13th International Conference on Automotive Safety (Automotive Safety 2022)

IOP Conference Series: Materials Science and Engineering
Volume 1247

Kaluza, Slovakia
26 - 29 April 2022

ISBN: 978-1-7138-5811-9
ISSN: 1757-8981

Printed from e-media with permission by:

Curran Associates, Inc.
57 Morehouse Lane
Red Hook, NY 12571

Some format issues inherent in the e-media version may also appear in this print version.

Institute of Physics
Dirac House, Temple Back
Bristol BS1 6BE UK

Phone: 44 1 17 929 7481
Fax: 44 1 17 920 0979

techtracking@iop.org

Additional copies of this publication are available from:

Curran Associates, Inc.
57 Morehouse Lane
Red Hook, NY 12571 USA
Phone: 845-758-0400
Fax: 845-758-2633
Email: curran@proceedings.com
Web: www.proceedings.com

TABLE OF CONTENTS

doi:10.1088/1757-899X/1247/1/011001

XIII International Scientific and Technical Conference

Automotive Safety 2022

Problemy bezpieczeństwa w pojazdach samochodowych

27-29 April 2022 Kaluža (SK)

Politechnika Świętokrzyska
Kielce University of Technology

Łukasiewicz
Przemysłowy Instytut Motoryzacji

UNIVERSITY OF ŽILINA

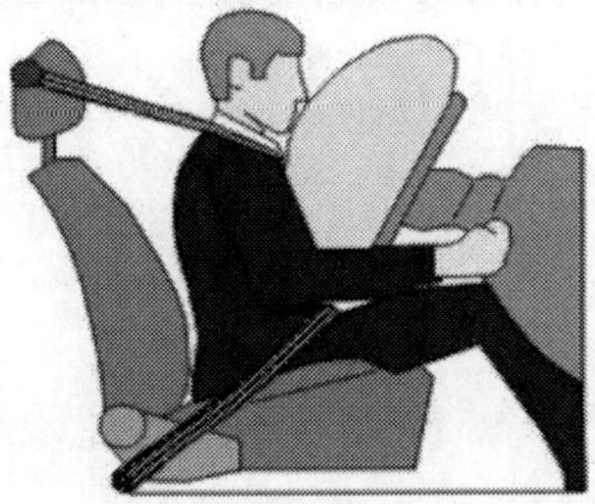

International Science-Technical Conference AUTOMOTIVE SAFETY

is organized by:

The Department of Automotive Engineering and Transport of the Kielce University of Technology,

The Faculty of Transport of the Warsaw University of Technology,

The Department of Road and Urban Transport at the Faculty of Operation and Economics of Transport and Communications of the University of Žilina,

Łukasiewicz Automotive Industry Institute

The International Science-Technical Conference AUTOMOTIVE SAFETY brings together scientists, experts and entrepreneurs dealing with the issues of vehicle safety and road safety. The AUTOMOTIVE SAFETY conference is held every two years and is becoming more and more popular among academics, scientists, entrepreneurs and other experts professionally involved in improving road safety.

The first edition of the Conference was held in Cedzyna near Kielce in 1998. Initially, it was an event gathering only scientists, experts and entrepreneurs from Poland. The first editions of the conference, entitled Safety problems in motor vehicles, were held in Poland each time. Since 2012, the Conference has become an international event.

In 2014, the Technical University in Žilina (Slovakia) became the co-organizer of the AUTOMOTIVE SAFETY Conference. Since 2014, the AUTOMOTIVE SAFETY Conference has been held alternately in Poland and Slovakia.

http://automotivesafetyconference.com

The 13th International conference on Automotive Safety (Automotive Safety 2022) IOP Publishing
IOP Conf. Series: Materials Science and Engineering 1247 (2022) 011001 doi:10.1088/1757-899X/1247/1/011001

XIII International Scientific and Technical Conference

Automotive Safety 2022

Problemy bezpieczeństwa w pojazdach samochodowych

27-29 April 2022 Kaluža (SK)

Politechnika Świętokrzyska
Kielce University of Technology

Łukasiewicz
Przemysłowy Instytut Motoryzacji

UNIVERSITY OF ŽILINA

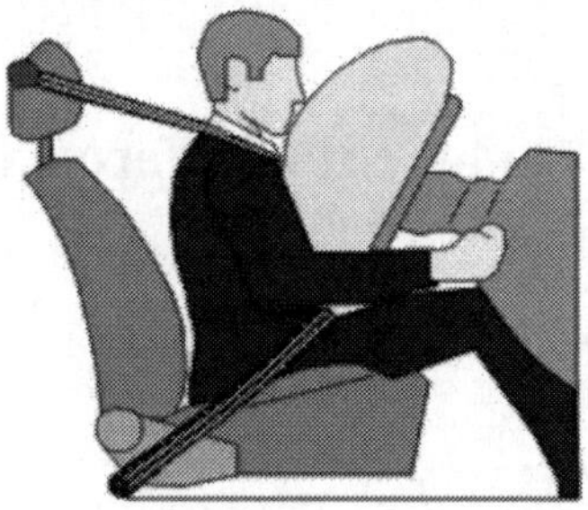

International Science-Technical Conference AUTOMOTIVE SAFETY

The aim of the Conference is to create an opportunity to exchange and share research results, experiences and new ideas on all aspects of vehicle and road safety.

The main topics of the conference include:

- **diagnostics for automotive safety assurance,**
- **braking systems: design and development,**
- **vehicle dynamics modelling and testing,**
- **road accident reconstruction,**
- **passive and active safety,**
- **safety in road transport,**
- **safety issues for electric and hybrid vehicles,**
- **intelligent transport infrastructure systems.**

Important issues raised by the speakers, substantive discussions that follow the presentations and a friendly atmosphere contribute to the growing interest in the Conference.

http://automotivesafetyconference.com

IOP Conf. Series: Materials Science and Engineering 1247 (2022) 011001 doi:10.1088/1757-899X/1247/1/011001

XIII International Scientific and Technical Conference

Automotive Safety 2022

Problemy bezpieczeństwa w pojazdach samochodowych

27-29 April 2022 Kaluža (SK)

Politechnika Świętokrzyska
Kielce University of Technology

Łukasiewicz
Przemysłowy Instytut Motoryzacji

UNIVERSITY OF ŽILINA

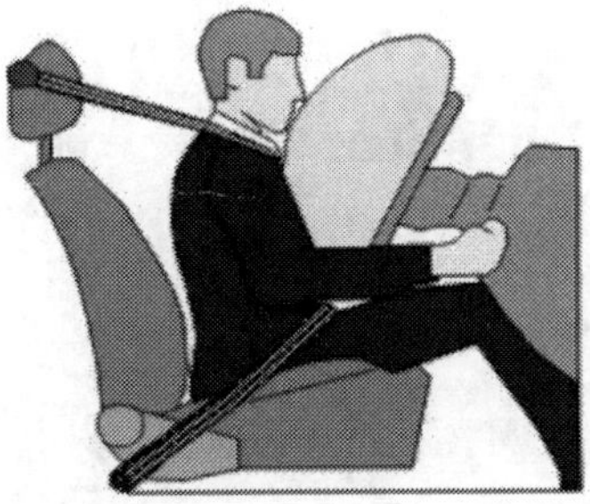

International Science-Technical Conference AUTOMOTIVE SAFETY

SCIENTIFIC COMMITTEE

Chairman:
- prof. dr hab. inż. Tomasz L. Stańczyk (Kielce University of Technology, PL)

Vice-Chairmen:
- prof. dr hab. inż. Zbigniew Lozia (Warsaw University of Technology, PL)
- prof. Ing. Milos Poliak, PhD (University of Žilina, SK)
- prof. dr hab. inż. Leon Prochowski (Military University of Technology, PL)
- prof. dr hab. inż. Jerzy Wicher

Members:
- doc. Ing. Dalibor Barta, PhD (University of Žilina, SK)
- prof. hab. dr Marijonas Bogdevičius (Vilnius Gediminas Technical University, LT)
- dr hab. inż. Paweł Droździel, prof. PL (Lublin University of Technology, PL)
- dr hab. inż. Mirosław Gidlewski, prof. WAT (Military University of Technology, PL)
- prof. dr hab. inż. Witold Grzegożek (Cracow University of Technology, PL)
- dr hab. inż. Marek Guzek, prof. PW (Warsaw University of Technology, PL)
- dr hab. inż. Jerzy Jackowski, prof. WAT (Military University of Technology, PL)
- prof. dr hab. inż. Marianna Jacyna (Warsaw University of Technology, PL)
- dr hab. inż. Marek Jaśkiewicz, prof. PŚk (Kielce University of Technology, PL)
- dr hab. inż. Rafał Jurecki, prof. PŚk (Kielce University of Technology, PL)
- prof. Ing. Alica Kalašová, PhD (University of Žilina, SK)
- doc. Ing. Rudolf Kampf, PhD (Institute of Technology and Business in Česke Budějovice, CZ)
- prof. Ing. Anna Križanová, CSc. (University of Žilina, SK)
- dr hab. inż. Przemysław Kubiak, prof. PŁ (Lodz University of Technology, PL)
- doc. Ing. Ján Ližbetin, PhD (Institute of Technology and Business in Česke Budějovice, CZ)
- dr hab. inż. Witold Luty, prof. PIMOT (Łukasiewicz Automotive Industry Institute, PL)
- dr hab. inż. Zbigniew Matuszak, prof. AM (Maritime University in Szczecin, PL)
- Ing. Ľubomír Moravčík, PhD (Slovakia's Ministry of Transport, Construction and Regional Development - MDV SK)
- prof. dr hab. inż. Andrzej Niewczas (Motor Transport Institute, PL)
- dr hab. inż. Krzysztof Olejnik, prof. PO (Opole University of Technology, PL)
- dr hab. inż. Marek Pawełczyk, prof. PŚk (Kielce University of Technology, PL)
- dr inż. Wiesław Pieniążek
- prof. Ing. Roman Rak, PhD (IRIS IDENT, CZ)
- dr hab. inż. Andrzej Reński prof. PW (Warsaw University of Technology, PL)
- doc. Ing. Branislav Šarkan, PhD (University of Žilina, SK)
- dr hab. inż. Ondrej Stopka (Institute of Technology and Business in Česke Budějovice, CZ)
- dr hab. inż. Andrzej Szosland, prof. PŁ (Lodz University of Technology, PL)
- dr hab. inż. Grzegorz Ślaski (Poznan University of Technology, PL)
- dr hab. inż. Wojciech Wach, prof. IES (Institute of Forensic Research in Cracow, PL)
- dr hab. inż. Dariusz Więckowski, prof. PW (Warsaw University of Technology, PL)
- dr inż. Karol Zielonka (Łukasiewicz Automotive Industry Institute, PL)
- dr hab. inż. Dariusz Żardecki, prof. WAT (Military University of Technology, PL)

IOP Conf. Series: Materials Science and Engineering 1247 (2022) 011001 doi:10.1088/1757-899X/1247/1/011001

XIII International Scientific and Technical Conference

Automotive Safety 2022

Problemy bezpieczeństwa w pojazdach samochodowych

27-29 April 2022 Kaluža (SK)

Politechnika Świętokrzyska
Kielce University of Technology

Łukasiewicz
Przemysłowy Instytut Motoryzacji

UNIVERSITY OF ŽILINA

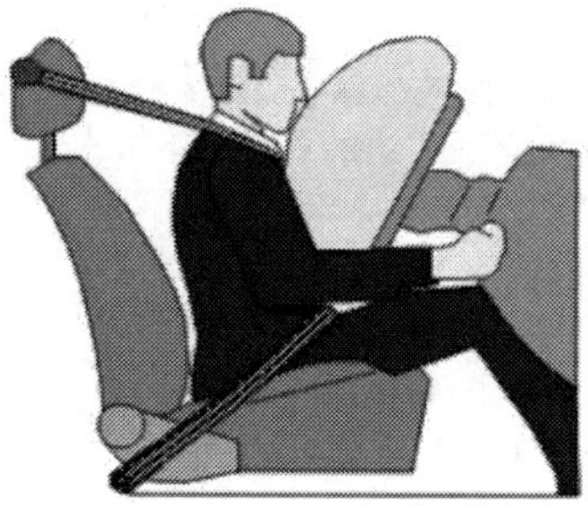

International Science-Technical Conference AUTOMOTIVE SAFETY

ORGANIZING COMMITTEE

Contact us

Kielce University of Technology
Department of Automotive Vehicles
and Transportation,
Aleja Tysiąclecia Państwa Polskiego 7,
25–314 Kielce

dr hab. inż. Marek Jaśkiewicz, prof. PŚk
tel. (+4841) 34–24–344
e-mail: m.jaskiewicz@tu.kielce.pl

http://automotivesafetyconference.com

IOP Conf. Series: Materials Science and Engineering 1247 (2022) 011002 doi:10.1088/1757-899X/1247/1/011002

Peer Review Statement

All papers published in this volume have been reviewed through processes administered by the Editors. Reviews were conducted by expert referees to the professional and scientific standards expected of a proceedings journal published by IOP Publishing.

- **Type of peer review:** Double Anonymous
- **Conference submission management system:** Morressier
- **Number of submissions received:** 79
- **Number of submissions sent for review:** 79
- **Number of submissions accepted:** 55
- **Acceptance Rate (Submissions Accepted / Submissions Received × 100):** 69.6
- **Average number of reviews per paper:** 3
- **Total number of reviewers involved:** 23
- **Contact person for queries:**
  Name: Emilia Szumska
  Email: eszumska@tu.kielce.pl
  Affiliation: Kielce Univeristy of Technology - Department of Automotive Engineering and Transport

Analysis of factors affecting the energy consumption of an EV vehicle - a literature study

A Skuza[1], **R S Jurecki**[1]

[1] Kielce University of Technology, Faculty of Mechatronics and Mechanical Engineering, Department of Automotive Engineering and Transport, Al. Tysiąclecia Państwa Polskiego 7, 25–314 Kielce, Poland

E-mail: askuza@tu.kielce.pl

Abstract. One of the most requested information about an electric vehicle is its energy consumption. This parameter is particularly important for electric vehicles users who drive them in real conditions. This paper aims to conduct literature research on the factors affecting the energy consumption of a vehicle. The article contains a set of information concerning how the temperature, traffic conditions, or properties of an electric vehicle translate into energy consumption. The results of studies presented in various publications were compared. The literature analysis allowed for the creation of a list that can be used as a compendium of knowledge on the energy consumption of electric cars.

Key words: hybrid electric vehicle, vehicle, energy consumption

1. Introduction

Electric vehicles are becoming more and more common. According to the publication [1], in 2020 the number of newly registered electric vehicles in Europe amounted to 11% of all vehicles, of which 6% were BEVs (Battery Electric Vehicle) and 5% were PHEVs (Plug-In Hybrid Electric Vehicle). Compared to the previous year, this share increased almost tripled, because in 2019, 3.5% of electric cars were registered.

Due to the increase in the number of electric vehicles, and the still limited access to charging stations (especially fast ones) for drivers traveling on the roads, the phenomenon of range anxiety, known in the literature, has emerged. The problem in their work was taken up by the authors [2] who decided to estimate energy consumption using neural networks. A similar topic also appeared in [3], where they analyzed how speed management affects range anxiety.

Energy consumption in electric vehicles appears as the topic of many publications and determines the range achieved. Various models have been created to determine energy consumption, allowing to predict energy consumption depending on many factors. In [4] a statistical model based on environmental conditions was presented. Its purpose is to determine the energy consumption of a vehicle using multiple linear regression. The authors in [5] showed an analytical model that used engine and drivetrain efficiency information. The publication [6] proposes three stages of modeling based on real driving profiles, simulated energy consumption, and driver behavior. Their purpose is to determine the primary energy consumption. The influence of individual factors on changes in the energy intensity of an electric vehicle also appears in the development of models for estimating energy consumption. In the work [7], in order to build the model, components were collected under real conditions. The change of slope, the use of auxiliary devices or the type of roads and traffic conditions were taken into account. This allowed the construction of the ECR (Energy Consumption Rate) and the fitting of its components.

Energy consumption is also related to the traffic conditions on the road. Therefore, driving cycles are created, i.e. a speed profile, which has been supplemented with information on braking, acceleration, driving with constant speed, and idling periods. The most popular cycle is NEDC (New European Driving Cycle), WLTC (Worldwide Harmonized Light-Duty Vehicles Test Procedure), HWFET (Highway Fuel Economy Test Cycle), FTP (EPA Federal Test Procedure) etc. The cycles mentioned were used in [8-11] to test electric vehicles, e.g. over energy consumption.

For many areas, driving cycles have been created that better represent the conditions of the roads there. The authors of the paper [12] created a cycle for the Chinese city of Xi'an to determine energy consumption and emissions under the conditions prevailing on the roads there. In publication [13], the electric vehicle test drive cycle was created for the Dublin area. It was shown that it differs from the legislative cycles and reliably reflects the real road conditions. The paper [14] shows how the energy consumption changes when variable speeds are applied to a highway in Perth, Australia. The authors of [15] created a real driving cycle based on the road driving in Shanghai to determine energy consumption and economic benefits.

2. Analysis of factors affecting energy intensity

One of the basic factors analyzed in the study of the energy consumption of electric vehicles is the ambient temperature. It affects the battery life and, consequently, the possible range of the vehicle. Energy consumption studies are most often carried out by comparing various temperature values and are the subject of numerous publications. This is because batteries are cooled down at high temperatures and heated up at temperatures below zero.

In [16], the authors conducted simulation studies for 5 selected temperature values: −20°C, −10°C, 0°C, 25° C, 45°C, at which the vehicle moved. 25°C was indicated as the nominal temperature. The simulation was performed for the NEDC test cycle. While driving at temperatures: -20°C, -10°C and 0°C, the vehicle range decreased by respectively: 36.7%, 28.76% and 24.96%, compared to the nominal temperature. During this time, energy consumption increased by 58.02%, 40.40% and 33.31%, respectively. The inverse relationship was recorded at an ambient temperature of 45°C. Then the range of the vehicle increased slightly, by 2.02%, and consumption decreased by 1.93%.

In publication [17] examined energy consumption based on real driving conditions in the Dutch city of Nieuwegein. The temperature range from 0°C to 35°C was analyzed with steps of 5°C. Driving was carried out in calm, normal and aggressive driving and speeds of 30 km/h and 130 km/h. was shown that at lower speeds, the lowest energy consumption occurs at 20°C and the highest at 0°C. The difference in this case is about 0.04 kWh. At 130 km/h the consumption is lower the higher the ambient temperature is.

Research on the influence of temperature on vehicle range is directly related to the use of auxiliary devices, i.e. air conditioning and heater. They are used to maintain the optimal temperature inside the vehicle and the battery.

In [18], a study was carried out comparing vehicle energy consumption in four test cycles, ie: NEDC, WLTC (Worldwide Harmonized Light Vehicles Test Cycle), WMTC (Motorcycle emission Test Cycle), and MAC (Mobile Air Conditioning). In each of the cycles, the vehicle was operated at a temperature of 23-25°C and -7°C. Additionally, the measurement was carried out with both the HVAC (Heating Ventilation and Air-Conditioning) system turned on and off (battery heating at negative temperature and cooling at positive temperature). The results showed that with the auxiliary system switched off, energy consumption is higher at -7°C than at +23°C. The difference varies from 6% to 14% depending on the cycle. It was also shown that energy consumption increases from 11% to 17% when using cooling and from 52% to 94% when using heating.

The authors in [19] discussed the influence of temperature on energy consumption and the range of an electric vehicle based on the Nissan Leaf. The simulation was also presented in four driving cycles, in the temperature range from -15°C to 20°C. It was shown that the range of the electric vehicle decreased from 150 to 85 km, i.e. by more than 40% at 0°C and to 60 km (by 60%) at -15°C. This is due to the influence of the cabin interior heating. On the other hand, the authors in [20] examined the annual variability of energy consumption during driving in the French city of Lille. The results of the analysis show that when the vehicle interior is heated, energy consumption increases by up to 33%. In summer, when cooling is used, the energy consumption increases to about 15%.

The authors of the paper [21] determined the energy consumption during heating at 18%, and at 14% during cooling. This analysis was based on the research of the Renault Zoe vehicle for daytime and nighttime journeys. On the other hand, in the work [22], a model was created to determine the energy consumption, the state of SOC or the influence of auxiliary devices. Popular driving cycles such as NEDC or WLTC were used to validate it. The results showed that using a heating system at -5°C can increase energy consumption by between 10% and 32%. For cabin interior cooling, energy consumption increases between 1% and 3% at an ambient temperature of 25°C and between 3% and 11% at an ambient temperature of 35°C. This is dependent on the cycle in which the vehicle is driven, but no actual cycle was used in the study.

The weather parameter that also affects the energy consumption of an electric vehicle is wind. This factor is the direct cause of the additional aerodynamic drag. In publication [23], the authors discussed in detail the influence of wind on energy consumption. For this purpose, they chose two routes: the first one was flat, with high exposure to wind, marked as LBR. The second was hilly and longer, marked as UBR, where wind gusts were much lower. At a wind speed of 18.5 km/h, energy consumption increased by 14% on the LBR route and by 5% on the UBR. At a wind speed of 50 km/h, the increase in energy consumption is already 31% and 15%, respectively. These results apply to free movement along the route.

Variable energy consumption also occurs with varying inclination. The energy intensity varies depending on whether the vehicle moves along the route as the altitude increases when ascending a hill or as the vehicle descends. Detailed research on this subject is presented by the authors in [24]. They carried out a study based on an analysis of the journeys of 492 electric vehicles over two years in Japan. Out of these, 68 cars were selected that traveled in the Aichi area. Twelve gradient ranges were extracted, from which a logarithmic model and a linear model were determined to predict energy consumption. According to the models, the higher the inclination increases the energy consumption. At the same time, the study showed that due to regenerative braking (energy recuperation), electric vehicles are more efficient on mountainous routes than conventional vehicles. In the work [25], a study of energy consumption with various road inclinations showed that a vehicle driving on a road with an inclination of 3% increases the specific energy consumption by 50%. When going downhill with a gradient of -3%, it reduces it by 80%. In contrast, in the work [26] the energy consumption of an electric vehicle was determined mathematically. A series of equations were used to determine how the energy consumption changes with a change in the angle of inclination. Based on these, general conclusions were made. It was determined that energy consumption increases with an increase in the angle of inclination and decreases with a decrease in the angle of inclination

In addition to the inclination of the route, it is also important to take into account other information related to the route from the point of view of energy intensity, such as traffic conditions along the route, type of road, speed values, etc. In [25] it was shown that the road conditions under which a journey takes place can reduce or increase energy consumption. Attention was paid to the length of the route over which the trip takes place. If the trip is made on a road shorter than 4 kilometers, the energy consumption is higher than for a 16-kilometer route. When driving on a motorway, the difference can be as much as 29%. In urban traffic, reducing the average speed can increase consumption by 19% on a route of fewer than 4 kilometers and by 15% on a longer route. When the proportion of stopping time increases from 12%-18% to 24%-34%, the energy intensity of the vehicle increases by 20%.

The authors in [14] used research of electric vehicles (Nissan Leaf and Mitsubishi i-Miev) moving at speeds in the range of 60-110 km/h on a section of a motorway in Perth, Australia, with a step of 10 km/h. It was determined that energy consumption increased with increasing speed. Then the authors repeated the measurements in four scenarios in which the load, battery charging protection and headwind changed. The increase in load contributed to an increase in energy consumption and a reduction in range for both vehicles. Also, the occurrence of a 20 km/h headwind increased the energy intensity of the vehicle.

In [27], the authors made predictions of energy consumption based on road conditions. To this end, they created models to determine how energy consumption develops as a function of driving speed, acceleration or route topography. They supported their analysis with graphs presenting energy consumption on the urban, suburban and motorway routes. The first two routes were driven by four drivers according to their driving style. On the motorway, measurements were taken at 90, 100 and 110 km/h. It can be seen that the energy consumption is different in each case. On each route, the vehicle was the most energy-consuming when it reached the maximum permitted speed.

The authors of the paper [28, 29] studied, among others, the influence of driving style on the energy consumption of an electric vehicle. From the results presented, it can be seen that the more aggressive the driving, the more energy-consuming the vehicle becomes. According to [29], the difference between aggressive and eco-driving at high speeds is relatively even 17%. The occurrence of speed oscillations can increase consumption. For 0.3 m/s it is 14% for eco-drivers, 37% for normal and 54% for aggressive ones.

In the paper [30], the authors investigated the influence of road conditions on the energy consumption of an electric vehicle. For this purpose, the AMESIM program was used. The results from the simulation were compared with real data obtained for the WLTC and RDE (Real Driving Emissions) cycle on a chassis dynamometer. Then, after establishing the consistency of the results, more than 100 driving cycles were used as input data for the program. The trips were grouped into congested, normal, uncongested and congested traffic conditions. The results showed that a change in traffic conditions causes a change in energy consumption. Changing conditions from congested to normal at low speeds reduces energy consumption by 15% and at high speeds increases it by 2%. Conversely, changing traffic conditions from normal to congested at low speeds reduces energy consumption by 6% at low speeds and 20% at high speeds. In addition, the authors determined that the use of traffic control measures, traffic calming measures, or roundabouts also changes energy consumption. Depending on the measure used, energy consumption decreases between 2% and 28%. This is due to the occurrence of regenerative braking.

In [31, 32], a vehicle-related factor was taken up as a part of research on energy consumption and vehicle range. The authors of [31] investigated, among others, the influence of load on energy consumption. The load was selected in the range from 50 to 250 kg with steps of 50 kg. It was shown that an increase in load causes an increase in the discharge of the vehicle battery. Energy consumption increases by up to 7% at the highest load compared to the lowest. At the same time, according to [32], the weight of the vehicle and the battery affects the range of the vehicle. The heavier the vehicle is, the less distance it can cover. It has been shown that the use of higher capacity batteries does not necessarily have a positive effect on energy consumption. This is due to the weight of the battery, which is much higher when using one with a higher capacity. The energy consumption per kilometer is then higher.

According to the authors of the papers [31, 33], attention should also be paid to factors such as SOC (State Of Charge) battery charge status or regenerative braking. According to [31], the state of charge of the battery at the start of the journey is not the most significant factor. The way to travel the route is more important. Regardless of whether the battery was 100% or 60% charged, the depth of discharge was the same. The authors in [33] showed that regenerative braking can improve energy transfer by over 40%, and range - even by almost 25%.

A summary of the energy consumption of the different factors is presented in Table 1. It includes the name of the most important factor, a diagram showing its value and the increase or decrease in energy consumption.

IOP Conf. Series: Materials Science and Engineering 1247 (2022) 012001 doi:10.1088/1757-899X/1247/1/012001

Table 1. Summary of energy consumption by the factors analysed.

Type of factor	Value of factor	Energy consumption
Negative temperatures	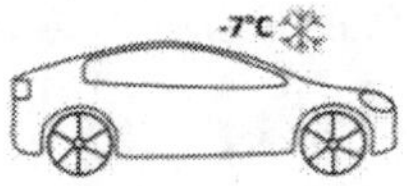	~6% -14% ↑
	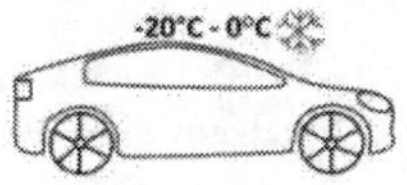	~33% - 58% ↑
High temperature	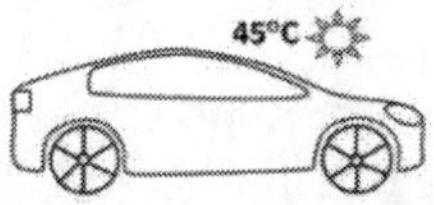	~2% ↓
HVAC heating (in winter)	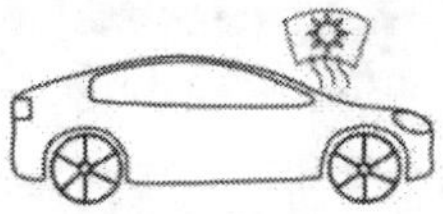	~52% - 94% ↑
HVAC cooling (in summer)		~11% - 17% ↑
Wind	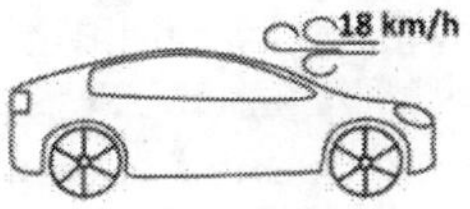	~5% - 14% ↑
	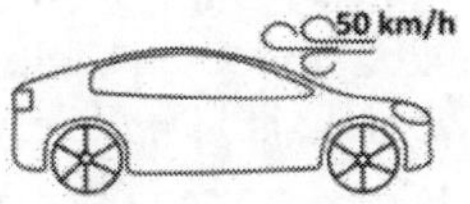	~15% - 31% ↑
Inclination	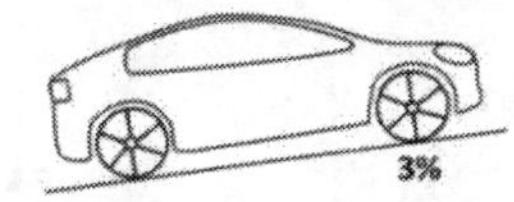	~50% ↑
	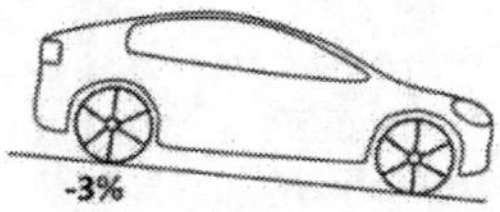	~80% ↓
Route length	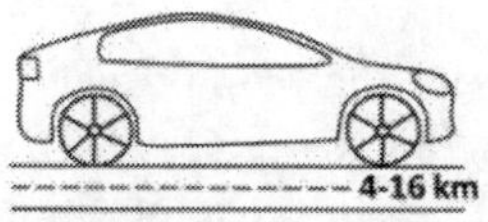	~15-29% ↑
Traffic conditions (proportion of stopping time)	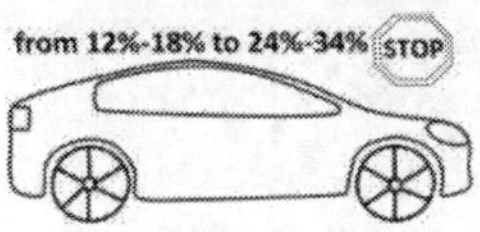	~20% ↑
Load	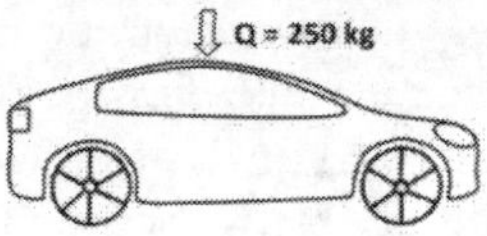	~7% ↑
Driving style	agresive 	~17% ↑

The paper [34] uses data on Nissan Leaf journeys in Sydney. Based on this, an experimental design was created, the results of which determine which factors have the greatest importance. It was shown that among: topography, climate, road traffic and infrastructure, it is the first two factors that have the greatest impact on energy consumption.

The authors [35] decided to determine the influence of driving conditions, driver aggressiveness and the use of auxiliary devices on the energy vehicle energy consumption. It was shown that the influence of auxiliary devices varies from 15% to 40% at a speed of 20 km/h. At higher speeds, this range is reduced from 5% to 15%. Driving style has a different effect on energy consumption. For aggressive drivers, it can be as much as 40% at low speeds (20 km/h) and 15% at higher speeds (60 km/h).

The authors of the study [36] analysed the ECR index and the factors influencing its change. They showed that energy consumption is higher in winter, as ECR increases by up to 34%. In addition, according to the study, the vehicle range is lower when trips are made over short distances. It was also shown that the effect of specific driving conditions on ECR is polynomial (e.g. initial battery charge level) or quadratic (e.g. driving speed).

3. Conclusions

The factors affecting the range of an electric vehicle are a very extensive subject for research and analysis. Hence the great interest of researchers in these issues, taking into account the variety of variable parameters whose influence on energy consumption varies

It is apparent that weather conditions can have a significant impact on the energy consumption of a vehicle. Negative temperatures can reduce it by up to 37%, while at a temperature of 40°C, it is possible to extend the range by about 2%. A similar relationship applies to the occurrence of headwinds. The greater the wind, the more energy a vehicle needs to overcome resistance to motion.

In addition, weather conditions are related to other factors, e.g. the use of auxiliary systems. In studies, it has been shown that heating a vehicle is more energy intensive than cooling its interior.

The analysis of other factors, such as vehicle load, battery weight, and driver's driving style showed variability in energy consumption. Both an increase in load and battery weight increase the energy consumption of the battery. Aggressive drivers should also be aware of the reduced range. Sudden acceleration and frequent acceleration of the vehicle have a negative effect, causing the battery to discharge faster.

Furthermore, some of the authors decided to identify and evaluate the factors that influence energy consumption on a comparative basis. By comparing several factors, they calculated which one has a greater influence on a vehicle's energy consumption.

The energy consumption of a vehicle is related to many factors. To maximise the battery range, many variables need to be taken into account that can reduce the range. An electric vehicle can't run only at temperatures close to 20°C, on a windless day, on a flat road, without passengers and during smooth traffic hours. However, drivers, taking into account the different influences of the parameters mentioned in this paper, among other things, plan their route to ensure its completion without unloading the vehicle. This will avoid range anxiety and stress while driving.

4. Bibliography

[1] European Environment Agency – New registrations of electric vehicles i Europe 2021 Available on-line: https://www.eea.europa.eu/ims/new-registrations-of-electric-vehicles ,

[2] Modi S, Bhattacharaya J, Basak P 2020 Estimation of energy consumption of electric vehicles using Deep Convolutional Neural Network to reduce driver's range anxiety. *ISA Transactions*, 98, 454-70,

[3] Kondo Y, Kato H, Ando R, Suzuki T, Karakama Y 2013 To what extent can Speer management alleviate the range anxiety of EV?. *World Electric Vehicle Symposium and Exhibition (EVS27),*

[4] De Cauwer C, Van Mierlo J, Coosemans T 2015 Energy consumption prediction for electric vehicles based on real-world data. *Energies* **8(8)** 8573-93.

[5] Fiori C, Ahn K, Rakha H.A 2016 Power-based electric vehicle energy consumption model: model development and validation. *Appl Energy* **168** 257-68.

[6] Muratori M, Moran M.J, Serra E, Rizzoni G 2013 Highly-resolved modeling of personal transportation energy consumption in the United-States. *Energy 58* 168-77,

[7] Qi X, Wu G, Boriboonsomsin K, Barth M J, 2018 Data-driven decomposition analysis and estimation of link-level electric vehicle energy consumption under real-world traffic conditions *Transp. Res. Part D Transp. Environ* **64** 36-52

[8] Szumska E, Jurecki S 2020 The Effect of Aggressive Driving on Vehicle Parameters. *Energies* **13(24)** 6675,

[9] Wahono B, Santoso W B, Nur A, Amin 2015 Analysis of Range Extender Electric Vehicle Performance using Vehicle Simulator *Energy Procedia*, **68** 409-18, ,

[10] Pavlovic J, Tansini A, Fontaras G, Ciuffo B et al, 2017 The Impact of WLTP on the Official Fuel Consumption and Electric Range of Plug-in Hybrid Electric Vehicles in Europe. *SAE Technical Paper* 2017-24-0133,

[11] Gis W, Żółowski A, Bocheńska A 2012 Testing of the electric vehicle in driving cycles. *J. of KONES*, **19(4)** 207-21,

[12] Zhao X, Ma J, Shi P, Chen H 2020 Construction of electric vehicle driving cycle for studying electric vehicle energy consumption and equivalent emissions. *Environ. Sci. and Poll. Res.* **27** 37395-409,

[13] Brady J, O'Mahony M, 2016 Development of a driving cycle to evaluate the energy economy of electric vehicles in urban areas. *Appl. Energy* **177** 165–78,

[14] Wager G, Whale J, Braunl T 2016 Driving electric vehicles at highway speeds: the effect of higher driving speeds on energy consumption and driving range for electric vehicles in Australia. *Renew. and Sust. Energy Rev.* **63** 158–65 ,

[15] Sun D, Zheng Y, Duan R 2021 Energy consumption simulation and economic benefit analysis for Urban electric commercial-vehicles. *Transp. Res. Part D Transp. Environ* **101** 103083,

[16] Xie Y, Li Y, Zhao Z, Dong H, Wang S, Liu J, Guan J, Duan X 2020 Microsimulation of electric bvehicle energy consumption and driving range. *Appl. energy*, **267** 115081,

[17] Donkers A, Yang D, Viktorović M 2020 Influence of driving style, infrastructure, weather and traffic on electric vehicle performance. *Transp. Res. Part D Transp. Environ*, **88** 102569,

[18] De Gennaro M, Paffumi E, Martini G, Manfredi U, Scholz H 2014 Experimental Investigation of the Energy Efficiency of an Electric Vehicle in Different Driving Conditions. *SAE Technical Paper 2014-01-1817*

[19] Iora P, Tribioli L 2019 Effect of Ambient Temperature on Electric Vehicles' Energy Consumption and Range: Model Definition and Sensitivity Analysis Based on Nissan Leaf Data. *World Electr. Veh. J.* **10(1)**,

[20] Desreveaux A, Bouscayrol A, Castex E, Trigui R, Hittinger E, Sirbu G M 2020 Annual Variation in Energy Consumption of an Electric Vehicle Used for Commuting. *Energies* **13(18)** 4639,

[21] Doyle A, Muneer T 2019 Energy consumption and modelling of the climate control system in the electric vehicle. *Energy Expl. & Expl* **37(1)** 519-43,

[22] Fiori C, Ahn K, Rakha H A, 2016 Power-based electric vehicle energy consumption model: Model development and validation *Appl. Energy* 168 257-268

[23] Burgess S C, Choi J M J 2003 A parametric study of the energy demands of car transportation: a case study of two competing commuter routes in the UK. *Transp. Res. Part D Transp. Environ*, **8** 21-36,

[24] Liu K, Yamamoto T, Morikawa T 2017 Impact of road gradient on energy consumption of electric vehicles. *Transp. Res. Part D Transp. Environ*, **54** 74-81,

[25] Al.-Wreikat Y, Serrano C, Sodré J R 2021 Driving behaviour and trip condition effects on the energy consumption of an electric vehicle under real-world driving. *Appl. energy* **297** 117096,

[26] Yang S C, Lin M, Lin Y, Posmak T Q, 2014 Electric vehicle's energy consumption on a road with different slope. *Phys. A: Stat. Mech. Appl.* **402** 41-48

[27] Wang J, Besselink I, Nijmeijer H 2015 Electric Vehicle Energy Consumption Modelling and Prediction Based on Road Information. *World Electr. Veh. J.* **7(3)** 447-58,

[28] Galvin R 2017 Energy consumption effects of speed and acceleration in electric vehicles: Laboratory case studies and implications for drivers and policymakers. *Transp. Res. Part D Transp. Environ* **53** 234-48,

[29] Donkers A, Yang D, Viktorović M 2020 Influence of driving style, infrastructure, weather and traffic on electric vehicle performance. *Transp. Res. Part D Transp. Environ* **88** 102569,

[30] Mamarikas S, Doulgeris S, Samaras Z, Ntziachristos L 2022 Traffic impacts on energy consumption of electric and conventional vehicles. *Transp. Res. Part D Transp.* **105** 103231

[31] Szumska E, Jurecki S 2021 Parameters influencing on electric vehicle range. *Energies* **14(16)** 4281,

[32] Mruzek M, Gajdáč I, Kučera L, Barta D 2016 Analysis of parameters influencing electric vehicle range. *Procedia Engineering* **134** 165-74,

[33] Qiu C, Wang G 2016 New evaluation methodology of regenerative braking contribution to energy efficiency improvement of electric vehicles. *Energy Conv. and man.* **119** 389-98,

[34] Li W, Stanula P, Egede P, Kara S, Herrmann C 2016 Determining the main factors influencing the energy consumption of electric vehicles in the usage phase. *Procedia CIRP* **48** 352-57,

[35] Badin F, Le Berr F, Briki H, Dabadie J C, Petit M, Magand S, Condemine E 2013 Evaluation of EVs energy consumption influencing factors, driving conditions, auxiliaries use, driver's aggressiveness. *World Elect. Veh. J.* **6(1)** 112-23,

[36] Fetene G M, Kaplan S, Mabit S L, Jensen A F, Prato C G, 2017 Harnessing big data for estimating the energy consumption and driving range of electric vehicle *Transp. Res. Part D Transp. Environ* **54** 1-11

Risks of the insurance in road transport

J. Kubáňová[1*], I. Kubasáková[1], M. Poliak[1]

[1] University of Zilina, Faculty of Operation and Economics of Transport and Communications, Department of Road and Urban Transport

* jaroslava.kubanova@fpedas.uniza.sk

Abstract. The growing trend of road freight transport also brings many negative factors, such as an increase in crime in the sector. In 2021, semi-trailers took the second place in the number of registered vehicles in Slovakia. This is also reflected the crime in the road freight transport, which has had a very slow declining trend in recent years. On the contrary, what is growing is the number of insurance frauds recorded by the Slovak insurance company. We analyse in the article individual types of criminal offenses in road freight transport. We are also trying to find measures by which the risk of crime can be reduced. Whether measures in the form of insurance or options verification of information before the start of the shipment itself. Insurance does not have to be just a means of compensating for any damages, it can at the same time serve as a tool to make the services offered more attractive, confirmation complexity of the services offered, but it can also be one of the motivational tools employer.

Key words: insurance, road freight carrier, risks, insurance frauds, vehicle

1. Introduction

A number of procedures and methods for working with risk, which are successfully used in practice almost in in all areas of human activity which, by its very nature, require some opportunity to influence existing risks. In road transport, we can only assess risks if we know identify the sources of risks and know their effects. Risk reduction is the process that aims to take measures to prevent losses. By identifying problems, it is possible perform adverse state modeling. If we group two or three methods, we achieve objective view of the whole problem. New security approaches are being developed combined methods. For risk assessment, these methods consist of two separate procedures, which complement each other. The first group includes mainly focused methods to identify the risk. Catalog sheets are a characteristic representative of this group. The second group includes methods for risk assessment.

Europe has removed borders within the EU for the movement of goods, services, capital and labor. This also allowed organized crime to operate more freely, which had already been extremely flexible in the previous period. However, the borders were not lifted for police and justice. Europe does not have an organization that operates at the federal level, something like the FBI. In Slovakia, issues such as phantom carriers are dealt with by police units at the district level.

Crime in road freight transport has many forms. The most common are:

- theft of semi-trailers and their often armed raids,
- drug smuggling and its distribution,
- smuggling and distribution of infringing products,

- customs infringements and customs fraud,
- violation of the tax regulation in the field of value added tax and excise taxes,
- illegal migration,
- illegal arms trade,
- international terrorism,
- in the past, there have also been cases of driver murders for the purpose of stealing trailers and their subsequent sale,
- phantom carriers. [1]

1.1. Schemes of activities of phantom carriers

As a rule, the perpetrators concentrate on high-value goods and goods that are not difficult to dispose of profitably. The loss is generally massive. These are some of the methods used by the perpetrators in order to acquire contracts.

An important category in crime is the phantom carrier:

- Phantom 1 – fake documents,
- Phantom 2 – duplicate of an existing company,
- Phantom 3 – tucked carrier,
- Phantom 4 – big bang,
- Phantom 5 – insurance fraud,
- Phantom 6 – freight forwarding company,
- Phantom 7 – trailer thieves,
- Phantom 8 – cash on delivery fraud,
- Phantom 9 – fake consignee [1].

Phantom 1 – fake documents

The carrier logs in to the consignor, loads the goods and disappears with the goods. After his disappearance, it is often found that the semi-trailer set was stolen, the vehicle registration number was stolen and the driver's documents, fuses and documents from the carrier were false, or stolen. Drivers report from mobile phones, telephone lines and faxes to petrol stations, freight forwarders, etc. It is the most frequently used and so-called basic scheme. Very often, they report via internet transport databases to companies that have missed loads on the Internet or unloaded goods for several days [1].

Phantom 2 – duplicate of an existing company

The carrier logs in to the consignor, loads the goods and disappears with the goods. After the inspection, it is found that the carrier who was supposed to load the goods knows nothing about the transport, even though the loading vehicle had the vehicle registration number identical to the existing one from the transport company's fleet. Duplicate - the vehicle has all the basic features and the vehicle registration number the same as the original vehicle. Pre-existing vehicles are sometimes used for the actual transport companies and the documents (concessions, copies of insurance) of these companies, although the vehicles no longer belonged to them and these documents were obtained legally by criminals at a time when the vehicles still belonged to the companies [1].

Phantom 3 – tucked carrier

An unknown forwarding company logs in to the sender, reports the vehicle and the carrier. The carrier is checked by the sender and it is an existing company and the vehicle, the insurance, the driver's documents and the documents from the vehicle are in order. The carrier will unload the loaded goods at a place other than that specified in the CMR document and the goods will disappear. Subsequently, it is found that the carrier was hired by the forwarder for a different transport than the carrier should have performed according to CMR. The carrier refers to the fulfillment of the obligation according to the order received from the forwarder. In cases where the carrier refused to obey the phantom's instructions, threats of violence followed [1].

Phantom 4 – big bang
The carrier who has been operating on the market for a long time is checked in with the consignor, the documents from the trailer set, the driver's documents and the insurance policies are in order. The carrier loads the goods and disappears with the goods. After his disappearance, it is found that this carrier recently changed owners, part of the employees and within a few days loaded several lucrative goods with various vehicles and the goods disappeared. During the crisis, several transport companies could get into the hands of criminals [1].

Phantom 5 – insurance fraud
The forwarder will procure a semi-trailer set for the consignor in a critical situation or under particularly advantageous conditions. The consignor then notifies the forwarder of the disappearance of the semi-trailer set with the goods. For the insurance fraud variant, the following testifies:

- Atypical delivery composition (inventory differences).
- A customer who does not take delivery of goods or deliveries made up of parts of other orders throughout the year.
- Problems and rejections of some previous semi-trailers.
- Carrier contacts and consignor's employees from the past. [1]

Phantom 6 – freight forwarding company
Forwarding companies procure transport orders at relatively advantageous prices, and thus quickly get to large volumes of transport. It offers carriers high prices, even higher than it receives from the customer. From a certain moment, they stop paying carriers, and as long as they find out what they are about, they stop cooperating with the company, they have extremely high overdue receivables. The company disappears. Neither the staff nor the owner are available. [1]

Phantom 7 – trailer thieves
This is a case several years old when the phantom-forwarder repeatedly lured the vehicles of the carriers for loading abroad in order to obtain the vehicles themselves, not the goods. For completeness, drivers were murdered and semi-trailers continued to the former Yugoslavia. The indiscretion of the Austrian police (media information) must have led to premature arrests and the case was not fully settled. At least one driver murder occurred while police were already watching the gang [1].

Phantom 8 – cash on delivery fraud
This type of crime is in the groupage delivery service, which collects cash on delivery for the delivered goods. Before the shipment is delivered from the foreign partner, or at a greater distance from the domestic service, the information received by the delivering partner about the cancellation of the need to select cash on delivery will appear. If it is not checked by another channel (another mail, fax, telephone) and delivered without picking up the cash on delivery, it will be found that the information was false. It is possible that hackers got into the forwarder´s information systems and modified the data directly. The beneficiary is a natural person, subsequently undetectable [1].

Phantom 9 – fake consignee
An unknown forwarding company logs on to the consignor, reports the trailer and the carrier. The carrier is checked by the consignor and it is an existing company and the semi-trailer set, the insurance, driver's documents and vehicle documents are in order. The carrier will take the loaded goods to their destination and there they will report to the consignee, who was added a phantom to the order. He will instruct him to unload at a place other than the one indicated in the CMR document, often with the justification "second warehouse - around the corner" and the goods will disappear. The carrier refers to the fulfillment of the obligation according to the order received from the forwarder. In addition, according to CMR document, he delivered correctly [1].

IOP Conf. Series: Materials Science and Engineering 1247 (2022) 012002 doi:10.1088/1757-899X/1247/1/012002

1.2. Checking information

First of all, at the beginning of every business relationship, we must try to check as much as possible the company with which we are going to cooperate. And this can be done in many ways, for example:

- Check in the list of carriers in the unified information system in road transport.
- Company control in the commercial register/ trade register.
- Checking the carrier´s insurance contract.
- Licence verification.
- Driver verification.
- Verification of the carrier´s vehicle registration number. [2]

In 2021, 3 436 018 vehicles were registered in Slovakia. Of these, passenger cars have the largest representation and semi-trailers are right behind them. (Figure 1)

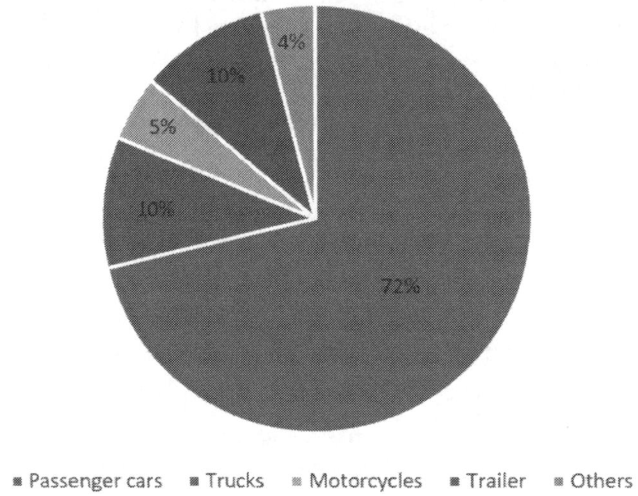

Figure 1. Registered vehicles by type in the Slovak Republic in 2021 [authors].

The compulsory contractual insurance is compulsory for motor vehicle owners. It has been on the rise in recent years as a result of the overall growing number of registered vehicles. (Figure 2)

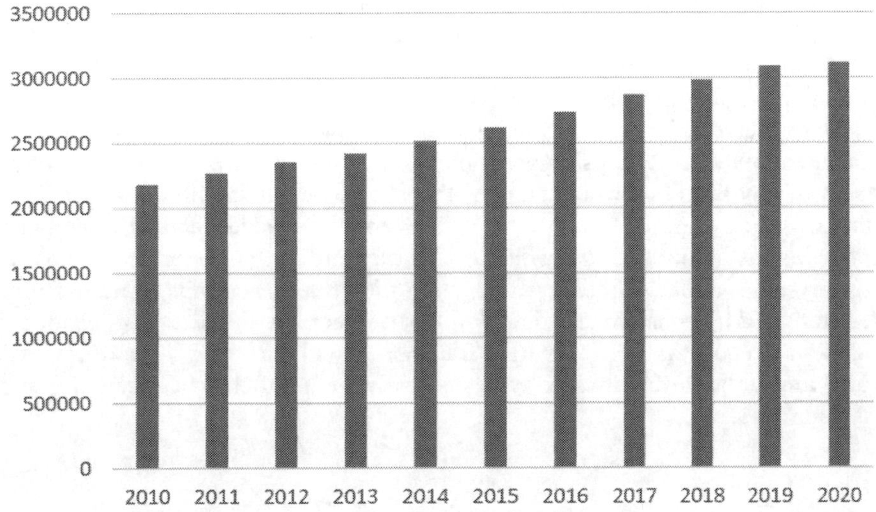

Figure 2. Number of insured vehicles by motor insurance [authors].

Insurance companies have created a register of insurance events (RIE) since 1 January 2017. It is mandatory for all members of the Slovak Insurance Association. The data is only entered into the register when you have an insured event. Then the name and surname will be entered in the register. In the case of a legal entity, the name of the company, the registered office of the company and the date of occurrence of the insured event will be published. In the case of motor vehicles, the registration number of the motor vehicle, the serial number of the motor vehicle and the date of occurrence of the insured event will be stated. Only the date of its occurrence and the number assigned by the insurer are recorded for the insured event. Anyone who has had an insured event and is registered can find out once a year at the written request of the insurer what information is kept about him. [3]

2. CMR neutralization

According to Article 12 (1) of the Convention on Contracts for the International Carriage of Goods by Road: The consignor is entitled to dispose of the consignment, in particular he may require the carrier to stop the shipment, change the place of delivery or issue the consignment to a consignee other than that indicated in the consignment note. [2]

In Article [4], the author seeks to better understand the problematic provisions of the Convention concerning the Contract for the International Carriage of Goods by Road (CMR). This analysis makes it possible to assess the CMR in order to decide whether it is poorly designed or misinterpreted. By showing different interpretations, this research highlights the hypothesis that CMR was not designed badly, but on the contrary that differences in the interpretation of the provisions are due to different approaches of national courts. [5]

At present, carriers quite often face customer requests to neutralize the CMR consignment note. Neutralization of transport documents is not a scientific concept; it is not defined anywhere. Neutralization in transport means the exchange of original consignment notes or transport documents for other consignment notes or transport documents. It is therefore a question of invalidating the first consignment note. Based on the analysis of commercial law in road transport, this is an act of the carrier which is not regulated by the CMR Convention. [6]

The carrier cannot know whether the neutralization of the CMR is not being used to cover up fraud on a shipper who is circumventing the regional pricing policy or are receiving unduly rebates. Or, for example, in VAT fraud, a chain of commercial transactions is declared, ending with fictitious exports to another EU country or to a third country, and in fact the goods remain in the country of origin. The consignor will claim a VAT refund and will sell the goods in the country of loading. These are often completely fictitious exports and fictitious business transactions, and the goods do not even exist or are

grossly overvalued. Unauthorized VAT refunds on timber transport have also been confirmed. One loaded semi-trailer will cross the border several times with the same load. They will claim an unauthorized VAT refund at each transfer. [2]
A possible solution against neutralizing the CMR is to have the driver sign a declaration as follows:
"I declare that I have received transport instructions from my employer or transport orderer that are fully in line with the content of the CMR consignment note number .... I have not received any instructions, and I do not know of any facts that would suggest that the shipment should not be carried out exactly according to this CMR consignment note. I am aware that the loaded goods are the property of the consignor and he alone is authorized to change the instructions for the shipment. I undertake to carry out the transport myself, without reloading or changing the driver, according to the instructions of the consignor given in the CMR consignment note and I will reject any instruction to change the route or place of unloading and report to: .... During transport, I will park in the secure car parks listed on the IRU website. I am further aware that I may be complicit in the offense of Fraud or Fraud according to the Criminal Code" [1,2].

3. Analysis of criminal activities in road freight transport

Road transport is a significant part of the EU economy and will remain so for a long time to come. The increase in road freight transport in the EU has been reported for the seventh consecutive year. This is the main argument for ensuring road safety in freight transport. [7]
The author of Article [8] found that attacks on foreign drivers and trucks, previously considered South Africa's internal problem due to the country's domestic social and economic challenges, raised concerns about South Africa's future as an economic power, its role as a transit. landlocked countries and its relationship with regional neighbours. For the landlocked countries in the region, any instability in South Africa can have a serious impact on logistics that support business and international trade.
Anecdotal evidence suggests that harsh social conditions in the road freight sector are affecting traffic crime. This article analyzes traffic crime and shows how to use a combination of official statistics and crowdsourcing data in the process. The findings show that transport is more vulnerable to antagonistic threats in geographical areas where low-wage carriers are more likely to operate. [9]
The authors focused on the analysis of traffic accidents classified as criminal offenses. The analysis also examines the structure of traffic accidents and their clarity. The aim of the paper is to find out the development of the traffic safety situation in Slovakia and its impact on the quality of life of the population. [10]
The risk analysis performed by the authors in Article [11] was performed using original algorithms developed on the basis of artificial neural networks, which, among other things, take into account the probability of the risk of cargo theft at a particular stage of the order for different types of cargo. The authors specify the values of expected losses, which relate to the type of stolen cargo, including penalties depending on the stage of transport.
The sustainability of road freight transport is becoming increasingly important due to the ever-increasing movement of freight, globalization and operational flexibility. It is difficult for transport managers around the world to manage the growing risks in operating and implementing risk mitigation strategies for economic sustainability. This study found that risk mitigation strategies must be chosen with regard to risk criticality along with a limited budget. In addition, we found that considering subjective assessments of expert judgment and justification of costs and benefits provides a convincing result in calculating the number of risk mitigation in the FMEA approach. [12]
The authors of this article [13] have focused on some strategic risks for road freight companies. A review of some Bulgarian and European policies and guidelines has been carried out to show the development of road freight transport, which is the new "White Paper 2020" of EU transport to business.
The total number of crimes in road freight transport is shown in Figure 3 below.

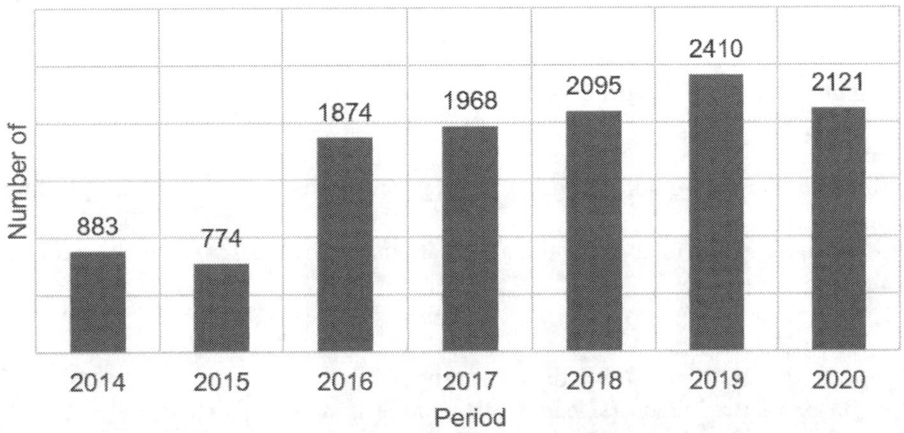

Figure 3. Total number of crimes in road freight transport [14]

Even the measures introduced to prevent criminal crime do not help to increase the volume of traffic, to reduce crime in EU road transport. The graph below (Figure 4) shows the total number of road freight offenses in the EU, in particular in the categories of vehicle theft, object theft, tractor / set theft, kidnapping, onset of an oncoming vehicle, container theft and phantom carrier. The number of criminal offenses is on the rise. Year-on-year, between 2015 and 2016, this increase was even more than double.

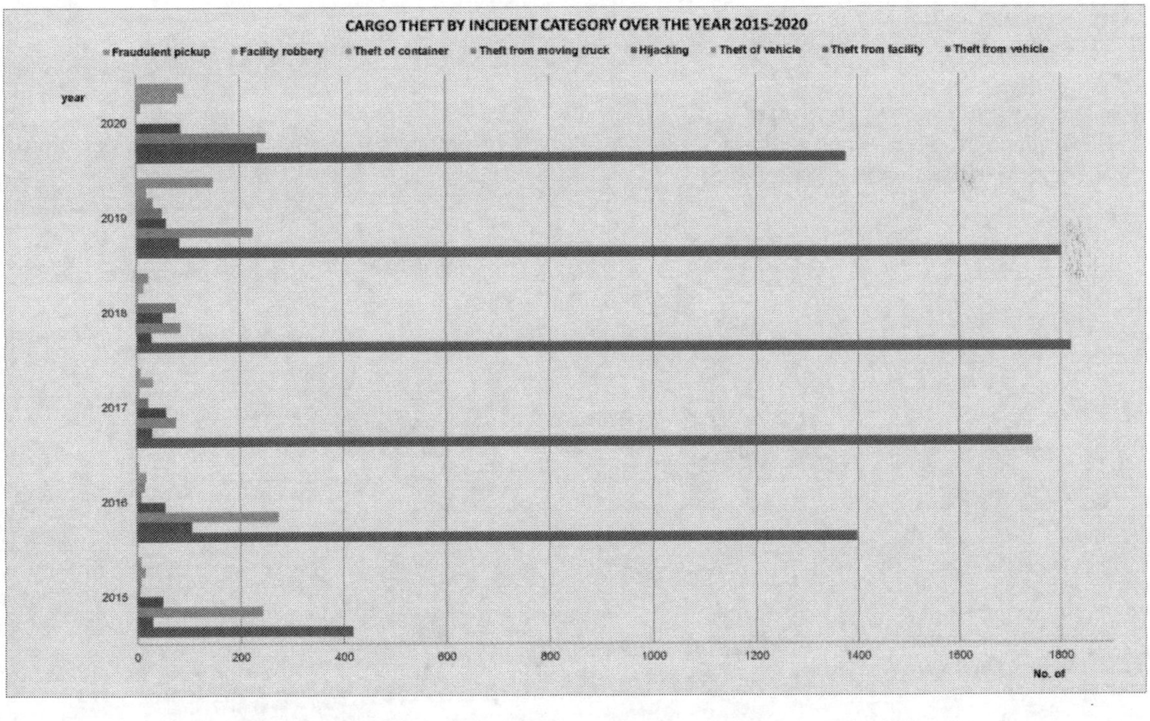

Figure 4. Cargo theft by incident category over the year 2015 – 2020 [14]

The largest volume of crime in road transport is vehicle theft, which, with the exception of 2020, is still growing year on year. The decline in 2020 may be due to a reduction in population movements during the COVID-19 pandemic. On the contrary, in 2020 the number of thefts from buildings increased (Figure 4).

4. Insurance options in road freight transport

The carrier's responsibilities are governed by international transport contracts. The risks associated with the transport of goods fall into the group of commercial risks in international trade.

These risks can:

1. be limited to the control of the goods being transported (can be accurately quantified and estimated)
2. extend to consequential damages if the goods are not in the right place at the right time (difficult to quantify and estimate).

Three stages of transport risk solution:

1. Risk transfer (transport and forwarding insurance).
2. Prevention or risk reduction (selection of suitable transport, suitable and safe packaging and marking of goods, national requirements for packaging of goods can be found in the International Chamber of Commerce).
3. Elimination of risk by transfer to business partners - depends on the applicable conditions governing liability:
   a) of the carrier for the goods in the individual international transport contracts,
   b) of the freight forwarder for goods according to freight forwarding contracts usually national.

The main products of the insurance company for road carriers are:

- road carrier liability insurance,
- shipment insurance.

4.1. Raod carrier liability insurance

The insurance product, which is characterized by Slovak insurance companies, has 3 forms depending on the nature of the transport: (Figure 5)

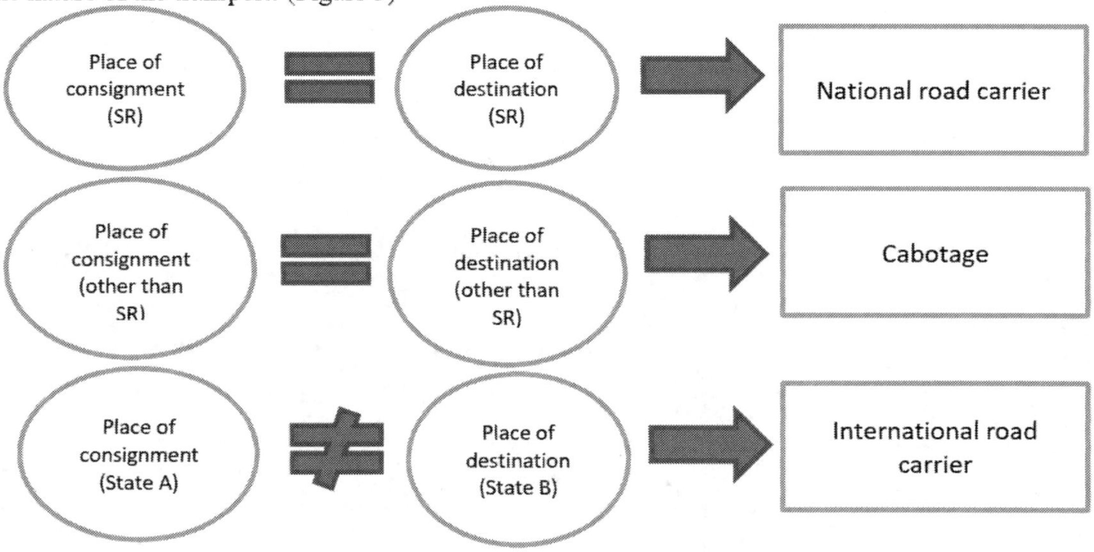

Figure 5. Determination of the carrier's character [authors]

4.1.1. Coverage range

The insurance covers the liability of the national road carrier in the territory of the Slovak Republic and the insurance of the liability of the international road carrier covering:

- the carrier's liability for damage caused directly to the consignment during transport,
- rescue costs,

- costs associated with securing the rights of the insured and the insurer against third parties; the precondition for reimbursement of costs is that they were incurred in connection with damage caused as a result of a damage event, which is an insured event under these insurance conditions,
- costs of investigation of the insured event by persons authorized by the insurer. [15]

4.1.2. Cabotage insurance also covers:

- actual and provable other property damage caused to a third party other than damage to the shipment, if this damage occurred due to exceeding the delivery time of the shipment due to a demonstrable mistake or negligence on the part of the insured. The maximum limit of indemnity for property damage caused by exceeding the delivery time for one transport may not exceed three times the transport agreed for the transport in question,
- transport costs if incurred in connection with damage to the shipment, which is the subject of insurance benefits under these insurance conditions and the insured is obliged to reimburse them, costs under this point will be reimbursed by the insurer up to 10% of the sum insured in the insurance contract, total costs is limited to € 10,000. [15]

In the following Figure 6 you can see the number of reported insurance events in Slovakia. These are insurance events aimed at insuring liability for damage caused by a motor vehicle.

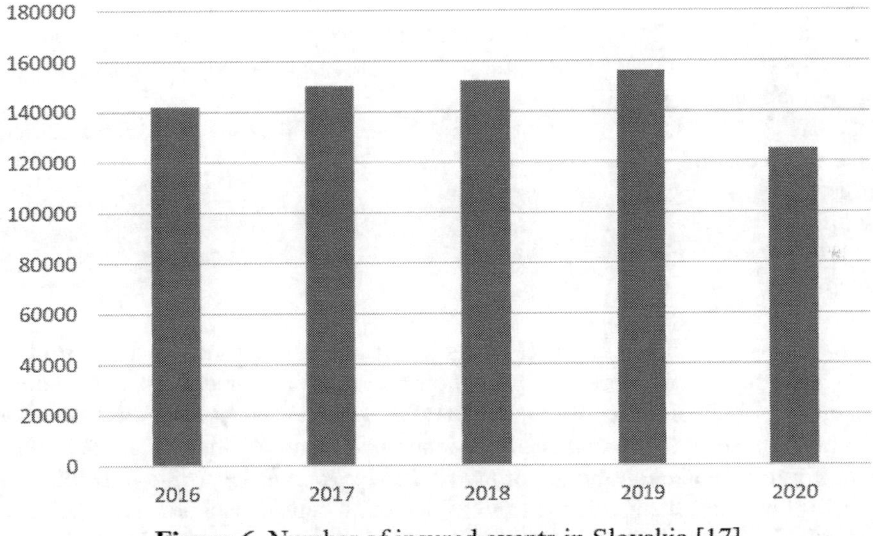

Figure 6. Number of insured events in Slovakia [17]

4.1.3. Insurance amount

In the liability insurance of a national road carrier performing transport on the territory of the Slovak Republic and in the liability insurance of an international road carrier, the sum insured is agreed in the insurance contract, which represents the upper limit of the insurance indemnity in the following options:

- Option A - the sum insured is the highest limit of indemnity for one and all insured events in one insurance period,
- Option B - the sum insured is the highest limit of indemnity for one insured event that occurred during the insurance period. [15]

In the case of an arrangement of Option B of the sum insured, the sum of paid insurance benefits from all claims incurred in one insurance period may not exceed ten times the agreed sum insured. [16]
Unless a specific variant of the sum insured has been agreed in the insurance contract, it is assumed that Option A of the sum insured applies to the insurance. Cabotage insurance is always covered by Option

IOP Conf. Series: Materials Science and Engineering 1247 (2022) 012002 doi:10.1088/1757-899X/1247/1/012002

A of the sum insured, so the sum insured represents the highest limit of indemnity for one and all insured events in one insurance period.

There are the following additional insurances for the product Liability of the road carrier:

- unsecured car parks,
- specialized vehicles,
- loading / unloading,
- used motor vehicles,
- live animals,
- theft of a transported consignment, including a vehicle,
- disposal costs,
- legal costs,
- simple damages,
- Phantom Carrier. [15]

4.2. The consignment insurance

Scope of coverage Insurance against all risks - ALL RISKS insurance. The insurance is arranged in case of damage, destruction, theft or loss of insured consignments as a result of all risks that may arise during the transport of consignments, carried out in accordance with the insurance contract and these insurance conditions.

This insurance covers the costs of a joint accident and rescue costs. There are additional insurances for this insurance product:

- loading / unloading,
- disposal costs,
- additional transport and storage costs,
- winter costs,
- expected profit. [15]

5. Insurance frauds

The Slovak Insurance Association publishes statistics on insurance fraud in the period from 2016 to 2020. Figure 7 shows significant increases in the number of investigated insurance frauds in 2016 and 2020. This suggests a cyclical situation, which may also be caused by the way in which fraudulent activity is detected by insurance companies. If an insurance company identifies a new type of fraud, it will incorporate it into its detection methodology (either by training or using algorithms in automated systems), which will increase the identified frauds. Detection can act as a precautionary measure and at the same time leads to the search for more sophisticated ways to commit new frauds. This creates space for cyclic repetition. This could be verified by longitudinal data collection and analysis.

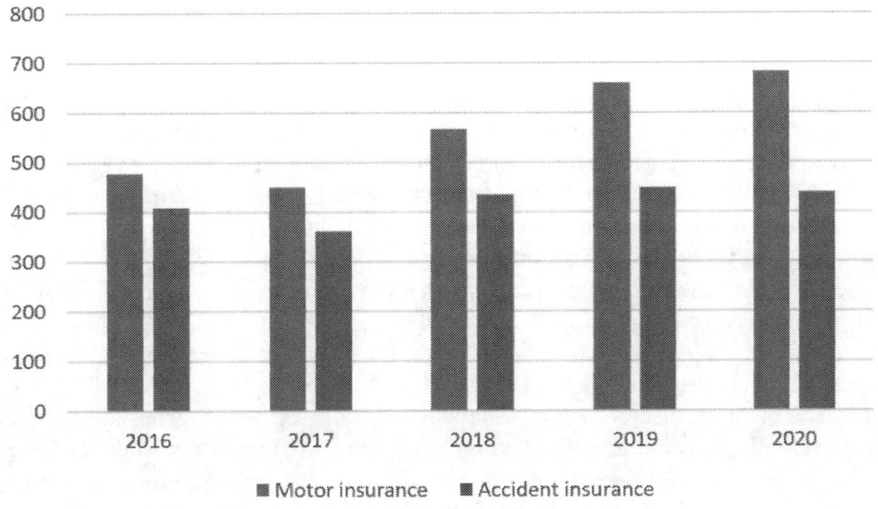

Figure 7. Number of insurance frauds [17]

The quantified value of insurance fraud in Slovakia by Allianz Insurance represented the following values (Figure 8).

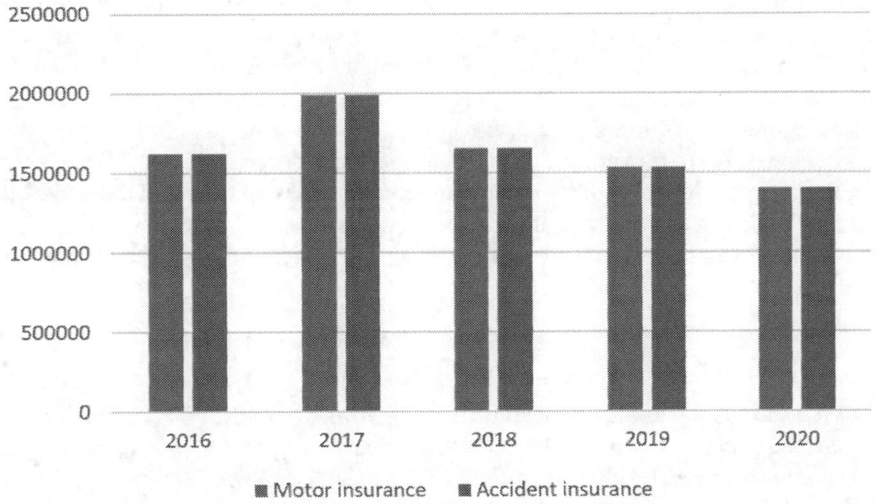

Figure 8. Value of insurance frauds by Allianz Insurance in Slovakia [18]

The primary objective of this study was to examine whether road traffic accidents are directly linked to the purchase of car insurance, and their examination of data on car insurance policyholders revealed that those drivers who purchase more insurance cover are more likely to be involved in accidents, as a result, they will tend to submit more applications. [19] This suggests that insurance coverage may contain information that can be used to assess the levels of probability of RTA risk. We have also found that less safety-conscious drivers will tend to buy more cover and that those who buy more cover will have more accidents and more applications. [20, 21]

6. Conclusion

Measuring something as complex and elusive as organised crime requires a clear understanding of the object of study. It takes a great deal of conceptual clarity and deep insights into the processes and mechanisms at play to know what empirical referents to choose for the measurement and how to interpret whatever data are collected. The scientific process is framed as much by what we know (or believe to know) as by what we do not know. Strategic planning and policy development, in contrast, have to rely on whatever knowledge is available and often appear to be inclined to disregard how narrow a basis of knowledge there is. Today's business environment is made up of business entities that carry out their activities in order to ensure their economic stability. Their goal is to make a profit. In order to achieve these goals, it is necessary to prepare in advance for situations and factors that would lead to destabilization or loss in a particular company. There are currently several options for reducing and eliminating risks. The most preferred type of financial risk coverage is insurance. Businesses have the right to choose, at their own discretion and financial capabilities, the appropriate type of insurance product that will meet their requirements. Insurance is one of the best ways to cover the risks arising from business activities. For businesses, it represents aid in the form of eliminating or reducing losses caused by adverse environmental effects. Today, a number of different insurance products can be found on the insurance market. It only depends on the carrier whether, in order to protect his business, he decides to take out only those types of insurance which are compulsory for him, or whether he also chooses to cover possible damages with one of the voluntary insurances. Insurance helps to solve basic problems that arise in connection with an extraordinary and unexpected event. The reason for the insurance is the existence of risks that endanger any human activity and can cause incalculable damage to the carrier, or even its complete extinction. At the beginning of the 21st century, a comprehensive increase in the level of security is one of the most important tasks facing the entire company, the governments of individual countries, but also the management of each company. On the other hand, the issue of achieving a safe life and continuously improving the standard of living is the goal of every human being. The criminals disguise their activities well. It is nevertheless possible to deploy a number of relatively simple measures to prevent the type of transport loss described here. The objective is always to make sure of the integrity and reliability of any potential business partners. This makes it possible to distinguish any black sheep from the reliable transport companies. In particular, if you are intending to establish new business contacts, it is crucial to subject the potential business partner to a rigorous examination.

References

[1] Association of Logistics and Forwarding of the Slovak Republic. 2012. *Falošná identita a jej zneužívanie pri páchaní trestnej činosti v kamiónovej deprave.* Conference e-Logistika 2021, Žilina. Available at: falosna_identita.ppt (live.com), (20.3.2022).
[2] Association of Logistics and Forwarding of the Slovak Republic. 2012. *Neutralizácia CMR.* Conference e-Logistika 2021, Žilina. Available at: neutralizacia_cmr.ppt (live.com), (20.3.2022).
[3] Daras M. 2018. Register poistných udalostí pomáha odhaľovať podvody. O poistení.sk Technológie. Poistný trh. Available at: Register poistných udalostí pomáha odhaľovať podvody - oPoistení.sk (opoisteni.sk). (18.3.2022)
[4] Defossez D. *CMR: what if the courts got it wrong?* Uniform Law Review, Volume 21, Issue 1, March 2016, Pages 75–100, https://doi.org/10.1093/ulr/unw003
[5] Denes AT. *Considerations on the contract for the international carriage of goods by road.* SO REVISTA DE LA FACULTAD DE DERECHO. 2018, 44, 253-285, DOI: 10.22187/rfd2018n44a9.
[6] Poliak M., Tomicová J. and Jaśkiewicz M. 2020. *Identification the Risks Associated With the Neutralization of the CMR Consignment Note.* Transportation Research Procedia. 44. 23-29. DOI: 10.1016/j.trpro.2020.02.004.
[7] Road freight transport statistics - Statistics Explained (europa.eu). Available at:

https://ec.europa.eu/eurostat/statistics-explained/index.php?title=Road_freight_transport_statistics (28.3.2022)

[8] Mlepo A. 2022. *Attacks on road-freight transporters: a threat to trade participation for landlocked countries in Southern Africa.* Journal of Transportation Security. 1-18. DOI:10.1007/s12198-021-00242-6.

[9] Sternberg He. and Lantz B. 2018. *Using crowdsourced data to analyze transport crime.* International Journal of Logistics Research and Applications. 1-15. DOI:10.1080/13675567.2018.1431211.

[10] Soltes V., Kubas J. and Stofkova Z. 2017. *The impact of a transport safety situation on crime situation and quality of life.* Transport Means. Lithuania, 963-967, https://www.webofscience.com/wos/woscc/full-record/WOS:000576601500171

[11] Lorenc, A. and Kuźnar, M. 2018. *An intelligent system to predict risk and costs of cargo thefts in road transport.* International Journal of Engineering and Technology Innovation. 8. 284-293. ISSN 2226-809X. Available at: (PDF) An intelligent system to predict risk and costs of cargo thefts in road transport (researchgate.net)

[12] Dadsena K., Sarmah S.P. and Naikan, V. 2019. *Risk evaluation and mitigation of sustainable road freight transport operation: a case of trucking industry.* International Journal of Production Research. 57. 1-23. DOI:10.1080/00207543.2019.1578429.

[13] Gagova, P. and Pencheva, V. 2012. *Strategic risks of the road freight forwarding companies working with international cargo.* Transport Problems: An International Scientific Journal. 7. 87-94. Available at: Strategic risks of the road freight forwarding companies working with international cargo | Request PDF (researchgate.net)

[14] https://database.tapa-global.org/incident/index (12.1.2022)

[15] Cunninghama E. Vehicle insurance. The University of Arizona. Risk Management. Available at: Vehicle Insurance | Risk Management (arizona.edu). (12.3.2022)

[16] Risboot P. 2022. Using risk rating in insurance customer acquisition, Digital Insurance. Telematics. Available at: Using risk rating in insurance customer acquisition | Digital Insurance (dig-in.com). (28.3.2022)

[17] Slovak Insurance Association. *Aggregated data for members of the Association on the number of claims*. Databses 2016-2020. Avalable at: Štatistiky SLASPO od roku 2016 | Slovenská asociácia poisťovní. (28.3.2022)

[18] Allianz Slovak Insurance. Allianz Societas Europaea. Databases of value of insurance frauds by Allianz Insurance Slovakia. Avalable at: https://www.allianz.sk/sk_SK/sukromne-osoby/ine-stranky/tlacove-centrum.html. (25.3.2022)

[19] Hsu Y.CH., Shiu Y.M., Chou P.L. and Chen, Y.M.J. 2015. *Vehicle insurance and the risk of road traffic accidents.* Transportation Research Part A: Policy and Practice. Volume 74, 2015, Pages 201-209, ISSN 0965-8564, DOI:https://doi.org/10.1016/j.tra.2015.02.015.

[20] Safarpour H., Khorasani-Zavareh D. and Mohammadi R. 2020. *The common road safety approaches: A scoping review and thematic analysis*. Chinese Journal of Traumatology. Volume 23, Issue 2, 2020, Pages 113-121, ISSN 1008-1275, DOI:https://doi.org/10.1016/j.cjtee.2020.02.005.

[21] Blows S., Ivers R.Q., Connor J., Ameratunga S. and Norton R. 2003. *Car insurance and the risk of car crash injury.* Accident Analysis & Prevention, Volume 35, Issue 6, 2003, Pages 987-990, ISSN 0001-4575, DOI:https://doi.org/10.1016/S0001-4575(02)00106-9.

Acknowledgments
This work was developed under the support of project: MSVVS SR – VEGA No. 1/0245/20 Poliak, M.: Identification of the impact of a change in transport related legislation on the competitiveness of carriers and carriage safety.

Identifying the Age of a Vehicle for Commercial Safety and Forensic Purposes Using "Window Stamps"

R. Rak[1*], V. Sulc[1], F. Vlach[1], D. Kopencova[1], V. Hudecova[2]

[1] Department of Security and Law, Ambis University, Lindnerova 1, 180 00 Prague 8, Czech Republic
[2] Department of Administrative Law, Police Academy, Sklabinska 8414/1, 835 17 Bratislava, Slovak Republic

*Rak.Roman@ambis.cz

Abstract. This paper deals with the issue of rapid, indicative identification of the age of a vehicle using time stamps on various components of the vehicle for technical, commercial, safety and forensic purposes. The main aim of the paper is the possibility of using markings of individual vehicle windows or "window stamps", in which the times of production of the windows are encrypted in various ways. An analysis of time markings on vehicle windows was conducted on the basis of an extensive collection of photographs of windows from 980 basic vehicle models manufactured between 1991 and 2021, and acquired between 2015 and 2021 as part of the Europe-wide initiative eCall. Research found that 98.8% of all common motor vehicle window panes on the European market come with one of only three basic types of window stamps, which were named Euro-American, Japanese and Korean. The paper provides a detailed description of the structure and decoding of vehicle window stamps of the three specified basic types.

Keywords: vehicles, safety, automotive safety

1. Introduction

For various reasons, in commercial, administrative, security (police) and forensic practice, we need to identify vehicles and their age (date of manufacture) in order to make decisions on further procedures. For that purpose, there is a wide range of identifiers and possibilities that require access to technical and administrative documentation, either in a limited form in vehicle documents, or in an extensive form, where we obtain the required information by connecting vehicle identifiers (registration numbers or VINs (*Vehicle Identification Number*), or other unique identifiers of individual production components) to various information systems (national vehicle registers, third-party computer records, security and police information systems, or the information systems of vehicle manufacturers themselves and their brand's servicing networks [1, 2, 3, 4]).

In practice, we may quite often encounter a lack of important information, when the vehicle documents or access to the relevant information systems is not immediately available, and we have to find the age of the vehicle (or its year of manufacture) in the field on the spot. The same problem arises in situations where, even though we have the vehicle documents, we have doubt about their accuracy and need to verify the information in the field on the spot without any other documents or access to remote information systems.

IOP Conf. Series: Materials Science and Engineering 1247 (2022) 012003 doi:10.1088/1757-899X/1247/1/012003

When faced with a lack of information in the field, we can ascertain a vehicle's year of manufacture (or the model year of manufacture) [5, 6, 7]:

- by decoding the VIN;
- from markings on the seat belts;
- from the dates or years of individual components;
- from the window markings ("window stamps").

1.1 Determining a Vehicle's Age by Decoding its VIN

In the structure of the VIN, specifically in the 10th position, the manufacturer can locate the year of manufacture (see Table 1), according to the applicable standard ISO 3779:1983 – *Road Vehicles -- Vehicle identification number (VIN) -- Content and structure*; ISO 3780:1983 – *Road vehicles -- World manufacturer identifier (WMI) Code* [8]. This is not an obligation, however, but merely a recommendation. Some manufacturers make use of this option, and specify the model year in the 10th position of the VIN. In order to decode this information ourselves, we need to be familiar with the particular conversion table, whether the manufacturer has made use of this option for the specific vehicle model, and the approximate time period in which the vehicle was made, because the codes in Table 1 are periodically repeated after 30 years.

1.2 Determining a Vehicle's Age from Seat Belt Markings

Some manufacturers (especially from Asia and Europe) mark seat belts with cloth or paper identification labels, which are attached to the seat belt in a suitable way around the area of its lower fastening. Selected manufacturers tend to mark these labels at the very least with the year of manufacture, and very often with the exact date. Seat belts are produced slightly before the vehicles themselves are manufactured (completed). In the case of lean manufacturing, numerous observations and analyses have found that the maximum time difference between the production of the belt and the assembly of the vehicle is up to 3 months, and in practice often many times less.

Figure 1. Examples of labels displaying the MFD (manufactured) date on seat belts. Source: Roman Rak

1.3 Determining a Vehicle's Age from Markings on Components

Much like seat belts, markings can be found on many components of vehicles [9]. In older vehicles they are marked with day/month/year by embossing or pressing. Modern-day vehicle components are marked using laser technology, engraving, scoring or stickers, with dates marked using digits, bar codes or 3D codes [10]. In this case too, there is no significant difference in the times between the manufacture of individual components and the vehicle. Not all vehicle manufacturers always mark seat belts with a time stamp. However, by examining a vehicle thoroughly we will find a considerably larger number of components marked with the time of their production, but not necessarily accurate to the day [11].

1.4 Determining a Vehicle's Age from Window Stamps

In accordance with current standards, all vehicle windows are marked with "window stamps", in which we can find a wide range of information, including the time of the window's manufacture. Unlike the markings of other vehicle components, window stamps can be found on all vehicle windows, observable from the outside without having to open the vehicle's cabin, engine space or luggage compartment. Window stamps are therefore the first easily accessible places for ascertaining the time of a window's production and, thereby, the approximate age of a vehicle that has been fitted with such a window. A key disadvantage of window stamps is that the date of manufacture is visually encoded, and in different ways for different makes.

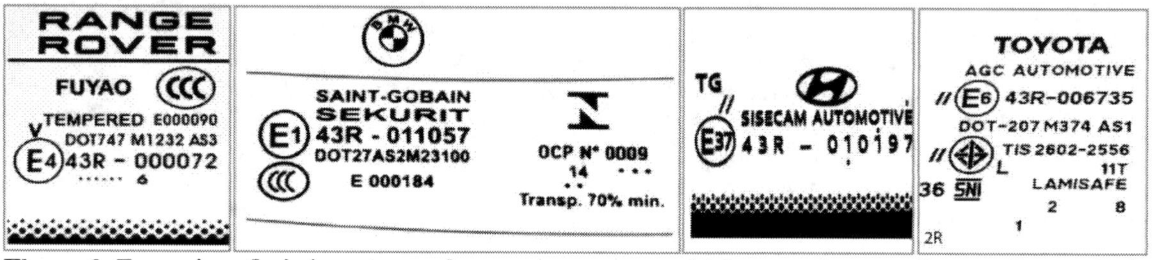

Figure 2. Examples of window stamps from various motor vehicle manufacturers Source: Roman Rak

2. Materials and Methods

An analysis of the structures and encoding of window stamps was conducted with a sample of 980 basic passenger car models covering the European market from 1991 to 2021. The research project was conducted at the Academy of the Police Force in Bratislava (Slovakia) from 2020 to 2021. The sample of basic models (980 in total) was acquired for the task of forensic identification of motor vehicles for the needs of investigative and forensic practice, and for the needs of the units of the Integrated Rescue System as part of the international eCall (emergency call) [12] initiative, which covers all EU member states. For the purposes of identifying vehicles and following correct technological processes when freeing persons from crashed vehicles, a set of photographs was taken for every model in an average quantity of 110-130 photographs per model. The photos captured the exterior and interior of the vehicles, their engine space and luggage compartment, cabin, basic vehicle identifiers, methods of opening the engine space and luggage compartment, the engine itself, the battery, etc. This photo-documentation also included images of all individual windows of the vehicles, including their visual identifiers. On

Table 1. Code identification of a vehicle's model year or year of manufacture in the 10th position of its VIN

Model year	Code	Model year	Code	Model year	Code
1980	A	1990	L	2000	Y
1981	B	1991	M	2001	1
1982	C	1992	N	2002	2
....	......	......	.....	....	.....
1989	K	1999	X	2009	9
2010	A	2020	L	2030	Y
2011	B	2021	M	2031	1
...	....	*2022*	....	....	....
2019	K	2029	X	2039	9

average, 10 photos were taken of the window stamps on various parts of all of the vehicles' windows. All of the images were analysed and compared with the basic technical standard. The conducted analyses established basic types of window stamps, their content and means of decoding the information stored in them.

3. Dates of Manufacture of Car Windows and Means of Determining them

One of the most complicated aspects when decoding window stamps is identifying a window's date of manufacture, because there are no unified norms and standards establishing means of dating car windows.

A detailed analysis ascertained that there are currently three basic methods used in practice, which can slightly differ. These three methods of dating can be named after the countries or continents where they are primarily used:

- **Euro-American**
- **Japanese**
- **Korean**

All three types of dating share a common principle, specifically the use of dots for decoding dates (except for the Japanese method, which has replaced dots with numbers, but the principle remains the same). They differ in the methods and locations in which dots are used.

3.1 Euro-American dating

Euro-American dating is based on dots and numbers, which appear, in most cases, near the bottom of the window stamp. The number located either below, next to or in between the dots identifies the year. Individual months are then identified by the dots next to or above the number (the location of the dots determines the method of decoding). The last places where dots can occur may be in the area under the number. These dots identify individual weeks in which windows are produced.

Figure 3 An example of one of the most frequently used forms of encoding the date of manufacture in a window stamp

Generally speaking, the method depicted in Figure 3 is the most commonly used form of encoding. The image shows the most common means of encoding the date in the Euro-American window stamp type. In this example, the figure 19 represents the year 2019. The number of individual dots on the left and right-hand sides identify the month. The dots underneath the figure 19 represent the number of the week within the particular month of manufacture. For example, three consecutive dots to the left of the number 19 (the year) mean the month of March.

IOP Conf. Series: Materials Science and Engineering 1247 (2022) 012003 doi:10.1088/1757-899X/1247/1/012003

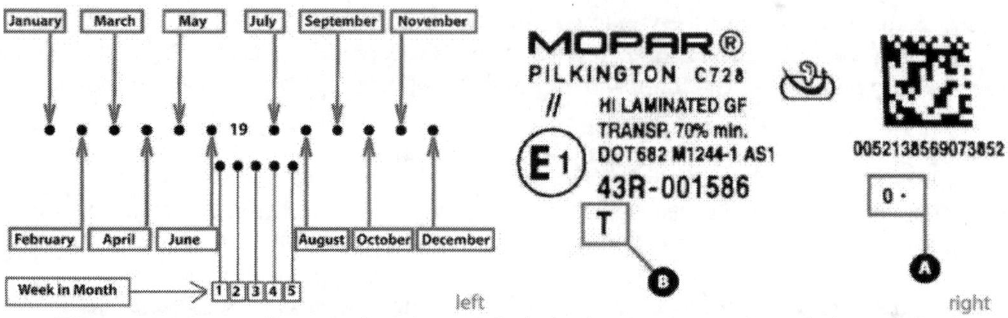

Figure 4. Means of expressing time periods in the Euro-American dating method

The box marked (A) in a window stamp from a Fiat 500 (Figure 4, right) contains the number 0 and one dot (meaning July of either 1990, 2000, 2010 or 2020); the letter T in the red box marked (B) then helps us to identify the precise year of manufacture as 2020 (T is the 20th letter in the English alphabet, see Table 2).

3.1.1 First Euro-American dating type

The first most frequently used method for vehicle windows made in Europe and America is shown in Figure 3. This type appears relatively simple to decode. A problem occurs when, instead of the number 19 (identifying the year), we find the number 1 (or any number from 0 to 9). In such cases, 1 could mean 2001, 2011 or 2021. To find the exact decade of production, we must be able to determine it based on the vehicle's model year, according to the known period of a minor model change, markings of the vehicle's components, identification labels and stickers, etc. Automotive specialists can easily ascertain such information.

Another way of more precisely decoding a window's date of manufacture is by using the English alphabet, or a number that can occasionally be found near the basic date markings. The year of manufacture can be identified according to the sequence of letters in the English alphabet and the corresponding numbers as shown in Table 2.

Table 2 The sequence of letters in the English alphabet and the corresponding numbers for identifying years of manufacture

A=1	**H**=8	**O**=15	**V**=22
B=2	**I**=9	**P**=16	**W**=23
C=3	**J**=10	**Q**=17	**X**=24
D=4	**K**=11	**R**=18	**Y**=25
E=5	**L**=12	**S**=19	**Z**=26
F=6	**M**=13	**T**=20	
G=7	**N**=14	**U**=21	

Some stamps use English letters that do not represent an exact number, but rather supplement another number, see Figure 4, right.

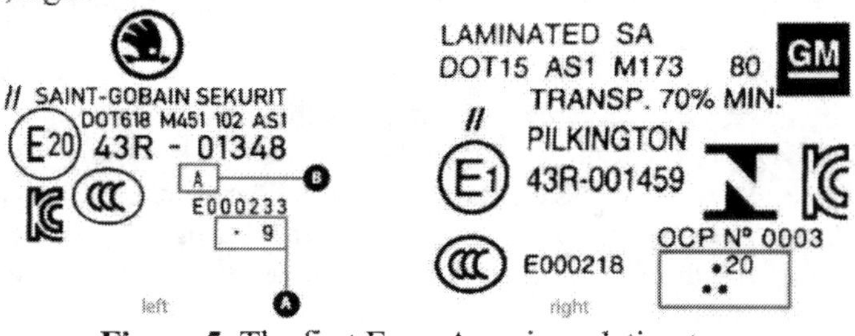

Figure 5. The first Euro-American dating type

Figure 5 (left) shows a window stamp from a Škoda Kamiq made in January 2019 (one dot before the number 9 under the letter (A) represents the month, see Figure 4, left, and the number 9 located below the letter (A) identifies the possible year as either 1999, 2009 or 2019. The letter A in the box marked (B) identifies the decade).

The examples in Figure 4 (left) and Figure 5 (right) are less commonly used methods, but it is important to be able to recognise them. This first dating type more commonly only uses a number and dots representing the month, and possibly the week of the window's manufacture.

The window stamp in Figure 5 (right) with a red box highlighting the method of encoding the date of the window's manufacture is taken from a Cadillac XT4. The date of manufacture is the second week (two dots at the bottom of the box) in January (one dot before the number 20) of 2020 (the number 20 itself).

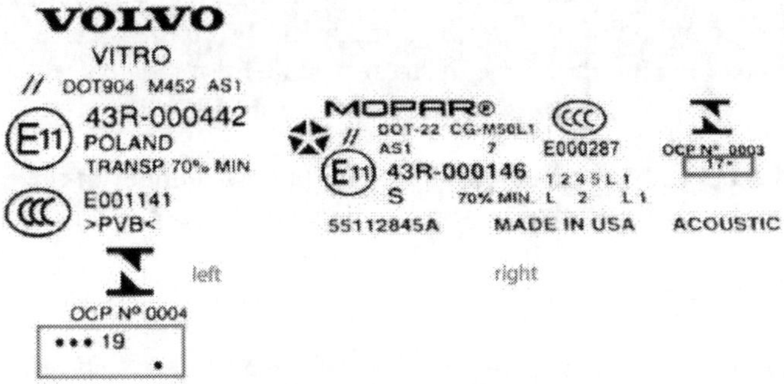

Figure 6. Window stamp from a Volvo V60 Cross Country (left) and a Dodge Durango (right)

Figure 6 (left) shows a window stamp from a Volvo V60 Cross Country highlighting the encoded date of manufacture (framed in red). The car window was manufactured in the first week (one dot at the bottom of the red frame) in March (three dots to the left of the number 19 represent the third month) of 2019 (the number 19 marks the year).

The example window stamp in Figure 6 (right) is taken from a Dodge Durango, again highlighting the window's date of manufacture (framed in red). In this case, the car window was made in July (one dot to the right of the number 17) of 2017 (the number 17 marks the year).

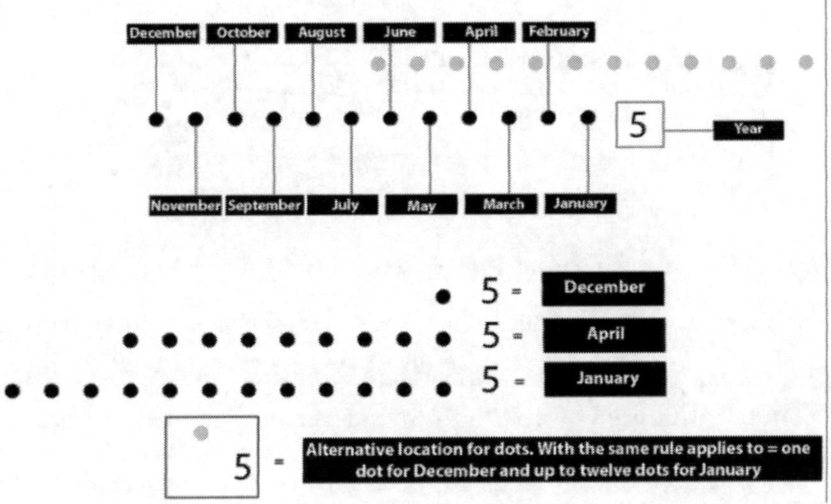

Figure 7. An alternative way of expressing the month of manufacture by subtraction

3.1.2 Second Euro-American dating type

The second type differs from the first in the way that the dots are positioned or the number of dots on one side of the number. This method uses subtraction. If a dot is in the 12th position (see Figure 7), this does not mean December as in the first case, but rather that the month of manufacture is January. In this case the dots are located either in the section above the number, or just to one side of the number. A problem occurs in cases where the window manufacturer uses precisely this subtraction method, and on one side we find, for example, two dots. It is not then possible to tell whether the month is February or November.

Figure 7shows an example for decoding the date of manufacture from the second Euro-American window dating type. Black dots are located next to the number five (framed in red) and grey dots are located above it. The black dots represent one way of locating the month in the window stamp, and the grey dots represent the second possibility for locating them. From the picture, it can be deduced that 12 dots either before or above the number 5 mean January, and 1 dot means December. For a complete list of the months and the corresponding number of dots, see Table 3.

Table 3 Distribution of individual months according to the number of dots in the second Euro-American car window dating type

Number of Dots		Month
••••••••••••	12	January
•••••••••••	11	February
••••••••••	10	March
•••••••••	9	April
••••••••	8	May
•••••••	7	June
••••••	6	July
•••••	5	August
••••	4	September
•••	3	October
••	2	November
•	1	December

Figure 8. Window stamps from a BMW series 7 (left) and a Škoda Kodiaq (right)

Figure 8 (left) shows a window stamp from a BMW series 7, highlighting the section where the window's date of manufacture is encoded. The window was produced in October (four dots above the number 4) of either 1994, 2004 or 2014. According to an analysis of the car's VIN (WBAYB01040D126525), the car was made in 2014, and therefore the window was also made in 2014.

Figure 8 (right) shows a window stamp with the encoded date of the window's manufacture framed in red; it contains the number 1, identifying the window's year of manufacture as either 1991, 2001, 2011 or 2021. Above the number there are 12 dots, identifying the month of manufacture as January. We cannot ascertain any further information about this window stamp (source of image[1]), so the exact date of manufacture cannot be identified with certainty. From the information available in the window stamp, we can identify the month of manufacture as January (see Table 3). The window's year of manufacture is either 1991, 2001, 2011 or 2021. Information which could help us to identify a more precise date of the window's manufacture includes, for instance, the technical properties of the glass (HUD, for example, would almost certainly eliminate 1991 and 2001), or the model of the car (for example, if the window was fitted in a Škoda Kodiaq, its year of manufacture would be 2021, because this car has only been in production since 2016).

Figure 9. Examples of encoding the year of manufacture using the subtraction method (left) and the Japanese dating method with an imaginary decoding table

Figure 9 (left) shows another car window stamp. The red frame contains the number 8, which identifies the possible year of manufacture as 1998, 2008 or 2018. You can see 11 dots to the left of the number 8, identifying the month as February (see Table 3). As we do not have any further information for this kind of window stamp, all we can say with certainty is that the window was made in February of either 1998, 2008 or 2018. We can determine the decade of manufacture based on further knowledge of the examined vehicle.

3.2 Japanese dating type

The Japanese dating type is distinguished by the fact that, at a first glance, the dots or numbers seem to make no sense. In order to decode them, we need to draw an imaginary grid, which will help us to identify the window's date of manufacture. This dating type can take the form of dots or numbers, where it is more complicated to decode the meaning of the dots in some cases (for example, when there are not enough dots for us to create a decoding grid in the area where the date is encoded).

The data in the window stamp in Figure 9 (right) can be decoded with the help of the outlined decoding grid.

The numbers in the rows from top to bottom mean:

1) years
2) months
3) days

There are two Japanese dating types.

3.2.1 First Japanese dating types

The first Japanese dating type uses dots. In order to decode the information, we need to create an imaginary calculation grid (see Figure 9, right) to help us determine when the car window was made. In the individual cells on each row, we enter powers of the number 2 (1, 2, 4, 8 and 16).

In the row to the left of the symbol (1) (in Figure 9, right), you will see a row with dots above the numbers 1 and 4. These numbers are added together (1+4=5), identifying the year of manufacture as 2005. The row to the left of the symbol (2) identifies the month of manufacture. Once again, we add the numbers together, revealing the month of May (1+4=5 and May is the 5th month of the year). The last row to the right of the symbol (3) helps us to ascertain the day of manufacture, in this case the second day of the month. From the information in the window stamp, we can therefore date the production of the window to 2 May 2005.

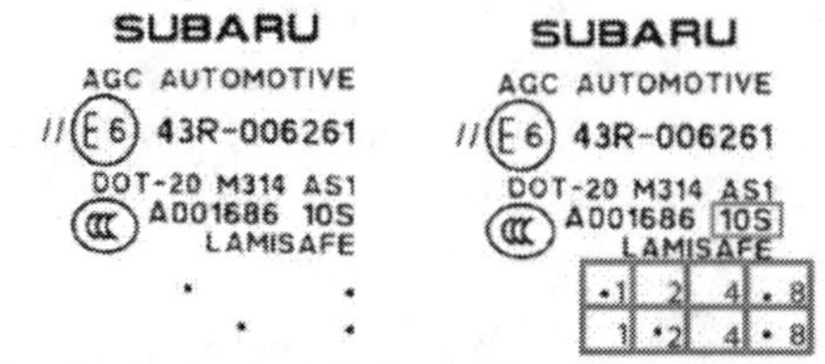

Figure 10. Window stamp from a Japanese Subaru Forester with an example of the first Japanese encoding type

The left side of Figure 10 shows a standard window stamp located on a Subaru Forester window. The right-hand side of the image presents an example of how to decode the first Japanese dating type using an imaginary table. A table has been drawn up around the dots at the bottom of the image (see left-hand side), with each cell marked with a number. When added together, the numbers in the cells containing the dots (1 and 8) in the top row (indicating the year of manufacture) equal the number 9 (identifying the year of manufacture as either 2009 or 2019). Adding up the numbers in the bottom row (2 and 8) produces the number 10, corresponding to the 10th month of the year (October). The red frame above the table highlights the data "10S". Adding the number 10 to the earlier possible year of manufacture (2009) gives us 2019, and S is the 19th letter of the English alphabet (see Table 2). The letter S helps us to determine the year of manufacture as 2019, not 2009, according to the relevant table. This confirms that the window was manufactured in October 2019.

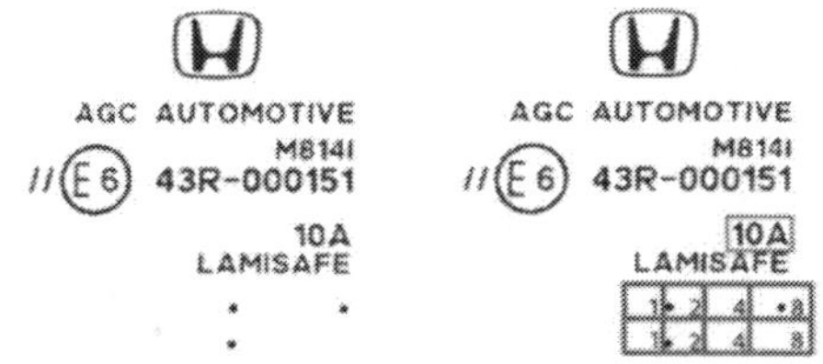

Figure 11. Window stamp from a Honda Jazz

The left-hand side of Figure 11 shows a window stamp from a Honda Jazz. The right-hand side shows a modified version with a table and a frame for decoding. The top row of the table (serving to decode the year of manufacture) contains dots in positions 2 and 8. The sum of these numbers is 10 (identifying the year of manufacture as 2010 or 2020), while the bottom row contains a dot in position 2. This indicates the second year of the month (February). The red frame above the table containing "10A" helps us to determine the exact date, just like in the previous example. Adding the number 10 from the frame to the previous sum of 10 (2+8) gives us the year 2020. The letter A here, just like in Figure 5, helps us to clarify the decade. The first decade ends in 2010 (the sum 2+8 in the first row of the table). The second decade is indicated by the letter A = 1 (see Table 2). The resulting decade is 2010 + 10, i.e., 2020. This example does not specify the day of production (there are only two rows in the table).

A problem with this kind of window dating arises when there are only two dots, one above the other, in the space for decoding the date (it is not then possible to draw up a decoding grid with the corresponding numbers with certainty and to obtain the date of manufacture).

3.2.2 Second Japanese dating type

Instead of dots, the second Japanese dating type uses numbers. It uses the same method as with dots, but in the created grid there are only numbers where the dots would be. This method is easier to decode. The same rules as in the previous method apply. The numbers can be arranged in up to three rows, where the top row indicates the year, the second row identifies the month, and the last row indicates the exact day of the window's manufacture (the third row indicating the precise day is less common and is not required).

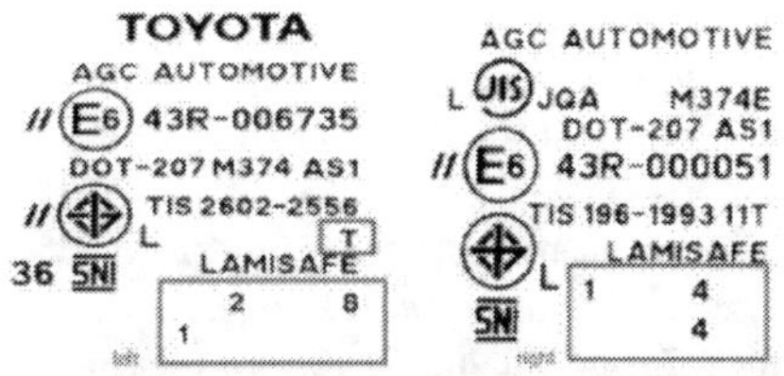

Figure 12. Window stamps from a Toyota Hilux (left) and Suzuki Celerio (right)

Figure 12 (left) shows a window stamp from a Toyota Hilux, with two areas marked for decoding the date of manufacture. The lower box highlights the second Japanese method of encoding the date of manufacture. This method is simpler, because we do not have to form a grid to ascertain the numerical values in the rows (the first Japanese type contains dots, which are located in similar places to the numbers in this method). The top row produces a sum of 10 (2+8), which gives us a value indicating that the window was manufactured in 2010 or 2020. The bottom row contains only the number 1, corresponding to January (the first month of the year). Finally, the letter T can be found in the upper box (helping us to ascertain the year of production using the English alphabet), representing the number 20 (see Table 2). From this information, we can deduce that the window was made in January 2020.

Figure 12 (right) shows a window stamp from a Suzuki Celerio with boxes highlighting the data encoding the date of manufacture. The top row contains the numbers 1 and 4, which produce a sum of 5 (possible years of manufacture are 1995, 2005 and 2015), while the bottom row contains the number 4, indicating the 4th month (April). The model year for this particular car is 2015, meaning that the window was made in April 2015.

3.3 Korean dating

Korean car window dating uses dots, which are most often located above and below the name of the window manufacturer in a **window stamp**. In this encoding system, the dots above the window manufacturer's name indicate the year of manufacture, while the dots below it indicate the month. There are two Korean dating types:

- The first Korean dating type uses the window manufacturer's name to encode the date of manufacture.
- The second Korean dating type uses a DOT code to encode the date of manufacture.

3.3.1 First Korean dating type

The first Korean dating type is distinguished by dots above and below the window manufacturer's name. While there are no clearly set encoding rules for individual manufacturers, as a general rule the dots placed above the manufacturer's name are used to identify the year of manufacture, and the dots below the name are used to indicate the month.

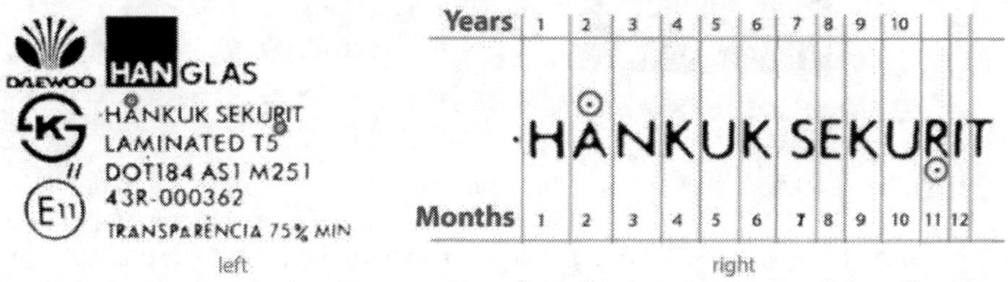

Figure 13. Window stamp from a Daewoo Evanda (left) and the means of decoding the time of manufacture (right)

IOP Conf. Series: Materials Science and Engineering 1247 (2022) 012003 doi:10.1088/1757-899X/1247/1/012003

Figure 13 (left) shows a window stamp from a Daewoo Evanda. The red circles contain dots that are used to decode the window's date of manufacture. The upper dots (in this case one above the letter A) help us to decode the year of manufacture, and the lower dots (in this case one below the letter R) help us to decode the month.

As mentioned above, the dots above the manufacturer's name serve to identify the year; a table can be drawn around the letters in the name to make it easier to decode the year and month of manufacture.

In Figure 13, a grid for decoding Korean dating is shown on the right. The top section contains data for decoding the year of manufacture, and the lower section for the month. It can therefore be said that the dot for the year of manufacture in this window stamp is located above the letter A, which corresponds to the position in auxiliary Table 2 (possible year of manufacture 2002 or 2012). This example does not give us enough information to determine the exact year of manufacture, but the possibilities are 2002, 2012, 2022, etc. In this case, we proceed in the established manner and determine the decade of manufacture based on other information about the vehicle that is available for us to examine.

By decoding the VIN, we ascertained that the model year was 2003. The lower dot serving to identify the month is located under the letter R and is marked with the number 11. The window was therefore made in November 2002 (the lower dot under the letter R is marked with the number 11). So the window was made in November 2002.

This dating type has already been shown in the first two pictures document. It is a method where dots are used above and below the name of the window's manufacture. The upper dots serve to decode the year of production, while the lower dots indicate the month.

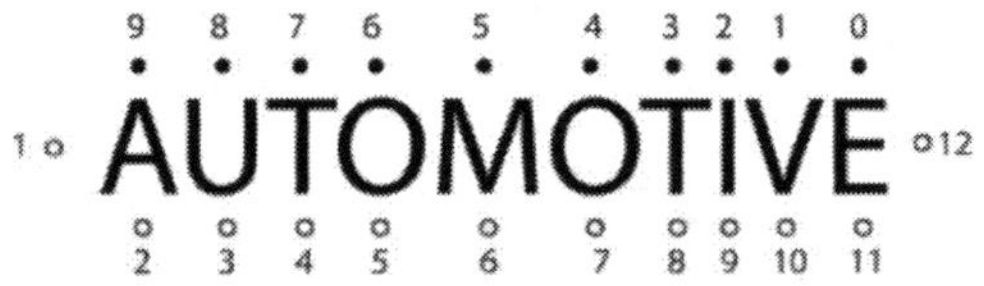

Figure 14. Example of decoding the time of manufacture of a window from a Lexus

Figure 14 provides an example of the procedure for decoding the first Korean dating type in the case of a window made by AGC AUTOMOTIVE (found primarily in the Lexus factory). The full dots in the top row indicate the year of manufacture, where 0 encompasses all whole numbers. The rings by the side of and under the letters of the word AUTOMOTIVE indicate the month of the window's production, where 1 means January and 12 means December.

Figure 15. Window stamp from a Lexus LC500 (left) and a Daewoo Evanda (right)

Figure 15 (left) shows a window stamp from a Lexus LC500 with dots circled in red above and below the manufacturer's AUTOMOTIVE brand for decoding the date. The year of manufacture is indicated in the upper circle (the dot above the letter E). The value above the letter E in this case corresponds to the number 0, meaning either the year 2000, 2010 or 2020. The dot in the bottom row designating the month of manufacture is located under the letter M, corresponding to the number 6 or June (the 6th month). The final and additional information for decoding the date is the make and type of motor vehicle itself. The model year is 2021. We can therefore say that the window was made in 2020.

Figure 15 (right) shows a window stamp from a Daewoo Evanda. The places for decoding the date of manufacture are located above and below the manufacturer's HANKUK SEKURIT brand. The same rule applies as for all Korean dating types, where the dots above the name denote the year, and the numbers below indicate the month. According to the sample image, the letter A corresponds to the number 2 for decoding the year of manufacture (2002, 2012 or 2022). The lower dot indicating the month of production is located under the letter R, which marks the 11th month (November). The Daewoo Evanda's model year is 2003. From the information that we know from the window stamp and the vehicle itself, we can deduce that the window was made in November 2002.

3.3.2 Second Korean dating type

The second Korean dating type uses dots above and below a DOT code. The method of decoding remains the same. The upper dots indicate the window's year of manufacture, while the lower dots identify the month.

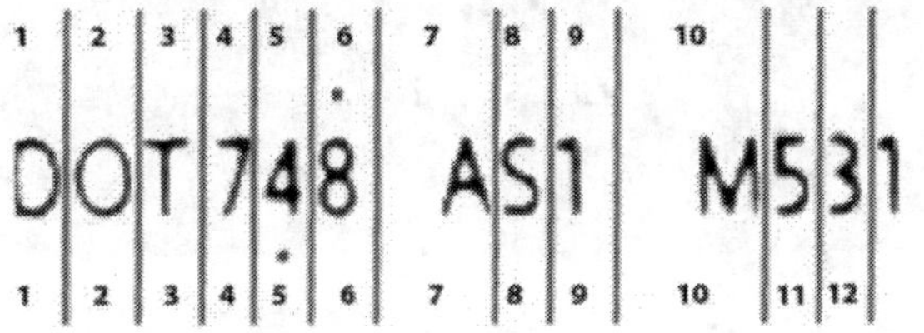

Figure 16. Basic method of decoding the second Korean type

Figure 16 shows the method of decoding the second Korean type. The spaces for decoding the date are located below and above the DOT code. The dots above the DOT code indicate the year of manufacture, where the code starts with the letter D, representing the number 1 for the years 1991, 2001, 2011 or 2021, and carries on until the letter M, indicating the number 10 (1990, 2000, 2010 or 2020). The bottom row indicates the month, where the code starts with the number 1 (January) and ends with the number 12 (December, in this image located under number 3 in the DOT code).

Figure 17. Window stamps from a Kia Sorento (left) and a Genesis G90 (right)

Figure 17 (left) shows a window stamp from a Kia Sorento. The spaces for decoding the date are located above and below the DOT code. The year of manufacture can be found in the red circle above the letter M in the DOT code, representing the number 10 (possible years 1990, 2000, 2010 and 2020). The month of the window's manufacture is located under the number 8, which in this case means June, the 6th month of the year (the number 8 is in the 6th position of the code). The model year of the Kia Sorento is 2021, so according to this information, the window was made in June 2020.

Figure 17 (right) shows a window stamp from a Genesis G90. The areas for decoding the date of manufacture are located below and above the DOT code, where the upper dots indicate the year, and the lower dots identify the month of the window's manufacture. The month of manufacture can be identified as May, because the lower dot is in the 5th position of the DOT code under the number 4. The year of manufacture is indicated by the upper dot, which is in the 6th position under the number 8. The sixth position could represent the years 1996, 2006 or 2016. The car's model year is 2017, so the window was made in May 2016.

3.4 Less common ways of encoding dates of manufacture in window stamps

Because there is no uniform regulation for dating car windows, in practice we may also encounter other window stamp dating types, which are in the minority in the European Union.

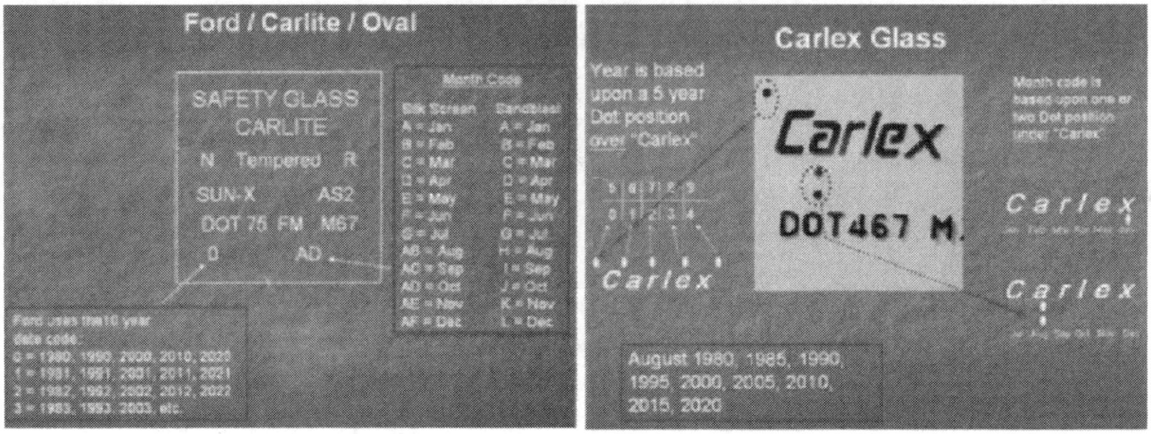

left right

Figure 18. Examples of Carlite codes

Figure 18 (left) shows an example of how to decode a Carlite window stamp. The box in the bottom left-hand corner shows the method of decoding the year of manufacture with an arrow pointing towards the figure 0 (same method of decoding the year as in the Euro-American type). The right-hand box contains two methods of determining the month of the window's production.

Figure 18 (right) shows a way of decoding Carlex window stamps. On the left-hand side (arrow from the dot pointing downwards to the left), you can see the method of decoding the year of manufacture. Above the word "Carlex", we create a table, and decode the year according to the position of the dots, e.g., the one dot above the letter C in this example represents either 0 or 5 (possible years of manufacture are 1980, 1985, 1990, 1995, 2000, 2005, 2010, 2015, 2020, and so on). The arrow pointing to the right-hand side shows us how to decode the month. A single dot underneath one of the letters in the word "Carlex" indicates a month from January to June (one dot under the letter C means January, and one dot under the letter X means June). Two dots under one of the letters indicate the months from July (C) to December (X).

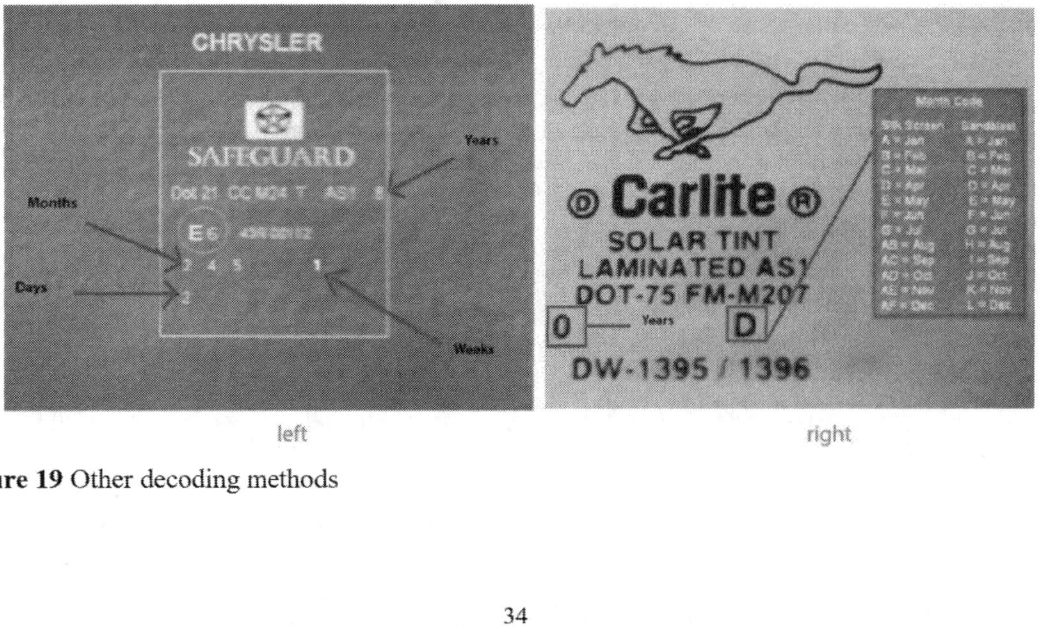

left right

Figure 19 Other decoding methods

Figure 19 (left) shows a Safeguard window stamp on an unidentified Chrysler model. The method of decoding is similar to the second Japanese type. The arrows show us how to decode the year of manufacture (the top right arrow pointing to the number 8) and the month (top left arrow pointing to the numbers 2, 4 and 5). In the bottom right, the number 1 indicates the week of manufacture, while the bottom left arrow points to a number indicating the day of the week. The decoded date on this window stamp is the second (day = 2) of November (month 2+4+5=11) of either 1988, 1998, 2008 or 2018 (years = 8).

Figure 19 (right) shows a window stamp of the manufacturer Carlite. The method of decoding is demonstrated by the red boxes. On the left-hand side, you will find the number 0, indicating the year of manufacture (same method of decoding as in the Euro-American type). On the right-hand side, the month is decoded using letters. In this example, you will find the letter D, which identifies the precise month (April) according to the table on the right. This window stamp was therefore produced in April of either 1980, 1990, 2000, 2010 or 2020.

4. Discussion

Identification of a vehicle, including its age, is a common task in automotive, commercial, safety and forensic practice, and can be put to practical use in various ways [13, 14]. There are many other ways to determine a vehicle's exact age. The most accurate method, although not always the fastest, simplest, or most readily available, is identifying a vehicle based on its VIN (global identifier) in the manufacturer's information system [15]. From there, it is possible to obtain the exact date of production of any given vehicle. Identifying a vehicle this way requires access to the above-mentioned manufacturer's database and the "willingness" of the manufacturer to provide the required information to interested parties from various institutions (individuals, private companies, third-party partners, and government authorities and institutions, including the police, investigating bodies, courts and security forces) [16]. As a general rule, access to a manufacturer's systems is limited to a restricted group of potential enquirers, based especially on applicable legislation [17, 18].

Situations arise, however, where we need to very quickly determine at least a vehicle's approximate age using various secondary identifiers during an initial examination of the vehicle at the site of an inspection or, in extreme cases, the scene of a crime, and only then, in case of further need and possibly after some time, can we identify the vehicle more precisely [19]

Vehicles generally contain a range of various components marked with the time of manufacture. Some can be identified to a specific day, while others can only be identified to a precise month or year of manufacture. For primary orientation, such time periods are more than sufficient. Today, vehicles are assembled on production lines "just in time", meaning that the date of their assembly in the modern globalised world does not differ fundamentally and extensively from the date of their completion with individual components marked with various time stamps. In five years of research and examination, we have found maximum time differences of just 1-3 months between the manufacture of individual components and the completion of a whole vehicle.

Another factor that decides on the method of determining a vehicle's age is whether we have the opportunity or the right to examine the interior, i.e., to open the vehicle by some means, and to provide access to spaces such as the cabin, engine space, luggage compartment and other spaces. We cannot always do so, and it is in precisely such situations that window stamps, which are a legitimate source of information for any observer from the outside of the vehicle, provide a suitable indication as to the approximate time of manufacture.

5. Conclusions

This paper has analysed the encoding of dates of manufacture in window stamps on 980 basic passenger car models manufactured between 1991 and 2021. The analysis found that there are three basic methods of encoding time data in window stamps: the Euro-American, Japanese and Korean

IOP Conf. Series: Materials Science and Engineering 1247 (2022) 012003 doi:10.1088/1757-899X/1247/1/012003

methods. Each type uses two other, more detailed methods of logically indicating a window's date of manufacture with dots, letters and numbers. We found that the three above-mentioned basic methods of indicating the date of manufacture of passenger car windows cover 98.8% of all commonly found vehicles on European roads within the past 30 years.

A fundamental problem when attempting to determine a vehicle's date of manufacture based on its windows or window stamps is the fact that there is no standardised, uniform procedure, and a person who wants to use this method must have good knowledge and experience of the three basic approaches that have been described above. In addition, this person must have at least a basic knowledge or an idea of which vehicle models (including versions with minor model changes) were produced within which time frame, so that it is possible to determine the correct decade, e.g., 2001, 2011 or 2021, in cases where the manufacturer encodes the year with the number 1.

A fundamental question remains: why do car window manufacturers today use such sophisticated or coded methods of indicating the year of manufacture, instead of marking windows with a simple date of manufacture such as 25/12/2021?

References

[1] Matuszak Z, Jaskiewicz M, Ludwinek K et al. 2015 Special characteristics of reliability for serial mechatronic systems. *Selected problems of electrical engineering and electronics* (WZEE). DOI: 10.1109/WZEE.2015.7394039.

[2] Böhm K, Kubjatko T, Paula D and Schweiger H G 2020 New developments on EDR (Event Data Recorder) for automated vehicles. *Open Engineering*, 10(**1**), pp. 140–146 ISSN: 23915439, DOI: 10.1515/eng-2020-0007.

[3] Mutawa N A, Bryce J, Franqueira V N L, Marrington A and Read J C 2019 Behavioural Digital Forensics Model: Embedding Behavioural Evidence Analysis into the Investigation of Digital Crimes, *Digital Investigation*, Volume **28**, pp. 70-82, ISSN 1742-2876, https://doi.org/10.1016/j.diin.2018.12.003.

[4] Matuszak Z, Jaskiewicz M, Wieckowski, D. et al. 2017 Remarks to the Reliability Assessment and to Human Actions. Especially Car Driver. *18th International Scientific Conference on LOGI Location: Book Series: MATEC Web of Conferences*. 134, Article Number: 00036. DOI: 10.1051/matecconf/201713400036.

[5] Kolitschova P, Rak, R and Kerbic J. 2018 Forensic and technical aspects of vehicle identification labels 11th International Scientific and Technical Conference on Automative Safety, Casta Papernicka, Slovakia, Apr 18-20, 2018. *Proceedings Paper IEEE Xplore, INSPEC Accession Number: 17823676*, DOI: 10.1109/AUTOSAFE.2018.8373340, UT WoS 000435296000046.

[6] Kopencova D, Rak R. 2020 Issues of Vehicle Digital Forensics, *XII International Science-Technical Conference AUTOMOTIVE SAFETY*, Kielce, 2020; pp. 1-6, EID: 2-s2.0-85099212235 doi: 10.1109/AUTOMOTIVESAFETY47494.2020.9293516.

[7] Jurecki R S and Jaskiewicz M. 2012 Analysis of road accidents in Poland over the last ten years. *Scientific journals of the maritime university of szczecin-zeszyty naukowe akademii morskiej w szczecinie*. 32(**104**):65-70.

[8] Rak R, Kopencova D and Felcan M. 2021 Digital vehicle identity – Digital VIN in forensic and technical practice, *Forensic Science International: Digital Investigation*, Elsevier, Volume **39**, 301307, ISSN 2666-2817, https://doi.org/10.1016/j.fsidi.2021.301307, https://www.sciencedirect.com/science/article/pii/S2666281721002328

[9] Soltani S and Hosseini Seno S A. 2019 A formal model for event reconstruction in digital forensic investigaton, *Digital Investigation*, Volume **30**, pp. 148-160, ISSN 1742-2876, https://doi.org/10.1016/j.diin.2019.07.006

[10] Kubjatko T, Görtz M, Macurova L and Ballay M. 2018 Synergy of forensic and security engineering in relation to the model of deformation energies on vehicles after traffic accidents. Conference Paper. Transport Means - *Proceedings of the International ConferenceVolume 2018-*

October, pp. 1342-1348, 22nd International Scientific on Conference Transport Means 2018; Trasalis - Trakai Resort and SPAGedimino pp. 26, Trakai; Lithuania; 3 October 2018 through 5 October 2018; Code 140271, ISSN: 1822296X

[11] Moravcik L, Jaskiewicz M. 2018 Boosting Car Safety in the EU. *11th International Scientific and Technical Conference on Automotive Safety.* Location: Casta Papiernicka, SLOVAKIA Date: APR 18-20

[12] Rak R, Zrubak R. 2012 Project eCALL – Car in Emergency Situation. 7th Scientific International Conference Crisis Management: Envirenmental Protection of Population – *Conference Proceedings. Edited by Horak, R; Juricek, L; Schwarz, R., 2012*; pp. 251-258, 2012. Proceedings paper, ISBN 978-80-86710-57-0.

[13] Reedy P. 2020 Interpol review of digital evidence 2016 – 2019, 2020, *Forensic Science International: Synergy,* Volume **2**, pp. 489-520, ISSN 2589-871X, https://doi.org/10.1016/j.fsisyn.2020.01.015.

[14] Cruz-Jesus F, Oliveira T and Bacao F. 2018 The Global Digital Divide: Evidence and Drivers. *Journal of Global Information Management (JGIM),* 26(**2**), 1-26, doi:10.4018/JGIM.2018040101.

[15] Arshad H, Jantan A B and Abiodun O I. 2018 Digital Forensics: Review of Issues in Scientific Validation of Digital Evidence, *Journal of Information Processing Systems*, Vol. **14**, No. 2, pp. 346-376, DOI: 10.3745/JIPS.03.0095.

[16] Jaskiewicz M, Lisiecki J, Lisiecki S. 2015 et al. Facility for performance testing of power transmission units. *Scientific journals of the maritime university of szczecin-zeszyty naukowe akademii morskiej w szczecinie.* 42(**114**):14-25 https://repository.am.szczecin.pl/handle/123456789/765

[17] Kopencova D, Felcan M and Rak R. 2020 Equilibrium and Limit States in Technical and Social Disciplines as Part of Risk Analysis and Management. In: Abdel Wahab M. (eds) *Proceedings of 1st International Conference on Structural Damage Modelling and Assessment. Lecture Notes in Civil Engineering,* vol **110**. Springer, Singapore. ISBN 978-981-15-9120-4. https://doi.org/10.1007/978-981-15-9121-1_22

[18] Mansor H, Markantonakis K, Akram R, Mayes K and Gurulian, I. 2016 Log your car: The non-invasive vehicle forensics. *2016 IEEE International Conference Trustcom/BigDataSE/ISPA, Trust, Security and Privacy in Computing and Communications*, [online, cit, 2019-11-26], Available at https://ieeexplore.ieee.org/document/7847047/authors#authors.

[19] Le-Khac, Nhien-An, Jacobs D L, Nijhoff, Bertens K and Choo, R K K. 2018 Smart vehicle forensics: Challenges and case study. Future *Generation Computer Systems.* DOI: 10.1016/j.future.2018.05.081.

Acknowledgments
This research was addressed in IGA No. 245/2019 "Modern technologies in the commissioning, detection, documentation, proof and prevention of crime in ensuring public order, safety and fluidity of road traffic (aspects of technical, forensic, criminological, penelogical, legal, public administration and security)" addressed at the Academy of the Police Force in Bratislava, Slovakia.

IOP Conf. Series: Materials Science and Engineering 1247 (2022) 012004 doi:10.1088/1757-899X/1247/1/012004

Influence of road conditions on the values of longitudinal acceleration during braking of a tourist coach

P Łagowski*, T Stańczyk, E Szumska, A Zuska, P Grabski, M Jaśkiewicz, R Jurecki, D Kurczyński,

Department of Automotive Vehicles and Transportation, Faculty of Mechatronics and Mechanical Engineering, Kielce University of Technology, Kielce, Poland

* p.lagowski@tu.kielce.pl

Abstract. The paper aims to analyse the longitudinal acceleration of a tourist coach on various road pavements. As part of the experimental research, braking tests were carried out at a speed of 30 km/h and acceleration to a speed of 30 km/h. The tests were carried out on four types of pavement, i.e. dry and wet asphalt and dry and wet concrete. The article describes the impact of road conditions on the braking and acceleration performance on selected road pavement types. To increase the reliability of the test results, ten tests were carried out for each tested pavement. Based on the test results, graphs of braking decelerations and accelerations during accelerating the vehicle as a function of time. This allowed for statistical analysis of the recorded decelerations and accelerations.

Keywords - acceleration, braking tests, automotive safety, road conditions

1. Introduction

The braking system is one of the most important active safety systems in motor vehicles. The system is designed to reduce the driving speed and immobilise the vehicle in the place specified by the driver [1-3]. Under emergency braking conditions, the type of pavement and the vehicle cargo have a significant impact on the deceleration value and the braking distance [4]. Moreover, technical condition of the braking system and the type and condition of tyres [5,6] also affect braking efficiency. The maximum value of deceleration during emergency braking is limited by the force of adhesion of the vehicle tyres to the ground. There are many methods used to test the performance of braking systems. One of them is the measurement of braking deceleration.

Two methods can be distinguished to determine the maximum value of braking deceleration [3, 7]:

– braking test with direct or indirect braking deceleration measurement,

– braking test with measurement of the stopping/braking distance.

To use the first method, it is necessary to install the equipment that records the course of braking deceleration, distance or speed [8,9]. The significant advantage of this method is that there is no need to determine the initial braking speed in a precise way.

As reported in the specialised literature, the simplest available devices for recording the speed or deceleration of the vehicle during the braking test can provide satisfactory results, allowing to determine the mean fully developed deceleration index (*MFDD*) or braking efficiency. These devices include acceleration sensors or speed recorders using the GPS signal [3]. Another parameter that allows assessing the braking performance is the value of the mean fully developed deceleration a_H.

Braking deceleration can also be determined by analysis of the braking marks left on the pavement but such analysis may lead to significant errors. This is due to difficulties in determining the actual length of the braking marks and the speed of the vehicle at the beginning of the braking marks [10, 11]. In addition, it should be taken into account that testing vehicles equipped with ABS system may be difficult in this case, as the lack of skidding may make it impossible to find braking traces [12-14].
A more accurate method of determining the braking deceleration is the registration of the vehicle speed during harsh braking [15-18]. By recording the course of speed as a function of time, it is possible to determine the mean value of deceleration of the vehicle braking by linearising the course of speed.

2. Object of research

An Apollo 970.28 coach was used to carry out the measurements. The coach was manufactured by Automet company from Sanok based on MB Atego 1222L. The coach is equipped with a Mercedes OM 904 LA (BlueTec 4) engine with a maximum power of 130 kW (177 HP). Basic technical data of the coach are presented in Table 1 [19].

Table 1. Technical data of the Apollo 970.28 coach [19]

Length	9,760 mm
Width	2,440 mm
Height	3,130 – 3,380 mm
Gross vehicle weight (GVW)	11,990 kg
Number of seats	36
Wheelbase	4,220 mm
Gearbox	6 - speed, manual
Maximum speed	130 km/h
Type of the engine	Mercedes OM 904 LA
Engine power	75-130kW at 2,200 rpm
Torque	400-670Nm at 1,200-1,600 rpm
Engine displacement	4,249 cm^3

The coach is equipped with a dual-circuit brake system with disc brakes on the front and rear axles, ABS system and parking brake. In addition, the vehicle is equipped with an electromagnetic retarder coupled with the gearbox. The photo of the coach is shown in Figure 1.

Figure. 1. The object of research: Type 970.28 Apollo coach

3. Experimental research and their analysis

An Apollo coach equipped with ABS system was used for the tests. The wheels of the vehicle were fitted with 245/70 R17.5 91H (Fulda Regioforce) tyres with a tread depth of approx. 8 mm and the nominal pressure recommended by the vehicle manufacturer. Road tests and the measurement of the maximum longitudinal acceleration values were carried out on dry and wet asphalt pavement and the dry and wet concrete pavement during harsh braking. The braking testing procedure was carried out at a speed of 30 km/h. There were three people on the coach during the measurements. For measuring and recording the results, research equipment was used, including S-350 Aqua Datron® optoelectronic sensor for measuring longitudinal speed and Datron uEEP-12® data acquisition station with ARMS® data acquisition and analysis software. In addition, TAA™ Datron and TANS™ Datron triaxial linear acceleration sensors were used to measure linear and angular accelerations.
Based on the curves of acceleration recorded in 10 tests during harsh braking on various pavements, the maximum, minimum and mean acceleration values were determined, as well as standard mean deviation SD and quantiles 0.1 and 0.9.
Figure 2 presents examples of recorded acceleration values during the process of harsh braking at a speed of 30 km/h on dry asphalt. The results of the measurements and calculations are presented in Table 2.

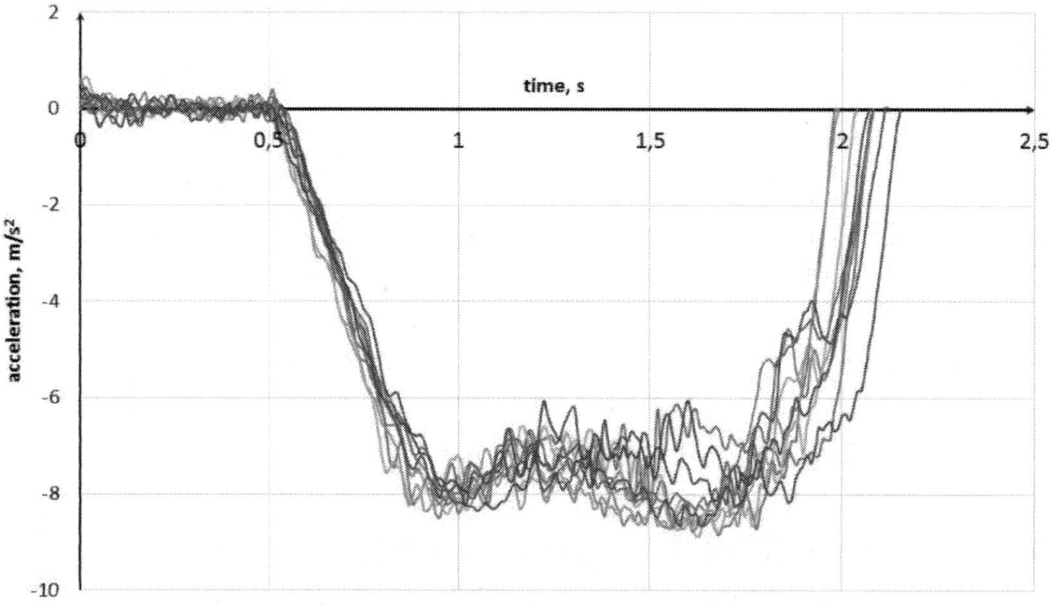

Figure. 2. Diagrams of longitudinal acceleration during harsh bus braking at a speed of 30 km/h on a dry asphalt pavement performed during 10 tests

Table 2. The acceleration statistics during the harsh braking of the coach at a speed of 30 km/h on different pavements

Measurement No.	acceleration, a [m/s²]			
	type of pavement			
	dry asphalt	wet asphalt	dry concrete	wet concrete
1	-8.688	-7.926	-7.44	-7.497
2	-8.221	-8.742	-7.284	-6.232
3	-8.863	-7.88	-8.018	-6.543
4	-8.835	-8.577	-8.119	-5.755
5	-8.697	-7.825	-7.788	-5.947
6	-8.807	-7.944	-8.018	-6.213
7	-8.642	-8.724	-7.954	-6.241
8	-8.349	-7.633	-7.77	-6.094
9	-8.294	-7.923	-8.247	-6.617
10	-8.239	-8.036	-7.724	-5.755
min	-8.221	-7.633	-7.284	-5.755
max	-8.863	-8.742	-8.247	-7.497
mean	-8.564	-8.121	-7.836	-6.289
SD	0.259	0.402	0.300	0.513
Q 0.1	-8.838	-8.726	-8.132	-6.705
Q 0.2	-8.237	-7.806	-7.424	-5.755

Figure 3 shows a graphical view of the average acceleration values for the same initial braking parameters, calculated based on ten tests carried out on different pavements.

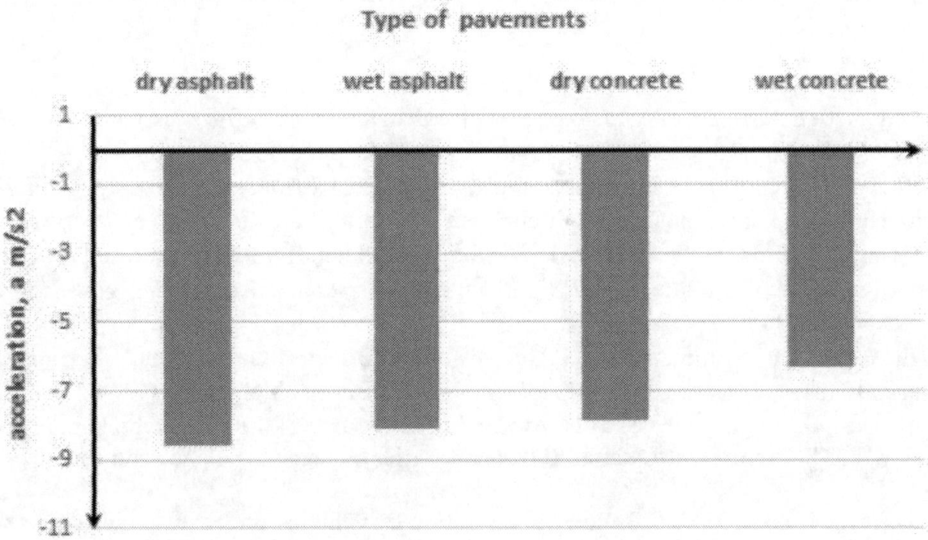

Figure 3. Average values of longitudinal acceleration during harsh braking of the coach at a speed of 30 km/h on different pavements

When analysing the values of the maximum accelerations during the braking of the coach at a speed of 30 km/h, it is clearly visible that the greatest acceleration values occurred for most measurements of braking on dry asphalt, which results from the greatest adhesion of tyres to this pavement a_{max}= -8.863 m/s^2. The smallest acceleration values were obtained on the wet concrete pavement, which may be related to the structural characteristics of concrete pavement a_{max} = - 7.497 m/s^2. The average values of the maximum accelerations during braking shown in Figure 3 confirm the nature of the change in acceleration values depending on the type of pavement. Moreover, when braking on the pavement of wet concrete, the highest value of the mean standard deviation SD = 2.846 was obtained.

4. Conclusions

Based on the conducted tests, the maximum and average values of braking accelerations were obtained for a tourist coach on various pavements at a speed of 30 km/h. The test results and their analysis showed the dependency between the type of pavement and the braking efficiency, which was determined by the acceleration statistics. As research shows, obtaining a high value of braking acceleration depends not only on weather conditions but also on the condition and the type of road pavement. The acceleration values specified in the paper were obtained on asphalt and concrete in good technical condition. In the case of asphalt and concrete of lower quality and greater wear, the acceleration will be lower, which should be borne in mind in further research and analysis of the influence of the pavement type and weather conditions on the braking efficiency.

The test results, especially the maximum acceleration values, will be used in further work related to the analysis of active safety during extreme braking. To limit the adverse impact of road conditions, as well as the type and condition of the pavement on the braking efficiency, the speed should always be adjusted to the existing road conditions, but also the vehicle load must be taken into account, which may directly affect the aforementioned active safety of the vehicle.

Acknowledgements

The research was carried out as part of the Innovative system research project supporting the motor vehicle insurance risk assessment dedicated to UBI (Usage Based Insurance) No. POIR.04.01.04 00 0004/19 00 financed by the National Centre for Research and Development.

References

[1] Jurecki R S 2016 Analysis of collision avoidance maneuvers in traffic emergencies. *Archiwum Inżynierii Motoryzacyjnej - Archiwum Motoryzacji* **72(2)** pp. 73-93 1

[2] Luty W 2018 Simulation-based analysis of the impact of vehicle mass on stopping distance. Eksploatacja i Niezawodność – Maintenance and Reliability **Vol. 20, No. 2,** pp. 1-19 2

[3] Oţăt V Oţăt O V and Oţăt D A 2022 Road safety-related determining of time and space braking under different travel conditions. *IOP Conf. Series: Materials Science and Engineering* **1220** pp. 1-8 3

[4] Sharizli A. et al. 2014 Simulation and Analysis on the Effect of Gross Vehicle Weight on Braking Distance of Heavy Vehicle. *Applied Mechanics and Materials* Vol. **564** pp. 77-82 4

[5] Szczypiński-Sala W and Lubas J 2016 Evaluation the course of the vehicle braking process in case of hydraulic circuit malfunction. IOP Conf. Series: Materials Science and Engineering **148** pp 1-16 5

[6] Tarkowski S., Mikus D 2016 Braking deceleration of vehicles under different traffic conditions. *Autobusy, Bezpieczeństwo i Ekologia*, **6** pp 438-441 6

[7] Wach W, Unarski J and Duś A 2005 Tachograph diagram - analysis of heavy braking recording, *SAE Technical Paper* **2005-01-1185** pp 1-15 7

[8] Szumska E, Młodzińska D and Jurecki R 2014 Vehicle braking effectiveness in different road conditions. *Logistyka* **6** pp 10430-39 9

[9] Witkiewicz M, Owczarz M and Taryma S 2016 Comparative analysis of measured values of braking deceleration of a passenger car with an efficient and a faulty braking system. *Autobusy, Eksploatacja i Testy,* **8** pp 154-157 8
[10] Botzer A Musicant O and Mama Y 2019 Relationship between hazard-perception-test scores and proportion of hard-braking events during on-road driving - An investigation using a range of thresholds for hard-braking. *Accident Analysis & Prevention,* Vol. **132**, pp. 1-16
[11] Stańczyk T L and Jurecki R 2014 Driver's reaction time in a complex road situation (braking with driving around an obstacle). *The Archives Of Automotive Engineering* **66** pp 69 – 82
[12] Huang R Zhao X Yuan Y Yu O Zhou Ch Daamen W 2021 Experimental study on evacuation behaviour of passengers in a high-deck coach: A Chinese case study. *Physica A: Statistical Mechanics and its Applications* Volume 579, pp 1-20.
[13] Silvanoa Ary P Ohlina M 2019 Non-collision incidents on buses due to acceleration and braking manoeuvres leading to falling events among standing passengers. *Journal of Transport & Health* **14**, pp. 1-9
[14] Bryant D Day A 2022 Chapter 7 - Electronic braking systems. *Braking of Road Vehicles (Second Edition).* pp. 227
[15] Janczur R and Zawałeń J 2021 Braking parameters of trucks and buses in reconstruction and analysis of road accidents. XVII Konferencja Problemy Rekonstrukcji Wypadków Drogowych. Kraków, Wydaw. Instytutu Ekspertyz Sądowych, Paragraf na Drodze, pp. 105-120
[16] Popa M. F., Capătă D. M. S. and Burnete N 2019 Aspects of wear braking process on brake discs and brake pads in the construction of commercial vehicles. *IOP Conf. Series: Materials Science and Engineering* **568** pp 1-8
[17] Stańczyk T L Jurecki R S Walczak S and Pieniążek W 2011 Analysis of the times of force build-up on the brake pedal and of the build-up of braking deceleration obtained during tests of accident risk situations simulated on a track. *Autobusy, Eksploatacja i Testy* **12** pp 349-359.
[18] Żardecki D and Dębowski A 2021 Non-smooth models and simulation studies of the suspension system dynamics basing on piecewise linear luz(…) and tar(…) projections. *Applied Mathematical Modelling* **94** pp. 619–634
[19] Automet. Technical data of the Apollo 970.28 coach.

IR spectroscopy analysis of skid marks for vehicle equipped with ABS considering slip ratio

E Sys[1], Ł. Gosławski[1], A Mrowicki[1]*, M Głogowski[2], S Madziara[1], B Droś[1], W Zakrzewicz[1], P Kubiak[3]

[1] Faculty of Automotive and Construction Machinery Engineering,, Warsaw University of Technology, Poland;
[2] Faculty of Process and Environmental Engineering, Lodz University of Technology, Poland;
[3] Institute of Marketing and Sustainable Development, Lodz University of Technology, Poland

* adam.mrowicki.dokt@pw.edu.pl

Abstract. Skid marks are an important source of information for road accident investigators. Up till now, analyzing them was a difficult task when ABS equipped vehicles are considered. Authors managed to solve this issue using IR spectroscopy. There are several factors to consider when investigating brake marks and one of them is slip ratio that was taken into account in this paper. Results show that for slip of 10% and more, IR spectroscopy technology was successful at revealing skid marks.

Keywords: IR spectroscopy, ABS, brakes, tire marks, vehicle, vehicle testing

1. Introduction

The problem of brake marks visibility has been known from the very beginning of the ABS system and it is unsolved to this day [1-3]. Authors decided to use IR spectroscopy to solve this problem. IR spectroscopy is a technology used to identify substances [4-6] and, according to author's opinion, it will be possible to use it to identify the braking trace at the scene of a road accident[7-10]. Preliminary studies have been performed. The results were presented in [11]. After first review and promising results, authors prepared following studies. Authors checked the influence of different types of asphalt on possibility of re-vealing braking marks. The results were published in [12]. In this article authors would like to present research on influence of different slip ratios for visibility of braking marks, using IR spectroscopy.

2. Literature review

Braking system equipped with the ABS improves the road safety [13-15], but on the other hand ABS caused difficulties in the reconstruction of road accidents [16-18].There has been a problem with the visibility of skid marks since vehicles with ABS were mass produced [19-21]. So far, two different ways to reveal skid marks have been proposed. Those methods are Image Refinement Methods and Thermovision method. Both methods were precisely described in first article, where it was proposed completely new method based on IR spectroscopy [11]. As part of the research presented in this article, it was decided to investigate the impact of another parameter on the possibility of revealing braking traces at the scene of a road accident - the slip ratio. Slip is the difference between the angular velocity of the wheel minus the angular velocity of the free-rolling wheel [22 -25]. Slip ratio is defines as:

$$\mathbf{S} = \frac{\mathbf{V_F}}{\mathbf{V_H}} = \frac{\mathbf{V_H} - \mathbf{R} \cdot \boldsymbol{\omega}}{\mathbf{V_H}}$$

Where:

S – slip
V_F - velocity of the tire at the tire patch

V_H - velocity of the wheel hub
ω - angular velocity of the wheel
R - rolling radius

While braking the vehicle, the slip ranges from 0% (free rolling) to 100% (wheel locked). When locking the wheels, the loss of steerability is observed. It is a dangerous situation in road traffic that makes it impossible to undertake additional defensive maneuvers, i.e. avoiding in an emergency. Therefore, it has been mandatory to install ABS systems in cars for over 20 years. ABS is a system that regulates the force acting on the vehicle brakes in such a way that the slip at the contact of the wheel with the asphalt is in the range of 10% -30%. This range ensures that on most surfaces the maximum available braking force is used. It also enables the transfer of lateral forces arising during a turning maneuver performed simultaneously with the braking maneuver. As a result, the vehicle remains steerable during braking. Relations between slip rate and adhesion coefficient is shown in Figure 1.

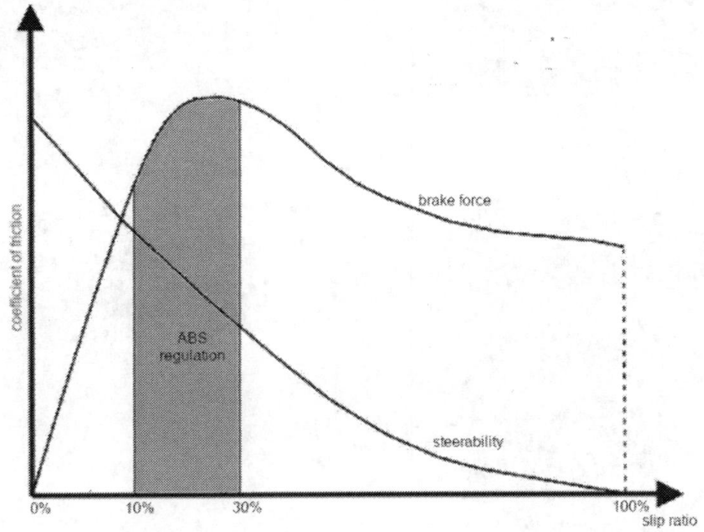

Figure 1. Relationship between braking coefficient and wheel slip [3]

3. Samples preparation

140 samples were prepared for the planned research. The samples were prepared from one type of asphalt. For each of 20 samples, braking traces were applied at various slips, i.e. 0%, 10%, 20%, 30%, 40%, 100% respectively and 20 samples were left without a trace. The exact method of sample preparation and research position where the trace was applied was presented in the previous article. It is worth noting that the traces were applied on the sample in only one layer, i.e. according to the method of applying the trace in real traffic conditions. The tire pressure on the sample was 4000N and the wheel speed was 40 km / h. Traces were applied with a summer tire of a budget class. Samples photos are presented in Table 1.

IOP Conf. Series: Materials Science and Engineering 1247 (2022) 012005 doi:10.1088/1757-899X/1247/1/012005

Table 1. Surfaces of prepared samples.

Slip ratio percentage	Surface	Slip ratio percentage	Surface
Without braking	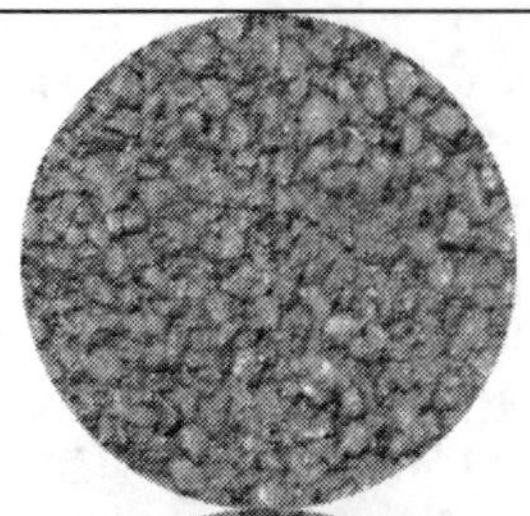	30	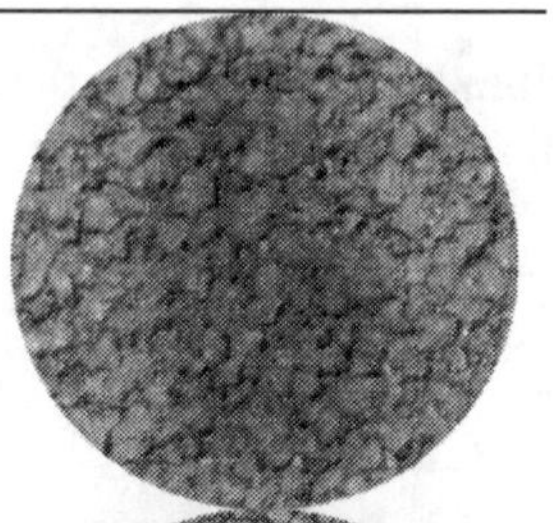
0		40	
10		100	
20	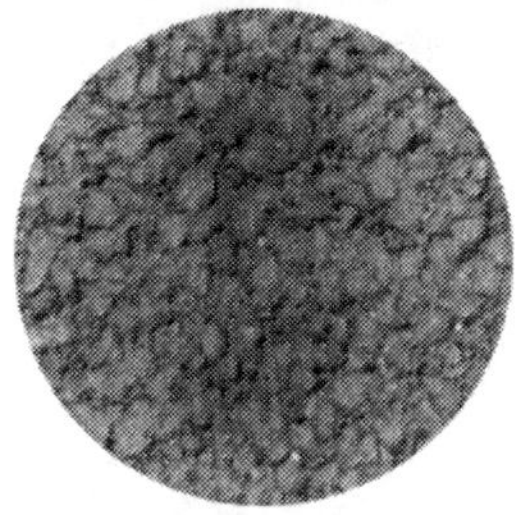		

Table 1 shows the surface of each samples. Samples with braking marks laid down with 0% slip ratios - traces are invisible. For 10 and 20 % slip ratios, braking traces were barely seen, between 30% - 40% traces are clearly visible and for 100 % we can observe almost black surface. The lower slip value, the more difficult organoleptic evaluation of the braking trace is. A series of measurements were made at 3 randomly selected points of the sample. The sample was mounted to the measuring table by using the standard spectrophotometer equipment, i.e. a pressure arm with a screw. The downforce was limited by a clutch in the screw handle. Measurements were made with a resolution of 128 and a measuring gap width of 2 mm. Such parameters lengthened the time of the measurement, but allowed to obtain high accuracy. The relation between the absorption and the electromagnetic wavelength was obtained from each measurement

Authors prepared 120 samples with traces and different slip ratios and we compared this to 20 samples without skid marks. All operations were performed by using the specialized software SpctraGryph [26], which is used to process spectra data. The results were normalized, i.e. the scale was adjusted to be able to compare the obtained values. It was a necessary procedure due to the fact that the strength of the measurement signal was different for each sample and largely dependent on the quality of the adherence of the sample to the measuring table. After that results were averaged to compare spectra for each slip ratios.

4. Results

Figure 2 presents the obtained results in form of a graph.

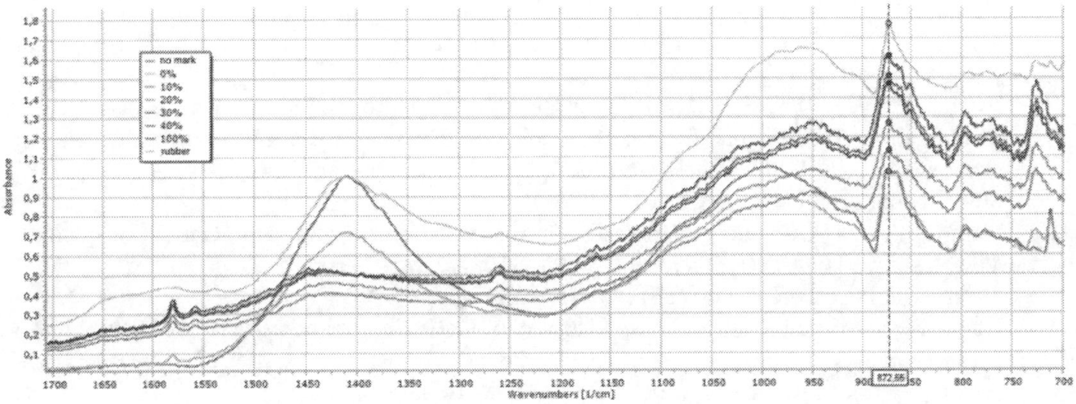

Figure 2. Graph presenting the absorbance as a function of wavenumber for different slip ratio percentages.

The braking trace was revealed on all tested samples. It was visible in the mid-infrared range, at the same wavelength as in the previous tests, i.e. 872 cm-1, which corresponds to approximately 11 500 nm. At this wavelength, local absorption extremes can be noticed on the graph. The absorption value increases with the increasing slip ratios. Already at the slip ratio 10% (lower range of adjustment ABS system) local extreme is visible. It means that break trace is revealed. It is also important that for the free rolling wheel, i.e. slip ratio 0%, the absorption curve corresponds to the absorption curve for the asphalt itself. Thanks to this, in real-life conditions it should be possible to indicate the beginning of the braking trace.

5. Conclusions

Based on the conducted research, it can be stated that for all tested slip ratios it is possible to reveal braking traces. For the free rolling wheel, i.e. slip ratio 0%, the absorption curve corresponds to the absorption curve for the asphalt itself. Thanks to this, in real-life conditions it should be possible to indicate the beginning of the braking trace. Future research will investigate the effects of other factors such as tire type and pressure force. If positive results are obtained, it should be possible to construct a device enabling investigation the analysis of braking traces. Due to the high potential of the method and the market requirements themselves, the authors of the article intend to continue the research.

References

[1] Beauchamp G, Hessel D, Rose NA, Fenton SJ, Voitel T. Determining vehicle steering and braking from yaw Mark striations. SAE Int J Passeng Cars - Mech Syst [Internet]. 2009;2(1):291–307. Available from: http://dx.doi.org/10.4271/2009-01-0092

[2] Tseng WK, Liao SX. Estimation of vehicle pre-braking speed. Appl Mech Mater [Internet]. 2012;151:165–9. Available from: http://dx.doi.org/10.4028/www.scientific.net/amm.151.165

[3] Žuraulis V, Levulytė L, Sokolovskij E. Vehicle speed prediction from yaw marks using photogrammetry of image of traffic accident scene. Procedia Eng [Internet]. 2016;134:89–94. Available from: http://dx.doi.org/10.1016/j.proeng.2016.01.043

[4] Theophile T. Infrared spectroscopy: Materials science, engineering and technology. BoD-Books on Demand; 2012. DOI: 10.5772/2055

[5] Brady DJ. Optical imaging and spectroscopy. John Wiley & Sons; 2009. DOI: 10.1002/9780470443736
[6] Sarkissian G. The analysis of tire rubber traces collected after braking incidents using pyrolysis-gas chromatog-raphy/mass spectrometry. Journal of forensic sciences. 2007;52(5):1050–6. DOI: 10.1111/j.1556-4029.2007.00529.x
[7] Ginzburg G, Evtiukov S, Brylev I, Volkov S. Reconstruction of road accidents based on braking parameters of category L3 vehicles. Transp res procedia [Internet]. 2017;20:212–8. Available from: http://dx.doi.org/10.1016/j.trpro.2017.01.054
[8] Evtiukov S, Kurakina E, Lukinskiy V, Ushakov A. Methods of accident reconstruction and investigation given the parameters of vehicle condition and road environment. Transp res procedia [Internet]. 2017;20:185–92. Available from: http://dx.doi.org/10.1016/j.trpro.2017.01.049
[9] Vangi D, Virga A. Evaluation of emergency braking deceleration for accident reconstruction. Veh Syst Dyn [Internet]. 2007;45(10):895–910. Available from: http://dx.doi.org/10.1080/00423110701538320
[10] Nagurnas S, Mitunevičius V, Unarski J, Wach W. Evaluation of veracity of car braking parameters used for the analysis of road accidents. Transport [Internet]. 2007;22(4):307–11. Available from: http://dx.doi.org/10.3846/16484142.2007.9638146
[11] Gosławski L, Kubiak P, Mrowicki A, Soghabatyan T, Sys E, Wang YW, et al. Analysis of braking marks left by vehicles equipped with ABS with IR spectroscopy. Archiwum Motoryzacji. 2019;84(2). DOI: https://doi.org/10.14669/AM.VOL84.ART3
[12] Sys E, Kubiak P, Gosławski L, Zakrzewicz W. Analysis of braking marks left by vehicles equipped with ABS with IR spectroscopy-different types of asphalt. Archiwum Motoryzacji. 2021;(3). DOI: https://doi.org/10.14669/AM.VOL93.ART1
[13] Darkunov S, Ozguner U. ABS control using optimum search via sliding mode. In: Proceeding of the 33rd Conference on Decision and Control. 1994. DOI: 10.1109/CDC.1994.411013
[14] Guntur RR. Design considerations of adaptive brake control systems. In: SAE Technical Paper Series. 400 Commonwealth Drive, Warrendale, PA, United States: SAE International; 1974. DOI: https://doi.org/10.4271/741082
[15] Aly AA, Zeidan E-S, Hamed A, Salem F. An antilock-braking systems (ABS) control: A technical review. Intell Contr Autom [Internet]. 2011;02(03):186–95. Available from: http://dx.doi.org/10.4236/ica.2011.23023
[16] Seipel G, Baumann F, Hermanutz R, Winner H. Analysis of the influence of vehicle dynamic parameters on tire marks. Tire Sci Technol [Internet]. 2013;41(3):196–213. Available from: http://dx.doi.org/10.2346/tire.13.410302
[17] Nešić M, Lipovac K. Analysis of traffic safety of vehicles equipped with ABS. European Automobile Engineers Cooperation-10th EAEC European Automotive Congress, EAEC 2005. 2005;1:294–306.
[18] Prochowski L, Unarski J, Wach W, Wicher J. Podstawy rekonstrukcji wypadków drogowych. Wydawnictwa Komunikacji i Łączności. 2008;276–7.
[19] Oppenheimer P. Comparing stopping capabiity of cars with and without antilock braking systems (ABS). In: SAE Technical Paper Series. 400 Commonwealth Drive, Warrendale, PA, United States: SAE International; 1988. DOI: https://doi.org/10.4271/880324
[20] Todoruţ A, Cordoş N, Barabas I, Bălcău M. Possibility of evaluation the pre-collisions speed and space crossing by vehicle within process of braking. ACTA TECHNICA NAPOCENSIS-Series: APPLIED MATHEMATICS, MECHANICS and ENGINEERING. 2014;(3).
[21] Grosch KA. Rubber abrasion and tire wear. Rubber Chem Technol [Internet]. 2008;81(3):470–505. Available from: http://dx.doi.org/10.5254/1.3548216
[22] Wellstead PE, Pettit NBOL. Analysis and redesign of an antilock brake system controller. IEE Proc - Control Theory Appl [Internet]. 1997;144(5):413–26. Available from: http://dx.doi.org/10.1049/ip-cta:19971441

IOP Conf. Series: Materials Science and Engineering 1247 (2022) 012005 doi:10.1088/1757-899X/1247/1/012005

[23] Batterman SD, Batterman SC. Introduction to Forensic Engineering and Accident Reconstruction. In: The Forensic Laboratory Handbook Procedures and Practice. Totowa, NJ: Humana Press; 2011. p. 539–61. DOI: 10.1007/978-1-60761-872-0_20
[24] Jurecki RS, Jaskiewicz M, Zuska, A The variety of the behaviour of drivers at risk accident situations Scientific Journals of the Maritime University of Szczecin-Zeszyty Naukowe Akademii Morskiej w Szczecinie. 2013, 35(107) p. 38-46.
[25] Jurecki RS, Stanczyk TL, A Methodology for Evaluating Driving Styles in Various Road Conditions, ENERGIES. 2021, 14 (12). DOI: 10.3390/en14123570
[26] Menges F. Spectragraph–optical spectroscopy software. 2020

The 13th International conference on Automotive Safety (Automotive Safety 2022) IOP Publishing
IOP Conf. Series: Materials Science and Engineering 1247 (2022) 012006 doi:10.1088/1757-899X/1247/1/012006

Motor vehicles' problems with keeping the straight driving direction; analysis of selected case

T Pusty [1*], R Lewiński [1], H K Kowieski [1]

[1] Łukasiewicz Research Network – Automotive Industry Institute, Jagiellońska 55, 03-301 Warszawa, Poland

*tomasz.pusty@pimot.lukasiewicz.gov.pl

Abstract: The article presents issues related to the testing and verification of the keeping of straight driving direction by motor vehicles. The subjective feelings of drivers were assigned specific values of physical quantities describing the rectilinear path of motion, i.e. the steering wheel angle and the force vector applied to the steering wheel of the vehicle under test while driven along a straight path. This made it possible to determine the limit values enabling the assessment whether the object under test has a tendency to move straight ahead without any unusual steering corrections. The cases analysed relate to the situations where measurements of wheel geometry settings do not show significant deviations from the standard but, in spite of that, the vehicle requires unusual steering corrections by the driver to maintain a straight path.

Key words: vehicle dynamics, vehicle testing, steering wheel

1. State of the art

The maintaining of the straight driving direction is one of the elements of motor vehicle's active safety. As stated by J. Wicher [4], the elements of kinematics of the road-vehicle system determine the vehicle's path on a specific road. This means that the vehicle should keep the straight driving direction on a flat horizontal road when accelerating, moving with a constant speed, and braking.

According to authors' experience, problems with maintaining the preset driving direction may happen even in the case of new vehicles; nevertheless, they occur much more often in result of operational damage. In most cases, they manifest themselves in a repeated and marked turn of the steering wheel when driver's hands lie freely on it and, in consequence, in a change in vehicle's direction of motion. Such a phenomenon may occur with different intensity at various driving speeds in vehicles of various types (passenger cars, off-road vehicles, motor trucks, buses, etc.); anyway, the matter of importance is its intensity, repeatability, and question whether it can affect the road traffic safety.

One of the requirements that must be met by vehicles' steering systems (according to UN ECE Regulation No 79) is that the system shall ensure vehicle's capability of traveling along a straight road section without unusual steering correction by the driver and without unusual vibration in the steering system at the maximum design speed of the vehicle. This means that the assessment whether the necessary steering corrections and the vibrations in the steering system are "unusual" or not has been left to discretion of the technical unit carrying out the tests. This work will make it possible to determine the limit values that will provide grounds for the assessment whether the vehicle has a tendency to travel straight ahead without any unusual steering corrections.

2. Assessment whether the vehicle can maintain the straight driving direction

2.1. Vehicle examination on test stands and at Vehicle Testing Stations

The assessment of current actual vehicle's technical condition begins from vehicle identification and verification of its completion (set of components used) for conformity with the vehicle conformity certificate issued by the vehicle manufacturer. This makes it possible to assess whether any design changes not meeting the specifications of the type approved have been made in the vehicle. Then, the tyre inflation pressure is measured and adjusted if necessary and the vehicle undercarriage is visually inspected with special attention being paid to the condition of vehicle's rims and tyres, wheel guiding parts, and suspension spring and damping elements. The inspection also includes preliminary measurements of: wheel toe-in on a sideslip plate, shock absorber damping performance, tyre rolling resistance, and tyre tread depth. During the inspection, signs of mechanical damage and excessive plays in the suspension and steering systems are sought. The inspection should be carried out in compliance with the latest relevant guidelines provided e.g. in publication [1]. During the measurements, attention is particularly focused on toe-in measurement results, differences in the damping coefficients of the left and right shock absorbers, and differences in the left and right tyre rolling resistance.

When reasonable, vehicle wheels may additionally be balanced, including the "road-simulation test", where the tyre-road contact is simulated during the balancing by means of a special roller pressed onto the tyre with a force of 5 kN. The complete balancing of all the wheels and appropriate selection of the location of individual wheels in the vehicle is to reduce the impact of tyres on vehicle sideslip.

As the next step, the linear and angular dimensions of the geometry of vehicle wheels and suspension components are precisely measured, i.e. the wheel toe-in, camber, castor, and kingpin inclination angles, as well as wheel track and wheelbase are determined. The measurement results are compared with the values specified by the vehicle manufacturer as acceptable. The parameters of particularly high importance for the vehicle stabilisation when driven straight ahead are the castor angle and the toe-in; however, the other parameters may indicate mechanical defects, such as damage to suspension components or, in extreme situations, to the undercarriage as a whole.

Then, the linear dimensions of the positions of the base undercarriage points, i.e. the main points of vehicle's longitudinal members and points of fastening the suspension system in relation to the said main points, are measured. The measurement results are compared with the values specified by the vehicle manufacturer as acceptable.

For the above rig tests and measurements, the equipment of Vehicle Testing Stations or garages may be used. Such operations, not very much time and cost consuming, may provide grounds for formulating or rejecting some hypotheses as regards the reasons for a possible vehicle fault.

2.2. Road tests

The tests that enable quantitative verification of vehicle's susceptibility to autogenous changing of the direction of motion are much more labour-consuming. They are carried out in specific dynamic road conditions, by means of specialist testing and measuring equipment to record the steering wheel angle and the value of the force vector applied to the steering wheel, referred to as a "dynamometer steering wheel" (or "strain-gage steering wheel"), Figure 1. In the tests described herein, an inertial and satellite navigation system RT 3000, installed and positioned in the vehicle in accordance with normative requirements, was also used. Less sophisticated test equipment, the application of which for tests of similar nature has been presented in publications [2] and [3], e.g. the VBOX Sport system to record time histories of accelerations and velocities in rectilinear motion as well as instantaneous locations in geographic coordinates or other apparatus to measure vehicle velocities, locations, and positions, may also be used.

It is recommended to carry out road tests on a dry, straight, flat and horizontal lane of a road section with bituminous or concrete surface, 300-500 m long, closed for public traffic. When analysing the measurement results, the direction and value of the slope of the road lane as well as the wind velocity should be taken into account. During the measurements, the side wind velocity should not exceed 5 m/s and the road slope should not exceed 2 %; the impact of the possible occurrence of these factors should be compensated by driving the vehicle on the same lane and road section in the opposite direction. The vehicles' ability to maintain the preset driving direction will vary depending on their body shapes, design of suspension and steering systems, and alignment of road wheels, especially in case of disturbances e.g. kinematic excitations caused by the road profile (surface roughness, cavities, etc.), or blasts of wind, translating into the necessity of steering corrections made by the driver.

Figure 1. Dynamometer steering wheel installed in a vehicle under test.
Photo: Ryszard Lewiński

Tests of two types may be carried out in road conditions. In one of the types, the vehicle is driven along a rectilinear track and a force necessary for the vehicle to be kept on the predefined path is applied to the steering wheel (closed-loop tests). In the other one, the vehicle having entered the test road section moves without any correction of its driving direction, i.e. any deviation from its predefined path is made possible (open-loop tests). The tests are carried out at different constant total vehicle velocity values in both driving directions.

2.3. Analysis of road test results

When the road test results are analysed, the arithmetic mean value of the force applied to the steering wheel and the arithmetic mean value of the steering wheel angle are calculated for individual tests.

In the open-loop tests, not only the steering wheel angle and the force applied to the steering wheel are measured but also the vehicle's deviation from the rectilinear trajectory (i.e. the distance between the vehicle and the preset vehicle path) is determined. In the analysis of the test results, special attention is paid to a possible predominance of the situations where a force has to be applied in a specific direction to the steering wheel for the desired straight driving direction to be maintained or where the vehicle autogenously deviated from the rectilinear path leftwards or rightwards. The authors of this publication make an attempt to establish specific numerical values of the steering wheel force and angle as criteria of assessment of vehicle's capability of maintaining the straight driving direction based on subjective drivers' opinions. The test results obtained for a selected case have been presented in Section 3.

3. Analysis of a selected case

The methodological approach described in the preceding Section was used for the analysis of a civil law dispute concerning a sport utility vehicle (SUV) of category M1G. The subject matter of the analysis was the ascertaining whether the vehicle had a physical fault that caused an autogenous change in the vehicle's driving direction.

A visual inspection of the SUV did not reveal any marks of impacts, wheel rim deformations, or excessive plays in, or damage to, the suspension and steering systems. The tyre tread depth was uniform, amounting to 6.3-6.7 mm and no dentation marks or cavities were noticed. The EUSAMA value (EV = the percentage of remnant vertical tyre-road contact force) for the left and right front axle wheels was 70 % and 75 %, respectively; for the rear axle wheels, it was 72 % on both sides. The values obtained did not show any malfunction of the suspension system. The tyre rolling resistance measured on a roller tester were: 310 N for the front left wheel, 210 N for the front right wheel, 130 N for the rear left wheel, and 150 N for the rear right wheel. The total tyre rolling resistance for the left wheels exceeded that for the right wheels by 80 N.

The difference found to occur between the left and right tyre rolling resistance forces, although quite small, was recognized as potentially capable of pulling the vehicle to the left. To verify this observation, the wheels were additionally balanced on a balancer supporting the "road-simulation test" function. This test did not confirm the differences indicated by the measurements carried out on the roller tester; on the contrary, the occurrence of a force of 30 N suggesting the possibility of pulling the vehicle in the opposite direction, i.e. to the right, was detected. The above discrepancies indicated that the total difference in the tyre rolling resistances detected by the roller tester at the Vehicle Testing Station came not from the wheels but, probably, from the power transmission or suspension system of the vehicle.

Measurements of the linear and angular dimensions showing the wheel alignment settings and locations of suspension system parts revealed some faults:

- The castor angle of the front left wheel was 3°18′ while the maximum acceptable value is 2° 15′; for the front right wheel, the castor angle was 2° 49′ (at the same acceptability limit of 2° 15′).
- There was a difference between the steering angles of the front wheels: the values of the angles were 3° 00′ for the right wheel and 2° 45′ for the left wheel (at the acceptable value of 2° 30′ ± 20′), i.e. the acceptability limit was exceeded for the right wheel.
- There was an excessive offset (of 10 mm) between the front wheel axes of rotation (the vehicle axles were not parallel to each other), while the tolerance is ± 8 mm.
- The maximum steering angle of the left wheel inwards (to the left) was 36° 54′, at the acceptability limit of 35° 30′. For the right wheel, the corresponding angle value was within the limit.

The excessive offset of the front wheels might have resulted from the difference between the castor angles of the left and right wheel or from incorrect fastening of the front axle to the vehicle body by means of the radius arms (reaction rods). The measurements of locations of the base undercarriage points did not reveal any deviations from the norm in the geometry of the vehicle body frame. During road tests carried out on a test road section on theUłęż airfield apron to check the maintaining of the straight driving direction, a difference of 5.1 N was found to occur between the moduli of the arithmetic mean values of the forces applied to the steering wheel. This meant that a force of 5.1 N (leftward) had to be applied to the steering wheel for the vehicle to keep the straight driving direction on a flat horizontal road section. An example measurement result has been presented in Figure 2. An example of the measuring trajectory apparatus fastening point when the vehicle was driven westwards has been presented in Figure 3. This trajectory was determined from the moment when the resultant value of the force was 0 N.

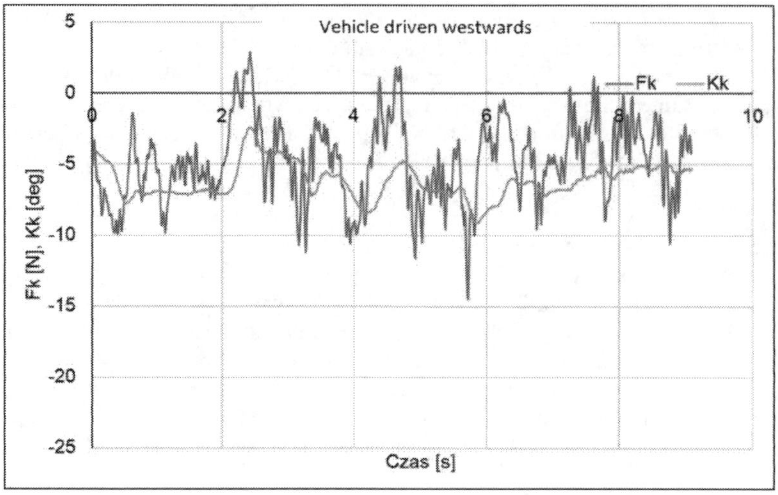

Figure 2. Time histories of the force applied to the steering wheel (Fk) and the steering wheel angle (Kk) during a road test on a test road section: vehicle driven westwards

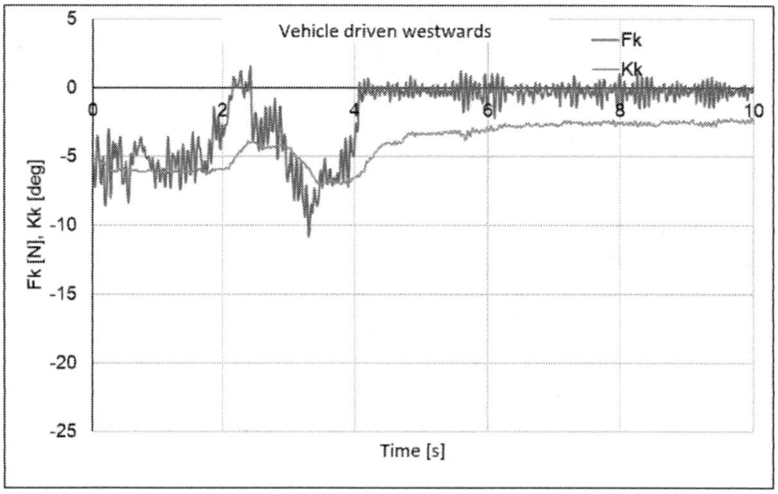

a)

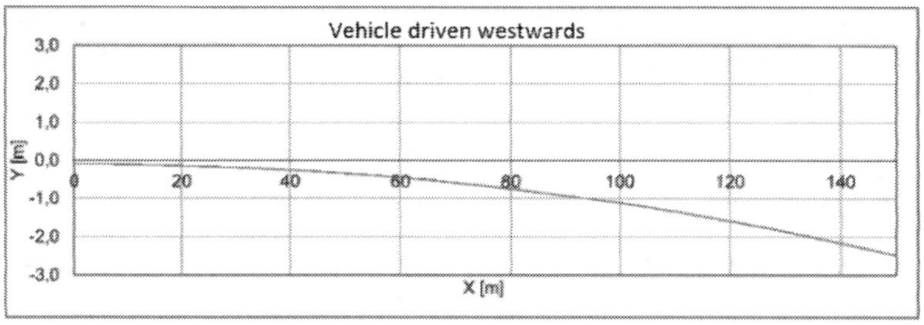

b)

Figure 3. An example of the analysis results, a) time histories of the force applied to the steering wheel (Fk) and the steering wheel angle (Kk, b) the trajectory of the vehicle

On a flat horizontal road section 150 m long, the vehicle deviated from its initial position on the straight driving path by 0.76÷1.22 m (this is the maximum and minimum difference between the moduli of the final and initial vehicle position coordinates measured in the lateral direction).
Based on measurements of the force applied to the steering wheel and the steering wheel angle, the vehicle was found to have a tendency to turn autogenously the steering wheel by an angle of about 2.0°÷2.5°, which will translate into a considerable rightward change in the driving direction when the resultant force applied to the steering wheel is 0 N.

4. Summary

The assessment of keeping the straight driving direction is a complex issue, for which various measuring instruments must be used. The most reliable element of the tests is the finding that the vehicle when subjected to road tests shows a tendency to pull to one side. Such a finding should in principle be based on measurements of the forces acting on the steering wheel and the steering wheel angle.

References

[1] Sitek K, *Badania techniczne pojazdów. Poradnik diagnosty (Motor vehicle inspection. Diagnostician's handbook)*, WKŁ, Warszawa 2020.
[2] Mikulec R, *Analiza przemieszczeń poprzecznych pojazdu na mokrej nawierzchni (Analysis of vehicle's lateral displacements on wet road surface)*, Paragraf na drodze, IES Kraków 2021, R pp. 149-158.
[3] Wolak St, Pieniążek W, Olszewski A, Dylewski S, *Badania intensywnego hamowania motocykla na mokrej nawierzchni jezdni (Research on the intensive motorcycle braking on wet road surface)*, Paragraf na drodze, IES Kraków 2021. R pp. 255-265.
[4] Wicher J, *Some Aspects of Vehicle Active Safety, International Journal of Occupational Safety And Ergonomics* 1995, VOL. 1, No 3, 244-251.

IOP Conf. Series: Materials Science and Engineering 1247 (2022) 012007 doi:10.1088/1757-899X/1247/1/012007

The estimation of frequency response of nonlinear quarter car model and bilinear model of damper characteristics

Z Klockiewicz[1] and G Ślaski[1]
[1] Poznań University of Technology, ul. Piotrowo 3, 60-965 Poznań, Poland

Abstract. The paper presents proposed and tested methods of estimation of frequency response of nonlinear quarter car suspension model and comparison with method for linear model used for research of vehicle vertical dynamics. Linear quarter car suspension model is widely used for estimation of comfort and safety performance of passive and semiactive suspensions. The real suspension - especially shock absorbers - are non-linear in three aspects: static characteristics, hysteresis and the presence of friction. Testing linear suspension model is possible with the use of analytical transfer function formulas but testing real suspension on a testing stand or even virtual but nonlinear suspension needs to use methods of transfer function estimation. It was necessary to design appropriate input signal allowing to get useful response signals. With the use of obtained frequency responses a method of linear estimation of nonlinear suspension for a given range of working condition was proposed. Also the bilinear model of damper characteristic estimation was proposed and tested proving to be good alternative to nonlinear characteristic.

Key words: vehicle dynamics, suspension, shock absorbers, vehicle

1. Vertical vehicle dynamics and its performance evaluation

Car suspension from a point of view of vertical dynamics is an oscillatory system, affected by two types of excitations - kinematic excitation z_r caused by the road surface profile and force excitation related to inertial forces acting on vehicle body.

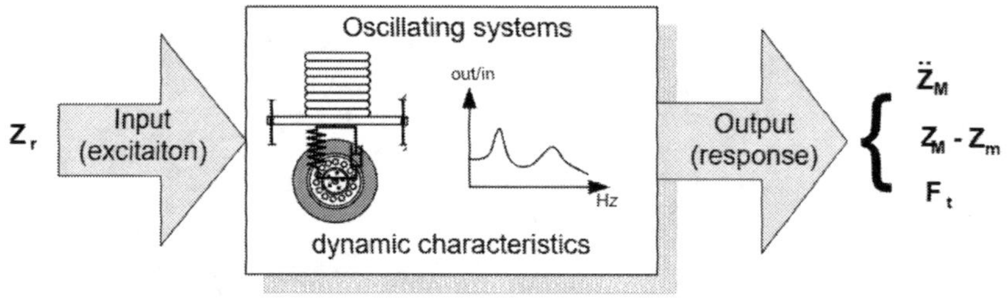

Figure 1. A quarter car suspension as a dynamic system.

These excitations processed by suspension dynamic structure result in changes in such variables as bounce absolute displacement of sprung mass z_M and unsprung mass z_m and the relative displacement of both masses, which describe the suspension deflection (rattle space) $z_M - z_m$, the acceleration of both these masses ($\ddot{z}_M$,$\ddot{z}_m$) and also tire-road contact forces F_t. The relation between road excitation and these variables treated as suspension system outputs (responses) in a function of excitation frequency are the dynamic characteristics of suspension, also called suspension transmissibility functions [1], frequency response functions or magnitude-frequency characteristics (Bode magnitude plot) [2,3,4].

The knowledge of frequency responses is essential to assess suspension performance according to such criteria as comfort, safety and technical limitations concerning limited suspension travel space. Ride comfort is often assessed by the use of sprung mass acceleration transfer function (acceleration amplification function), $\ddot{z}_M(\omega)$, safety – with an analysis of dynamic wheel load amplification function $F_t(\omega)$ and by analysis of wheel rattle space amplification function. This variable is important also from a point of view of kinematic performance of suspension (possible changes of wheel camber and steer angles) and the vertical dynamics influence on lateral dynamics.

2. Linear and non-linear quarter car models

There are many different kinds of vehicle suspension models (understood as a vertical dynamics vehicle models) of varying degree of complication. The one most frequently used is a simple linear quarter-car model, which can be found for example in old [2,3] and in newer publications [5, 6]. As the years went by, more and more research and simulations started using non-linear models, especially in the last decade [7,8].

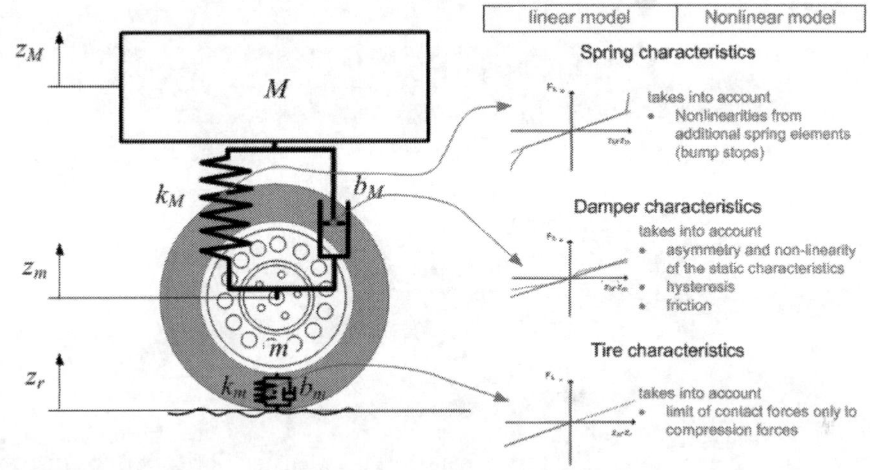

Figure 2. A quarter car suspension model with linear and nonlinear elements.

In most conditions of vehicle use simulation made to test ride safety or comfort can be conducted on simpler linear models. Roads of classes A and B rarely cause suspension to work in the non-linear range of its elements characteristics with the exception of vehicles being heavily loaded in the case of B class road, where the non-linear model might be worth considering. For worse quality roads, especially class D, linear models can yield somewhat satisfactory and realistic results only for best case scenarios, i.e. very low speeds and an almost empty vehicle. Otherwise, the suspension regularly enters non-linear working range [9].

Both the linear and nonlinear models use the same structure of suspension system, while the differences are found in the characteristics of spring and damping elements. Nonlinear spring characteristics takes into account big force increase when suspension enters the range of bump stop work. Nonlinear damper characteristic is even more complicated and takes into account asymmetry and nonlinearity of static shock absorber characteristics, hysteresis of damping force and friction. Nonlinear tire model can limit tire forces only to compression force (zero tire force when there is no road-tire contact).

3. Frequency responses of quarter car model and its interpretation

There are two main factors contributing to vehicle ride quality: the vehicle's suspension system dynamic characteristics and the road kinematic excitation resulting from the pavement surface and vehicle velocity. But in the process of assessing vehicle suspension performance other variables are taken into account as well - suspension deflection and tire dynamic load.

All of these are responses to input in the form of road excitation. As road excitation is time frequency signal vehicle suspension responses are also time frequency signals. The input and response signals can be expressed in the form of a Power Spectral Density (PSD) function. The response PSD function is

IOP Conf. Series: Materials Science and Engineering 1247 (2022) 012007 doi:10.1088/1757-899X/1247/1/012007

related to the excitation PSD function via the transmissibility Frequency Response Function (FRF) of the vehicle suspension under consideration. The transmissibility FRF acts as a weighting factor between the input and output. It is sometimes called also transmissibility function. Depending on the variable being analysed, various functions need to be used: sprung mass acceleration transmissibility, suspension deflection transmissibility, tire load (vertical force) transmissibility [10] and also sprung and unsprung mass displacement transmissibility (e.g.[11]).
The simplest way to experimentally determine FRF function is to excite suspension system with constant amplitude input signal with slowly and linearly changing frequency. The response signal can be then easily compared with excitation signal and multiplication factor of amplitudes for every frequency allow to determine FRF function values – Fig 3. The example of testing suspension deflection when excited by 0.01 m amplitude road excitation input signal is presented in figure 3. It is clearly visible that there are two resonances - one for sprung mass and one for unsprung mass. For frequencies near 0 Hz suspension almost does not move, the same can be said for frequencies much higher than unsprung mass resonance, where there is very little suspension deflection.

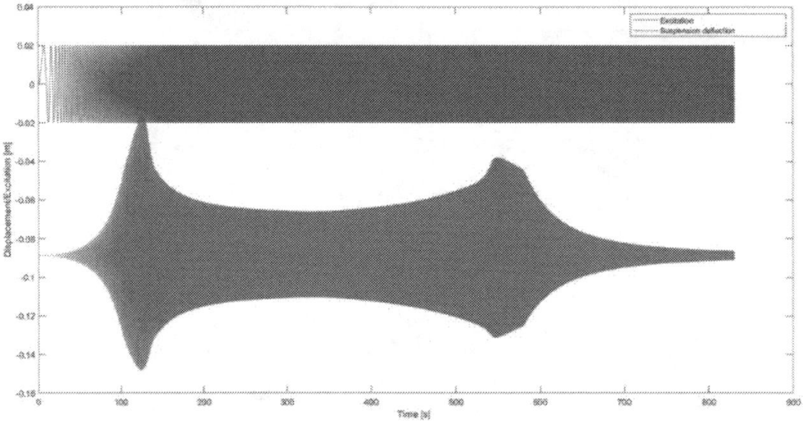

Figure 3. Comparison of excitation and response signals (suspension deflection) of suspension excited by constant amplitude sinusoidal signal with linearly changing frequency

An example of interpretation of FRF in practical consideration can be relation between EUSAMA test [12] and tire vertical force FRF function – figure 4. In this figure changes of tire vertical force are presented as the response to constant amplitude (0.003 m) kinematic excitation with frequency fluently changing in the range of 0 to 25 Hz, used on EUSAMA test stands during Periodical Vehicle Inspections. Yellow line presents the FRF assuming that frequency is linearly changing according to time.

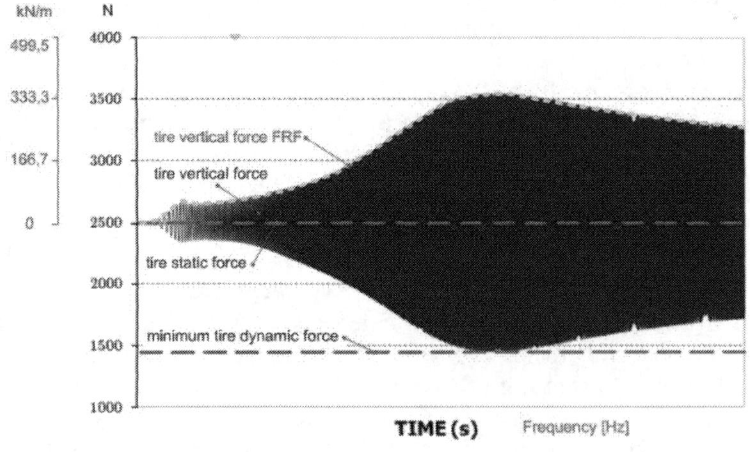

Figure 4. Relation between EUSAMA test and tire vertical force FRF function (red scale and yellow line)

One can note that FRF amplitudes in an unsprung mass resonance range can be used to calculate maximum and minimum dynamic loads - if it is for example 333.3 kN/m then multiplying 0.003 m excitation we obtain amplitude of 999 N of dynamic wheel load. That mean maximum wheel load of 3500 N and minimum 1500 N if the static load is equal to 2500 N. EUSAMA index will then be equal to EU=1500/2500=0.6.

4. Frequency response function estimation methods

4.1 Linear model frequency response function calculation

In the case of linear dynamic systems, which enable researchers to build relatively simple mathematical models, amplification functions can be evaluated using fundamental methods of control engineering (exemplary description of the use of these methods for a linear quarter car model can be found e.g. in [13,14]).

This is done by determining Fourier transforms of differential motion equations with zero initial conditions and then by determining frequency response functions for selected inputs and outputs. These frequency responses can be expressed in algebraic and exponential form [15]:

$$G(j\omega) = \frac{x(j\omega)}{u(j\omega)} = P(\omega) + jQ(\omega) = A(\omega)e^{j\varphi(\omega)} \qquad (1)$$

Gain of frequency response function expressed as:

$$A(\omega) = \sqrt{P^2 + Q^2}|_\omega = \frac{A_{output}(\omega)}{A_{input}(\omega)} \qquad (2)$$

physically means the gain between magnitude of input (excitation) and magnitude of output (response) in form of sinusoidal signal.

Frequency response function argument φ expressed as:

$$\varphi(\omega) = arctg\frac{Q}{P}|_\omega \qquad (3)$$

physically means the phase shift of sinusoidal signals - input and output.

4.2 Nonlinear model frequency response function estimation

Frequency characteristics of suspension can be determined also by analysing the relations between input signal and output signals. These signals can be measured and acquired during tests on a physical object or simulated with different types of suspension model - for example nonlinear model. It is worth pointing out that the responses of a non-linear system do not depend on the frequency alone, but on the input amplitude as well. The authors decided to focus their research on the frequency response for amplitudes in the range for typical road excitation levels and they acknowledge that the interpretation of the results acquired here might not be applicable for amplitudes of vastly differing values.

After preparing measured or simulated appropriate signals in the form of time histories, frequency response evaluation can be made using expression (4) [16]:

$$\hat{H}_{xy}(\omega) = \frac{\hat{G}_{xy}(\omega)}{\hat{G}_x(\omega)} \qquad (4)$$

where:

$\hat{G}_x(\omega)$ - estimator of power spectral density of input signal - kinematic excitation signal,

$\hat{G}_{xy}(\omega)$ - estimator of cross power spectral density of input and output signals.

The use of this formula requires knowledge of power spectral density - PSD (power spectrum value divided by frequency band for which a given PSD value was determined). The two-sided continuous power spectral density can be calculated as a limit [2]:

$$S_x(\omega) = \lim_{\Delta\omega \to 0} \frac{\Delta\langle f_n^2(t)\rangle}{\Delta\omega} \quad (5)$$

where:
$\Delta\langle f_n^2(t)\rangle$ - power share of function $f(t)$ near the frequency ω_n,
$\Delta\omega$ - frequency band for which power share is determined.
Measurements only help to estimate the power spectrum or power spectrum density, but do not allow to determine them exactly. This is more complex task, caused mainly by a limited set of samples [17].
The estimate of power spectral density can be found with various methods, for example squaring the magnitude of $f(t)$ signal Fourier transform and thus obtaining the periodogram or calculating discrete Fourier transform (DFT) of autocorrelation $f(t)$ signal function [17].
When calculating the estimates of power spectral density with few signal samples, unsatisfactory results are obtained which are shown as considerable PSD signal fluctuations. In order to obtain smoothed PSD estimates various method are used, for example calculating and averaging periodograms. One of the most common is the Welch method, which results in the so-called modified periodogram. It is a modified version of the Bartlett method [17].
Both methods involve dividing the signal (a series of samples) into several periods, for which shorter periodograms are calculated. Next an averaged periodogram over all the segments is calculated. The difference between the Bartlett and Welch methods of spectrum estimation is that the first one uses segments which do not overlap, while the other applies overlapping segments and time windows other than rectangular - e.g. the Hanning windows. As a result the Welch method ensures estimation with a smaller variance compared to the Bartlett method. This method is programmed amongst others in Matlab environment and allows to obtain the power spectral density estimate or cross-spectral density expressed in units of power per radian per sample (rad/Sa) or Hz.
Using calculated estimates of power spectral density we can calculate estimate of the frequency response function or one can use Matlab function for estimating the frequency response estimate using the presented procedure. Authors of the paper used the ***tfestimate*** function in Matlab for estimating frequency response function for linear, bilinear and nonlinear models of quarter car suspension.

5. Frequency response function estimation experiment

The experiment consist of two phases:

1. checking and design of input signal allowing to get good estimation of Frequency Response Function and compare it with FRF calculated analytically for fully linear model,
2. estimation of FRF for fully nonlinear suspension model (tire, spring and shock absorber) and comparing it with two models with different shock absorber models (fully linear, symmetrical bilinear) and possibility to identify their damping coefficients.

5.1 Suspension model used for experiment

The vehicle model used in simulation was a non-linear quarter-car model in three variants that differed in a damper model used. The variants were non-linear damper, linear damper and bilinear damper models. The tire forces were linearly dependant on the tire deflection and its velocity. The tire forces model had a non-linearity in the form of a possibility for a tire to loose contact with the road surface, which causes the tire forces to drop to 0 N. This meant the tire forces would never be negative – the tire is never pulled towards road surface. The characteristic can be seen in Figure 5

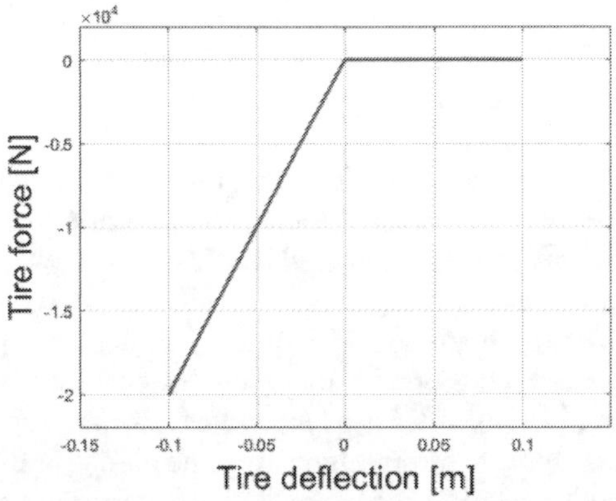

Figure 5. Tire force model used in research.

Other non-linearity in the model was the suspension spring, which used characteristic that was acquired in tests described in [18]. In this instance the characteristic was that of a front suspension spring that was a part of a MacPherson strut. The characteristic included also a rubber bump stop, which made the characteristic progressive. It also included an artificially added point of data for when the spring is completely compressed – in which case the force to compress it any further dramatically increases as now the metal coils of a spring themselves need to be compressed. The full characteristic can be seen in Figure 6.

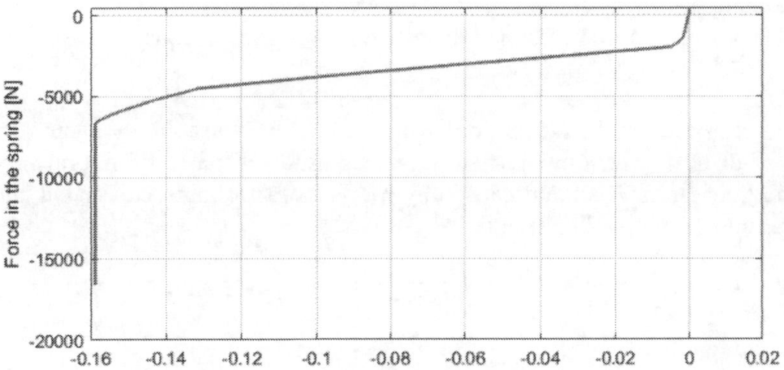

Figure 6. Spring characteristic used in research

The last important element to be addressed is the damper model, the non-linear version it used the front damper characteristic from [18], which can be seen in Figure 7. The model also had modules responsible for simulating hysteresis and internal friction of a damper, which could be disabled and enabled. More details on the damper model can be found in [19].

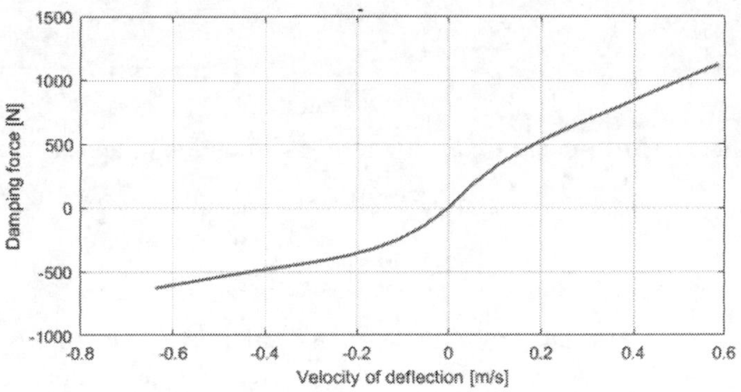

Figure 7. Damper characteristic used in non-linear model

5.2 Experiment first phase - input signal preparation for FRF estimation

The use of appropriate Matlab function **tfestimate** for FRF estimation is not the only condition to get good estimate. The other necessary operation is preparing appropriate input signal to obtain good input and output signals.

The kinematic excitation signal was prepared so that it would produce well represented, broad spectrum of frequencies for transfer function calculations. Generally so called chirp signal was used which is sinusoidal signal of a linear swept-frequency in a given time. However, normally generated "chirp" signal using Matlab or Simulink generator will not produce good enough results, as the automatic generator tends to go through low frequencies really quickly. It does not produce even one full cycle in the lowest frequencies while in the high there are many cycles present.

The **tfestimate** function in Matlab needs at least a few cycles in or around a given frequency to estimate it properly – around 10. That is why different signal was needed - one that was glued together from a few shorter samples. Connected signal consisted of three parts lasting enough time to consist of multiple cycles of similar frequencies occurring during simulation:

- the first one had only low frequencies up to 3 Hz changing linearly from 0 to 300 s,
- the second one consisted of medium-range frequencies from 3 to 10 Hz changing linearly from 300 to 580 s,
- the last one was of high frequencies from 10 to 35 Hz changing linearly from 300 to 580 s.

Another important condition to input signal was to produce response signals in a range of operational suspension deflections. It needed to control the response range. The overshoot of 35 Hz was implemented, because **tfestimate** function tends to give blurry results at the ends of examined frequency range. In order to acquire important data about frequencies around 25 Hz, the signal was artificially prolonged to 35 Hz to avoid aforementioned distortions for highest frequencies.

5.3 Experiment first phase - FRF estimation for linear model

First of all method of estimation FRF functions with use of **tfestimate** function and results of analytical calculation of this function were compared. In fig. 8 magnitude of transfer function (frequency response function) for sprung mass displacement caused by road kinematic input calculated and estimated with use of Matlab **tfestmate** function is presented.

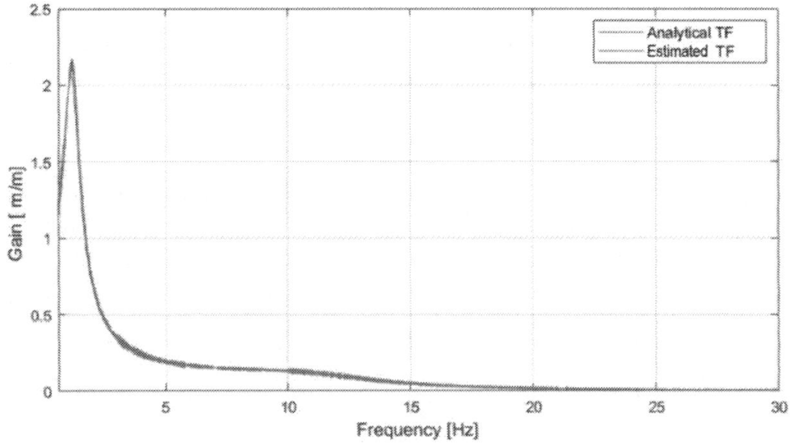

Figure 8. Analytical and estimated frequency response function of sprung mass displacement

Analytical FRF function is calculated with use of following formula

$$H_{z_M}(s) = \frac{(b_M s + k_M) k_m}{mMs^4 + (mb_M + Mb_M)s^3 + (Mk_M + mk_M + Mk_m)s^2 + b_M k_m s + k_m k_M} \quad (6)$$

The second function (red in Figure 8) was estimated in Matlab using **tfestimate** and their values are almost the same. The only difference is 5% lower value in a range of first resonance. In the rest - almost 99% of a frequency range the values of both functions are the same.

5.4 Experiment second phase - nonlinear suspension model FRF estimation and comparison with linear and bilinear damper model.

In the next phase of experiment FRF for five variables were estimated for nonlinear model of suspension and two other suspensions differing in a model of shock absorber. One was fully linear damping force model and the second one was symmetric bilinear model with two damping coefficients - one of a greater value for small suspension deflection velocities - up to 0.2 m/s and second one of smaller value for the higher deflection speed values. The results obtained during simulation were used to estimate the transfer functions between kinematic excitation and a given variable. These FRF functions were:

- sprung mass displacement,
- unsprung mass displacement,
- suspension deflection,
- sprung mass acceleration,
- cumulative tire force.

Those were then used to estimate the transfer functions between kinematic excitation and a given variable. In the case of suspension deflections it was deemed useful to calculate the density of probability of a given deflection to appear, in order to establish whether or not given excitation would cause the suspension to work in a normal operation range.

The first batch of experiments was done with simplest variant of damper model using just a single damping coefficient, which had its value changed in iterations in order to find the one that most closely resembled the non-linear model when it comes to vehicle responses. This was done in a broad spectrum of damping coefficients – starting from 1000 Ns/m every 200 Ns/m up to 3000 Ns/m. This phase was needed to try and gauge, what damping coefficient would give results most similar to those of a non-linear model. In this phase it became apparent, that there is no single value of a coefficient which would make the linear model act similar to the non-linear one – it was possible to find a coefficient which would make them act similarly near the first or the second resonant frequency, but not both at the same time - figure 9.

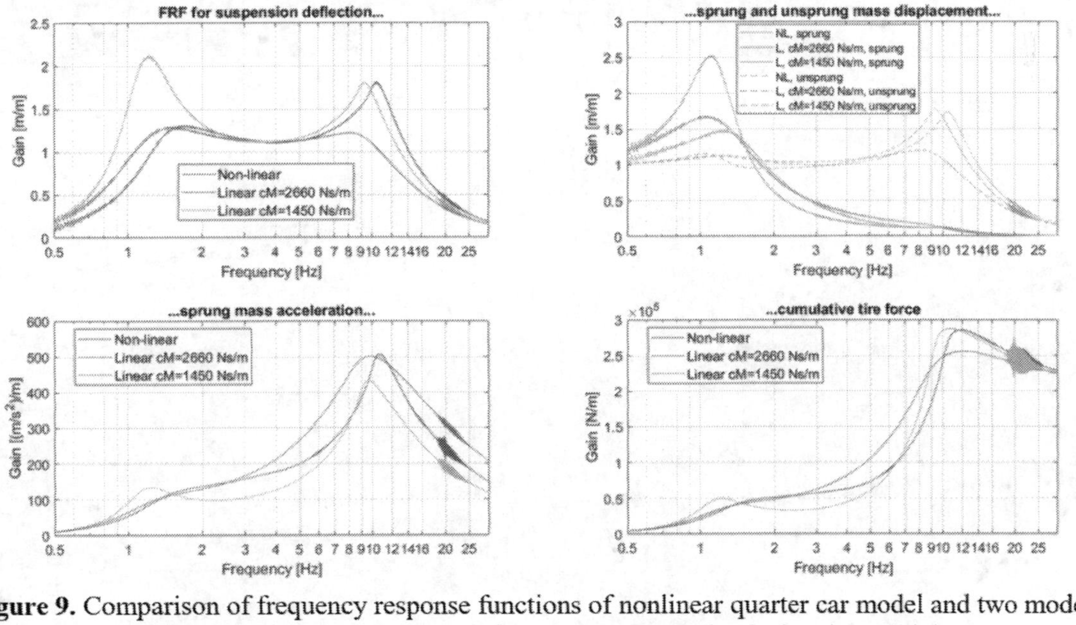

Figure 9. Comparison of frequency response functions of nonlinear quarter car model and two models with linear (b_M=2660 Ns/m and b_M=1450 Ns/m) shock absorber model.

In the second batch, previously recognized damping coefficients' values which bore the most resemblance to non-linear models in both frequencies were tested once again, this time with much smaller increases of 10 Ns/m per iteration. For sprung mass resonance, the testing started from

c_M= 1400 Ns/m, while for unsprung mass from c_M= 2600 Ns/m. Those two values were then also used to test if the bilinear model with them can approximate the behavior of the system better than that with values calculated based on the characteristic itself (which, to remind the reader, does not take into account hysteresis and friction in the damper).

Lastly, recognizing the shortcomings of the linear model, slightly more complex, but still easily implementable bilinear model was proposed. It consisted of two linear functions, connected at the joint point - Figure 10. The name bilinear comes from the fact, that there are two different linear damping coefficients – even if the characteristic consists of three lines, two of them share the same slope, while their variable independent part is different. It is worth noting, that this characteristic is still very simplified – for example, it is symmetrical, while the real characteristic differs depending on whether the damper is being compressed or stretched.

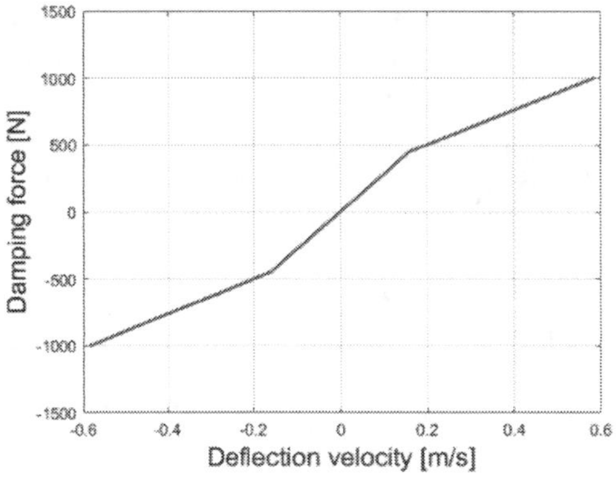

Figure 10. Damper characteristic in bilinear model.

Best matched FRF was obtained for bilinear model with two damping parameters $b_{M_1} = 2800$ Ns/m and $b_{M\_2} = 1220$ Ns/m - fig. 11. They differ from parameters really used in nonlinear model witch were determined to be about $b_{M_1} = 2815$ Ns/m and $b_{M\_2} = 1440$ Ns/m. The difference is caused by the fact that the nonlinear shock absorber is also nonsymmetrical.

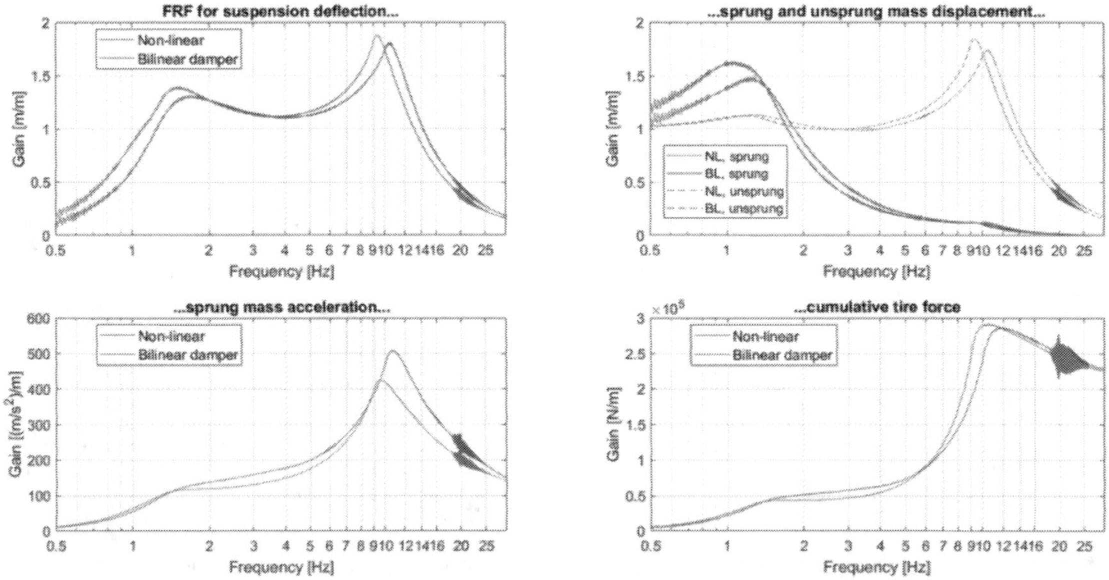

Figure 11. Best matched bilinear model frequency response functions compared with these functions of nonlinear model of shock absorbers.

The last part of the experiment was to evaluate which of nonlinear elements of shock absorber model - friction force or hysteresis - make it impossible to estimate ideally nonlinear model with linear or bilinear model.
In the Figure 12 FR functions of different versions of nonlinear shock absorber model (with friction and hysteresis, without friction, without hysteresis and without friction and hysteresis) and bilinear model are compared. It is visible that the best match with bilinear model is achieved using nonlinear damper without friction and hysteresis (but still with asymmetrical static characteristics).
Nonlinear model without friction behaves in low frequencies more similar to its version without friction and hysteresis and in a range of higher frequencies more like the version with friction and hysteresis.
Nonlinear model without hysteresis is more similar in a low frequencies range to fully nonlinear model (hysteresis and friction enabled) and in a range of higher frequencies it behaves like the version of a model without friction and hysteresis. This allows researchers to conclude that friction is more important in modelling behaviour of suspension in a range of low (first resonance) frequencies and hysteresis is more important in a range of second resonance at higher frequencies.

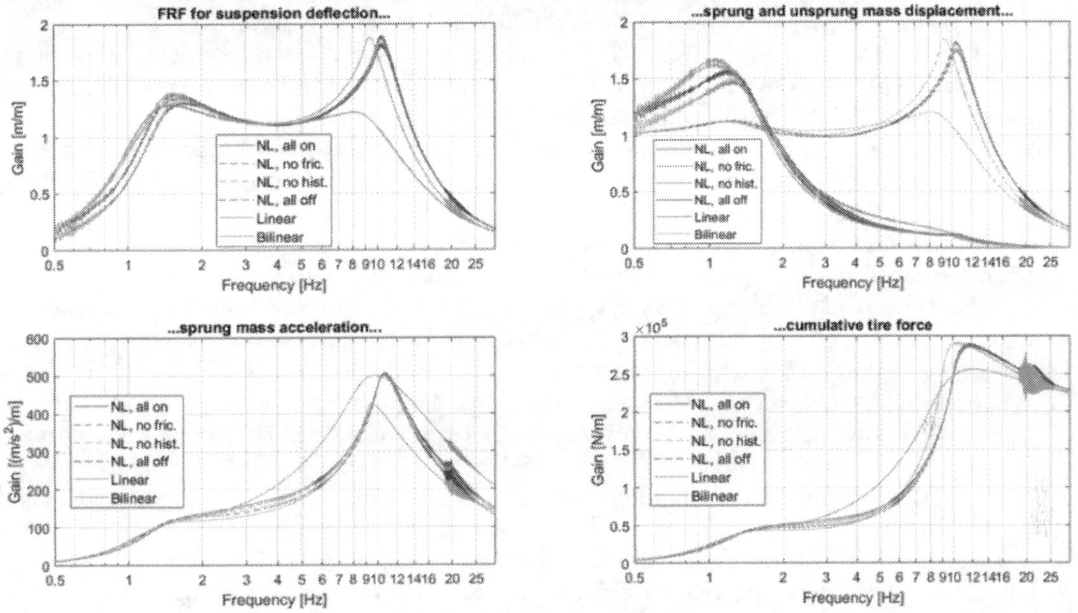

Figure 12. Comparison of FR functions for bilinear and different variants of nonlinear shock absorber model.

6. Conclusions

The paper presented the method of frequency response function estimation for nonlinear models of quarter car suspension, which can be used also for physical quarter car suspensions and measured wheel excitation and suspension responses.
The method is based on PSD function estimation and later FRF estimation with special preparation of the input signal (wheel kinematic excitation). Verification performed proved that for linear models it is possible to get almost the same frequency response function shapes.
In case of nonlinear models their ideal estimation using linear damper model is impossible. The reasons for this are as follows:

- due to nonlinearity of shock absorber characteristics (digressive) damping coefficient for the first (sprung mass) resonance is much higher than for the second (unsprung mass) resonance,
- due to presence of friction especially at low frequencies nonlinear shock absorber with friction active gives lower FRF gains due to different friction force behavior than viscous damping forces - friction force works for even the smallest deflection velocities when viscous damping force at the smallest deflection velocities are almost absent,
- due to presence of hysteresis of damping force there is a difference in all testing frequency ranges giving the biggest differences in a second (unsprung) mass resonance.

Performed experiments proved that use of bilinear model (even symmetrical) can give quite good estimation of fully nonlinear damper giving very close values of gain at the resonances with small shift in a value of resonance frequency. However those conclusions were tested only for a linear range of spring characteristic, which means that in order to extend them for other input amplitudes reaching non-linear spring work ranges, it would be necessary to run more simulations and create a function of two variables – both frequency and amplitude.

7. References

[1] J. Grajnert (red): Izolacja drgań w maszynach i pojazdach. Oficyna Wydawnicza Politechniki Wrocławskiej, Wrocław (1997).

[2] E. Kamiński, J. Pokorski: Dynamika zawieszeń i układów napędowych pojazdów samochodowych, Wydawnictwa Komunikacji i Łączności, Warszawa (1983),

[3] M. Mitschke: Dynamika samochodu: Drgania. Tom 2, Wydawnictwa Komunikacji i Łączności, Warszawa (1989).

[4] A. Reński: Bezpieczeństwo czynne samochodu. Zawieszenia oraz układy hamulcowe i kierownicze. Oficyna Wydawnicza Politechniki Warszawskiej, Warszawa (2011).

[5] J. Celko, M. Decky, and M. Kovac, An analysis of vehicle – road surface interaction for classification of IRI in the frame of Slovak PMS. Maintenance and reliability, Polish Maintenance society, Vol. 1, Iss. 41, 2009, pp. 15 - 21

[6] M. Doumiati, A. Victorino, A. Charara, and D. Lechner,: Estimation of road profile for vehicle dynamics motion: Experimental validation. Proceedings of the American Control Conference, pp. 5237–5242, 2011, DOI: 10.1109/ACC.2011.5991595.

[7] A. Mitura: Modelowanie drgań nieliniowego zawieszenia pojazdu samochodowego z tłumieniem magnetoreologicznym: rozprawa doktorska, Politechnika Lubelska, Lublin (2010).

[8] P. Zdanowicz and Z. Lozia: Wyznaczenie optymalnej wartości współczynnika asymetrii amortyzatora pasywnego zawieszenia samochodu z wykorzystaniem modelu „ćwiartki samochodu". Pr. Nauk. Politech. Warsz. Transp., no. 119, pp. 249–265, (2017).

[9] Z. Klockiewicz, G. Ślaski, and M. Spadło: The influence of the conditions of use and the type of model used on the vertical dynamic responses of a car suspension, The Archives of Automotive Engineering – Archiwum Motoryzacji, Vol. 85, No. 3, 2019, pp. 57–82.

[10] G. Reina, G. D. Rose: Active vibration absorber for automotive suspensions: a theoretical study. Int. J. Heavy Vehicle Systems, Vol. 23, No. 1, 2016.

[11] S. Hegazy, A. M. Sharaf: Ride Comfort Analysis Using Quarter Car Model. Proceeding of International Conference on Aerospace Sciences and Aviation Technology, May 2013

[12] A. Kuranc: Diagnostyczne badania zawieszenia pojazdu w aspekcie zmian parametrów eksploatacyjnych. Inżynieria Rolnicza, Vol. 117, No. 8, 2009.

[13] D. Karnopp: How significant are transfer function relations and invariant points for a quarter car suspension model? Vehicle System Dynamics, Vol. 47, Iss. 4, 2009.

[14] S. M. Savaresi, et al: Semi-Active Suspension Control Design for Vehicles. Oxford: Butterworth-Heinemann Ltd (Elsevier), 2010.

[15] A. Czemplik: Modele dynamiki układów fizycznych dla inżynierów. Warszawa, WNT, 2008.

[16] D. Borkowski: Symulacyjne badanie nieparametrycznej metody estymacji impedancji sieci energetycznej, Proceedings of XV Sympozjum Modelowanie i Symulacja Systemów Pomiarowych 18-22.09.2005.Krynica.

[17] D. Stranneby: Cyfrowe przetwarzanie sygnałów. Metody. Algorytmy. Zastosowania. Warszawa, Wydawnictwo BTC, Warszawa (2004).

[18] G. Ślaski: Studium projektowania zawieszeń samochodowych o zmiennym tłumieniu, Poznań: Wydawnictwo Politechniki Poznańskiej, (2012).

[19] D. Więckowski, K. Dąbrowski, G. Ślaski: Adjustable shock absorber characteristics testing and modelling, International Automotive Conference KONMOT 2018, 13 - 14 SEPTEMBER 2018, CRACOW, IOP Conf. Series: Materials Science and Engineering 421 (2018) 022039 IOP Publishing doi:10.1088/1757-899X/421/2/022039.

Airborne sound transmission in fluted PVC panels-properties of materials and structures

JM Kopania [1,*], W Zakrzewicz [2], P Kubiak [1,2], A Mrowicki [2], M Głogowski [1], J Gralewski [1], G Bogusławski [1], K Wójciak [3], P Gaj [3]

[1] Lodz University of Technology, Żeromskiego 116, 90-924 Lodz, Poland
[2] Warsaw University of Technology, Pl. Politechniki 1, 00-661 Warsaw, Poland
[3] Institute of Power Engineering, Mory Street 8, 01-330 Warsaw, Poland

* joanna.kopania@p.lodz.pl

Abstract. Authors present a study on transmission loss and absorption of corrugated PVC panels. Tests were conducted in reverberation room and it has been proven, both theoretically and experimentally, that there is a relationship between the panel type and its transmission loss and absorption. Panel size and thickness was also taken into consideration during testing procedure. Analyses also show that the Hansen model was not perfectly accurate when predicting the transmission loss for some of the tested samples.

Keywords: road accidents, sensors, automotive safety

1. Introduction

One of the main blocks of the European Green Deal is Circular Economy Action Plan (CEAP). The plan targets product design, circular economy processes and sustainable consumption. This causes industry to play a leading role in the ecological transition by reducing its carbon and material footprint. Further actions include implementing circularity across the economy [1], also by management of waste, especially plastics or polymers materials [2]. Due to this fact, the industry uses waste polymers materials more and more for automotive, aeronautics and space applications considering its mechanical properties and low weight [3]. Polyvinyl Chloride (PVC) is one of the most popular polymers in the industry, having a versatile nature and it is broadly applicable to technical solutions including in building, transport, packaging, electrical/electronic and healthcare applications, as well as everyday items. Although recent works have shown that PVC production harms the natural environment and human health [4,5], it is still universally used and will be used in the future.

A critical problem during the design, construction and exploitation of vehicles, buildings or machinery is not only the energy efficiency, but also good fire or sound insulation. Determination of the sound absorption (absorption coefficient - α) and sound transmission loss (TL, insulation parameter) of materials have become a necessity due to the technical progress and thus increased number of noise sources. Nowadays, noise pollution is one of the world's most leading environmental problems and about 40 % of the UE residents are always exposed to noise [6]. The latest research has shown, that long-term exposure to noise in urban areas has negative impact on health, causing 12 000 premature deaths and contributes to 48 000 new cases of ischaemic heart disease every year across Europe [7]. Different actions are taken to ensure reduction of noise pollution, for example by correcting the noise wave propagation to exposed people, by using panels, barriers and screens to block the direct path of sound.

Some of the earliest works focus on PVCs acoustic insulations in different forms: foam or fiber or composites or a mix with additives. Cheng et al., pointed out that the addition of flame retardants polymer composite to PVC does not influence the acoustic insulation[8]. Yang et al., studied the honeycomb weave fabric/PVC composite material, showing that repetitions of weave have great influence on the acoustic insulation [9]. Liu et al., noticed that adding fillers to PVC deteriorates mechanical properties and has little effect on the acoustics [10]. Huang et al. shows that PVC fiber panels can reduce the noise of the air conditioning more effectively, as compared to an only absorption material [11]. Xue et al., found that sound insulation cover made from PVC/PET fiber can effectively reduce the noise of the compressor [12]. Pulgern et al., studied the WPVC composite panels, indicating that those panels reduce sound pressure level by approximately 15-19 dB [13]. The other works focus on sound-absorption and sound insulation of the PVC in different building applications, like floor covering [14], blocks mixed with fine PVC [15], PVC foam materials [16] or PVC windows [17]. This resulted in several papers about PVC capacity for acoustic barriers for machineries or transport system. Fang et al., found that PVC panels caused noise reduction in refrigerators [18]. Agyeman-Prempeh [19] studied the reduction of noise levels of coconut fiber combined with polyvinyl chloride as sound insulating panels. Moreover, the noise insulation system for high-rise building by using PVC foam on wood, concrete and glass were studied [20]. The results for indoor showed that concrete mixed with PVC foam has the highest noise insulation effectiveness.

Roschke and Esche examined another acoustics parameter of materials - insertion loss (IL) and have studied various polymer barriers [21]. Macbain et al., [22] studied barrier from various recycled plastic and found its performance to be better at lower frequencies (100 to 250 Hz). Combination of tire rubber, PVC grains and nylon fibers with polyurethane as a binder resulted in good absorption coefficient [23]. Another way to improve the performance of noise barrier was to modify its shape by using additional elements on the top of barrier, made from different material, like T-shaped and Y-shaped caps [24,25], cylindrical caps [26,27], multiple-edge caps [28] and tilted caps [29]. All of these design variants changed the effective acoustic length of the barrier or modified the diffraction points of the screen. Also the application of scatters acoustic barriers in cities was studied. Its aim was to reduce the transmitted transport noise that affected buildings [30].

Corrugated or fluted panels have experienced increasing number of application as construction elements for roofs, claddings and walls of modern industrial buildings as well as acoustics road panels. The range of profile shapes may be large and intended for different situations. As corrugated and fluted panels are so widely used for industrial applications, its acoustics parameters are of considerable interest. Engineers need this information to estimate expected community noise levels at the building or construction design stage.

The corrugated or fluted panels are generally referred to as orthotropic because they are more stiff along one direction than the other. Heckl' [31,32] analyzed sound transmission through orthotropic panels. He derived some approximate solutions for transmission coefficient for panels of infinite extent with the effects of panel damping neglected. The earliest works focus on the calculation of moments of area of the profiled metal sheet [33]. on the sound absorption and TL of single metal skin cladding with varying profiling and thickness [34] or on the dependences of the profiling to critical frequency [35] or application the finite element and boundary element analysis to predict the TL of a corrugated and fluted steel sheet [36].

Due to the development of the plastics processing industry, those materials are used even more frequently. Manufacturing specifics and material additives caused the acoustical parameters for different type of plastics to remain in the cognitive phase. Therefore, the main goal of this paper is study the absorption coefficient and sound insulation of fluted PVC panels, taking into account the panel size and thickness. Measured results are also compared with the theoretical model. The acoustical analysis of a single fluted PVC panels could be the next step to find a method for improving their sound insulation capacity.

2. Materials and methods

2.1. Analitical model infinite and finite homogeneous orthotropic panel

According to Hansen [38] for thin, homogeneous, orthotropic panels, the wave impedance Z_{orth} can be expressed by:

$$Z_{orth} = j\omega m\left[1 - \left(\frac{f}{f_{c1}} cos^2\varphi + \frac{f}{f_{c2}} sin^2\varphi\right)^2 sin^4\theta(1 + jn)\right], \quad (1)$$

where:

$$f = \frac{\omega}{2\pi}, \quad (2)$$

$$f_{c1} = \frac{c^2}{2\pi}\sqrt{\frac{m}{B_x}}, \quad (3)$$

$$f_{c2} = \frac{c^2}{2\pi}\sqrt{\frac{m}{B_y}}, \quad (4)$$

The panel loss factor is n, the frequency of the panel is ω, the mass per unit area of the panel is m, the speed sound in air is c and the angles θ and φ are defined in Figure 1.

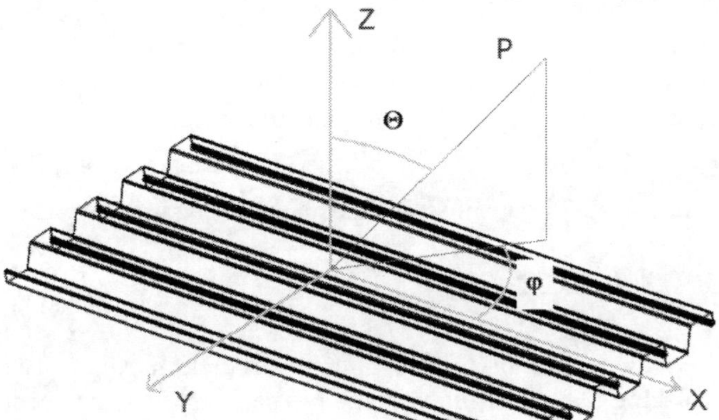

Figure 1. Panel coordinate system showing incident wave.

The f_{c1} and f_{c2} are known as the critical frequencies, at which the speed of bending wave propagation is equal to the speed of acoustic wave propagation in the surrounding medium. The corrugated panel commonly found in constriction engineering are stiffer along the direction of the ribs than across the ribs. Orthotropic panels are characterized by a range of bending wave speeds caused by the two different values of the cross-sectional second moment of area per unit width. Therefore, for thin corrugated trapezoidal panel the bending stiffness about X and Y axis, B_x and B_y, in Equation (3) and (4) are given by:

$$B_x = \frac{Eh}{(1 - v^2)L}\sum_n b_n\left[z_n^2 + \frac{h^2 + b_n^2}{24} + \frac{h^2 - b_n^2}{24} cos2\theta\right], \quad (5)$$

$$B_y = \frac{Eh^3}{(1 - v^2)L}\sum_n b_n, \quad (6)$$

where L and b_n are defined in Figure 2, and ν is Poisson's ratio, z_n is the distance from the neutral axis of the panel cross section to the center of the panel segment n, where neutral axis is defined such that:

$$\sum_n b_n z_n = 0, \quad (7)$$

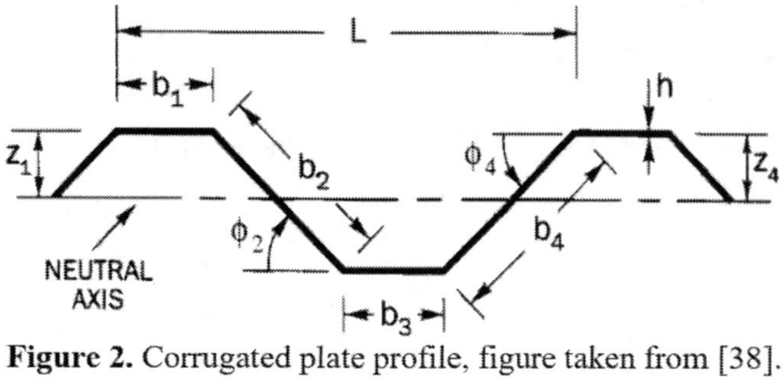

Figure 2. Corrugated plate profile, figure taken from [38].

The transmission coefficient for a wave incident at angle (θ, φ) to the panel is given in terms of the panel impedance by

$$\tau_{\theta\varphi} = \left|1 + \frac{Z_{orth}cos\theta}{2\rho_0 c}\right|^{-2}, \tag{8}$$

The transmission coefficient for normal incidence, τ_N, is found by substituting θ=0 in Equation (5). The diffuse field transmission coefficient, τ_d, is found by determining a weighted average for $\tau(\theta, \varphi)$ over all angles of incidences using the following relationship:

$$\tau_d = \frac{1}{\pi}\int_0^{2\pi} d\varphi \int_0^{\pi/2} \tau(\theta,\varphi)cos\theta sin\theta d\theta, \tag{9}$$

The $cos\theta$ term accounts for the projection of the cross-section area of the panel wave that is incident upon a unit area of wall at an angle, θ, to the wall normal. The $sin\theta$ term is a metrical coefficient, which arises from the use of spherical coordinates. For orthotropic panel Equation (3) becomes:

$$\tau_d = \frac{2}{\pi}\int_0^{\pi/2} d\varphi \int_0^1 \tau(\theta,\varphi)d(sin^2\theta), \tag{10}$$

Substituting Equation (5) into (7) gives for the diffuse field transmission coefficient for an infinite panel, like given by Heckl [33]:

$$\tau_d = \frac{2}{\pi}\int_0^{\pi/2}\int_0^1 \frac{d(sin^2\theta)d\varphi}{\left|1+\frac{Z}{2\varrho_0 c}cos\theta\right|^2}, \tag{11}$$

The transmission loss is commonly observed by means of testing in a reverberation room. The transmission loss is usually measured by placing the panel in a window between two adjacent reverberant rooms. Noise is introduced into one of the rooms (source room) and part of the sound energy is transmitted through the test panel into the second room (receiver room). So, during measurements, panels are not of infinite extent.

It is possible to perform numerical integration of the Equation (11) for various range of frequency and transmission loss can then be calculated by substituting τ_d for τ in Equation (12), considering that the angle θ is dependent on the size of the panel (and $\theta=0$ - the acoustic wave is in the direction angle θ to the normal to the panel surface). So, to reduce the extent of the numerical calculations the various approximation can be made [34]. At frequency $f = f_{c1}/2$, it may be assumed that the panel behaves as a limp mass and transmission loss TL can be calculated from:

$$TL = 20log_{10}f_{c1}m - 54, (dB) \tag{12}$$

For $f_{c1} < f < f_{c2}/2$ the following approximation can be used:

$$TL = 20log_{10}f + 10\,log_{10}(m) - 10log_{10}f_{c1} - 20log_{10}\left[log_e\frac{4f}{f_{c1}}\right] - 13.2, (dB) \tag{13}$$

For $f=2f_{c2}$ transmission loss TL is calculate from:

$$TL = 10log_{10}m + 10log_{10}f_{c2} - 5log_{10}f_{c1} - 17, (dB) \tag{14}$$

Above $2f_{c2}$ the TL is given by:

$$TL = 20log_{10}f + 10log_{10}m - 5log_{10}f_{c1} - -5log_{10}f_{c2} - 23, (dB) \tag{15}$$

The transmission loss between $0.5f_{c1}$ and f_{c1} and also between $0.5f_{c2}$ and $2f_{c2}$ is approximated by connecting points corresponding to frequency with a straight line. It is generally accepted that these formulas provide reasonably accurate results and are satisfactory for most commonly used orthotropic building panels, like corrugated or fluted panels.

2.2. Studied panels

The transmission loss between $0.5f_{c1}$ and f_{c1} and also between $0.5f_{c2}$ and $2f_{c2}$ is approximated by connecting points corresponding to frequency with a straight line. It is generally accepted that these formulas provide reasonably accurate results and are satisfactory for most commonly used orthotropic building panels, like corrugated or fluted panels.

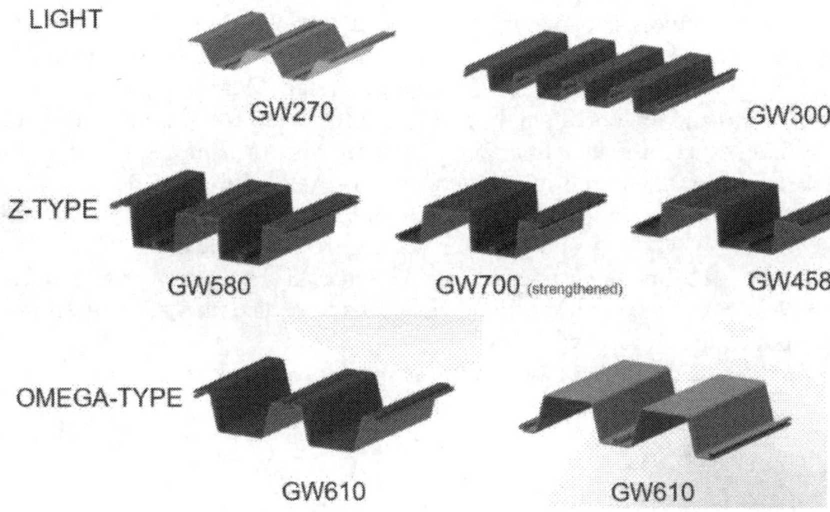

Figure 3. Panel's profiles studied in this work.

Table 1. The parameters of studied fluted PVC panels.

Panel	Panel thickness [mm]	Commercial mark -	Young's modulus [GPa]	Poisson's Ratio -	Cross-sectional panel height [mm]
	h	-	E	ν	Z
GW270	3,5	light	2,4	0,4	0,150
GW300	5,5	light	2,4	0,4	0,115
GW580	7,0	Z-type	2,4	0,4	0,240
GW700	9,0	Z-type Strengthened* *undisclosed trade information	2,4	0,4	0,250
GW610	9,0	Ω-type	2,4	0,4	0,230
GW448	12,0	Z type	2,4	0,4	0,254

Studied commercial fluted plastic panels are manufactured in the process of molding, otherwise known as co-extrusion of polyvinyl chloride with additive substances refining its parameters (toughness modifiers, UV and thermal stabilizers etc.). The sheet are manufactured using extrusion molding method as monolithic profiles or co-extrusion method, where core is made from a recycled construction type PVC with an outer layer of the default PVC material. Interlocking connections at the edges allow for producing panels of various width.

2.3. Transmission loss measurements

The studies were done in reverberation rooms of the Institute of Power Engineering, OTC – Thermal Technology Branch „ITC" in Lodz. The panels were tested by mounting in an aperture between two reverberant test chambers. The measurement were carried out in accordance with the current ISO standard: *10140-2:2021 Acoustics — Laboratory Measurement of Sound Insulation of Building*

Elements — Part 2: Measurement of Airborne Sound Insulation and *10140-4:2021 Acoustics — Laboratory measurement of sound insulation of building elements — Part 4: Measurement procedures and requirements*. Both of the source and receiving rooms were constructed to give a reverberant sound field with volumes of $V=237m^3$ and area of 231.5 m^2 for each.
The tested PVC panels (size 1,0 m × 1,0m) in a frame were fixed to an aperture opening between the two reverberation rooms. Panels had to be cut down to fit the aperture opening dimension. Edges of each specimen were sealed tightly by means of sound-proof silicone sealant.
A white noise was produced by using all-in-one DO 203 omnidirectional noise source, including a dodecahedron loudspeaker with a power amplifier and a noise generator. Measured parameters were sound pressure level-in both rooms and reverberation time (RT_{60}) in the receiver room. RT is defined as the time required for the sound source to reduce to a level of 60 dB when it is turned off in a closed room. Reverberation times have been measured in the frequency range of 50Hz to 10000Hz. Sound pressure levels in the rooms were sampled by means of two sets of l/2 inch microphones, each coupled with a pre-amplifier and fed to SVAN 958 analyzer for 1/3 octave band spectrum analysis. The background level was measured and calibrated using the Bruel & Kjaer 4231 calibrator before and after the measurement procedure. Moreover, the environmental conditions were also monitored by recording the temperature, relative humidity and atmospheric pressure.
Transmission loss was obtained using the equation for the transmission of sound between two reverberant spaces:

$$TL = \overline{L_s} - \overline{L_r} - 10log\left(\frac{S}{A}\right)\ (dB) \quad (16)$$

where:
$\overline{L_s}$ - spatial average sound pressure level in the source room, dB
$\overline{L_r}$ - spatial average sound pressure level in the receiver room, dB
$TL-$ reverberant field transmission loss, dB
$S-$ area of the transmitting surface, m^2
$A-$ room constant in the receiver room, using the Sabine equation: $A=0,16\times(V/RT_{60})$, m^2

2.4. Sound absorption measurements
Second phase of test considered examination of absorption coefficient of studied PVC panels. For every test around 10 m^2 of panel were placed on the floor of the reverberation chamber according to standard *ISO 354:2005 - Acoustics — Measurement of sound absorption in a reverberation room.* Interlocking connections allowed to build the PVC sheet of approximately 5m in width. Prior to testing, panels were inserted into reverberation room for 24 hours to acclimatize to the atmospheric conditions in chamber - Figure 4. The measurements were done using rotating microphone boom Nor265 with an integration time of 30s for each microphone position. The 1/3 octave bands were measured using Nor140. Microphone was calibrated before the acoustic test.
The reverberation time of the room was measured in two conditions:
- when the reverberation room was empty;
- when the studied panels were inside.

In general, once material is placed into the reverberation room, a lower reverberation time can be observed. The difference in reverberation times is a measure of the amount of absorption caused by the studied panels.

Figure 4. Reverberation room with studied panel.

3. Results and discussion

3.1. Transmission loss

Figure 5 shows the measured TL values for the studied fluted PVC panels. For single panels, the transmission loss is influenced by four factors: size, stiffness, mass, and damping. In Figure 5 similar TL spectrums are observed. In the low and medium frequencies, the three visible peaks are observed: around 100Hz, 160Hz and 315Hz and dips at 125Hz and 200Hz. These dips can be interpreted as specific vibrational mode of the panel or the lowest acoustic coincidence of the panels. At very low frequencies the transmission depends heavily on stiffness, as it can be observed for GW700 panel under 63Hz . With greater bending stiffness and shorter span, the transmission loss is the highest. But the measured transmission loss of all fluted panels shows no distinct lower critical frequency shift with the change of panel shape. The low-frequency response of panels of different depths is similar.

However, at the frequency of the incident, the wave increases and the panels will resonate mechanically at a series of frequencies, like resonance or eigenfrequencies. This is because of the fact that the panels have a finite boundary and edge fixing. The resonance frequencies consist of a fundamental frequency and integer multiples of this fundamentally called harmonics. At the fundamental resonance, $f_{1,1}$, the sound transmission through the panel is enhanced and the transmission loss drops significantly. Adding the damping material to the panel usually reduces the resonance amplitude. The fundamental resonance frequencies, $f_{1,1}$, of a supported rectangular orthotropic panel of width a and length b were calculated according to:

$$f_{i,n} = \frac{\pi}{2m^{1/2}}\sqrt{\left(\frac{B_x i^4}{a^4} + \frac{B_y n^4}{b^4} + \frac{B_{ab} i^2 n^2}{a^2 b^2}\right)}, \tag{17}$$

$$B_{ab} = \frac{1}{2}(B_a v + B_b v + Gh^3/3), \tag{18}$$

where, $G=E/[2(1+v)]$ is material modulus rigidity, E is Young's modulus, v is Poisson's ratio, a and b are panel dimension (Table 3).

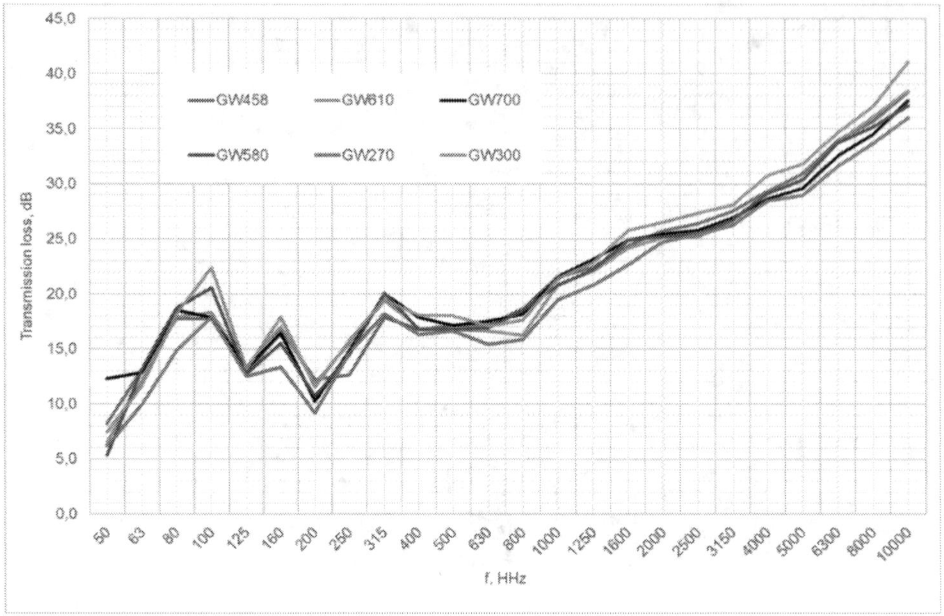

Figure 5. Measured transmission loss (TL) of studied PVC panels

Above 800 Hz the sound transmission depends on the mass inertia of panels and is independent of its stiffness. In this frequency region, part of the acoustic energy is transmitted through the panel and the remainder is reflected. A direct relationship exists between panels' weight and the resulting transmission loss. If the structure of panels is of considerable weight, the sound insulation could not be improved only by a small increase in mass. However, increasing panel mass lowers resonance frequencies and raises the critical frequencies. For the studied panels, as an orthotropic panel, two critical frequencies were calculated, according to Hansen, and shown in Table 2.

Table 2. Calculated properties of fluted (orthotropic) PVC panels.

Panel type	**Surface weight [kg/m²]**	**Bending stiffness [kg×m²/s²]**	**Effective bending stiffness [kg×m²/s²]**		**Critical frequencies [Hz]**		**First resonance frequency [Hz]**
	m	**B_x**	**B_y**	**B_{eff}**	**f_{c1}**	**f_{c2}**	**$f_{1,1}$**
GW270	7,2	44599	13,7	780,5	238	13594	136
GW300	14,3	35282	65,8	1523,8	377	8728	86
GW580	23,6	224863	142,45659,6		192	7621	168
GW700	29,4	255173	282,98496,1		201	6036	161
GW610	22,6	219878	266,87658,5		190	5450	170
GW448	30,7	529198	692,219139,6		143	3943	227

For these type of panels the critical frequency is dependent on the direction of the incident acoustic wave. The range of critical frequencies is bounded at the lower end by the critical frequency corresponding to a wave travelling in the panel most stiff direction and at the upper end by the critical frequency corresponding to a wave propagating in the least stiff direction.
Calculated panels properties allowed for a comparison of the theoretical values and measurements result – as shown in Figure 6. It can be seen that the Hansen's scheme provides a reasonable TL prediction in agreement with measured results for studied GW270 and GW300. For the rest of the panels, the predictions differ from the measurement overestimating at higher frequencies. There are also differences between theory and experiment at very low frequencies. The first resonance frequency of the studied panels varies from 80Hz to 230Hz. As this theory is only applicable to frequencies well above the first

resonance frequency of the panel, some discrepancies are expected. However, this approach that was developed by Hansen is unable to account for localized vibration effects. The following behavior is especially to be noted. A very stiff construction tends to move the first resonance to higher frequencies but, at the same time, the frequency of coincidence tends to move to lower frequencies. Thus, the extent of the mass law region depends on the stiffness of the panel.

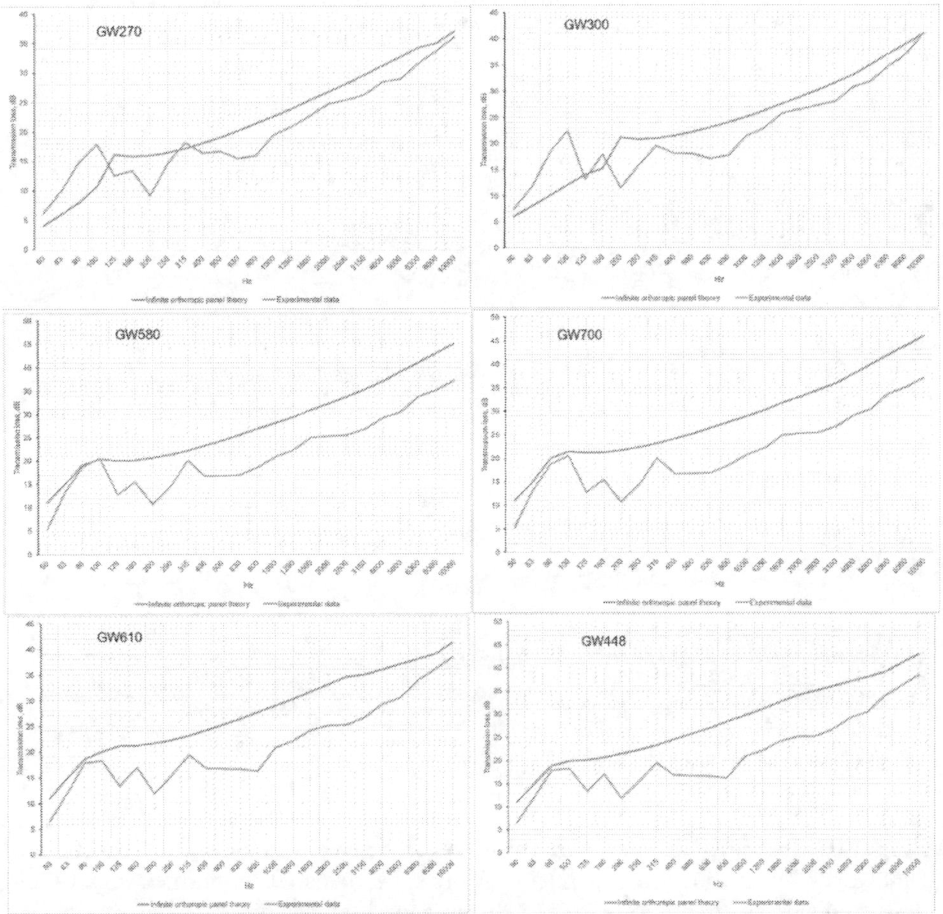

Figure 6. Measured and predicted transmission loss (TL) of studied PVC panels

The orthotropic plate theory predicts a shift of critical frequency (especially f_{c2}), which would make the TL below the critical frequency higher. Such case was not observed in experimental data.
The transmission loss of orthotropic panels is strongly affected by the existence of a critical frequency range. In this case the coincidence region may extend over two decades for common fluted panels. Therefore, this type of panels should be avoided where noise control is crucial. Although it can be shown that damping of these panels can improve by adding a damping material.

3.2. Absorption coefficient

The results of the absorption coefficients are given in at Figure 7 in 1/3-octave bands. The results are the arithmetic average of the three measurements and calculations. The studied panels have a good acoustic absorption properties in selected frequencies. If the material's sound absorption coefficient is above 0.4, it is considered to have good noise absorbing properties. For studied panel, some dependencies between absorbing coefficient and their construction were found. It can be argued that the shape of panel matters here. The similar spectrum for GW700, GW610 and GW580 were observed, with $\alpha \approx 0{,}4 \div 0{,}45$ at 125Hz. For GW610 panel, the second peak was observed at 200Hz ($\alpha \approx 0{,}3$), similar to

GW710 (α≈0,25 at 315Hz). For GW710 panel, an increase in the absorption coefficient is observed between 2000-8000Hz. For GW270, GW300 and GW448 poor absorption was observed below this range (α≈0,3). However, an advantage here is an increased absorption in the range of 100-400Hz. Higher absorption coefficient was observed in the frequency range with a decrease of transmission loss. The absorption properties of the studied fluted PVC panels are strictly connected with its resonance. In studied cases, this dip could correspond to the resonance frequency associated with the space between the ribs and wave propagation in the plane of the panel. This frequency is independent of panel size, and although it can be calculated reasonably accurately, the magnitude of the corresponding dip cannot be calculated at this research stage.

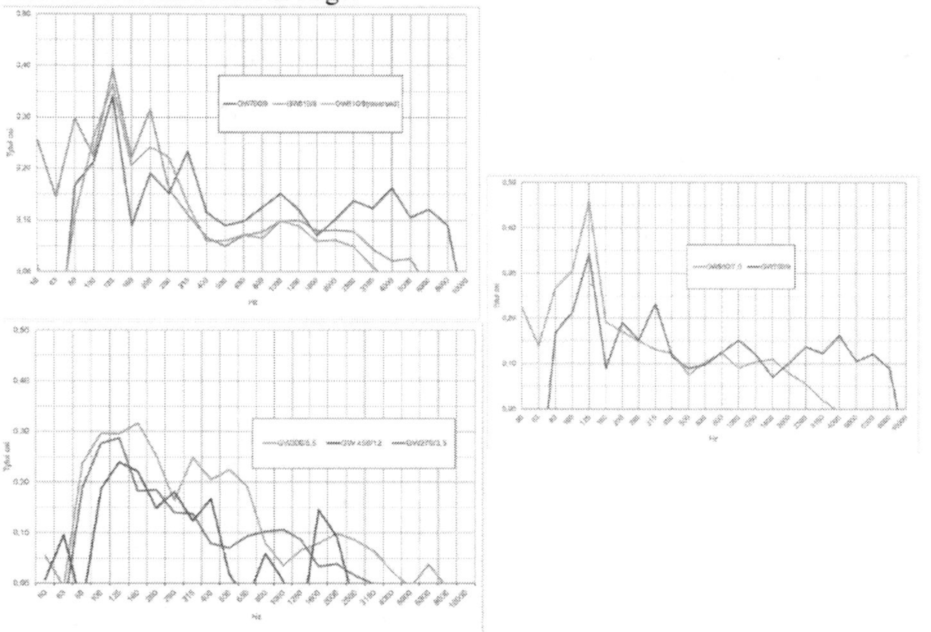

Figure 7. Measured absorption coefficient of studied PVC panels

4. Conclusions

This paper discusses the PVC fluted panels and its transmission loss as well as absorption coefficient. Authors proved that those two quantities are dependent on the type of panel, both theoretically and experimentally. At low frequencies, there were minor differences between the Hansen model and experimental data. Authors managed also to observe a weakness of the Hansen model when it comes to localized vibration effects—recorded sound spectrum was not in line with the orthotropic plate theory.

References

[1] European Comission, *A New Industrial Strategy for Europe*; 2020;
[2] Geyer, R.; Jambeck, J.R.; Law, K.L. Production, Use, and Fate of All Plastics Ever Made. *Science Advances* **2017**, *3*.
[3] Kirchherr, J.; Reike, D.; Hekkert, M. Conceptualizing the Circular Economy: An Analysis of 114 Definitions. *Resources, Conservation and Recycling* **2017**, *127*, 221–232, doi:https://doi.org/10.1016/j.resconrec.2017.09.005.
[4] Al-Malack, M.H. Migration of Lead from Unplasticized Polyvinyl Chloride Pipes. *Journal Hazardous Materials* **2001**, *82*, 263–274, doi:10.1016/s0304-3894(00)00366-6.
[5] Health Care Without Harm *The Polyvinyl Chloride Debate: Why PVC Remains a Problematic Material*; 2021;
[6] World Health Organization *Environmental Noise Guidelines for the European Region*; 2018;
[7] European Commission *Noise Impacts on Health*; 2015;
[8] Cheng, L.; Yao, Y.; Tang, X. On the Flame-Retardant Properties of Soft PVC-Base Composite Insulation. *Journal of Zhejiang Technology University* **2011**, *28 (2)*, 160–165.

[9] Yang, T.; Fu, Y. Sound Insulation of Honeycomb Weave Fabric Reinforced Polyvinylchloride Composite. *Journal of Textile Research* **2011**, 65–70.

[10] Liu, H. Study on Performance Impact of Filler to PVC Flexible Insulation Composite Material. *Zhejiang Technology University* **2012**.

[11] Huang, L.; Tan, L.; Xia, B. Noise Test and Control for Commercial Air Conditioner's Compressors. *Noise and Vibration Control* **2015**, *1*, 191–194.

[12] Xue, H.; Lai, X.; Zhang, S. Research on Deadening Felt of Inverter Compressor. *Electrical Appliances* **2014**, 76–78.

[13] Zhang, L. Research on the Acoustic Insulation Properties of PVC.; 2016.

[14] Pulngern, T.; Eakintumas, W.; Rosarpitak, V.; Sombatsompop, N. Compressive Load, Thermal and Acoustic Properties of Wood/Polyvinyl Chloride Composite Log-Wall Panels. *Journal of Reinforced Plastics and Composites* **2017**, *36*, 1183–1193, doi:https://doi.org/10.1177/0731684417699712.

[15] Mun, D.-H.; Song, G.-G.; Lee, C.-S.; Park H.-G. Reduction of Floor Impact Noise and Impact Force for PVC Floor Covering and Floor Mat. *Transactions of the Korean Society for Noise and Vibration Engineering* **2014**, *The Korean Society for Noise and Vibration Engineering*, doi:https://doi.org/10.5050/ksnve.2014.24.7.501.

[16] Hussein, H.H.; Eedan, O.A.; Ahmed, M.K. Mechanical, Thermal and Acoustical Properties of Concrete with Fine Polyvinyl Chloride (PVC). *Iraqi Journal of Civil Engineering* **2016**, *12*, 81–91.

[17] Xu, X.; Quan, J. Brief Analysis on Application of PVC Foam Materials in Building Material Industry. *Materials Science Forum, Trans Tech Publications, Ltd* **2019**, *944*, 729–735, doi:doi:10.4028/www.scientific.net/msf.944.729.

[18] Konstantinov, A.; Verkhovsky, A.; Lyabakova, E. Sound Insulation of PVC Windows at Negative Outdoor Temperatures.; 2020; Vol. 896.

[19] Fang, Z.; Jiang, M.; Zhang, Y. Experimental Research on PVC Sound Insulation Improving the Front Noise of Refrigerator. *Home appliance technology* **2014**, 2–4.

[20] Agyeman-Prempeh, N.O.W.O. Noise Abatement in Machine Shops Through the Use of Recycled Coconut Fibre and Polyvinyl Chloride Sheets As Noise Absorbing Materials, University of Ghana, 2017.

[21] Kuan, L.J. Analysis of Noise Barrier for Apartment, Tunku Abdul Rahman University College, 2017.

[22] Roschke, P.N.; Esch, S.T. Construction of Full-Scale Noise Barrier with Recycled Plastic. *Transportation Research Record* **1999**, *1656*, 94–101.

[23] Saageghvaziri, M.; Macbain, K. Sound Barrier Applications of Recycled Plastics. *Transportation Research Record* **1998**, *1626(1)*, 85–92, doi:10.3141/1626-11.

[24] Benkreira, H.; Khan, A.; Horoshenkov, KV. Sustainable Acoustic and Thermal Insulation Materials from Elastomeric Waste Residues. *Chemical Engineering Science* **2011**, *66(18)*, 4157–4171.

[25] May, D.N.; Osman, N.M. Highway Noise Barriers: New Shapes. *Journal of Sound and Vibration* **1980**, *71*, 73–101, doi:10.1016/0022-460X(80)90410-1.

[26] Murata, K.; Nagakura, K.; Tanaka, SI. Noise Reduction Effect of Noise Barrier for Shinkansen Based on Y-Shaped Structure. *Quarterly Report of RTRI* **2006**, *47(3)*, 162–168.

[27] Galińska, B.; Kopania, J. Hałas Drogowy a Skuteczność Ekranów z Oktagonalnymi Reduktorami Dźwięku. *Autobusy : technika, eksploatacja, systemy transportowe* **2016**, *6*, 168–171.

[28] Fujiwara, K.; Hothersall, D.C.; Kim, C. Noise Barriers with Reactive Surfaces. *Applied Acoustics* **1996**, *53*, 255–272.

[29] Watts, G.R. Acoustic Performance of Parallel Traffic Noise Barriers. *Applied Acoustics* **1996**, *47*, 95–119, doi:10.1016/0003-682X(95)00031-4.

[30] Jolibois, A.; Defrance, J.; Jean, P.; Duhamel, D.; Sparrow, VW. In Situ Measurement of the Acoustic Performance of a Full Scale Tramway Low Height Noise Barrier Prototype. *Applied Acoustics* **2015**, *94*, 57–68.

[31] Castiñeira-Ibañez, S.; Rubio, C.; Sánchez-Pérez, J.V. Environmental Noise Control during Its Transmission Phase to Protect Buildings. Design Model for Acoustic Barriers Based on Arrays of Isolated Scatterers. *Building and Environment* **2015**, *93*, 179–185, doi:10.1016/j.buildenv.2015.07.002.
[32] Heckl, M. Untersuchungen an Orthotropen Platten. *Acoustics* **1960**, *10*, 109–115.
[33] Cremer, L.; Heckl, M. *Structure-Borne Sound*; Springer Berlin Heidelberg: Berlin, Heidelberg, 1988; ISBN 978-3-662-10123-0.
[34] Bies, D.A.; Hansen, C.H. Engineering Noise Control. *Unwin Hyman Ltd, London* **1988**.
[35] Alexander, N.J.H. Measurement of Sound Absorption and Transmission of Corrugated Steel Plates. *Proceedings of the Institute of Acoustics* **1988**, *10(5)*, 39–50.
[36] Rowell, M.A.; Oldham, D.J. The Directivity of Orthotropic Factory Cladding Panels. *Proceedings of Institute of Acoustics* **1988**, *10(5)*, 39–51.
[37] Cordonnier-Cloarec, P. Contribution to the Study of Sound Transmission and Radiation of Corrugated Steel Structures. *Journal of Sound and Vibration* **1992**, *157*, 515–530.
[38] Hansen, C.H. Sound Transmission Loss of Corrugated Panels. *NOISE CONTROL ENGINEERING JOURNAL* **1993**, *40*, 187–197.

IOP Conf. Series: Materials Science and Engineering 1247 (2022) 012009 doi:10.1088/1757-899X/1247/1/012009

Lateral acceleration tests during circular driving maneuvers and double lane change

P Grabski*, M Jaśkiewicz, R Jurecki, D Kurczyński, P Łagowski, TL Stańczyk, E Szumska, A Zuska

Kielce University of Technology, Aleja Tysiąclecia Państwa Polskiego 7, 25-314 Kielce, Poland

* pgrabski@tu.kielce.pl

Abstract. The paper presents the analysis of lateral acceleration based on a vehicle traveling in a circle with a double lane change. The maneuvers of driving in a circle and double lane changes were chosen not by accident. Tests were carried out on a popular brand passenger car on wet and dry surfaces. The vehicle is equipped with high-class electronic equipment that records vehicle motion parameters such as lateral acceleration, longitudinal acceleration, speed and others. In the study, the results of maneuvers were statistically analyzed and tabulated in order to compare and influence the comfort of travellers.

Keywords: vehicle dynamics, vehicle movement, acceleration

1. Introduction

One of the key elements faced by modern vehicle systems is passenger comfort and safety. Side rollover accounts for about 2.5% of the total number of road accidents. This type of event occurs when the vehicle rotates ninety degrees or more about its longitudinal axis. In this case, it is worth paying attention to the fact that the side rollover of the vehicle is influenced by the design and operational properties. The most important of them are the turning characteristics, traffic parameters, wheel track and tire characteristics [1]. Unfortunately, the course of such accidents is becoming more and more popular, and their course is complex. The constructed models are necessarily based on the approximation of the nonlinear elastic-plastic characteristics of the impact processes, especially during the transition from the compression phase to the deformation restitution phase. Obtaining such characteristics requires a lot of experimental research, which is also not an easy process and leads to obtaining specific answers to specific situations [9]. A modern car is a complex device, built with the use of the latest achievements in such fields as: mechanics and machine construction, electrical engineering, electronics, information technology, automation and chemistry. Along with the development of technology, the share of electronic systems controlling the work of assemblies, subassemblies and individual devices is constantly growing. There is a rapid, very drastic evolution in the automotive world, which consists in providing the best devices or components that will be primarily an element of active safety, but at the same time will be an element of protection in the event of the consequences of already existing road incidents. This fundamental change in the technical parameters of motor vehicles and their optimization has a significant impact on today's selection of the optimal vehicle. Particularly important issues related to active safety are steering and vehicle stability.

Many years of steering and stability tests contributed to the recognition and improvement of the vehicle properties and were the basis for determining the directions of modernization and development of vehicle structures [7]. Taking into account the methods of testing controllability, developed over many years in various research centers, they focus mainly on the properties of the steering system, especially the vehicle as a dynamic element. In this case, it is worth paying attention to the issue related to the tires. As you know, the wheel in the vehicle is the element through which the vehicle comes into contact with the ground. The characteristics associated with the appropriate selection of tires are very wide. It depends primarily on the grade, type, size and performance of the tire. The undisputed advantage of pneumatic tires is, of course, their resistance to flattening, which is typical for pneumatic tires and often leads to an accelerated need to replace them due to the rapid and extreme wear of the tread. Most of today's tires are made of rim, structure and tread components. Currently, tubeless tires are used in most passenger vehicles, which are able to ensure maximum safety in road traffic with the appropriate selection of parameters [8]. Along with the development of simulation studies, new research possibilities were created that allow to learn about various processes of phenomena affecting the behavior of the vehicle in various road conditions, even extremely unfavorable or unfavorable, and taking into account the actions of the driver. The aim is to equip the vehicle with assistance systems so as to be able to compensate for the defects and wrong decisions of the driver [2].

The introduction of newer and newer driver assistance systems is associated with solving problems with dynamics. In terms of control of the so-called longitudinal dynamics, these elements are already well developed and widespread in the latest vehicle designs. Such systems include, among others ABS, ASR, ICC, Lane Assistant and others. Unfortunately, the development of curvilinear motion control systems is still quite difficult. It is true that there are more and more concepts, developments of new systems (e.g. crosswind assistant, ESP), which are already introduced as additional equipment during the production of vehicles, unfortunately they are far from the level that would allow them to be thoroughly researched, modernized and permanently introduced to the production system as part of the basic equipment [2]. Among the solutions to this problem, two directions should be distinguished and presented. The first is where the classic system components of controlling the deviation need to be improved. The second direction could be the model of the driver, using the so-called optical flux method. It consists in the fact that while driving, the driver adjusts the track to the course of the road, compares the position of the car and its movement with the ideal state assumed by him. However, in critical situations, he can act in panic, make bad decisions, act too abruptly. In such situations, the car, and in principle the systems installed in it, should support or possibly correct the driver's actions, but its operation should not be predictable for him and ensure the ability to accurately follow the assumed trajectory. In situations leading to accidents, extreme road conditions are the most common. The double lane change maneuver was selected for the analysis. This maneuver is basic test performed during vehicle evaluation. It consists in driving the tested vehicle over a certain section of the test track compliant with the appropriate standard and at a constant speed. Currently, tests are carried out on the basis of ISO standards [5].

The main aim of the research was to achieve the maximum lateral acceleration during the tests, to achieve the limit values of lateral acceleration, the inability to pass the track without damaging the track infrastructure elements (cones, columns). The data will depend on how a given driver feels about his safety limit. It is enough for the driver to behave dangerously or in a dangerous situation and the issue of driving changes drastically. During the maneuvers driving in a circle, the driver did not have a specific steering wheel angle and the radius on which the vehicle was to move was not set. Table 1 below summarizes the track dimensions for each vehicle type. Passenger cars are most often intended for private use and transport up to five people. They have a luggage space, either directly or indirectly separated from the passenger compartment, depending on the vehicle structure. In each of these cases, the most important factor will be comfort and both positive and negative feelings during the journey, thanks to which passengers will be able to classify the journey as successful or less or unsuccessful in a given environment, vehicle. Therefore, it was decided to carry out research on a passenger car, the type of which has the highest percentage in terms of usability in city traffic, with the possibility of transporting up to five people and the possibility of loading a fairly large amount of luggage.

Table 1. Track dimensions for individual standards [5]

Norm	Section number	Lenght	Width	Traverse shift	Remarks
ISO 3888-1	1	15 m	1,1 x B+0,25 m		
	2	30 m			
	3	25 m	1,2 x B+0,25 m	3,5 m	The total length of the test track is 125m
	4	25 m			
	5	15 m	1,3 x B+0,25 m		
	6	15 m	1,3 x B+0,25 m		
ISO 3888-2	1	12 m	1,1 x B+0,25 m		
	2	13,5 m			Total measurement track length 61 m. For vehicles up to 3500 kg.
	3	11 m	B+1 m	1 m	
	4	12,5 m			
	5	12 m	1,3 x B+0,25 m		

The track according to ISO 3888 - 1 standard, the diagram of which is presented below, was selected for testing the vehicles.

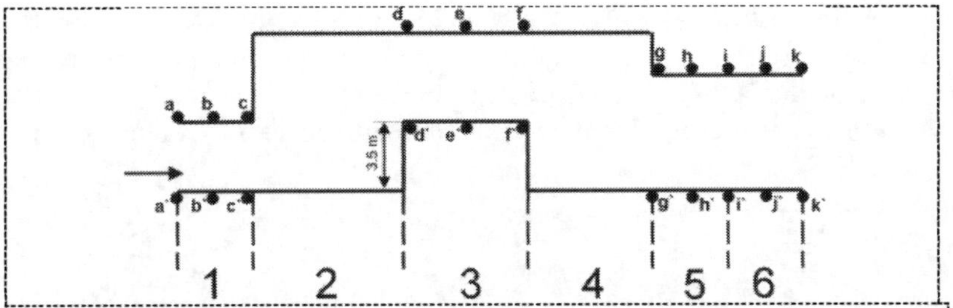

Figure 1. Double lane change track according to ISO 3888 - 1.

2. Methodology

The test object was an Audi A6 Avant passenger car Fig. [2] tab. [2]. The power unit used for the in-line cylinder is a diesel engine with a six-cylinder 2957.00 cm^3 and a power of 176 kW, which was produced in 2009. The vehicle is equipped with a turbocharger and an intercooler. Injection takes place directly via the Common Rail system. The car is equipped with a gearbox, permanent all-wheel drive, as well as air suspension. The vehicle ran on tires for years. In fig. 2 The vehicle used for the test is shown together with its technical data.

IOP Conf. Series: Materials Science and Engineering 1247 (2022) 012009 doi:10.1088/1757-899X/1247/1/012009

Figure 2. Audi A6 Avant passenger car

Table 2. Motor vehicle dimensions.

Parameter [m]	Car
Lenght	4.93
Width	1.85
Height	1.46
Weight	1920
Drive	quattro

The vehicle was moving on paved asphalt both when driving in a circle and when changing lanes twice. Depending on the needs, the surface was dry or wet. Driving in a circle is arranged to the left and right, and on dry and wet surfaces. The two-lane shift was performed on a dry and wet surface at a speed of 50 km / h. The speed was determined on the basis of road traffic regulations for vehicles moving in city traffic. During the measurement of vehicle, motor vehicle, access control, access control, from which the control was carried out. The vehicle is equipped with a clock device consisting of: a) Corrsys Datron S-350® Aquare on-board optoelectronic system, which registers elements of motion, including the clock measurement of the vehicle, b) control of TAA® clock traffic, registration of the measurement system in the X, O direction, Y, Z, TANS® sensor for measuring the linear and angular measurement of the X, Y, Z axis; c) DatronuEEP12® data acquisition station with tablet, connection and ARMS®. The measurement system recorded data at a frequency of 100 Hz. Data.

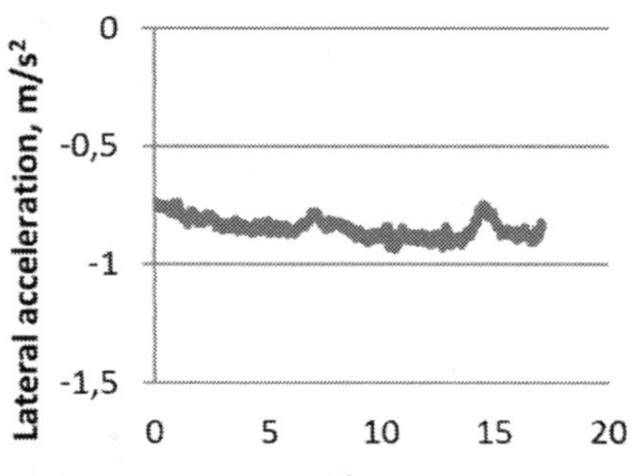

Figure 3a) Lateral accelerations recorded during the maneuver of driving in a circle on a dry surface to the left

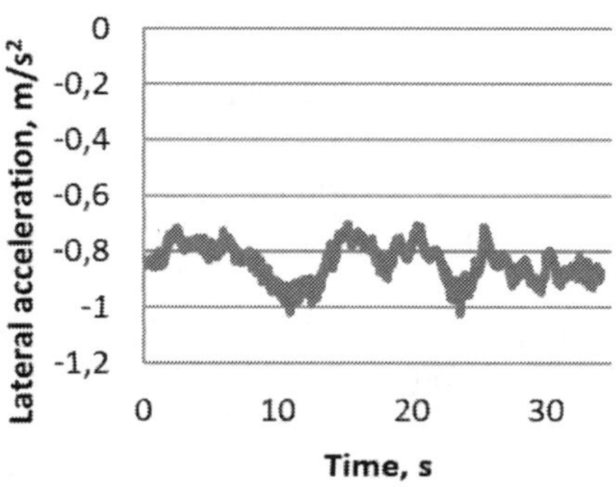

Figure 3b) Lateral accelerations recorded during the maneuver of driving in a circle a on a wet surface to the right

3. Results

Taking into account the exemplary curves of lateral acceleration during the maneuver of driving in a circle, it is worth noting that when the vehicle is moving on dry asphalt, the range of values is between 0.5 and 0.98 m / s2. However, when driving the vehicle on a wet surface, the range of values is in the range of 0.55 to 1.05 m / s2. The wide range of ranges is mainly due to the fact that in the first case the driver moved to the right - it is harder to hold your own body during sudden maneuvers, because there is no greater point of support - than to the left, where our body rests on the door upholstery. In addition, it can also be concluded that the driver feels safer when moving the vehicle to the left, because he has a greater range of visibility in relation to the vehicle and the road.

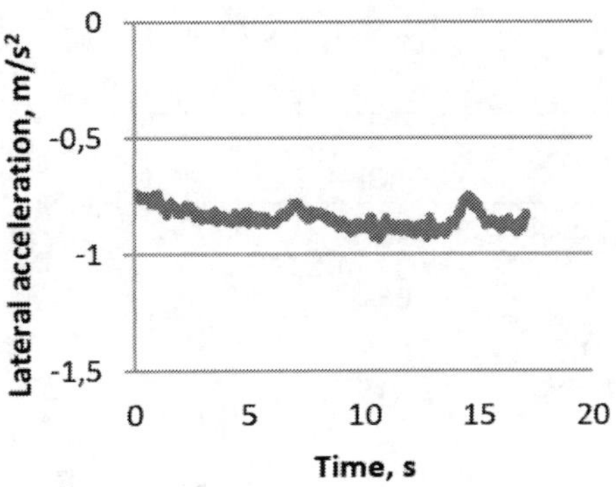

Figure 4a) An example of the course of lateral accelerations dry road to the right

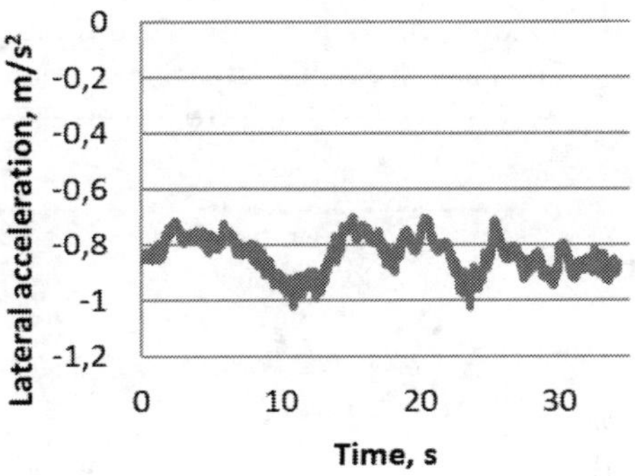

Figure 4b) An example of the course of lateral accelerations wet road to the right

According to the recorded minimum values of a vehicle moving on a wet surface, during the maneuver driving in a circle, they reach the value of 7.17 m / s^2 to the left, and the minimum values to the right reach 7.36 m / s^2. As you can see, the minimal lateral accelerations of a vehicle moving left and right on a wet surface are very close to each other. When the vehicle moves in a circle to the left, the maximum values of lateral accelerations are 9.19 m / s^2 and to the right - 9.12 m / s^2. The difference between the maximum and minimum values is, respectively, 2.02 m / s^2 for driving in a circle to the left and 1.76 m / s^2 for driving in a circle to the right. The situation is quite similar when it comes to driving in a circle on dry asphalt. To the left, the minimum values are 7.91 m / s^2, and to the right, 7.55 m / s^2. For the maximum values of lateral accelerations for driving in a circle to the left, the values are respectively 9.6 m / s^2 for driving to the left and 9.42 m / s^2 for driving to the right. The difference between the maximum and minimum values is 1.69 m / s^2 for driving to the left and 1.87 m / s^2 for driving to the right, respectively. As you can easily notice, higher differences between the maximum lateral accelerations are indicated by driving in a circle on a wet surface to the left, while when driving to the right, greater differences in values are indicated by driving in a circle on a dry asphalt to the right.

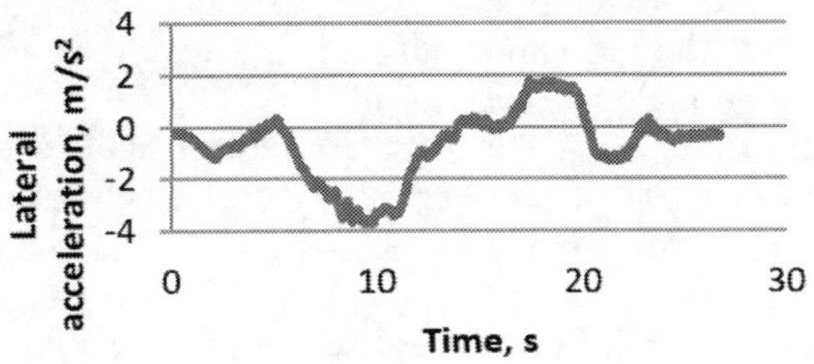

Figure 5a) An example of the course of lateral acceleration on dry asphalt

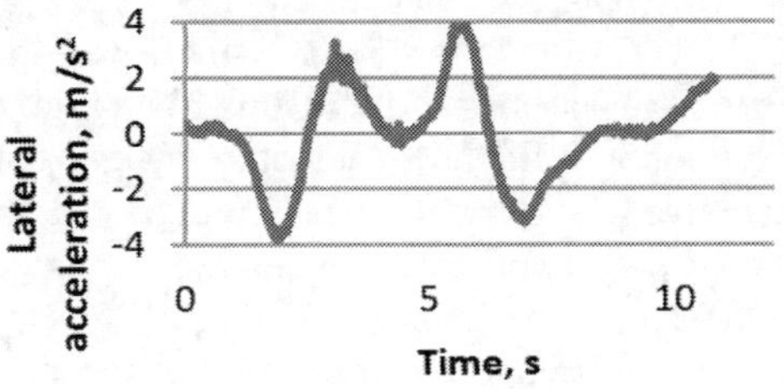

Figure 5b) An example of the course of lateral acceleration on wet asphalt

IOP Conf. Series: Materials Science and Engineering 1247 (2022) 012009 doi:10.1088/1757-899X/1247/1/012009

The results of the presented research show that the values of lateral acceleration are mainly influenced by the type of surface on which the vehicle is traveling. The values presented in the article clearly show that changes in the maximum lateral acceleration significantly affect the concept of comfort and discomfort for passengers of given vehicles. The data obtained in this way were subjected to statistical analysis, the results of which are presented in Table 3. It presents the minimum and maximum values of lateral accelerations y depending on the type of maneuver and the surface on which a given vehicle was driving while driving in a circle.

Table 3. Summary of minimum and maximum accelerations for circular maneuver, double lane change and road surfaces

Car Surface	Dry asphalt	Car	Wet asphalt	
Acceleration	min	max	min	max
Double lane change	4.47	4.87	4.75	6.11
Circle left	7.91	9.6	7.17	9.19
Circle right	7.55	9.42	7.36	9.12

For a passenger car, the minimum values of lateral acceleration were recorded while driving in a circle on the left side on a wet surface. They are 7.17 m / s^2. The maximum acceleration values were recorded while driving in a circle to the left on a dry surface of 9.6 m / s^2. The minimum values obtained during the double lane change maneuver on a dry surface were 4.47 m / s^2, and on wet asphalt the minimum values were 4.75 m / s^2. The maximum values achieved on the dry surface were 4.87 m / s^2, and on the wet surface - 6.11 m / s^2. The values of lateral acceleration recorded during driving in a circle and during a double lane change were characterized by large differences. The minimum acceleration with a double lane change was 4.47 m / s^2, while the maximum was 9.19 m / s^2 while driving in a circle. The difference between the minimum and maximum values is quite large and amounts to 4.72 m / s^2.

4. Conclusion

The performed tests of lateral acceleration were divided according to the type of surface and the type of maneuver. Analyzing the obtained data while driving in a circle, both to the left and to the right, as well as in terms of the type of surface - wet or dry, it results that the highest values of lateral acceleration are obtained on a dry surface. Minimum values for double lane change maneuvers and the maximum values of a vehicle running on dry asphalt are similar. However, in the case of a vehicle running on wet asphalt, the difference between the minimum and maximum values is much greater and amounts to 1.36 m / s^2. When analyzing the process of driving in a circle or double lane change and the maximum acceleration values achieved during this maneuver, many factors should be taken into account. One of the main ones is the implementation of the drive in a passenger vehicle. If the vehicle is not equipped with four-wheel drive, the obtained acceleration values may differ significantly from the obtained ones. In addition, the vehicle used for the tests has active safety systems in the form of the ABS or ESP system, which can start working at the right moment and in a given situation, which translates into the results of lateral acceleration (vehicle stabilization). To a large extent, such behavior is influenced not only by the type of surface on which the vehicles traveled, the type of maneuver, but also the characteristics of the passenger car and its equipment. Lateral acceleration is one of the parameters influencing not only the driving comfort but also the stability of the vehicle. Extensive research and tests have a significant impact on the method of production of vehicles and the safety of passengers moving in it. Exceeding certain barriers related to lateral acceleration may have a significant impact on the trajectory of a moving vehicle, and thus the safety of passengers. Therefore, it is very important to carry out laboratory tests with the use of the highest quality equipment and to obtain real values that can be used in an appropriate way for a cognitive and analytical approach to vehicle safety issues.

5. References

[1] K. Parczewski, H. Wnęk, *Analiza wpływu parametrów masowych na stateczność ruchu samochodu ciężarowego w oparciu o badania mobilnego modelu pojazdu*, Postępy Nauki i Techniki 2012/2014
[2] T. L. Stańczyk, D. Więckowski, *Computer simulatoion testing of a double traffic lane – change manoeuvre,* Journal of KONES Powertrain and Transport, Vol. 13, No. 1, 2006
[3] Yoshimoto, K., Obawa, H., Kubota, H., *Course tracking control algorithm using visual information*, Vehicle System Dynamics 28(1997).
[4] Yoshimoto, K., Iwatani, K., Kokubo, T., *Automatic driving using image informations*, Technical Notes/JSAE Review 18(1997).
[5] *ISO 3888-1 Passenger cars – Test track for a severe lane-change manoeuvre – Party 1: Double lanechange*. First edition 1999-10-01;
[6] ISO 3888-2 *Passenger cars - Test track for a severe lane-change manoeuvre - Part 2: Obstacle avoidance* First edition 2002. First edition 2002-11-15;
[7] W. Pieniążek, D. Więckowski – *Badania kierowalności i stateczności pojazdów samochodowych,* PWN, Warszawa 2020.
[8] Jackowski, J., Żmuda, M., Wieczorek, M., Zuska, A.*Quasi-static research of ATV/UTV non-pneumatic tires*. Energies, 2021, 14(20), 655.
[9] Prochowski, L., Gidlewski, M., Ziubiński, M., Dziewiecki, K. *Kinematics of the motorcar body side deformation process during front-to-side vehicle collision and the emergence of a hazard to car occupants*. Meccanica, 2021, 56(4), pp. 901–922.

IOP Conf. Series: Materials Science and Engineering 1247 (2022) 012010 doi:10.1088/1757-899X/1247/1/012010

Study of lateral acceleration of a Ford Transit car on various road surfaces

D Kurczyński*, P Łagowski, T Stańczyk, E Szumska, A Zuska, P Grabski, M Jaśkiewicz, R Jurecki

Department of Automotive Vehicles and Transportation, Faculty of Mechatronics and Mechanical Engineering, Kielce University of Technology, Kielce, Poland

* kdarek@tu.kielce.pl

Abstract. The measurement of car acceleration in time can be used to assess driving styles and safety behaviours of drivers. The values of lateral acceleration of the car can be an indication of the driver's aggressive driving style and tendency for risky behaviour. If the lateral acceleration is too high, it may affect the car's stability and potentially cause it to roll over. The paper outlines the results of the lateral acceleration analysis of a Ford Transit car driving in a circle and in the attempt to change two driving lanes. The tests were conducted in driving practice areas. Measurements were taken for the test vehicle driving in circles with maximum attainable velocity, and changing two driving lanes with pre-specified velocity. The tests were conducted on asphalt and concrete surfaces, in dry, wet and icy conditions. The purpose of the tests was to determine the maximum lateral acceleration of the analysed vehicle. The impact of the surface condition on the lateral acceleration of the test vehicle was also determined. The obtained results can be used as threshold values to assess driving style, and to analyse causes of accidents taking into consideration the condition of the road surface.

Key words: lateral acceleration, vehicle testing, vehicle movement, vehicle dynamics

1. Introduction

Lateral acceleration of the vehicle can affect driving safety, causing the vehicle to become unstable or, in extreme cases, even to roll over. The outcome depends on a number of factors, including vehicle velocity and acceleration, its type, size and weight, but particularly its centre of gravity, the type and condition of the road surface, and the changes thereof, and sudden and unexpected lateral forces, such as strong wind gusts, as well as other possible factors. In some cases, strong wind gust can tip over a coach or a truck, especially when the vehicle is leaving the cover of a noise wall. Road tankers are particularly exposed to the risk of tipping over, because their centre of gravity is located high on the vehicle. The measured lateral acceleration values can be an indication of the driver's driving style, for instance, whether they drive safely or take risks, make aggressive and dangerous manoeuvres, or how often do they find themselves in perilous situations. This might be of particular importance for coach and truck drivers, especially those carrying hazardous materials. The methodology of driving style assessment is discussed in [1].

In [2], the results of the analysis of lateral acceleration of a fire truck during single lane change are presented. Fire trucks are particularly exposed to the risk of tipping over during sudden manoeuvres. The purpose of the analysis was to determine the most advantageous sites for installing accelerometers which could signal the driver that the truck is at risk of tipping over. In [3-5], the stability of armoured personnel vehicles during double lane change was analysed. It was shown that the values which affect the stability of those vehicles the most were longitudinal velocity and lateral acceleration. The vehicle movement parameters are also impacted by test conditions, such as the wheel load, tyre pressure, road conditions and weather. Also, the driver's driving style depends on their stress level. In [6], the results of road tests of double lane change by a car are presented. The results of the road tests were compared with the simulations for various models of tyred wheels. In [7], the simulation of double lane change is discussed.

Ali G. et al. analysed the impact of factors such as vehicle velocity, driver's age and sex, vehicle class and location on the incidence of longitudinal and lateral acceleration [8]. It was shown that velocity was the key factor for the incidence of all acceleration values. Another important factor for vehicle acceleration turned out to be the driver's age. It was shown that both younger and older drivers caused higher longitudinal acceleration when accelerating and braking. On the other hand, the incidence of lateral acceleration decreased with the age of the drivers. The vehicle class also plays a role in the incidence of sudden accelerations. Duan K. et al. developed a model for lane changing process in a roadworks zone [9], which they based on experimental tests conducted in a driving simulator. The authors showed that in heavy traffic, drivers are more prone to aggressive behaviour, which translates into higher values of longitudinal and lateral acceleration. Mourad L. et al. discussed the issue of driving narrow urban vehicles with tilting body which have the tendency to tip over due to lateral acceleration during turns [10]. The authors proposed a method of body tilt control that would reduce the lateral acceleration felt inside the vehicle, using accelerometer and gyroscope measurements.

The authors of this paper measured the lateral acceleration of a passenger minivan driving in circles and during double lane change. The authors showed the impact of different road surfaces on the measured lateral acceleration values.

2. Methodology

The tests were conducted in the driving practice areas of the Laboratory of the Department of Automotive Vehicles and Transportation at the Kielce University of Technology, and in Jarząb Autodrome, on asphalt and concrete roads. The test vehicle was a nine-passenger version of Ford Transit Mark VI with the curb weight of 2.070 kg. The gross vehicle weight of this model is 3.050 kg. The test vehicle was powered by a 2.198 cm^3 92 kW engine and had Continental ContiVanContact 200 205/65 R16C summer tyres with very little wear. The tyre tread depth was just 1 mm below the nominal depth. The test vehicle is shown on Figure 1. The test vehicle was equipped with traction control systems (ABS, ASR, ESP) which were turned on during the test. Before commencing the test, the vehicle's technical condition was checked, focusing in particular on the braking system and tyres. During the test, the vehicle carried the test apparatus and two persons conducting the test activities.

During the circular test drive, the vehicle was accelerated to the maximum attainable velocity which could then be maintained for at least a dozen seconds. The driver had to maintain the specified vehicle trajectory while keeping it stable. The first test cycle was conducted on dry asphalt and dry concrete. Test drives on each road surface were performed clockwise and counterclockwise. During the tests, the lateral acceleration of the vehicle and its velocity were measured. The test drives were repeated in identical weather conditions on wet asphalt and wet concrete. Additionally, a test drive on thin ice was also performed in temperature slightly below 0°C.

For the circular test drives with maximum attainable velocity while maintaining the vehicle stability, the vectors of lateral acceleration of the vehicle at stable acceptable velocity in the given traffic conditions were determined. The following values were derived from the test: the minimum measured acceleration a_{omin}, maximum measured acceleration a_{omax}, difference of accelerations $a_{omax} - a_{omin}$, and the mean measured acceleration a_{om}.

The tests on the track also included double lane changes, which are performed in order to assess the steerability and stability of the vehicle. During the test, the vehicle response to the driver turning the steering wheel is assessed. The test of the Ford Transit passenger van was conducted in accordance with ISO 3888-1:2018 standard [11]. For safety reasons, the double lane change was performed at 50 km/h. The test was conducted on dry and wet asphalt. The manoeuvre was repeated 10 times for each road surface in identical weather conditions.

Figure 1. Ford Transit test vehicle

3. Test apparatus

For the tests, specialised measuring apparatus was used including:

- Correvit S-350 Aqua® optical sensor for measuring longitudinal velocity,
- μEEP-12 data acquisition system with ARMS™ data analysis software,
- TAA® 3-axis linear accelerometer.

The Correvit S-350 Aqua sensor makes it possible to precisely measure the distance, longitudinal and lateral velocity, and angle in dynamic tests of the vehicle, e.g. during steady circular motion in accordance with ISO 4138 standard [12]. The sensor ensures high accuracy of measurement, also in difficult conditions, on all standard measurement surfaces [13]. Also, the μEEP-12 data acquisition and analysis system dedicated for mobile vehicle testing apps was used [14]. During the tests, the μEEP-12 system was connected to a laptop. The measurement system was operated by ARMS™ software. Furthermore, the test apparatus included the TAA linear accelerometer which enabled 3-axis dynamic acceleration measurement [15]. The measurement range of the sensor was ±3 g.

4. Test results and analysis

Circular driving tests were performed with maximum attainable velocity while maintaining the stability of the test vehicle. During the tests, the vehicle accelerations were recorded, including the lateral acceleration. Figure 2 shows the measured lateral acceleration of the test vehicle while in circular counterclockwise motion. The acceleration values were determined for dry, wet and icy road surfaces. The lateral acceleration values of the test vehicle driving on the aforementioned road surfaces in clockwise motion is presented on Figure 3. The highest lateral acceleration was recorded for the vehicle driving on dry asphalt. The acceleration values for wet asphalt were slightly lower, while the lateral acceleration of the test vehicle driving on surface covered by thin layer of ice was considerably lower.

The analysis of the lateral acceleration values shown on Figure 2 and Figure 3 shows that the amplitude of acceleration values measured for the vehicle driving clockwise on dry and icy asphalt is higher than for the test vehicle driving counterclockwise. Driving the vehicle in circles clockwise on dry and icy surface, it was more difficult for the driver to maintain constant velocity than when driving in circles counterclockwise.

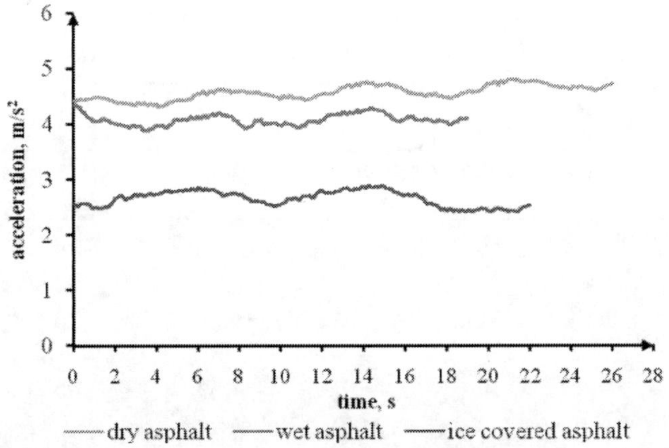

Figure 2. Lateral acceleration of Ford Transit during circular counterclockwise motion (left) with the maximum attainable velocity, on dry, wet and icy asphalt surface.

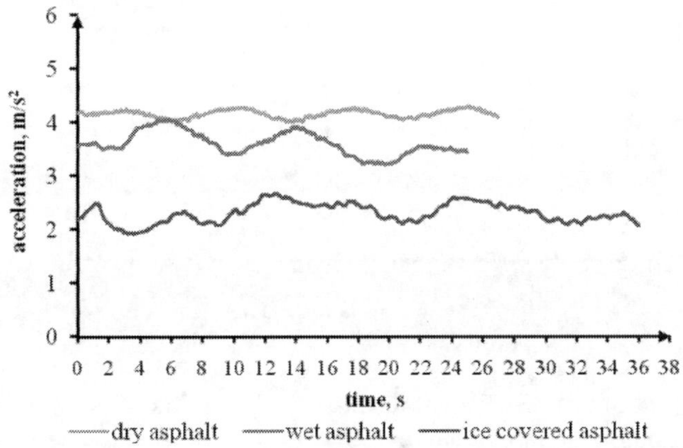

Figure 3. Lateral acceleration of Ford Transit during circular clockwise motion (right) with the maximum attainable velocity, on dry, wet and icy asphalt surface.

For each circular test drive, the acceleration of the vehicle was measured at least a dozen seconds after the vehicle had stabilised its maximum attainable velocity. The table lists the mean measured acceleration a_{om}, the minimum measured acceleration a_{omin}, and the maximum measured acceleration a_{omax}. The difference of the maximum and minimum measured acceleration $a_{omax} - a_{omin}$ is also specified. The acceleration values a_{om}, a_{omin} and a_{omax} were highest for dry asphalt, followed closely by wet asphalt, and considerably lower for icy asphalt. The calculated $a_{omax} - a_{omin}$ for clockwise circular motion of the vehicle on dry and icy surface was significantly lower than for counterclockwise circular motion. For the clockwise circular motion, the measured accelerations a_{om}, a_{omin} and a_{omax} were lower than for counterclockwise circular motion.

The accelerations of the Ford Transit test vehicle were also measured during circular motion on concrete surface. Figure 4 and Figure 5 show the measured lateral accelerations of the test vehicle on dry and wet concrete surface, driving counterclockwise (Figure 4) and clockwise (Figure 5).

Table 1. Lateral acceleration of Ford Transit during circular motion with the maximum attainable velocity, on dry, wet and icy asphalt surface.

Parameter	Asphalt					
	dry		wet		icy	
	driving direction		driving direction		driving direction	
	left	right	left	right	left	right
Mean measured acceleration a_{om}, m/s^2	4.57	4.16	4.08	3.60	2.65	2.30
Minimum measured acceleration a_{omin}, m/s^2	4.31	4.00	3.88	3.20	2.40	1.92
Maximum measured acceleration a_{omax}, m/s^2	4.80	4.30	4.35	4.05	2.88	2.67
$a_{omax} - a_{omin}$, m/s^2	0.49	0.30	0.47	0.85	0.48	0.75

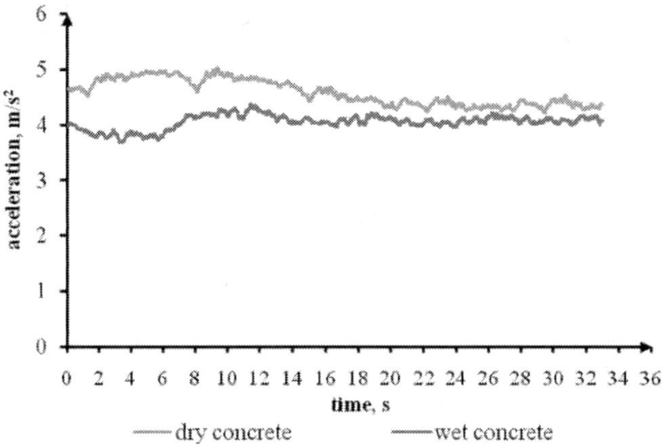

Figure 4. Lateral acceleration of Ford Transit during circular counterclockwise motion (left) with the maximum attainable velocity, on dry and wet concrete surface.

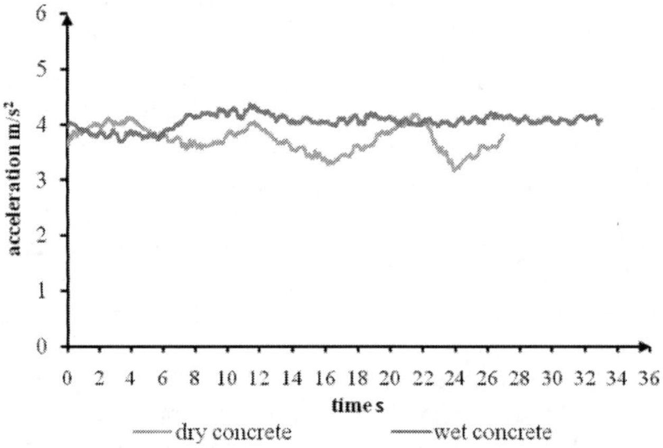

Figure 5. Lateral acceleration of Ford Transit during circular clockwise motion (right) with the maximum attainable velocity, on dry and wet concrete surface.

Table 2 presents the measured lateral acceleration values of the vehicle driving in circles on concrete. When driving in circles counterclockwise, the lateral acceleration was lower on wet concrete as compared to dry concrete. When driving in circles clockwise, the a_{om}, a_{omin} and a_{omax} values were slightly higher on wet concrete than on dry concrete. On dry concrete, the measured lateral acceleration values fluctuated significantly when driving clockwise. This might have been caused by the presence of concrete surface wear residues which might have affected the test results. This possibility is further highlighted by the values of $a_{omax} - a_{omin}$ which were higher for circular motion on dry concrete as compared to circular motion on dry asphalt, and also higher than on dry and icy asphalt.

Table 2. Lateral acceleration of Ford Transit during circular motion with the maximum attainable velocity, on dry and wet concrete surface.

Parameter	concrete			
	dry		wet	
	driving direction		driving direction	
	left	right	left	right
Mean measured acceleration a_{om}, m/s^2	4.58	3.73	4.36	4.05
Minimum measured acceleration a_{omin}, m/s^2	4.22	3.15	3.97	3.68
Maximum measured acceleration a_{omax}, m/s^2	5.04	4.18	4.82	4.36
$a_{omax} - a_{omin}$, m/s^2	0.82	1.03	0.85	0.68

Figure 6 presents a comparison of the mean measured lateral accelerations of Ford Transit during circular motion, with the maximum attainable velocity, on different road surfaces. The lowest acceleration was measured for test drives on icy asphalt. The mean lateral acceleration measured on wet concrete surface were slightly higher than on wet asphalt surface. The mean lateral acceleration measured on dry asphalt and dry concrete are virtually identical. The mean lateral acceleration measured for driving in circles clockwise was lower than for driving in circles counterclockwise.

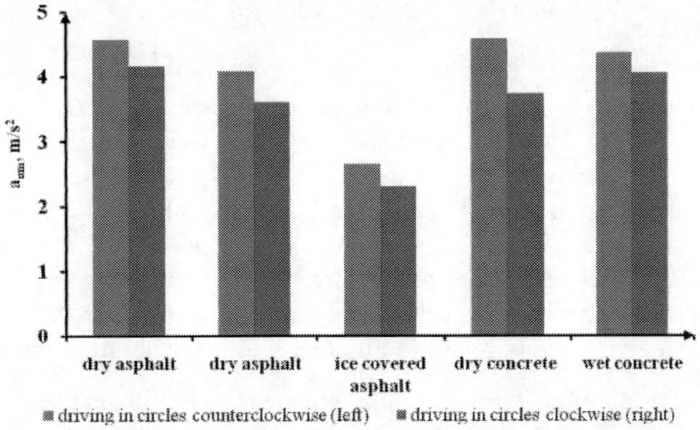

Figure 6. Comparison of the mean measured lateral accelerations of Ford Transit during circular motion, with the maximum attainable velocity, on different road surfaces.

The second phase of the tests involved measuring lateral acceleration of the test vehicle during double lane change. The measurements were conducted in Jarząb Autodrome, on dry and wet asphalt. On each road surface type, 10 double lane change manoeuvres were performed at 50 km/h. Figure 7 shows the lateral acceleration measured during double lane change by the test vehicle on dry asphalt, while Figure 8 shows the acceleration on wet asphalt. For wet asphalt, fluctuations of the lateral acceleration values are clearly visible. In each test on wet asphalt, lateral acceleration fluctuations were observed in exactly the same phase. For tests on dry asphalt, no such fluctuations of the measured acceleration were found. This might have been caused by the onboard systems preventing the test vehicle from losing stability which would be more likely on wet asphalt.

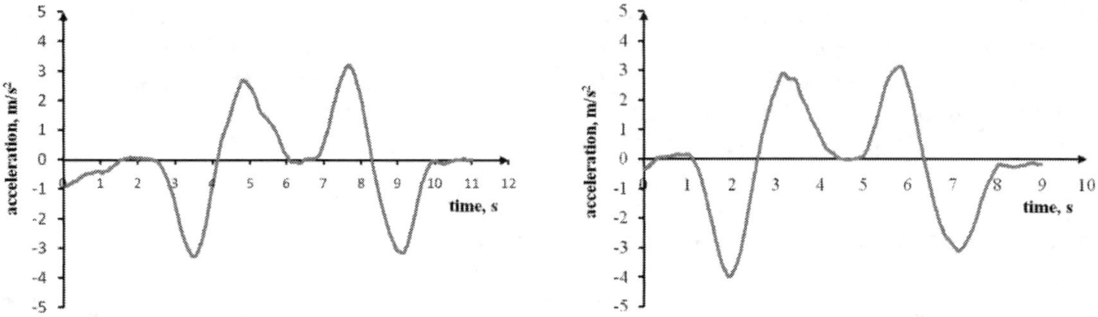

Figure 7. Examples of lateral acceleration values during double lane change by the test vehicle on dry asphalt at 50 km/h.

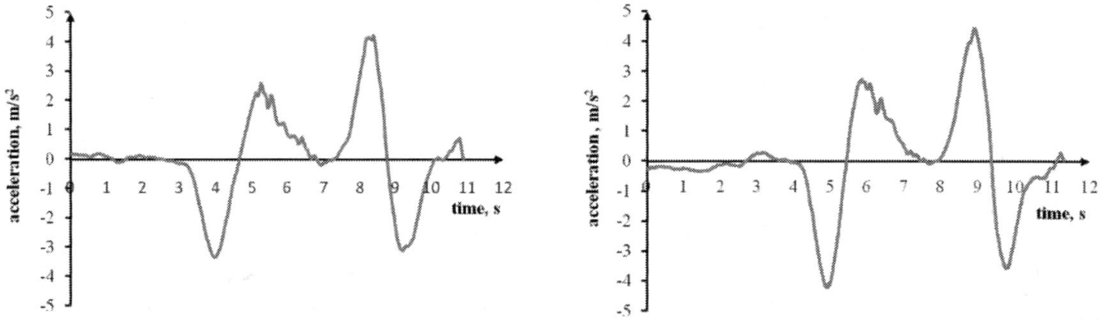

Figure 8. Examples of lateral acceleration values during double lane change by the test vehicle on wet asphalt at 50 km/h.

For each lateral acceleration during double lane change, the following values were determined: the maximum lateral acceleration a_{max}, the minimum lateral acceleration a_{min} and the sum of absolute maximum and absolute minimum accelerations $|a_{rmax}| + |a_{min}|$. The mean values of lateral accelerations from 10 test attempts were calculated, along with the standard deviations. The values of those parameters are listed in table 3 and table 4. The lateral accelerations of the test vehicle were higher for all double lane changes on wet asphalt, as compared to the accelerations measured for the vehicle driving on dry asphalt. This was likely caused by the ESP system which prevented the vehicle from losing traction. The ESP system operates by applying the brakes to the appropriate wheels and reducing the engine torque, which may result in higher vehicle acceleration values.

Table 3. Lateral acceleration values during double lane change by the test vehicle on dry asphalt at 50 km/h.

Attempt no.	a_{max} m/s^2	a_{maxm} m/s^2	SD	a_{min} m/s^2	a_{minm} m/s^2	SD	$|a_{rmax}| + |a_{min}|$ m/s^2	$(|a_{rmax}| + |a_{min}|)_m$ m/s^2	SD
1	2.91	3.07	0.23	-3.10	-3.31	0.31	6.01	6.38	0.47
2	2.75			-2.86			5.62		
3	2.99			-3.09			6.09		
4	3.02			-3.19			6.21		
5	3.20			-3.27			6.47		
6	3.27			-3.33			6.60		
7	2.79			-3.28			6.07		
8	3.50			-3.46			6.96		
9	3.09			-3.54			6.62		
10	3.14			-4.00			7.15		

Table 4. Lateral acceleration values during double lane change by the test vehicle on wet asphalt at 50 km/h.

Attempt no.	a_{max} m/s^2	a_{maxm} m/s^2	SD	a_{min} m/s^2	a_{minm} m/s^2	SD	$\lvert a_{rmax}\rvert + \lvert a_{min}\rvert$ m/s^2	$(\lvert a_{rmax}\rvert + \lvert a_{min}\rvert)_m$ m/s^2	SD
1	3.91	4.06	0.39	-3.47	-3.72	0.33	7.38	7.78	0.62
2	4.01			-3.56			7.57		
3	4.25			-3.70			7.96		
4	3.33			-3.52			6.84		
5	4.22			-3.35			7.57		
6	3.97			-3.42			7.39		
7	4.15			-3.73			7.88		
8	3.65			-4.00			7.65		
9	4.41			-4.22			8.64		
10	4.70			-4.25			8.94		

5. Conclusions

The study analysed the lateral acceleration of a Ford Transit passenger van. The study tests were conducted on the test vehicle driving in circles with the maximum attainable velocity on dry and wet asphalt, and dry and wet concrete surfaces. Furthermore, tests were also conducted for the asphalt surface covered by a thin layer of ice. In addition, the lateral acceleration of the vehicle during double lane change was tested on dry and wet asphalt.

The measured lateral acceleration values of the test vehicle during circular motion were slightly higher on dry asphalt as compared to wet asphalt. For circular motion on asphalt covered by a thin layer of ice, the acceleration values were considerably lower. Similar correlation was found for counterclockwise and clockwise circular motion. On dry concrete, the lateral acceleration for counterclockwise circular motion was also slightly higher than on wet concrete. For clockwise circular motion, the results were different. On wet surface, the mean measured lateral acceleration was slightly higher than the acceleration measured on dry surface. This might have been caused by the presence of the concrete surface wear residues. For all counterclockwise circular test drives, higher lateral acceleration values were measured than for clockwise test drives.

For double lane change tests on wet asphalt, higher lateral acceleration of the test vehicle was measured as compared to the acceleration measured on dry asphalt. This was most likely caused by the ESP system which applies traction control to stabilise the vehicle on the trajectory chosen by the driver.

The obtained results may be useful for the assessment of driving styles. Drivers, who in normal traffic conditions record lateral acceleration values similar to those presented herein, might be considered aggressive and potentially dangerous. The results presented in this study may also be used in the analysis of road accidents in which Ford Transit vans are involved.

References

[1] Jurecki R S and Stańczyk T L 2021 A Methodology for Evaluating Driving Styles in Various Road Conditions. *Energies* **14** 3570

[2] Dzida J 2020 Badania przyspieszenia poprzecznego samochodu specjalnego pod katem wykorzystania w sygnalizatorze niebezpieczeństwa bocznego wywrotu. *Pomiary Automatyka Kontrola* **56** pp 233-235

[3] Motrycz G and Stryjek P 2010 Problems of evaluation of parameters of wheeled armored vehicles the method a severe lane-change manoeuvre. *Logistyka* **4** CD-CD

[4] Motrycz G and Stryjek P 2011 Influence of wheeled APC design changes on lateral dynamic stability. *Zeszyty Naukowe WSOWL* **1(159) pp** 192-203.

IOP Conf. Series: Materials Science and Engineering 1247 (2022) 012010 doi:10.1088/1757-899X/1247/1/012010

[5] Gidlewski M, Jackowski J, Jemioł L, Żardecki D 2021 Sensitivity of a vehicle lane change control system to disturbances and measurement signal errors – Modeling and numerical investigations. *Mechanical Systems and Signal Processing* **147** 107081
[6] Janczur R, Świder P and Walczak S 2007 Double lane change maneuvers – experimental and simulations result done by means of a computer program V-sim, with applying two different tire model. *Logistyka* 3 CD-CD
[7] Stańczyk T L and Więckowski D 2006 Computer simulation testing of a double traffic lane-change manoeuvre. *Journal of KONES Powertrain and Transport* **13/1** pp 177-185
[8] Ali G, McLaughlin S and Ahmadian M 2021 Quantifying the effect of roadway, driver, vehicle, and location characteristics on the frequency of longitudinal and lateral accelerations. *Accident Analysis & Prevention* **161** 106356
[9] Duan K, Yan X, Ma L, Hang J and Li X 2022 A multistage analytic model of the longitudinal and lateral acceleration during lane changing in work zone areas with the aid of a driving simulator experiment. *Transportation Letters* 14 pp 28-38
[10] Mourad L, Claveau F and Chevrel Ph 2011 A Lateral Control Strategy for Narrow Tilting Commuter Vehicle Based on the Perceived Lateral Acceleration. Proceedings of the 18th World Congress The International Federation of Automatic Control Milano (Italy) 6254-6259
[11] ISO 3888-1:2018 Passenger cars — Test track for a severe lane-change manoeuvre — Part 1: Double lane-change
[12] ISO 4138:2012 Passenger cars — Steady-state circular driving behaviour — Open-loop test methods.
[13] Correvit® S-350 Non-Contact Optical Sensors, Type CS350A... Kistler Patent No. DE 43 13 497 C2 https://www.kistler.com/files/document/000-807e.pdf
[14] μEEP-12 Data Acquisition and Evaluation Typ CMEP1A... Kistler. https://www.techtarget.com.br/site/wp-content/uploads/2016/09/CMEP12A_000-819e.pdf
[15] Acceleration modules SAA (1-axis), DAA (2-axis), TAA (3-axis), Type KCD15760, KCD15951, KCD15911. Kistler https://www.kistler.com/files/document/000-913e.pdf?callee=frontend

Acknowledgements
The research was carried out as part of the Innovative system research project supporting the motor vehicle insurance risk assessment dedicated to UBI (Usage Based Insurance) No. POIR.04.01.04 00 0004/19 00 financed by the National Centre for Research and Development.

The concept of the application of modelling and simulation to simplify and reduce the cost of durability tests of vehicles and their components

Zbigniew Lozia[1*]

[1]Łukasiewicz Research Network – Automotive Industry Institute. Jagiellońska Street 55, 03-301 Warsaw, Poland. Faculty of Transport, Warsaw University of Technology, Koszykowa Street 75, Warsaw, Poland.

* zbigniew.lozia@pw.edu.pl

Abstract. The paper deals with durability tests of road vehicle components. They may be performed in a real work environment or in the laboratory. New or modified structures of vehicles and their components should and are tested in an accelerated manner in real or simulated conditions, corresponding to the anticipated conditions of use. The work presents the concept of using modelling and simulation to determine the inputs during the tests of the car's structural components (assemblies, subassemblies, and parts), without the need to test the entire vehicle. The author proposes the use of physical and mathematical models of vehicles to determine the excitations affecting the assembly, subassembly, or element of interest during the research tests of the entire vehicle. This approach reduces the dimensionality of the experimental research and limits its scope to the structural node of interest, which in turn reduces the costs. The next, more extensive publication will present an exemplary application of the presented concept.

Keywords. durability tests, laboratory tests, modelling, simulation, costs reduction, vehicle

1. Introduction

New or modified structures of vehicles and their components should and are tested in an accelerated manner in real or simulated conditions, corresponding to the anticipated conditions of use [2, 4, 5, 14, 16, 17, 20, 21, 22, 23]. Granting a product of different types (e.g. a safety certificate) requires an appropriate cycle of tests to confirm their suitability under the target operating conditions [20]. Durability tests of the structure and validation tests of the prototype are also important [2, 4, 5, 14, 16, 17, 20, 21, 22, 23]. Tests in real conditions (traction or field) are very long-lasting, expensive, and the most dangerous. Simulation tests on simulation stand significantly accelerate the product testing process; it also becomes repetitive [16, 17]. Manufacturers of vehicle structural components (assemblies, subassemblies, and parts) are mainly interested in testing their products; fewer other elements of the vehicle structure - this is the domain of activity of the final vehicle manufacturer (OEM) [5, 22, 23].
The author presents a proposal for the use of modelling and simulation to determine the test conditions for structural components (assemblies, subassemblies, and parts), without the need to test the entire vehicle. The construction of a mathematical model of the car's motion under representative operating

conditions allows for simulation tests in the frequency and/or time domain [3, 6, 7, 8, 9, 10, 11, 15, 18, 19], leading to the determination of load time courses selected items. They can be used in the research of the vehicle structure nodes of interest to us: axle axles, wheel hubs, tires, elastic and damping suspension elements, couplings of vehicle assemblies (saddle - kingpin, towing hook - trailer drawbar towing hitch). The anticipated benefit is the limitation of the scope of research to the structural node of interest to us, which reduces the "dimensionality" of the research process and its costs.

2. Examples of stands for the simulation of operational loads of vehicles and their construction nodes

Simulation tests of vehicles, their assemblies, and components have been conducted for several dozen years all over the world [2, 4, 16, 17]. In Poland, the leading role (for many years) is played by the Automotive Industry Institute (currently Łukasiewicz - Automotive Industry Institute), both in the field of research methodology [16, 17] and practical applications [16, 17, 20]. In recent years, even private manufacturers of the automotive sector have built research centres and applied the methodology of durability testing and validation of prototypes [22, 23]. For this purpose, the available production offer of advanced test stands and devices together with control software (for example [14, 21]) is used.
However, methods of the most accurate reflection of the real load conditions of vehicles and their design nodes are still sought, allowing the accelerated and possibly low-cost checking of the tested structures.

3. Methods of conducting bench simulation tests of operational loads of vehicles and their construction nodes

It is worth quoting the words of the work, the authors of which have conducted simulation tests for many years [17]: "Laboratory test-band research gives better possibilities and, at the same time, the possibility of evaluating the vehicle in a relatively short time. (...). Generally, bench tests are poor in input signals and are also artificially shaped; However, they provide much greater possibilities of measuring output signals, both in terms of their number and type; moreover, they have a very important feature: repeatability of tests". The same authors note that each experimental research can be regarded as a certain simulation of real events, only its accuracy, scope, and purpose for which it is carried out differ.
Simulation test stands are built for various tests, both in terms of their purpose and completeness of the vehicle. There are stands for whole cars (fully equipped), partially disassembled (not fully equipped), individual assemblies, as well as selected parts of the vehicle.
The experimental simulation (on laboratory stands) is carried out (Fig. 1) in two variants [17]:

- ➢ in variant 1 (*simulation based on input signals*), the signals controlling the vibrators of the stand are predetermined, e.g. based on measurements of geometric unevenness of the road surface that we want to recreate;
- ➢ in variant 2 (*simulation based on output signals*), external inputs are introduced in such a way as to obtain the compliance of the values measured in the vehicle at the station and on the road (during previous measurements under imposed conditions).

The authors of the work [17] draw attention to the fact that in variant 1 there is a risk of distorting the output signals of the forcing devices if the conditions of the input (on the stand) on the vehicle differ significantly from the actual ones. The most frequently quoted example is the lack of rotation of road wheels during bench tests, which significantly changes the dynamic properties of the pneumatic wheel. It is worth noting, however, that in modern stands, forcing a rotating wheel is introduced. However, this significantly increases the cost of the stand and the test itself.
For similar reasons, in variant 2, the input signals are distorted, because the compliance of the selected signal during bench tests and in road conditions is important.

IOP Conf. Series: Materials Science and Engineering 1247 (2022) 012011 doi:10.1088/1757-899X/1247/1/012011

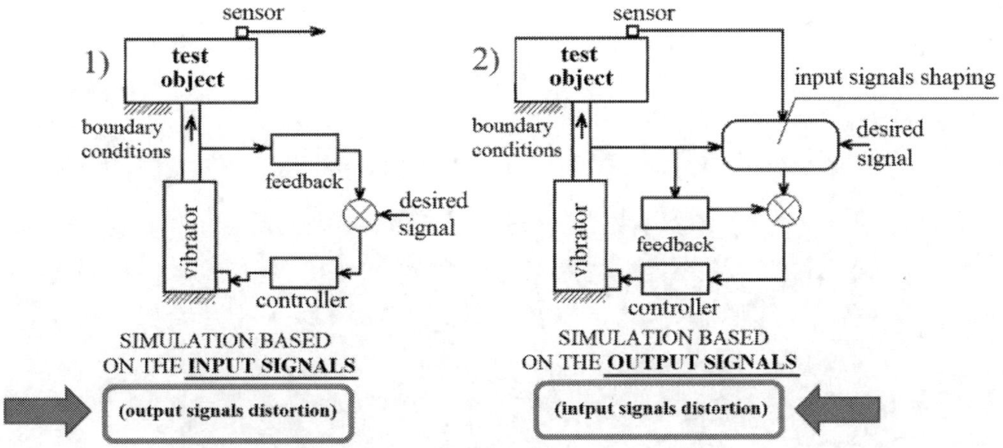

Figure 1. Variants of research in the experimental simulation (the figure was prepared based on Fig. 7.9 in the paper [17]).

4. The essence of the proposed method

A digital simulation will be used - on a computer, using a mathematical model of a real object, along with models of inputs and outputs. It is carried out in variant 1 (*simulation on the basis of input signals*) - with the use of input signal (excitation) models. These are road unevenness, aerodynamic effects of air, internal forces (driving and braking torques, turning angles of the road wheels), resulting from the activity of the vehicle driver.

The mathematical model/simulation model of the vehicle includes the model of the structural node of interest with the model of inputs and outputs (Figs. 2 and 3). Assuming an excitation representative for the operating conditions of the object, we use this model to determine the excitation acting on the vehicle structure node of interest to us. This extortion can be used in the experimental simulation of loads of the structural node on a laboratory stand, which reduces the "dimensionality" of the test process and their costs.

The determined time courses of loads (corresponding to representative conditions of their operation) can be used in the tests of the vehicle design nodes of interest: wheel axles, wheel hubs, tires, elastic and damping elements of suspensions, couplings of vehicle assemblies (saddle - kingpin, towing hook - hitch tow bar of the trailer).

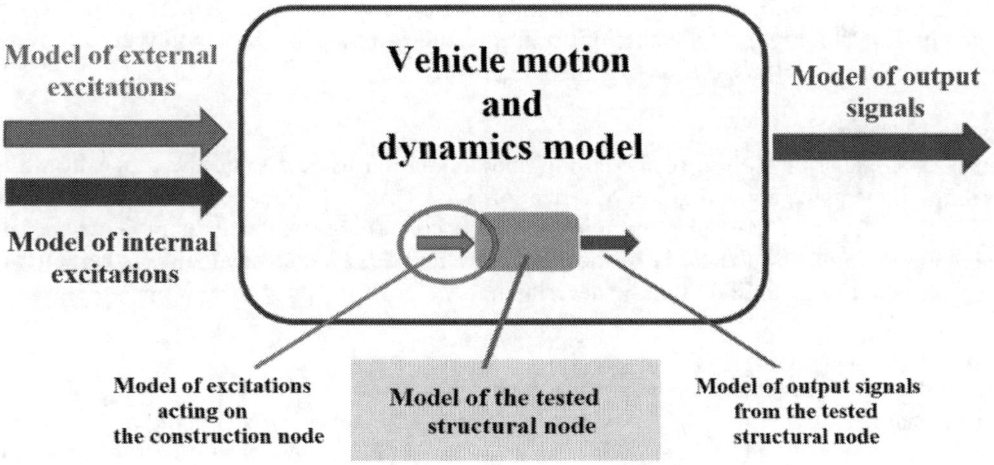

Figure 2. Determining the excitation acting on the selected vehicle structure node.

IOP Conf. Series: Materials Science and Engineering 1247 (2022) 012011 doi:10.1088/1757-899X/1247/1/012011

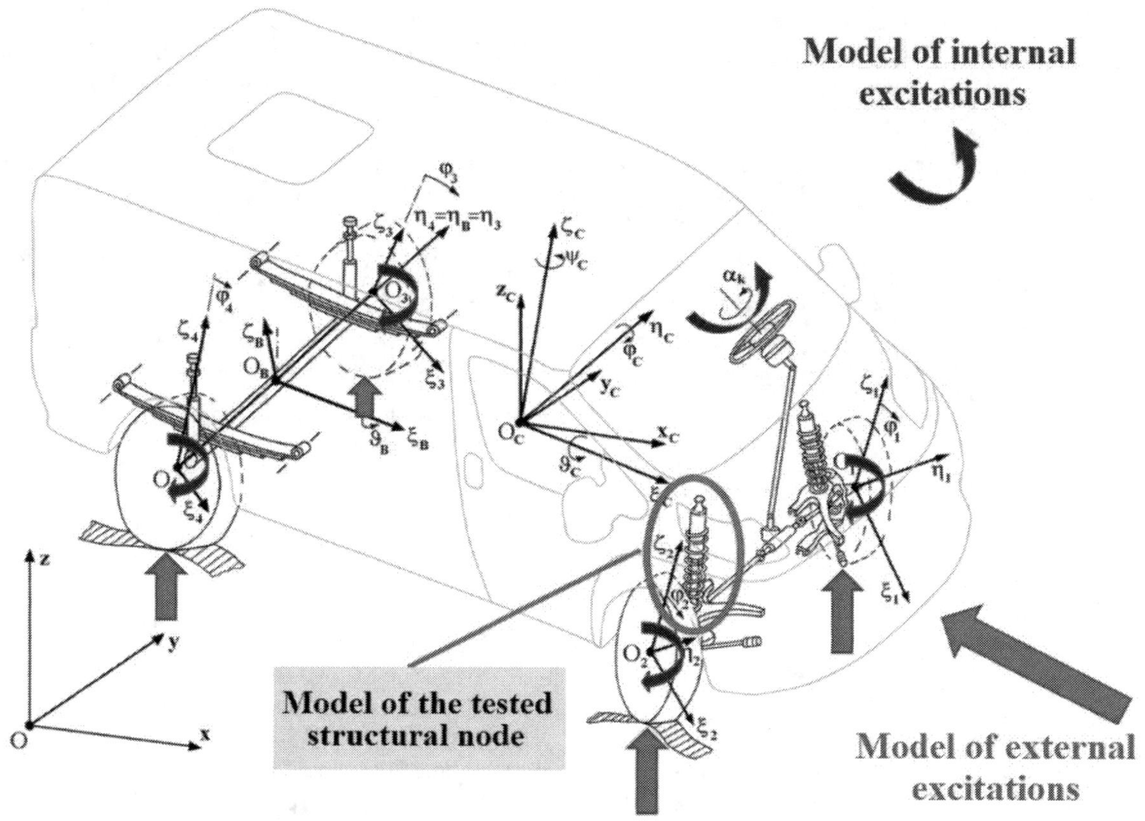

Figure 3. Determining the excitation acting on a selected construction node of the vehicle, using a digital simulation of working processes accompanying the vehicle movement (the own drawing presented in the paper was used [10]).

5. Examples of simulation models of motion, and dynamics of road vehicles

Models of this type have been used for several dozen years in the study of motion, and vehicle dynamics. They are widely presented in monographs [1, 4, 6, 7, 9, 12, 13, 19].

The author has built many own, original models and simulation programs, described in the works [8, 9, 10, 11]. They can be used to determine the input acting on a selected vehicle structure node, as schematically shown in Fig. 3, in the case of the elastic and damping element (coil spring and hydraulic damper) of the front wheel suspension of a passenger car.

6. Recapitulation

The presented concept of the use of modelling and simulation to determine the input during tests of structural components of a car requires verification for example applications. The author expects to present a much more extensive publication in the future, containing such an example - the use of a road vehicle simulation model in the study of the construction node. The physical and mathematical model of the vehicle under consideration will be described.

References

[1] Arczyński S 1993 *Mechanika ruchu samochodu* (Warszawa: WNT)
[2] Azrulhisham E (et al.) 2011 *Application of road simulator service loads in automotive component durability assessment.* (The Open Industrial & Manufacturing Engineering Journal vol 4) pp 1–7
[3] Dukkipati R (et al.) 2008 *Road vehicle dynamics* (Warrendale: SAE International USA)

[4] Gierulski W 1991 *Kształtowanie drgań w stanowiskowych badaniach symulacyjnych* (Kielce: Zeszyty Naukowe Politechniki Świętokrzyskiej Mechanika 46) PL ISSN 0239-4979
[5] Kaczor M and Januszka M 2019 *Optymalizacja konstrukcji naczep z zastosowaniem symulatora drogi MTS* (Mechanik No 1/2019) pp 52-54 (DOI: https://doi.org/10.17814/mechanik.2019.1.10)
[6] Kamiński E and Pokorski J 1983 *Teoria samochodu Dynamika zawieszeń i układów napędowych pojazdów samochodowych* (Warszawa: WKŁ)
[7] Kasprzyk T and Prochowski L 1990 *Teoria samochodu Obciążenia dynamiczne zawieszeń* (Warszawa: WKŁ)
[8] Lozia Z 1990 *Analiza wpływu podatnego zamocowania haka holowniczego na dynamikę zestawu samochód - przyczepa w ruchu prostoliniowym ze stałą prędkością* V Konferencja Naukowa Instytutu Transportu Politechniki Warszawskiej "Nauka i praktyka w transporcie". Warszawa, 19-21.09.1990 Materiały pp 116-123
[9] Lozia Z 1998 *Analiza ruchu samochodu dwuosiowego na tle modelowania jego dynamiki.* (Warszawa: Prace Naukowe Politechniki Warszawskiej. Transport Zeszyt 41)
[10] Lozia Z 2015 *Examples of authorial models for the simulation of motor vehicle motion and dynamics* (Warszawa: Zeszyty Naukowe Instytutu Pojazdów PW vol 04 No 4) pp 9-27
[11] Lozia Z 2016 *The use of a linear quarter-car model to optimize the damping in a passive automotive suspension system – a follow-on from many authors' works of the recent 40 years* (Warsaw: The Archives of Automotive Engineering – Archiwum Motoryzacji vol 71(1)) pp 39-71 http://dx.doi.org/10.14669/AM.VOL71.ART3
[12] Mitschke M 1977 *Teoria samochodu Dynamika samochodu* (Warszawa: WKŁ)
[13] Mitschke M 1989 *Dynamika samochodu Drgania* (Warszawa: WKŁ)
[14] *MTS Model 320 Tire-Coupled Road Simulators Versatile repeatable road simulation for the full range of vehicle sizes and application* (MTS booklet available as: https://www.mts.com/cs/groups/public/documents/library/ dev_002219.pdf) (Google attempt: 2020-02-1).
[15] Osiecki J W 1994 *Dynamika maszyn* (Warszawa: Wojskowa Akademia Techniczna)
[16] Osiecki J W 2004 *Badania pojazdów samochodowych i ich zespołów na stanowiskach badawczych* (Kielce: Politechnika Świętokrzyska Specjalne zagadnienia mechaniki i sterowania) ISBN 83-88906-16-x pp 19-49
[17] Osiecki J W Gromadowski T and Stępiński B 2006 *Badania pojazdów samochodowych i ich zespołów na symulacyjnych stanowiskach badawczych* (Radom: Wydawnictwo Instytutu Technologii Eksploatacji – PIB) ISBN 83-7204-510-0
[18] Osiński Z 1978 *Teoria drgań* (Warszawa: PWN)
[19] Prochowski L 2005 *Pojazdy samochodowe Mechanika ruchu* (Warszawa: WKŁ)
[20] Homepage Łukasiewicz PIMOT. Laboratorium BLY: https://pimot.lukasiewicz.gov.pl/index.php?option=com_content&view=article&id=40&Itemid=222 (Google attempt: 2022-03-07)
[21] Homepage MTS: https://www.mts.com/home/ oraz https://test.mts.com/industries/automotive (Google attempt: 2022-03-07)
[22] Homepage firmy Wielton: https://wielton.com.pl/ oraz https://wieltongroup.com/innowacyjnosc/ (Google attempt: 2022-03-07)
[23] Homepage 40ton.net (motoryzacja dla profesjonalistów): https://40ton.net/zobacz-jak-dziala-i-do-czego-sluzy-centrum-badawczo-rozwojowe-firmy-wielton-otwarte-wczoraj-w-wieluniu/ (Google attempt: 2022-03-07)

Quasi-static research of cornering stiffness for selected tires of UTV/ATV vehicles

J Jackowski[1], M Żmuda[1*], M Wieczorek[1]

[1] Department of Mechanical Engineering, Institute of Vehicles and Transportation, Military University o Technology (WAT), st. gen. Sylwestra Kaliskiego 2, 00-908 Warsaw, Poland.

* marcin.zmuda@wat.edu.pl

Abstract. The cornering stiffness is an important feature of the tire influences the dynamic behavior of the vehicle in curvilinear motion. The article presents results of quasi-static research of cornering stiffness. Two non-pneumatic tires (NPTs) of similar size and a pneumatic tire for utility terrain vehicles (UTVs) or all-terrain vehicles (ATVs) were selected for the research. The analysis of the research results showed a significant influence of the flexible structure of NPTs on the values of lateral forces.

Key words: pneumatic tire, vehicle, vehicle dynamics

1. Introduction

The tire is one of the most important elements of the vehicle. Its construction and the materials used affect, among others, traction and controllability of the vehicle, especially in critical conditions of its motion [1,2]). It is also an important, next to the suspension [3], an elastic-damping element that co-decides on the driving comfort and car movement of the car in various road conditions [4].

The pneumatic tire is the most common type of tire today. The performance of specific functions of a pneumatic tire is possible by appropriate materials selection of supporting structure and tread, as well as maintaining compressed air or gas under a certain pressure. Requirement to maintain compressed air (resilient factor) is unfortunately one of the major disadvantages of pneumatic tires. The air loss initially causes the deterioration of the tire's performance properties and, in the absence of any reaction from the driver, may lead to its irreversible damage [5]. An alternative solution to this problem are non-pneumatic tires (NPTs). Their modern construction correctly mimic the properties of compressed air thanks to the use of a supporting structure with often complex geometry [6 - 8]. In [6,8] detailed analysis of the normal load transfer mechanism depending on the supporting structure of NPTs are presented.

The structure of each NPT include the following common elements: rim, belt (shear beam/belt), tread, flexible structure (connecting the tread with the rim), e.g. single spokes or with a cell structure. The available test results for NPTs are mostly the results of numerical tests (finite element method). For example, [9] analyzed the effect of changes in the width and length of the spokes on the aerodynamic drag of the wheel based on numerical tests, supplemented by scaled model experimental tests in a wind tunnel. In [5], the NPT radial characteristics were analyzed using the finite element method for various geometries of the supporting structures, which were validated by a laboratory experiment. In [8] the influence of the NPTs spoke curvature on the radial characteristics of the wheel was analyzed. Increasing the spoke curvature resulted in an increased displacement of the NPTs center. The assessment of the

influence of wheels on the vehicle curvilinear motion (especially in transient non-steady-states) is always an important and carefully considered aspect. The vehicle course is determined by the cornering stiffness [10]. This parameter also provides information on the tire susceptibility to lateral deformation, which in dynamic conditions translates into a slower response of the vehicle to dynamic changes in the values of lateral forces transferred by the vehicle wheels. The lateral force loading affects the gradual increase in the deformation of the wheel resilient layer. The shape of the contact patch with the ground changes, while the lateral response increases until a certain constant value is reached, which is transmitted under the steady-state vehicle motion condition. This phenomenon is well described by the so-called tire relaxation length. Taking it into account is particularly important in the dynamic model of curvilinear vehicle motion.
The purpose of this paper is to measure and analyze the cornering stiffness of NPTs and pneumatic tire used in utility terrain vehicles (UTVs) or all-terrain vehicles (ATVs).

2. Methodology and research conditions

The quasi-static research of cornering stiffness were carried out with use of the Universal quasi-static indoor tyre test bench (figure 1). A detailed description of the constrution and measurement possibilities have been described earlier in [11, 12]. The authors are conducting comprehensive tests of NPTs. Therefore, the research of the cornering stiffness were carried out in the same conditions as those described earlier in [13] (the tests of the radial characteristics). The ATVs and the UTVs, during curvilinear motion, often move using large angles of wheels steering, which was taken into account in the selection of the cornering angles. The test conditions are presented in table 1.
Complementing the comprehensive research with the cornering stiffness was carried out for (figure 2):

- NPT_1 (non-pneumatic tire nr 1) - equipped with 24 pairs of spokes;
- NPT_2 (non-pneumatic tire nr 2) - equipped with a cellular (hexagonal – honey comb) flexible cell;
- PT (pneumatic tire) - diagonal tire.

Each measurement under specified conditions was repeated four times.

Figure 1. The universal quasi-static indoor tyre test bench.

Table 1. Research conditions.

Research objects	Normal load [N]	Cornering angle [°]	Inflation pressure [MPa]
NPT_1	1000	5	-
NPT_2	2000	10	-
PT	3000	15	0,3

a) b) c)

Figure 2. Badane koła: a) NPT_1, b) NPT_2, c) PT.

The tire cornering stiffness contains the stable value of lateral force as a function of tire cornering angle. During the measurement, the wheel was positioned at a certain angle with respect to the direction of movement of the measuring surface. After applying the determined value of the normal force, the measuring (mimics road) surface was set in motion (figure 3). During the test, the value of the lateral reaction and the displacement of the measuring surface were recorded. The obtained measurements allow to determine the stable value of the lateral force as a function of cornering angle. The procedure for marking the cornering stiffness is shown in Figure 4.

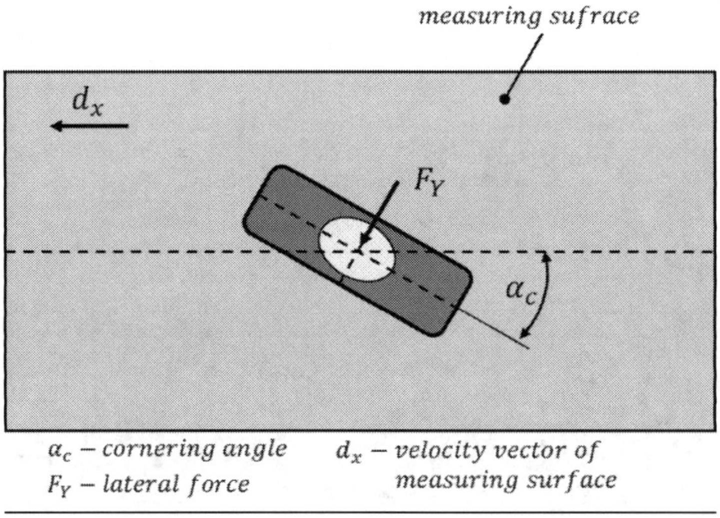

Figure 3. The method of measuring the cornering stiffness (top view).

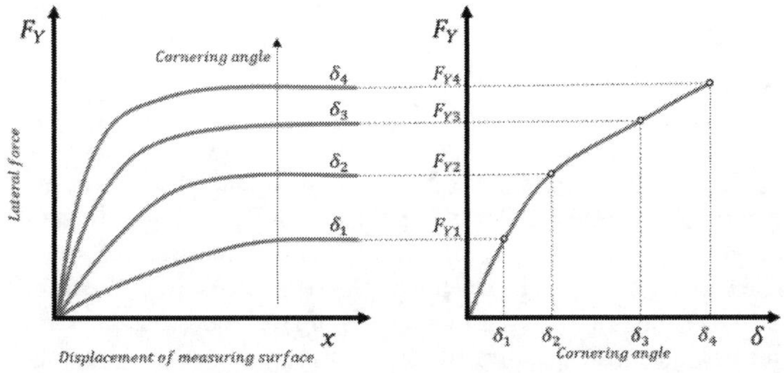

Figure 4. From left: example of trajectory of the lateral tire F_Y as a function of measuring surface displacement x, characteristic of cornering stiffness.

3. Results

Research results were shown on figure 5. Measurement results are marked with gray pluses. Next, the mean values for each cornering angle are connected by a line. Green, orange and blue colors indicate the characteristics of the cornering stiffness, respectively for the normal force of 1000, 2000, 3000 N.

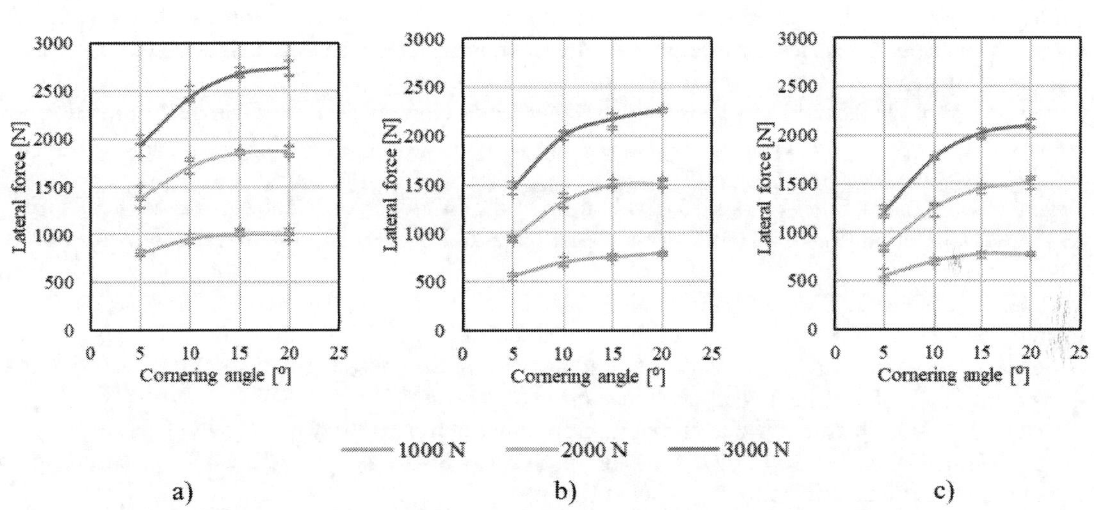

Figure 5. Characteristic of cornering stiffness: a) NPT_1, b) NPT_2, c) PT.

In general approach, the trajectory of the characteristics of the cornering stiffness for all research objects is similar. Increasing the cornering angle increases the lateral force, which stabilizes at the cornering angle of 15° or 20° (smaller value of angle for less load), depending on applied normal load. In the analysis, a pneumatic tire was used as a reference ~~point~~, which can be replaced by NPTs. This comparison shows that the use of NPT_2 allows to achieve higher values of lateral force even at the smallest cornering angle (especially for applied normal load 3000 N). Probably the first cause of this phenomenon is the stiffer NPT_2 tire caps under tread area with a simultaneously larger radius of curvature of the cross-section. The effect of the above-mentioned is greater stability of the shape and contact path of the wheel. Another reason may be the high torsional stiffness of the with honeycomb structure. This inference can be confirmed by the results obtained in the research of the NPT_1. The measured values of lateral force are definitely greater than obtained for the other research objects. The reduction in the torsional stiffness of the "spoke" support structure is observed while maintaining the advantages of a stiff tread with belt. As a result, the tread protrusions are less deformed, which results in an increase in the values of the lateral forces achieved (resulting from the adhesion force of the rubber forming the tread protrusions on the ground).

4. Conclusion

The quasi-static research of cornering stiffness were carried out for non-pneumatic tires and pneumatic tire used in UTVs/ATVs. During the research, the wheels were loaded with three different force values and were rolled at four cornering angles. NPT_1 (wheel equipped with single spokes) in each of the adopted test conditions was characterized by the highest value of lateral force. Obtaining higher values of the established lateral force may have an influence on the increase of the relaxation path compared to NPT_2 and PT. It may also have an impact on higher values of lateral deformation and a slower response of the vehicle to dynamic changes in the values of tire lateral force. Other research objects for normal loads of 1000 and 2000 N achieve similar cornering stiffness characteristics.
For the analyzed pair of wheels (NPT_2, PT) for the load of 3000 N, visible differences were obtained in the trajectory of cornering stiffness characteristics.
In the next stage of the research, it is planned to extend the test conditions with additional cornering angles and an additional value of inflation pressure of the pneumatic tire. The analysis of the test results will be enriched with the results of the relaxation path and the analysis aimed at explaining the observed results and the possible confirmation of the hypotheses formulated in the paper.

References

[1] Gidlewski, M., Jemioł, L., Żardecki, D. Sensitivity investigations of the automated lane-change manoeuvre – selected issues. Proceedings of the Institution of Mechanical Engineers. Part I: Journal of Systems and Control Engineering, 2019, 233(4), pp. 360–369.
[2] Gidlewski, M.J., Jankowski, K., Muszyski, A., Żardecki, D. Vehicle Lane Change Automation with Active Steering - Theoretical Studies and Numerical Investigations.SAE Technical Papers 2017-01-1555.
[3] Hryciów, Z., Krasoń, W., Wysocki, J. Evaluation of the influence of friction in a multi-leaf spring on the working conditions of a truck driver. Eksploatacja i Niezawodność, 2021, 23(3), pp. 422–429
[4] Jurecki, R., Stanczyk, T. Analysis of vehicle moving parameters in various road conditions. Communications - Scientific Letters of the University of Zilina, 2021, 23(3), pp. F58–F70.
[5] Rugsaj R, Suvanjumrat C 2019 Proper radial spokes of non-pneumatic tire for vertical load supporting by finite element analysis, *Int.J Automot. Technol.* 20, 801-812.
[6] Żmuda M 2021 Construction analysis of non-pneumatic tires, *Biuletyn WAT*, 70 (1), 113-128.
[7] Sim J, Hong J, Cho I, Lee J 2021 Analysis of Vertical Stiffness Characteristics Based on Spoke Shape of Non-Pneumatic Tire. Appl. Sci. 11, 2369.
[8] Hryciów Z, Jackowski J, Żmuda M 2020 The Influence of Non-Pneumatic Tyre Structure on its Operational Properties, International Journal of Automotive and Mechanical Engineering 17(3), 8168–8178.
[9] Li H, Xu Y, Si C, Yang Y 2021 A research on aerodynamic characteristics of non-pneumatic tire, *Mechanics & Industry* 22, 27.
[10] Luty W 2013 Nieustalone stany znoszenia bocznego ogumienia kół jezdnych w symulacji ruchu krzywoliniowego pojazdu, *Prace Naukowe Politechniki Warszawskiej. Transport* 98, 357-367.
[11] Jackowski, J Wieczorek M 2011 Analysis of interaction between tyre tread and road on the basis of laboratory test. In: 7th International Science Conference Transbaltica Vilnius, 292 – 296.
[12] Luty W, Prochowski L, Szurkowski Z 1999 Stanowisko do badań ogumienia dużego rozmiaru. In: VII Międzynarodowe Symp. IPM. Warszawa-Rynia 375 – 382.
[13] Jackowski J, Żmuda M, Wieczorek M, Zuska A 2021 Quasi-Static Research of ATV/UTV Non-Pneumatic Tires, *Energies* 14 (20), 1-12.

Yellow and brown grease—characteristics of compression-ignition engine

R Ciesielski[1], M Zakrzewski[1], O Shtyka[1], T Maniecki[1,*], A Rylski[1], M Wozniak[2], P Kubiak[3] and K Siczek[2,*]

[1] Lodz University of Technology, Lodz., Poland
[2] Department of Vehicles and Fundamentals of Machine Design, Lodz University of Technology, Lodz. Poland
[3] Warsaw University of Technology, Warsaw, Poland

* tomasz.maniecki@p.lodz.pl; ks670907i@p.lodz.pl

Abstract. This papers presents the results of analysis done on a compression-ignition engine supplied with methyl ester of rapeseed oil (Yellow Grease), methyl ester of goose fat (Brown Grease) and pure diesel. The analysis included the engine characteristics, emissions and fuel consumption. Results also include chromatographic analysis for all of the three fuels. Additional evaluation was done on a vehicle idling and under load.

Keywords: biodiesel, yellow grease, brown grease, diesel engine, fuel consumption, engine emission

1. Introduction

Both insufficient coverages of increasing energy demand and the harmful environmental effects of fossil fuels made biofuels both solid ones [1,2] and liquid ones [3] an interesting alternative to them. Liquid biofuels, particularly derived from inedible biomass are promising renewable energy sources for transport. It is due to they are usable within current infrastructures and also need less technological advances than electric vehicles and nuclear power. The negative effects of biofuels on food security and production strongly depend on the type of feedstock used in biofuel pro-duction and need further studies on methods decreasing their level of harmfulness [3].

Nowadays the demand to replace the commonly utilized diesel by alternative fuels, particularly the bio-derived ones, has become more and more visible [4,5].

It is necessary to notice that there are some differences between biodiesels and renewable bio-diesels, both being alternative to the fossil diesel sometimes also called petrodiesel [6-8].

Biodiesel is a renewable, biodegradable fuel manufactured domestically from vegetable oils, animal fats, or recycled restaurant grease. Biodiesel meets both the biomass-based diesel and overall advanced biofuel requirement of the Renewable Fuel Standard related to the a minimum volume of renewable fuels contained in transportation fuel [9]. Biodiesel in its pure form is known as "neat" biodiesel. A biodiesel blend is pure biodiesel blended with petrodiesel with the notation of Bxx where the xx indicates the volumetric amount of biodiesel in the blend, commonly, varying from B2 to B20 [6]. According to [10] biodiesel is often blended with petroleum diesel in ratio range from 5% to 20% biodiesel, or B5-B20. The Renewable Fuel Standard (RFS), a federal program that mandates the blending of biofuels into the nation's fuel supply generally encouraged higher biodiesel blends.

Renewable hydrocarbon biofuels comprise renewable gasolines, renewable diesels and sustainable aviation fuels (SAF) produced from biomass sources through a variety of biological, thermal, and chemical processes via various technology pathways including: traditional hydrotreating, biological sugar upgrading, catalytic conversion of sugars, gasification, pyrolysis and hydro-thermal processing [11].

The renewable diesel, so-called 'green diesel' is a biomass-derived transportation fuel suitable for use in diesel engines. It meets the ASTM D975 specification for petroleum in the United States and EN 590 in Europe. It is chemically identical to diesel and can be used in existing diesel engines [12].
Van Gerpen [13] stated that biodiesel is the form of biofuel obtained from vegetable oils and animal fats and is an important alternative to fossil diesel. It is renewable, non-toxic, biodegradable, sulfur, and benzene-free, can be applied in various diesel engines without any modification, and can be blended with fossil diesel at any given ratio [14-19].
Diaz-Felix et al. [20] stated that biodiesel is a renewable fuel comprised of fatty acid methyl esters (FAMEs).
However, biodiesel exhibited lower oxidative stability and cold flow properties, higher viscosity and density, and higher NOx emissions in comparison to fossil diesel [21-24]. Additionally, bio-diesel is characterized by higher production costs compared to fossil diesel [25-27]. Some activities, including the use of the cheaper technologies, are carried out to decrease such costs [28-34].
Huang et al. [35] reviewed various types of biodiesel, the characteristics, processing, and eco-nomics of industry related to biodiesel. They discussed particularly the application of biodiesel in the automobile industry.
The goal of our study was to compare the effect of a supply the engine with a pure diesel fuel and two blends including the one of diesel and 6% biodiesel in the form of the MEYG - methyl ester of yellow grease (rapeseed oil) and the one of MEBG - methyl ester of brown grease (goose fat) on the characteristics, fuel consumption and emissions of diesel engine applied in the vehicle chosen.

2. Feedstocks for Biodiesel

According to [36] there are six EPA-approved biodiesel and renewable diesel production feed-stocks are soybean oil, yellow grease, distillers corn oil, animal fats, canola oil and camelina. As of September 2017, their respective content the biomass-based diesel market were approximately 46%, 15%, 15%, 13%, 11% and below 1%.
Bartha-Vári et al. [37] noted that biodiesel can be produced from edible oils including soybean and palm oil, and also non-edible ones like cooking oils and algal or fungal oils.
Edible oils are cheaper for biodiesel production in comparison to the algal oils [38].
Triglycerides of edible oils comprise various fatty acids, including mostly (9Z)-9-octadecenoic acid (oleic acid, OA), followed by (9Z,12Z)-9,12-octadecadienoic acid (linoleic acid, LA), myristic, palmitic and stearic acids, respectively. Biodiesel can be also produced from the fatty phase of wastes (brown grease – BG) or palm fatty acid distillate (PFAD) [39].
The mentioned raw materials contain a high concentration of free fatty acids (FFAs). Due to their high content of the OA and LA, and palmitic acid in PFAD, such acids are used as model systems during investigations [40-42].
According to Gebremariam and Marchetti [43] fresh vegetable oils are used for biodiesel pro-duction at an industrial scale. However, Mansir et al. [44] stated that most biodiesel production costs resulted from utilizing fresh edible oils as feedstock.
Therefore, alternative, cheaper sources of lipids were proposed including waste cooking oil, non-edible Jatropha, algae, tamarind seed oil, municipal sewage sludge, and recycled grease trap waste [45-51].
Elgharbawy et al. [52] noticed that feedstocks of biodiesel varies between edible oil such as the oil of sunflower, soybean and coconut or non-edible oil such as the oil of Jatropha, jojoba and used cooking oil.
However, the high free fatty acid (FFA) content of mentioned feedstocks resulted in saponification and emulsification in the presence of homogeneous base catalysts, limiting biodiesel yield and making its separation from glycerol difficult [53,54].
According to [55] biodiesel is found as a nontoxic, biodegradable and renewable alternative fuel applicable with little or no engine modifications. Biodiesel can be produced from low cost oils (restaurant waste, frying oils, animal fats) comprising high levels of free fatty acids.

IOP Conf. Series: Materials Science and Engineering 1247 (2022) 012013 doi:10.1088/1757-899X/1247/1/012013

Grzelak and Zoltowski [56] proposed the use of waste oils to produce hydrogenated vegetable oils (HVOs) and described the existing biorefinery plant in Venice producing HVO fuel. They presented the parameters of such a biofuel and compared them with the parameters of diesel and biodiesel containing FAMEs applied as a 7% additive to diesel oil.

3. Technologies for Biodiesel Production

The most popular method of biodiesel production at an industrial scale is the alkali-catalyzed transesterification of the aforementioned fresh vegetable oils [43].

According to Elgharbawy et al. [52] biodiesel is often obtained from vegetable oil via transesterification reaction substituting the organic group (alkyl) of an alcohol by the organic group of a triglyceride producing FAME and glycerol. The authors reviewed four types of the transesterification processes, acid-catalyzed transesterification, alkaline catalyzed transesterification, enzyme catalyzed transesterification and non catalyzed supercritical methanol. They noticed that the alkaline catalyzed transesterification method is the most commonly utilized process due to the fastest reaction, the highest yield, the mildest reaction condition, the lowest cost, corrosive, and toxicity.

As reported in [57,58] biodiesel can be produced via transesterification and esterification. The transesterification resulted in formation of alkyl esters of FFAs and glycerol as a byproduct of the reaction between mono-, di- or triglycerides and alcohol. During the esterification, a reaction between alcohol and FFA takes place resulting in formation of alkyl esters.

Both esterification and transesterification belong to the group of equilibrium reactions [59].

As methanol and ethanol are used in such reactions [57], therefore both methyl and ethyl esters are found to be biodiesels [60].

The activation of transesterification and esterification reactions is realized by agitation and heating [61].

Ultrasonication can also be applied for activation of chemical reactions, as the esterification of FFAs under ultrasonic activation proceeds quickly and efficiently, without additional heating [62-64].

However, application of ultrasonication is expensive. Furthermore, unwanted direct contact between the treated corrosive media and ultrasonic devices is possible, which can be prevented by a decrease of the residence time of reagents in the ultrasonic reactor or by the use of contactless reactors [62].

Activation of reactions also needs a chemical or enzymatic catalyst [61]. Chemical catalysts can basic or acid. Basic catalysts are better than the latter, as allow reaction at relatively low temperatures (60–90 °C) and pressures (1.5–4 atm). However they are highly sensitivity to water and reactive with FFAs in saponification [65].

The acid catalysis are preferred [66] but have longer duration (over 12 h), need higher temperatures (120–250 °C) and a larger amount of catalyst (here of acid type) compared to the basic catalysis. The sulfuric acid insensitive to small amounts of water, it is usually applied as a catalyst in the esterification of FFAs [65].

As homogeneous aggressive catalysts cause corrosion of equipment they can be replaced by heterogeneous basic or acidic catalysts [39].

The wide application of solid-phase catalysts is limited by problems in mass transfer, resolvable using an oscillatory flow technique, microwave irradiation or ultrasonic technology [67].

Lewis acids can also be utilized as catalysts in biodiesel production [68-70].

Lewis acids are effective catalysts for both esterification and transesterification reactions [71,72], and in both homogeneous and heterogeneous forms [73].

Lewis acid boron trifluoride in methanol was proposed as a good catalyst for the esterification of fatty acids [74].

Heterogeneous catalysis by Lewis acids such as $AlCl3$, $AlF3$ and $ZnCl2$ was found effective in transesterification of canola oil for biodiesel production [75-77].

Various tin-based catalysts (Sn(II) acetate, Sn(II) chloride and Sn(II) 2-ethyl-hexanoate) exhibited high catalytic activity in esterification and transesterification of vegetable oils [78].

The use of Lewis acid catalysts allows proceeding the reaction under not harsh conditions, ambient temperature and in a short time [79].
A two-step biodiesel production process from waste cooking oil through recycling crude glycerol esterification catalyzed by alkali catalyst was described in [80].
The homogeneous base catalysts such as potassium hydroxide, sodium hydroxide, sodium methoxide, and potassium methoxide often utilized for transesterification are corrosive, difficult to be recycled, and are removed from the reaction mixture by water washing, causing enhanced amounts of wastewater [81-84]. According to Alaei et al. [85] biodiesel obtained using homogenous base catalysts can be contaminated with metal ions.

4. Research on Biodiesel from Yellow Grease

Several studies were conducted on biodiesels obtained from various Yellow Greases.
Panchal et al. [86] studied the one-step process of producing biodiesel from yellow fat by kinetic transesterification of yellow grease with dimethyl carbonate in a molar ratio (1: 5), using a CH_4O_3S catalyst, reaction temperature equal to 708 °C, time of 45 min, and mixing speed equal to 125 rpm. The maximum yield of biodiesel with yellow fat was achieved of 94%. The biodiesel properties tested, such as kinematic viscosity at temperature equal to 408 °C, specific gravity at temperature equal to 258 °C, flash point, pour point, cloud point, and corrosion grade of the copper strip were in accordance with the ASTM D6751–02 biodiesel quality standards.
Ortiz-Martínez et al. [87] investigated the transesterification reaction of Pongamia pinnata oil also known as Karanja oil in supercritical methanol in one-step catalyst-free process. Triglyceride (TG) conversion and the yield of FAMEs at an alcohol-to-oil molar ratio of 43:1 were investigated in the temperature range of 250-350 °C under pressure 12-43 MPa and in a time range of 15–90 min, respectively. They also studied the evolution of monoglycerides (MG) and diglycerides (DG) and the thermal decomposition of fatty acid chains for the mentioned reaction conditions. The optimal conditions for such a reaction related to temperature of 300 °C and reaction time of 90 min allowing almost complete triglyceride conversion. Then, the high level of thermal decomposition occurred from 325 °C, mainly due to the degradation of polyunsaturated fatty acid methyl esters. Maximum degree of thermal decomposition of 38% took place at 350 °C in the end of reaction time.
Goembira and Saka [88] studied biodiesel production from Pongamia pinnata oil using an interesterification process called one-step supercritical methyl acetate method for under reaction condition of 300 °C / 20 MPa / 45 min / 42 M ratio in methyl acetate to oil. The addition of 10 wt.% aqueous acetic acid allowed conducting the interesterification process under mentioned reaction condition. It was achieved the highest yield of 96.6 wt.% FAME and 11.5 wt.% triacetin (total 108.1 wt.%). Both products were miscible and their biodiesel properties were in agreement with biodiesel standards.
Özçelik et al. [89] investigated the effects of biodiesel (B100) obtained from safflower oil in the transesterification process and its two blends with Eurodiesel (10%:90% - B10) and (20%:80% - B20) on the characteristics of a four-stroke, common-rail fuel system, water-cooled, four-cylinder diesel engine. They found that engine power values obtained with Eurodiesel fuel, biodiesel, and its blends were close to one another at all engine speeds. The fuel consumption was slightly enhanced for the engine supplied with Biodiesel and its blends compared to the supply with Eurodiesel fuel.
Syamsuddin and Hameed [90] conducted synthesis of glycerol free-methyl esters (FAME) for the transesterification of Jatropha oil utilizing dimethyl carbonate (DMC) as methylating agent over heterogeneous Ca–La mixed-oxide catalyst. The optimal reaction conditions occurred at a 4:1 molar ratio of DMC to oil, 150 °C reaction temperature, 120 min reaction time, and 3 wt.% of catalyst loading allowing obtaining conversion and yield of 99.84%.
Canakci and Van Gerpen [55] studied the effect of the biodiesel produced from high free fatty acid feedstocks on engine performance and emissions. One biodiesel was prepared from animal fat-based yellow grease with 9% free fatty acids. The neat biofuel and its 20% blend with No. 2 diesel fuel were investigated at steady-state engine operating conditions in a four-cylinder turbocharged diesel engine. Biodiesel fuel allowed obtaining highly limited emissions of PMs, CO, and unburned HCs. However,

the NO_x emission were enhanced by 11% for the yellow grease methyl ester. The conversion of the biodiesel fuel energy to work was close to that of diesel fuel.
According to [20], the esterification reduced the FFA levels in Yellow Grease, but large amounts of methanol had to be employed to drive the reaction to high yield. The authors investigated the effect of FFA in the range of 2% to 32% (w/w) at three different molar ratios of methanol to FFA (4.5: 1, 9: 1, 18: 1), fiding the non-linear nature of the effect.
Gaurav et al. [91] suggested the use of the catalytic distillation (CD) combining a catalytic reaction and separation via distillation in the same distillation column for the biodiesel production from yellow grease containing both triglyceride and free fatty acid. They emphasized an advantage of CD in comparizon to a conventional biodiesel processes due to the limited number of distillation columns, waste streams and greenhouse gas emissions.
Duda et al. [92] studied the performance and emission characteristics of a compression ignition engine equipped with a Common Rail injection system. The engine was fueled with diesel and diesel-biodiesel blends containing 25% and 50% share (by volume) of methyl esters of rapeseed oil obtained from the single-stage transesterification process. The experiments were conducted on a medium-duty, turbocharged, inter-cooled, Common Rail Direct Injection (CRDI) diesel engine operating at speed of 1500 rpm and mid-load range from 100 Nm to 200 Nm. They found that the addition of the mentioned renewable component to a fuel blend positively affected performance parameters and weakened engine emissions, in the majority of cases.

5. Research on Biodiesel from Brown Grease

Less studies were conducted on biodiesel obtained from various Brown Greases.
According to [93], since 3 December 2021 the International Sustainability and Carbon Certification (ISCC) has stated that the category of brown grease or grease trap fat "shall exclusively be used for material that is removed from grease traps" and "this category shall not be used for material that is removed from the sewage system". Such a definition has excluded waste edible oil collected from sewers from being classed as brown grease. Oil removed from sewers was instead covered under the sewage sludge category expanded to include fats, oils and grease (FOG) from the sewage system, while the separate sewage system FOG category has been deleted.
Kolet et al. [94] stated that brown grease is a mixture of oils, fats, solids and detergents from food industry wastes that is captured in grease traps. Brown grease contains oils and fats convertable into biodiesel. However, the high concentration of free fatty acids in brown grease excluded the application of conventional biodiesel production schemes. The authors elaborated a scheme for biodiesel production from brown grease. They studied conditions for the effective separation of a fat phase from brown grease, and determined the composition of a fatty phase for several grease traps. They conducted esterification and transesterification of brown grease lipids with methanol, with accompanied catalysts of the Lewis acids BF_3 and $AlCl_3$. The reaction was activated by ultrasound. They found that biodiesel was obtainable from brown grease via esterification and transesterification within several minutes under ultrasonic activation at room temperature.
The researches from the same group also in [39] combined the application of Lewis catalysts with ultrasound activation to the scheme of biodiesel production from FFAs.
Tran et al. [95] reviewed recent achievements in the production of biodiesel from grease trap waste (GTW).
According to Hums et al. [96] a thickened and dewatered Grease Trap Waste (GTW) can be upgraded to brown grease via heating and settling. Using a bubble column reactor brown grease can be cinverted into crude biodiesel. Such a conversion process is realizable at ambient pressure and relatively low temperatures, using readily-available industrial chemicals. Purification of the reaction products is conducted via distillation and adsorption resulting in weakening sulfur content but such a scheme is still slightly effective. After purification, the biofuel typically contains approximately 25ppm – 50ppm sulfur, which is still over the required specification. Interestingly, it was mentioned that laboratory-scale desulfurization experiments using reactive desulfurization and enhanced adsorption strategies allowed reducing sulfur content to 10-15 ppm.

Zheng et al. [97] studied a potential biodiesel feedstock in form of larval grease extracted from yellow mealworm beetle (Tenebrio molitor L.) (YMB), a post-harvest scavenger, fed with decayed vegetables in 9 weeks before extraction of its grease. Interestingly, 34.2 g biodiesel was got from 234.8 g dried YMB larval biomass. The YMB biodiesel comprised various fatty acids including linolenic acid (19.7%), palmitic acid (17.6%), linoleic acid (16.3%) and stearic acid (11.4%). The YMB biodiesel features met the standard EN 14214, including ester content (96.8%), density (860 kg/m^3), flash point (127 °C), cetane number (58), water content (300 mg/kg), and methanol content (0.2%).
Spiller et al. [98] characterized two brown grease samples captured from local grease traps or wastewater facilities. Raw brown grease was fermented using Clostridium butyricum to produce butyric acid. It was obtained a yield of 0.55 butyric acid/g sugars confirming also the convertion of the glycerol and lactic acid. Hexane extraction of the fermentation broth allowed obtaining an 81.3% recovery of lipids.
Fan et al. [99] converted brown grease to biodiesel FAME. It was made in a two-stage reaction comprising esterification and transesterification using Novozym 435 in a solvent-free medium. They studied the effects of methanol amounts and enzyme concentration on the first-stage reaction, and the operational stability of Novozym 435 in the two-stage reaction. They found that the addition of biodiesel into the reaction system prior to two-stage reaction prevented the lipase from deactivation. The latter was induced by excess amounts of methanol. They have obtained contents of FAME above 95 wt.% even after 15 cycle reactions.
The research group Duda et al. [92] studied also the performance and emission characteristics of a mentioned earlier diesel engine supplied also with diesel-biodiesel blends containing 25% and 50% share (by volume) of methyl esters of turkey lard obtained from the single-stage transesterification process. They found that the addition of the mentioned renewable component to a fuel blend improved performance parameters and weakened engine emissions, in the majority of cases studied.

6. Materials and methods

During present study three kinds of fuel were utilized including:

- pure diesel fuel (as delivered),
- the blend of diesel with 6% of biodiesel obtained from yellow grease, more specifically in the form of methyl ester of rapeseed oil (as delivered),
- the blend of diesel with 6% of biodiesel obtained from a brown grease, more specifically in the form of methyl ester of goose fat obtained via transesterification process described further.

These fuels were utilized to supply a diesel engine of the vehicle chosen. Such a vehicle was investigated on its fuel consumption and emissions under both zero and full load of the engine.

6.1. Goose fat transesterification

6.1.1. Reagents and equipment

For the transesterification process the following reagents were used:

- goose lard,
- sodium methoxide solution.

For the transesterification process the following equipment was utilized:

- round bottom flask with a capacity of 100 and 250ml,
- reflux cooler,
- magnetic stirrer,
- water bath,
- thermometer,
- manifold with a capacity of 250ml,
- measuring cylinder,

• beakers,
• conical flask with a stopper,
• glass funnel.

6.1.2. Methodology
In order to prepare the methanolic catalyst solution, in a conical flask with a capacity of 100 ml of 3.36 g of solid NaOH were weighed and dissolved in 36 ml of methanol.
Next, for a rounded flask with a capacity of 250 ml were introduced with 120 g of goose lard, a magnetic rod and the flask was fitted with a reflux condenser. The whole was placed in a water bath at a temperature of about 50 ° C (monitored with a thermometer) and the catalyst solution prepared in advance was introduced. The blend was stirred with a magnetic stirrer for 60 minutes. The reaction blend was then transferred to a separating funnel, 50 ml of 10% HCl was added, mixed and allowed to separate (about 10-20 minutes). The lower water-glycerin layer was separated and the upper oily layer was washed twice with distilled water (2 x 50 mL). Washing was done carefully by slightly tilting the funnel to avoid the formation of stable emulsions.
The procedure was repeated to obtain amount of biodiesel needed for mixing with pure diesel to obtain the estimated amount of fuel necessary to supply the diesel engine during investigation. To be more specific, that amount of fuel had to be sufficient to supply the tested engine during two measurements of its fuel consumption during operation at a rotational speed of 1500 rpm and separately at its zero and at its full load. So it was twice the volume of an additional 1 L transparent plastic fuel tank. Since the amount of biodiesel in the blend with diesel was to be 6%, the required amount of produced methyl ester of goose fat was about 200 mL.

6.2. Chromatographic analysis of fuels used
Qualitative analysis of samples I - III (I – pure diesel fuel sample; II – diesel with 6% of methyl ester of rapeseed oil; III – diesel with 6% of methyl ester of goose fat) was performed using a Hewlett Packard series 5890 chromatograph (Agilent Technologies) coupled with a Hewlett Packard series 5972 mass spectrometer. A DB-5MS capillary column (30m, 0.250mm) was used. During the measurement, the following GC column temperature program was used: 40 ° C for 4 min, then the temperature was increased to 70 ° C at a rate of 4 ° C / min, then increased to 250 ° C at a rate of 20 ° C / min and held for 10 min. The injection port and detector temperatures were 250 and 200 ° C, respectively. Helium of purity 5.0 was used as the carrier gas. The range of scanned masses in the range of 20 - 700 m / z was used for qualitative determinations of the tested samples. Before the measurements, the samples were extracted in hexane in a ratio of 1: 100. The microliter phase dissolved in hexane was taken with a syringe for further analysis. The injection volume of the samples was 1 μl.

6.3. Measurement of engine characteristics, fuel consumption and emissions at full engine load
During the present study the engine characteristics were determined for the diesel engine applied in the vehicle Citroen C5 II 1.6 HDI 2008 100 kW operating under full engine load. Such an engine had nominally a cubic capacity equal to 1560 cc, maximum power equal to 80 kW at 4000 rpm and maximum torque equal to 240 Nm at 1750 rpm [100]. The measurement of fuel consumption and emissions were determined for this diesel engine under zero engine load and under full engine load. The way the fuel was supplied to the engine was slightly changed for the time of measurements. Instead of the classic method of supplying fuel from the vehicle's tank to the low-pressure fuel pump, an additional 1 L transparent plastic fuel tank was introduced with a metered amount of fuel connected with the said fuel pump. During measurement the time and the actual fuel level in the tank was recorded (using a camera of the mobile phone). The density of fuel was determined in the following manner. Using a graduated glass beaker, the amount V_i of i-th fuel was measured, equal to 50 ml. That beaker with fuel was weighed, and then, after the fuel was withdrawn and beaker was dried, it was weighed. The difference in the masses measured in this way allowed estimating the mass m_i of the i-th fuel. The density d_i of such a fuel was estimated from Equation 1.

$$d_i = \frac{(m_i+m_b)-m_b}{V_i};\ i = \begin{cases} 1, \text{for pure diesel} \\ 2, \text{for diesel} + 6\%\ \text{MEYG}, \\ 3, \text{for diesel} + 6\%\ \text{MEBG} \end{cases} \quad (1)$$

where:
$m_i + m_b$ – measured sum of the beaker mass and the i-th fuel mass,
m_b – measured mass of beaker,
m_i – mass of the i-th fuel, while: $i = 1$ for pure diesel, $i = 2$ for the blend of diesel and 6% of methyl ester of yellow grease (rapeseed oil), $i = 3$ for the blend of diesel and 6% of methyl ester of brown grease (goose fat),
V_i – volume of the i-th fuel equal to 50 mL.

It was assumed that during the measurements the density of the fuel delivered to the engine did not change, in particular, there was no suction of false air with impurities.

The i-th fuel consumption $g_{ei}(n)$ at the engine speed n set was estimated from Equation (2).

$$g_e(n) = \frac{d_i \cdot [V_{i0}(n) - V_{ie}(n)]}{P_i(n) \cdot t_i(n)}, \quad (2)$$

where:
$g_{ei}(n)$ – the i-th fuel consumption at the engine speed n set,
d_i – the i-th fuel density obtained from Equation (1),
$V_{i0}(n)$ – the i-th fuel volume registered in the beginning of the measurement process of the i-th fuel consumption at the engine speed n set,
$V_{ie}(n)$ – the i-th fuel volume registered in the end of the measurement process of the i-th fuel consumption at the engine speed n set,
$t_i(n)$ – time period registered for the measurement process of the i-th fuel consumption at the engine speed n set,
$P_i(n)$ – measured engine power at the engine speed n set under full engine load, as described further.

6.3.1. *Measurement under full engine load*

The engine characteristics including engine power P versus engine speed n and engine torque M versus engine speed n were obtained on the Dynomet Roller Dynamometer [101] for the engine operating under full load at engine speed n sequentially set to the value from the range of 1000-4000 rpm every 500 rpm. The measurements were realized as follows:

- First engine operated at idle speed until reaching the temperature value of the balance temperature;
- Then the additional transparent plastic tank was replaced by the other one with the same dose of the pure diesel;
- Engine characteristics were obtained for diesel engine supplied with pure diesel (fuel quality and content as delivered);
- Simultaneously for each engine speed value set the time of engine operation and volume of fuel consumed were determined;
- Additionally, for each engine speed value set the engine emissions were measured using the Elwico unit [102] comprising an exhaust gas analyzer and an opacimeter;
- Then the additional tank was replaced with the one filled with diesel with 6% of biodiesel in the form of the methyl ester of rapeseed oil (blend quality and content as delivered);
- Next the engine characteristics, time of engine operation, volume of fuel consumed and engine emissions were determined for the same engine speed n values as for the case of pure diesel supply;
- Then the additional tank was replaced with the one filled with pure diesel;

- Engine operated once again at idle speed until reaching the temperature value of the balance temperature.
- Next the additional tank was replaced with the one filled with diesel with 6% of biodiesel in the form of methyl ester of goose fat obtained from the transesterification process described earlier.
- Then the engine characteristics, time of engine operation, volume of fuel consumed and engine emissions were determined for the engine speed value n equal to 1500 rpm.
- Next the engine was supplied with fuel from the vehicle tank and operated once again at idle speed until reaching the temperature value of the balance temperature.

6.3.2. Measurement under zero engine load
The measurement of fuel consumption and emissions of diesel engine operating under zero engine load was done in the Diagnostic Station in the Vocational Automotive High School of Lodz in Poland. Figure 1 presents such a vehicle in this Diagnostic Station.

Figure 1. Measurement of fuel consumption and emissions of diesel engine applied in the vehicle Citroen C5 II 1.6 HDI 2008 100 kW in the Diagnostic Station in the Vocational Automotive High School of Lodz in Poland.

The engine speed was coarsely controlled using the engine speed indicator on the dashboard of the vehicle. More accurately, the engine speed was measured using measuring unit Tester XENON 310U WTM with piezoelectric sonde fastened on one of the high pressure fuel lines of engine.
The measurements were realized similarly as for the case of the full engine load. However, for each engine speed value set the engine emissions were measured using an exhaust gas analyzer Arcon ISC Oliver K9000 and an opacimeter ISC OLIVER D60.

7. Results and discussion
The obtained results comprised these related to chromatographic analysis of fuels, and engine characteristics determined under full engine load for three fuels studied. The other results related to the fuel consumption end engine emissions determined under both zero and full engine load for the mentioned fuels.

7.1. Results from chromatographic analysis of fuels supplied to the combustion engine
The obtained chromatogram for pure diesel was shown in Figure 2. The highest amount of about 19% related to 9-octadecenoic acid, methyl ester. The amounts over 5% were observed for eicosane, 2,6-dimethyl-heptadecane, 4-ethyl-tetradecane, hexadecane, pentadecane, tetradecane, and 2,3-dihydro-4,7-dimethyl-1H-indene. The other amounts were below 5%.

The obtained chromatogram for diesel with addition of 6% of methyl ester of Yellow Grease MEYG (rapeseed oil methyl ester) was presented in Figure 3. The highest amount of about 8% related to 2,6-dimethyl-heptadecane. The amounts of over 5% were observed for 3-methyl-dodecane, pentadecane, 4-methyl-tridecane, tetradecane, 2,6,11-trimethyl-dodecane, and 3,8-dimethyl-decane. The other amounts were below 5%.

The obtained chromatogram for diesel with addition of 6% of methyl ester of Brown Grease MEBG (goose fat methyl ester) was presented in Figure 4. The highest amount of about 8% related to 9-octadecenoic acid, methyl ester. The amounts of over 5% were observed for eicosane, 2,6-dimethyl-heptadecane, 3-methyl-dodecane, pentadecane,4-methyl-tridecane, tetradecane, and 2,3-dihyfro-4,7-dimethyl-1H-indene. The other amounts were below 5%. To our knowledge, this is the first report for the blend of diesel and 6% biodiesel in the form of methyl ester of goose fat.

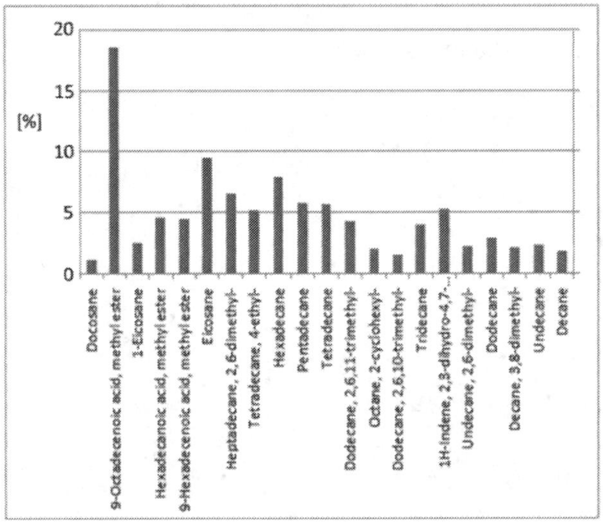

Figure 2. Chemical content of pure diesel suppled to the combustion engine studied.

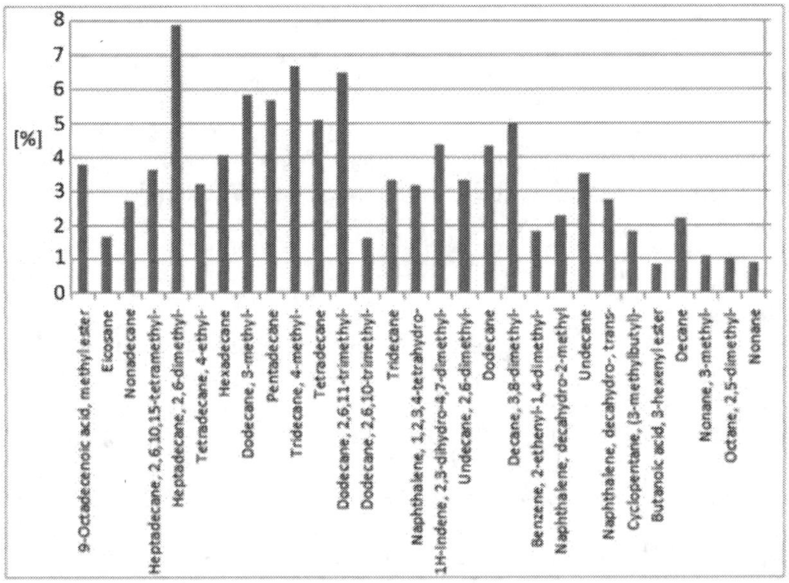

Figure 3. Chemical content of diesel with addition of 6% of methyl ester of Yellow Grease MEYG (rapeseed oil methyl ester).

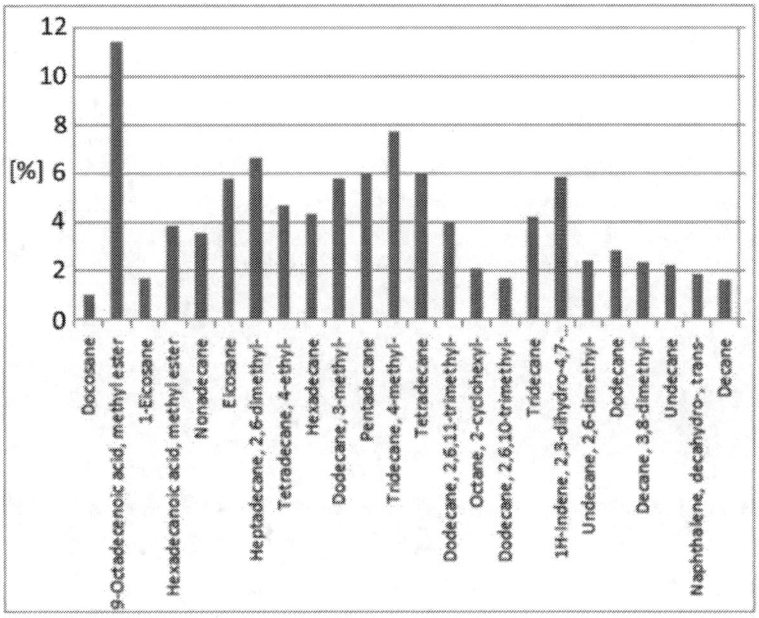

Figure 4. Chemical content of diesel with addition of 6% of methyl ester of Brown Grease MEBG (goose fat methyl ester).

7.2. Resulted Engine Characteristics

The engine characteristics obtained from research on dynamometer under full load of an engine were shown in Figure 5. They comprised engine torque versus engine speed M(n) and engine power versus engine speed obtained for pure diesel, diesel with 6% of rapeseed oil methyl ester and diesel with 6% of goose fat methyl ester. The latter is related by only one characteristic point in the waveform. For the pure diesel the maximum power was lower by about 3% compared to that presented in [103]. The maximum torque was lower by 4.3 % compared to that presented in [103]. These decreases were probably resulted from the wear of engine components. Under full load of engine the addition of 6% of rapeseed oil methyl ester to the diesel resulted in weakening of engine power by 8% and engine torque about 4%. Addition of 6% of goose fat methyl ester to the diesel resulted in an enhancement of an engine power about 1%. It was accompanied by no visible change of the engine torque.

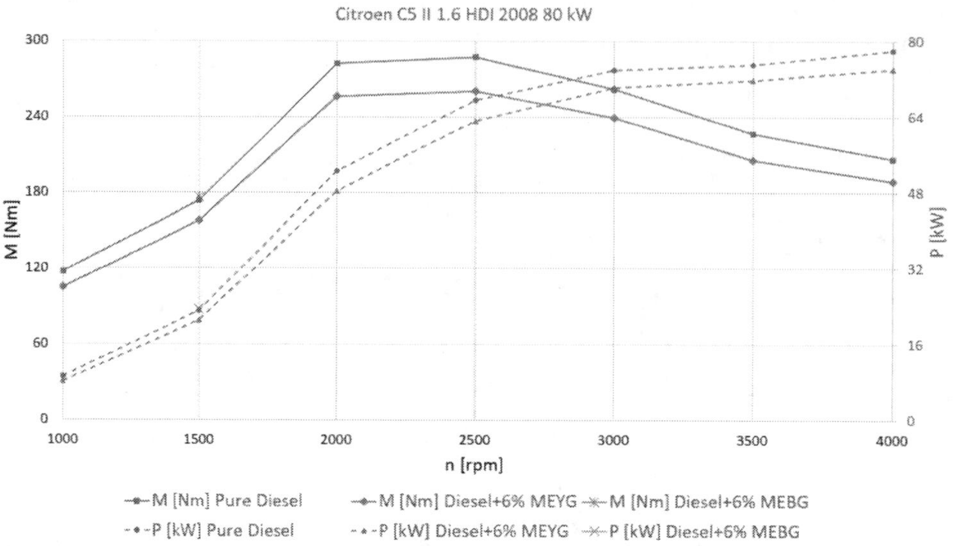

Figure 5. The engine characteristics: Torque M(n) and Power P(n) under full engine load for pure diesel, diesel with 6% of rapeseed oil methyl ester and diesel with 6% of goose fat methyl ester.

7.3. Resulted fuel consumption and emissions at zero engine load

The fuel consumption g_e obtained at zero engine load was presented in Table 1. Addition of 6% of rapeseed oil methyl ester into diesel resulted in as increase of the fuel consumption by 8.6%. Addition of 6% of goose fat methyl into diesel resulted in an enhancement by 3.4% of such a consumption.

Table 1. The fuel consumption g_e at zero engine load.

	g_e [g/kWh]	
Pure Diesel	**Diesel + 6% MEYG**	**Diesel + 6% MEBG**
580	630	600

The engine emissions of CO, CO_2, and PM obtained at zero load of engine supplied with pure diesel, diesel with 6% of rapeseed oil methyl ester and diesel with 6% of goose fat methyl ester, respectively were shown in Figure 6. The emissions of NO_x, O_2 and HC under such conditions were presented in Figure 7. The addition of 6% of rapeseed oil methyl ester to diesel caused a small enhancement of CO and HC emissions and a more clearly visible increase of CO_2 and decrease of PM emissions, compared to the engine supplied with a pure diesel. The addition of 6% of goose fat methyl ester to diesel resulted in a small increase of CO emissions and a more clearly visible enhancement of HC and CO_2 emissions, and a decrease of PM emission, in comparison to the case of pure diesel.

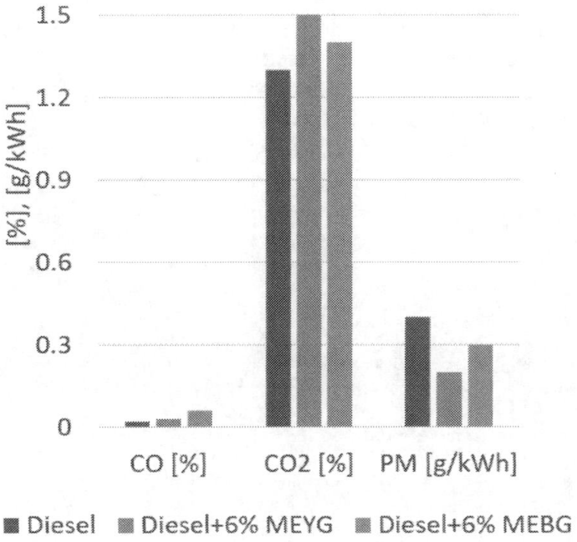

Figure 6. The engine emissions of CO, CO_2, and PM obtained at zero load of engine supplied with pure diesel, diesel with 6% of rapeseed oil methyl ester and diesel with 6% of goose fat methyl ester, respectively.

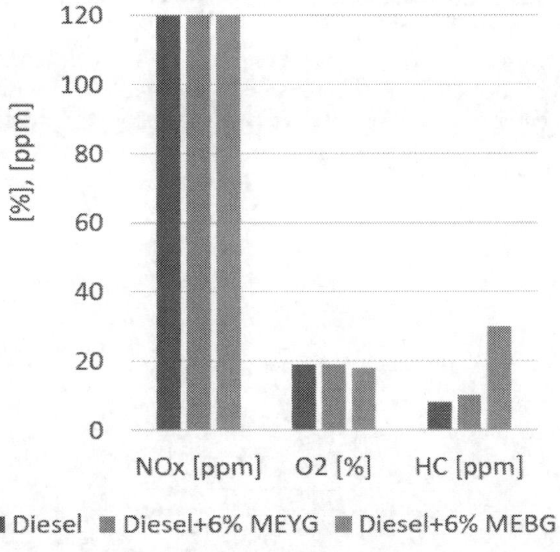

Figure 7. The engine emissions of NO_x, O_2 and HC obtained at zero load of engine supplied with pure diesel, diesel with 6% of rapeseed oil methyl ester and diesel with 6% of goose fat methyl ester, respectively.

7.4. Resulted fuel consumption and emissions at full engine load

The fuel consumption g_e versus engine speed obtained at full engine load for pure diesel, diesel with 6% of rapeseed oil methyl ester and diesel with 6% of goose fat methyl ester, respectively, was presented in Figure 8. For the pure diesel supply of engine, the obtained values of fuel consumption were lower by about 15%-17% compared to these presented in [89], which were obtained for four-stroke, common-rail fuel system, water-cooled, four-cylinder diesel engine with cubic capacity equal to 1910 cc, compression ratio of 5.18:1, maximum power equal to 79 kW at 4000 rpm and maximum torque equal

200 Nm at 1750 rpm. The addition of 6% of rapeseed oil methyl ester to diesel resulted mostly in an increase of the fuel consumption. Contrary, the slight resulted decrease in such a consumption was observed only for the engine speed equal to 3500 rpm. The addition of 6% of goose fat methyl ester to diesel resulted in a small enhancement of the fuel consumption at engine speed equal to 1500 rpm.

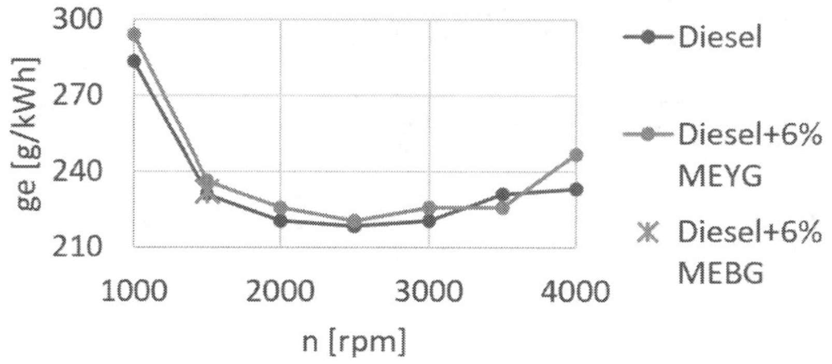

Figure 8. The fuel consumption g_e at full engine load.

The emissions of CO versus engine speed obtained at full engine load for pure diesel, diesel with 6% of rapeseed oil methyl ester and diesel with 6% of goose fat methyl ester, respectively, was shown in Figure 9. The addition of 6% of rapeseed oil methyl ester to diesel resulted in an decrease of CO emission up to 80 %. For the comparison, the decrease of CO emission when the blend of diesel and 5% methyl ester of rapeseed ester was used instead pure diesel was below 52% for a four stroke, four cylinder, water cooled, direct injection, naturally aspirated Diesel engine D-243 with maximum power equal to 59 kW and the splash volume equal to 4.75 L [104]. This decrease weakened with engine speed. The addition of 6% of goose fat methyl ester to diesel resulted in a small enhancement of the CO emissions at engine speed equal to 1500 rpm.

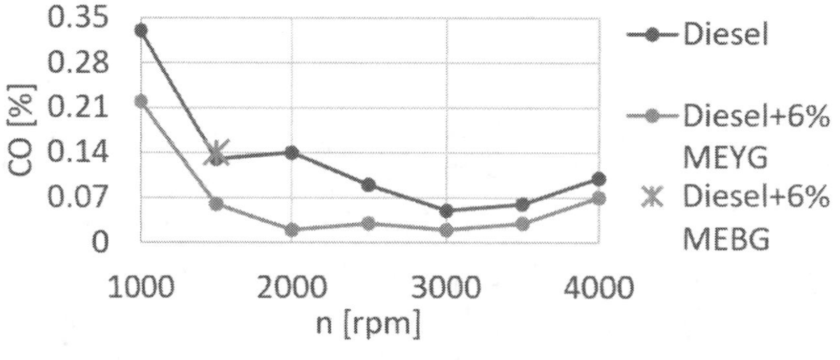

Figure 9. The CO emissions at full engine load.

The emissions of CO_2 obtained at full engine load for pure diesel, diesel with 6% of rapeseed oil methyl ester and diesel with 6% of goose fat methyl ester, respectively, was shown in Figure 10. The addition of 6% of rapeseed oil methyl ester to diesel resulted in an increase of the CO_2 emissions for engine speed below 2000 rpm and in a decrease of it for engine speed above 2000 rpm. The smallest change has occurred at engine speed equal to 2500 rpm. It was contrary to the increase of CO_2 emission from engine supplied by bland of diesel with 20 % rapeseed methyl ester by 10 % compared to the case of supply with diesel [105]. Such contrary feature can be affected by a difference in the mentioned contents of rapeseed methyl ester in the blends with diesel. The addition of 6% of goose fat methyl ester to diesel resulted in a decrease of the CO_2 emissions by a 4.5% at engine speed equal to 1500 rpm.

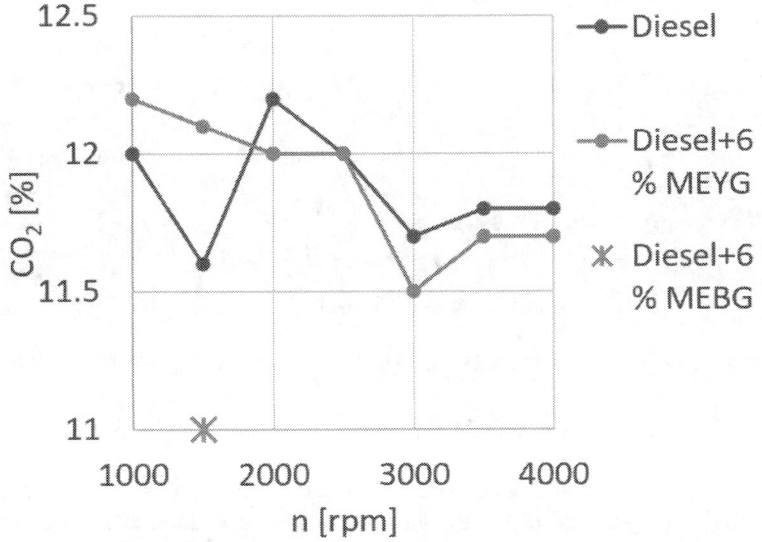

Figure 10. The CO_2 emissions at full engine load.

The emissions of HC versus engine speed obtained at full engine load for pure diesel, diesel with 6% of rapeseed oil methyl ester and diesel with 6% of goose fat methyl ester, respectively, was shown in Figure 11. The addition of 6% of rapeseed oil methyl ester to diesel resulted in an increase of the HC emissions for all engine speeds. The smallest change has occurred at engine speed equal to 3000 rpm. For the case of engine supply by pure diesel and by the blend of diesel with 6% of rapeseed oil methyl ester the HC emission were at least twice higher than HC emission from diesel engine supplied with the blend of diesel and neat rapeseed methyl ester for all engine speeds [104]. The higher HC emission during present study could result from the significant wear of components of diesel engine studied. The addition of 6% of goose fat methyl ester to diesel resulted in a triple increase of the HC emissions at engine speed equal to 1500 rpm.

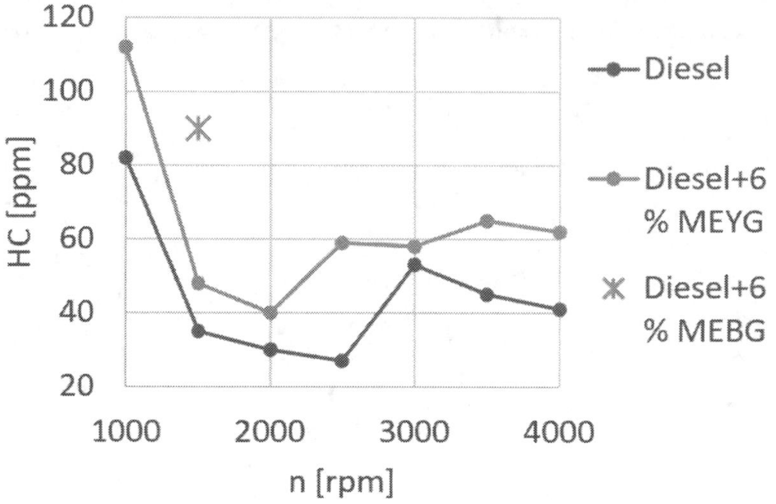

Figure 11. The HC emissions at full engine load.

The emissions of NO_x versus engine speed obtained at full engine load for pure diesel, diesel with 6% of rapeseed oil methyl ester and diesel with 6% of goose fat methyl ester, respectively, was shown in Figure 12. The emission of NOx was lower by 40% compared to that reported for the diesel engine presented in [104]. The addition of 6% of rapeseed oil methyl ester to diesel resulted in an increase of the NO_x emissions for all engine speeds. The smallest change has occurred at engine speed equal to 3000 rpm. The addition of 6% of goose fat methyl ester to diesel resulted in an increase of the HC emissions by 31% at engine speed equal to 1500 rpm.

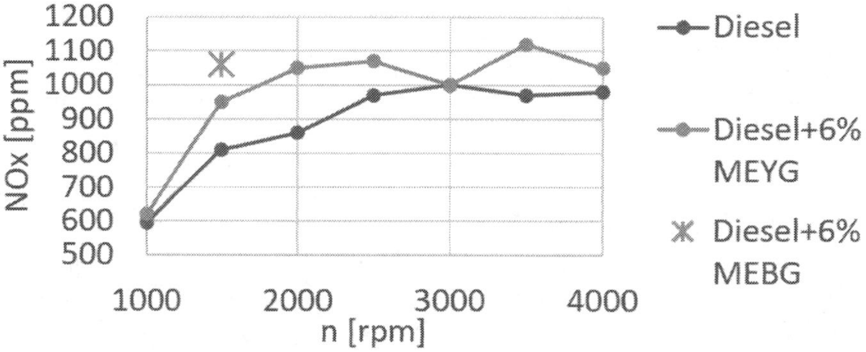

Figure 12. The NO_x emissions at full engine load.

The total emissions of PM versus engine speed obtained at full engine load for pure diesel, diesel with 6% of rapeseed oil methyl ester and diesel with 6% of goose fat methyl ester, respectively, was shown in Figure 13. The addition of 6% of rapeseed oil methyl ester to diesel resulted in an decrease of the TPM emissions for all engine speeds. The smallest change by has occurred at engine speed equal to 2000 rpm. The relative decrease of the TPM emissions were over 50%. It was much more compared to the average decrease of PM emissions for diesel engine supplied with a blend of diesel with 20% of methyl ester of rapeseed oil compared to the engine supply with pure diesel [106]. Such a difference was affected by a difference in amounts of methyl ester of rapeseed oil in the blends with diesel used to supply of engines. The addition of 6% of goose fat methyl ester to diesel resulted in an decrease of the TPM emissions by 3% at engine speed equal to 1500 rpm.

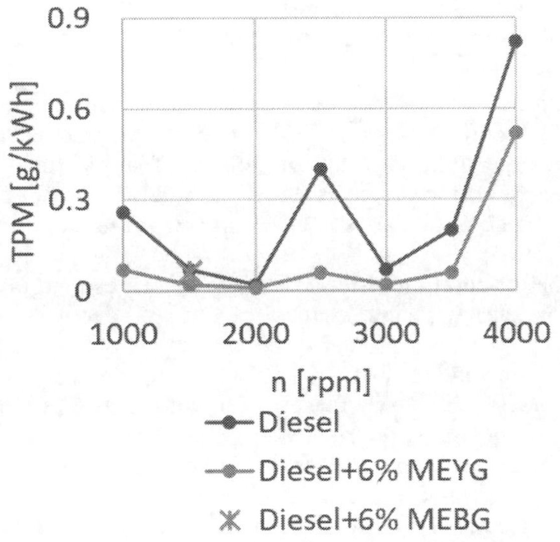

Figure 13. The total emissions of PM at full engine load.

The amount of O_2 in exhaust gas versus engine speed obtained at full engine load for pure diesel, diesel with 6% of rapeseed oil methyl ester and diesel with 6% of goose fat methyl ester, respectively, was presented in Figure 14. The addition of 6% of rapeseed oil methyl ester to diesel resulted in an decrease of the amount of O_2 in exhaust gas for engine speed above 1500 rpm. Below the latter it was observed an increase of such an amount. The smallest change has occurred at engine speed equal to 2000 rpm. The addition of 6% of goose fat methyl ester to diesel resulted in an increase of the amount of O_2 in exhaust gas by 2.7% at engine speed equal to 1500 rpm.

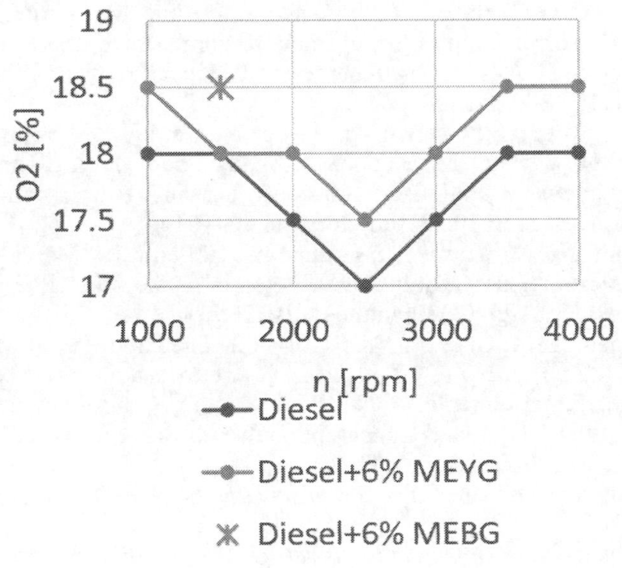

Figure 14. The amount of O_2 in exhaust gas at full engine load.

8. Conclusions

Based on the obtained results some conclusion was made.

1. The comparison of the effect of the diesel engine supply with three fuels including pure diesel, the blend of diesel with 6% of methyl ester of rapeseed oil (Yellow Grease), and the one with 6% of methyl ester of goose fat (Brown Grease) on the engine characteristics, fuel consumption and engine emissions has been done.
2. Obtained biodiesel containing 6% of Yellow Grease clearly differed not only from pure diesel but also much more from the biodiesel containing 6% of Brown Grease;
3. Under full load of engine the addition of 6% of Yellow Grease to the diesel allowed decreasing of engine power by 8% and engine torque about 4%, while addition of 6% of Brown Grease to the diesel allowed increasing of engine power about 1% and practically resulted in no change of the engine torque;
4. Under 0% load of engine the addition of 6% MEYG to diesel resulted in a small increase of CO and HC emissions and a more clearly visible increase of CO_2 and decrease of PM emissions, compared to the case of pure diesel;
5. Under 0% load of engine the addition of 6% MEBG to diesel caused a small enhancement of CO emissions and a more clearly visible increase of HC and CO_2 emissions, and a decrease of PM emission, in comparison to the case of pure diesel;
6. Under full load of engine the addition of 6% MEYG to diesel resulted in a clear decrease of CO and PM emissions for all engine speed and an enhancement of NO_x and HC emissions for all engine speed, compared to the case of pure diesel. The fuel consumption was a little higher but rather for the lower engine speed, the CO_2 emission varied with engine speed compared to the case of pure diesel;
7. Under full load of engine the addition of 6% MEBG to diesel resulted in a small enhancement of the fuel consumption, a clear increase of NO_x and HC emissions, a small decrease of PM emission and a clear decrease of CO_2 emision, while CO emision was not changed, compared to the case of pure diesel.

References

[1] Szufa, S.; Wielgosinski, G.; Piersa, P.; Czerwinska, J.; Dzikuc, M.; Adrian, L.; Lewandowska, W.; Marczak-Grzesik, M. Torrefaction of Straw from Oats and Maize for Use as A Fuel and Additive to Organic Fertilizers - TGA Analysis, Kinetics as Products for Agricultural Purposes. *Energies* 2020, ***13***, 2064. doi: 10.3390/en13082064.

[2] Szufa, S.; Piersa, P.; Adrian, L.; Czerwinska, J.; Lewandowski, A.; Lewandowska, W.; Sielski, J.; Dzikuc, M.; Wrobel, M.; Jewiarz, M.; Knapczyk, A. Sustainable Drying and Torrefaction Processes of Miscanthus for Use as a Pelletized Solid Biofuel and Biocar-bon-Carrier for Fertilizers. *Molecules* 2021, ***26***, 1014. doi: 10.3390/molecules26041014.

[3] Khan, M.A.H.; Bonifacio, S.; Clowes, J.; Foulds, A.; Holland, R.; Matthews, J.C.; Percival, C.J.; Shallcross, D.E. Investigation of Biofuel as a Potential Renewable Energy Source. *Atmosphere* 2021, ***12***, 1289. https://doi.org/10.3390/atmos12101289.

[4] Afshin, G.; Rahimpour, M.R; Ghasemi, Y.; Raeissi, S. The Biodiesel of Microalgae as a Solution for Diesel Demand in Iran. *Energies* 2018, ***11*****(4)**, 950. https://doi.org/10.3390/en11040950.

[5] Tito I.F., Sales, E.A.; Nascimento, A.N.; Padula, A.D. Optimizing the Cooperated "Multi-Countries" Biodiesel Production and Consumption in Sub-Saharan Africa. *Energies* 2020, ***13*****(18)**, 4717. https://doi.org/10.3390/en13184717.

[6] Knothe, G.; Krahl, J.; Van Gerpen, J. *The Biodiesel Handbook*; AOCS Press, Champaign, IL, USA, 2005.

[7] Brown, R.C. *Biorenewable Resources: Engineering New Products from Agriculture*; Iowa State Press, Ames, IA, USA, 2003.

[8] Pahl, G. *Biodiesel: Growing a New Energy Economy*; Chelsea Green, White River Junction, VT, USA, 2005.

[9] Alternative Fuels Data Center. Biodiesel Fuel Basics. Available online: https://afdc.energy.gov/fuels/biodiesel_basics.html (accessed on 14 March 2022).

[10] Hanson, S.; Agarwal, N. Biodiesels produced from certain feedstocks have distinct properties from petroleum diesel. *Today in Energy*, May 3, 2018. Available online: https://www.eia.gov/todayinenergy/detail.php?id=36052 (accessed on 14 March 2022).

[11] Alternative Fuels Data Center. Renewable Hydrocarbon Biofuels. Available online: https://afdc.energy.gov/fuels/emerging_hydrocarbon.html (accessed on 14 March 2022).

[12] Renewable Hydrocarbon Biofuels. Available online: https://afdc.energy.gov/fuels/emerging_hydrocarbon.html (accessed on 14 March 2022).

[13] Van Gerpen, J. Biodiesel processing and production. *Fuel Process. Technol.* 2005, ***86***, 1097-1107. doi: 10.1016/j.fuproc.2004.11.005.

[14] Demirbas, A. Biodiesel production via non-catalytic SCF method and biodiesel fuel characteristics. *Energy Convers. Manage.* 2006, ***47***, 2271-2282. doi: 10.1016/j.enconman.2005.11.019.

[15] Nabi, M.N.; Akhter, M.S.; Shahadat, M.M.Z. Improvement of engine emissions with conventional diesel fuel and diesel-biodiesel blends. *Bioresour. Technol.* 2006, ***97***, 372-378. doi: 10.1016/j.biortech.2005.03.013;

[16] Khan, S.A.; Rashmi, Hussain, M.Z.; Prasad, S.; Banerjee, U.C. Prospects of biodiesel production from microalgae in India. *Renew. Sustain. Energy Rev.* 2009, ***13***, 2361-2372. doi: 10.1016/j.rser.2009.04.005.

[17] Kumar, D.; Kumar, G.; Singh, P.C.P. Fast, easy ethanolysis of coconut oil for biodiesel production assisted by ultrasonication. *Ultrasonics Sonochem.* 2010, ***17***, 555-559. doi: 10.1016/j.ultsonch.2009.10.018;

[18] Kumar, G.; Kumar, D.; Singh, S.; Kothari, S.; Bhatt, S.; Singh, C.P. Continuous Low Cost Transesterification Process for the Production of Coconut Biodiesel. *Energies* 2010, ***3*(1)**, 43-56. https://doi.org/10.3390/en3010043.

[19] Mekhilef, S.; Siga, S.; Saidur, R. A review on palm oil biodiesel as a source of renewable fuels. *Renew. Sustain. Energy Rev.* 2011, ***15***, 1937-1949. doi: 10.1016/j.rser.2010.12.012.

[20] Diaz-Felix, W.; Riley, M.; Zimmt, W.; Kazz, M. Pretreatment of yellow grease for ef-ficient production of fatty acid methyl esters. *Biomass & Bioenergy - BIOMASS BIOENERG.* 2009, ***33***, 558-563. doi: 10.1016/j.biombioe.2008.09.009.

[21] Monyem, A.; Van Gerpen, J.H. The effect of biodiesel oxidation on engine per-formance and emissions. *Biomass Bioenergy* 2001, ***20***, 317-325. doi: 10.1016/S0961-9534(00)00095-7.

[22] Agarwal, D.; Sinha, S.; Agarwal, A.K. Ex-perimental investigation of control of NOx emissions in biodiesel-fueled compression ignition engine. *Renew. Energy* **2006**, ***31***, 2356-2369. doi: 10.1016/j.renene.2005.12.003.

[23] Canakci, M. The potential of restaurant waste lipids as biodiesel feedstocks. *Bioresour. Technol.* **2007**, ***98***, 183-190. doi: 10.1016/j.biortech.2005.11.022;

[24] Tesfa, B.; Mishra, R.; Gu, F.; Powles, N. Prediction models for density and viscosity of biodiesel and their effects on fuel supply system in CI engines. *Renew. Energy* 2010, ***35***, 2752-2760. doi: 10.1016/j.renene.2010.04.026.

[25] Hasheminejad, M.; Tabatabaei, M.; Mansourpanah, Y.; Khatamifar, M.; Javani, A. Upstream and downstream strategies to economize biodiesel production. *Bioresource Technol.* 2011, ***102***, 461-468. doi: 10.1016/j.biortech.2010.09.094.

[26] Yaakob, Z.; Narayanan, B.N.; Padikkaparambil, S.; Unni, K.S.; Akbar, P.M. A review on the oxidation stability of biodiesel. *Renew. Sustain. Energy Rev.* 2014, ***35***, 136-153. doi: 10.1016/j.rser.2014.03.055.

[27] Ullah, Z.; Bustam, M.A.; Man, Z. Biodiesel produc-tion from waste cooking oil by acidic ionic liquid as a catalyst. *Renew. Energy* 2015, *77*, 521-526. doi: 10.1016/j.renene.2014.12.040.

[28] Kumar, G.; Kumar, D.; Singh, S.; Kothari, S.; Bhatt, S.; Singh, C.P. Continuous Low Cost Transesterification Process for the Production of Coconut Biodiesel. *Energies* 2010, ***3*(1)**, 43-56. https://doi.org/10.3390/en3010043.

[29] Omotola, B.; Petrik, L.; Amigun, B.; Ameer, F. Low-Cost Feedstock Conversion to Biodiesel via Ultrasound Technology. *Energies* 2010, ***3*(10)**, 1691-1703. https://doi.org/10.3390/en3101691.
[30] Sunde, K.; Brekke, A.; Solberg, B. Environmental Impacts and Costs of Hydrotreated Vegetable Oils, Transesterified Lipids and Woody BTL - A Review. *Energies* 2011, ***4*(6)**, 845-877. https://doi.org/10.3390/en4060845.
[31] Deng, X.; Han, J.; Yin, F. Net Energy, CO_2 Emission and Land-Based Cost-Benefit Analyses of Jatropha Biodiesel: A Case Study of the Panzhihua Region of Sichuan Province in China. *Energies* **2012**, ***5*(7)**, 2150-2164. https://doi.org/10.3390/en5072150.
[32] Muthu G.R., Nagaraj, S.; Manjunath, M.; Boopathy, A.B.; Kurinjimalar, C.; Rengasamy, R.; Jayakumar, T.; Sheu, J.-R.; Li, J.-L.. Biofuel and Biochemical Analysis of Amphora coffeaeformis RR03, a Novel Marine Diatom, Cultivated in an Open Raceway Pond. *Energies* 2018, ***11*(6)**, 1341. https://doi.org/10.3390/en11061341.
[33] Ramos, M.; Dias, A.P.S.; Puna, J.F.; Gomes, J.; Bordado, J.C. Biodiesel Production Processes and Sustainable Raw Materials. *Energies* 2019, ***12*(23)**, 4408. https://doi.org/10.3390/en12234408.
[34] Osorio-González, Carlos S., Natali Gómez-Falcon, Fabiola Sandoval-Salas, Rahul Saini, Satinder K. Brar, and Antonio A. Ramírez. 2020. "Production of Biodiesel from Castor Oil: A Review" Ener-gies 13, no. 10: 2467. https://doi.org/10.3390/en13102467.
[35] Huang, D.; Zhou, H.; Lin, L. Biodiesel: an Alternative to Conventional Fuel. *Energy Procedia* 2012, ***16C***, 1874-1885. https://doi.org/10.1016/j.egypro.2012.01.287.
[36] Biofuel Feedstock Overview. Oil & Energy online 2022. Available online: https://oilandenergyonline.com/articles/all/biofuel-feedstock-overview/ (accessed on 14 March 2022).
[37] Bartha-Vári, J.-H.; Moisă, M.E.; Bencze, L.C.; Irimie, F.; Paizs, C.; Tosa, M.I. Efficient Biodiesel Production Catalyzed by Nanobioconjugate of Lipase from Pseudomonas fluorescens. *Molecules* 2020, ***25***, 651. doi: 10.3390/molecules25030651.
[38] Chua, S.Y.; Periasamy, L.A.; Goh, C.M.H.; Tan, Y.H.; Mubarak, N.M.; Kansedo, J.; Khalid, M.; Walvekar, R.; Abdullah, E.C. Biodiesel synthesis using natural solid catalyst derived from biomass waste - A review. *J. Ind. Eng. Chem.* 2020, ***81***, 41-60. doi: 10.1016/j.jiec.2019.09.022.
[39] Kolet, M.; Zerbib, D.; Molina, K.; Nakonechny, F., Nisnevitch, M. Biodiesel Production us-ing Lewis Catalysts under Ultrasonic Activation. *Isr. J. Chem.* 2020, ***60***, 1-9. doi: 10.1002/ijch.202000034.
[40] Canakci, M.; Sanli, H. Biodiesel production from various feedstocks and their effects on the fuel properties. *J. Ind. Microbiol. Biotechnol.* 2008, ***35*(5)**, 431–441. doi: 10.1007/s10295-008-0337-6.
[41] Ping, B.T.Y.; Yusof, M. Characteristics and Properties of Fatty Acid Distillatesfrom Palm Oil. *Oil Palm Bull.* 2009, ***59***, 5-11.
[42] Masri, A.N.; Mutalib, M.I.A.; Yahya, W.Z.N.; Aminuddin, N.F.; Leveque, J.M. Rapid esterification of fatty acid using dicationic acidic ionic liquid catalyst via ultrasonic-assisted method. *Ultrason. Sonochem.* 2020, ***60***, 104732. doi: 10.1016/j.ultsonch.2019.104732.
[43] Gebremariam, S.N.; Marchetti, J.M. Economics of biodiesel production: review. *Energy Convers. Manage.* 2018, **168**, 74-84. doi: 10.1016/j.enconman.2018.05.00
[44] Mansir, N.; Teo, S.H.; Rashid, U.; Saiman, M.I.; Tan, Y.P.; Alsultan, G.A., Taufiq-Yap, Y.H. Modified waste egg shell derived bifunctional catalyst for biodiesel production from high FFA waste cooking oil. A review. *Renew. Sustain. Energy Rev.* 2018, ***82***, 3645-3655. doi: 10.1016/j.rser.2017.10.098.
[45] Nagarajan, S.; Chou, S.K.; Cao, S.; Wu, C.; Zhou, Z. An updated comprehensive techno-economic analysis of algae biodiesel. *Bioresource Technol.* 2013, ***145***, 150-156. doi: 10.1016/j.biortech.2012.11.108.
[46] Yusuf, N.N.A.N.; Kamarudin, S.K. Tech-no-economic analysis of biodiesel production from Jatropha curcas via a supercritical methanol process. *Energy Convers. Manage.* 2013, ***75***, 710–717. doi: 10.1016/j.enconman.2013.08.017;
[47] Olkiewicz, M.; Torres, C.M.; Jimenez, L.; Font, J.; Bengoa, C. Scale-up and economic analysis of biodiesel production from munici-pal primary sewage sludge. *Bioresource Technol.* 2016, ***214***, 122-131. doi: 10.1016/j.biortech.2016.04.098.

[48] Abdurakhman, Y.B.; Putra, Z.A.; Bilad, M.R.; Nordin, N.A.H.M.; Wirzal, M.D.H. Techno-economic analysis of biodiesel production process from waste cooking oil using catalytic membrane reactor and realistic feed com-position. *Chem. Eng. Res. Des.* 2018, ***134***, 564–574. doi: 10.1016/j.cherd.2018.04.044.

[49] Tran, N.N.; Tisma, M.; Budzaki, S.; McMurchie, E. J.; Gonzalez, O.M.M.; Hessel, V.; Yung, N. Scale-up and economic analysis of biodiesel production from recycled grease trap waste. *Applied Energy* 2018, ***229***, 142-150. doi: 10.1016/j.apenergy.2018.07.106.

[50] da Silva, C.A., dos Santos, R.N.; Oliveira, G.G.; de Souza Ferreira, T.P.; de Souza, N.L.G.D.; Soares, A.S.; de Melo, J.F.; Colares, C.J.G.; de Souza, U.J.B.; de Araújo-Filho, R.N.; de Souza Aguiar, R.W.; dos Santos, G.R.; Gabev, E.E.; Campos, F.S. Biodiesel and Bioplastic Production from Waste-Cooking-Oil Transesterification: An Environmentally Friendly Approach. *Energies* 2022, ***15*(3)**, 1073. https://doi.org/10.3390/en15031073.

[51] Mushtaq, A.; Hanif, M.A.; Zahid, M.; Rashid, U., Mushtaq, Z.; Zubair, M.; Moser B.R.; Alharthi, F.A. Production and Evaluation of Fractionated Tamarind Seed Oil Methyl Esters as a New Source of Biodiesel. *Energies* 2021, ***14*(21)**, 7148. https://doi.org/10.3390/en14217148.

[52] Elgharbawy, A.S.; Sadik, W.A.; Sadek, O.M.; Kasaby, M.A. A Review on Biodiesel Feed-stocks and Production Technologies. *Journal of the Chilean Chemical Society* 2021, ***66*(1)**, 5098-5109. https://dx.doi.org/10.4067/S0717-97072021000105098.

[53] Ghiaci, M.; Aghabarari, B.; Habibollahi, S.; Gil, A. Highly efficient Bronsted acidic ionic liquid-based catalysts for biodiesel synthesis from vegetable oils. *Bioresource Technol.* 2011, ***102***, 1200-1204. doi: 10.1016/j.biortech.2010.09.095.

[54] Alegria, A.; de Arriba, A.L.F.; Moran, J.R.; Cuellar, J. Biodiesel production using 4-dodecylbenzenesulfonic acid as catalyst. *Appl. Catal. B Environ.* 2014, *160-161*, 743-756. doi: 10.1016/j.apcatb.2014.06.033.

[55] Canakci, M.; Van Gerpen, J.H. Comparison of Engine Performance and Emissions for Petroleum Diesel Fuel, Yellow Grease Biodiesel, and Soybean Oil Biodiesel. *Transactions of the ASAE* 2003, ***46*(4)**, 937-944. doi: 10.13031/2013.13948.

[56] Grzelak, P.; Zoltowski, A. Environmental assessment of the exploitation of diesel engines powered by biofuels. *Combustion Engines* 2020, ***180*(1)**, 31-35. https://doi.org/10.19206/CE-2020-105.

[57] Singh, D.; Sharma, D.; Soni, S.L.; Sharma, S.; Sharma, P.K.; Jhalani, A. A review on feedstocks, production processes, and yield for different generations of biodiesel. *Fuel* 2020, ***262***, 116553. https://doi.org/10.1016/j.fuel.2019.116553.

[58] Karmakar, B.; Sucharita, S.; Halder, G. Delonix regia heterogeneous catalyzed two-step biodiesel production from Pongamia pinnata oil using methanol and 2-propanol. J. Cleaner Prod. 2020, 255, 120313. https://doi.org/10.1016/j.jclepro.2020.120313.

[59] Basu, R. Evaluation of some renewable energy technologies. *Min. Miner. Depos.* 2017, ***11*(4)**, 29-37. https://doi.org/10.15407/mining11.04.029.

[60] Pascoal, C.V.P.; Oliveira, A.L.L.; Figueiredo, D.D.; Assunção, J.C.C. Optimization and kinetic study of ultrasonic-mediated in situ transesterification for biodiesel production from the almonds of Syagrus cearensis. *Renewable Energy* 2020, ***147*(P1)**, 1815-1824. doi: 10.1016/j.renene.2019.09.122.

[61] Raman, L.A.; Deepanraj, B.; Rajakumar, S.; Sivasubramanian, V. Experimental investigation on performance, combustion and emission analysis of a direct injection diesel engine fuelled with rapeseed oil biodiesel. *Fuel* 2019, ***246***, 69-74. https://doi.org/10.1016/j.fuel.2019.02.106.

[62] Ashokkumar, M. *Ultrasonic Synthesis of Functional Materials*; Springer International Publishing, Cham, Switzerland, 2016, 41-42.

[63] Gole, V.L.; Gogate, P.R. A review on intensification of synthesis of biodiesel from sustainable feed stock using sonochemical reactors. *Chem. Eng. Process.* 2012, ***53***, 1-9. https://doi.org/10.1016/j.cep.2011.12.008.

[64] Pukale, D.D.; Maddikeri, G.L.; Gogate, P.R.; Pandit, A.B.; Pratap, A.P. Ultrasound assisted transesterification of waste cooking oil using heterogeneous solid catalyst. *Ultrason. Sonochem.* 2015, ***22***, 278-286. doi: 10.1016/j.ultsonch.2014.05.020.

[65] Fukuda, H.H.; Kondo, A.; Noda, H. Biodiesel Fuel Production by Transesterification of Oils. *J. Biosci. Bioeng.* 2001, ***92***, 405-416. http://dx.doi.org/10.1016/S1389-1723(01)80288-7.

[66] Anguebes-Franseschi, F.; Abatal, M.; Bassam, A.; Soberanis, M.A.E.; Tzuc, O.M.; Bucio-Galindo, L.; Quiroz, A.V.C.; Ucan, C.A.A.; Ramirez-Elias, M.A. Esterification Optimization of Crude African Palm Olein Using Response Surface Methodology and Heterogeneous Acid Catalysis. *Energies* 2018, ***11*(1)**, 157. https://doi.org/10.3390/en11010157.

[67] Lam, M.K.; Lee, K.T.; Mohamed, A.R. Homogeneous, heterogeneous and enzymatic catalysis for transesterification of high free fatty acid oil (waste cooking oil) to biodiesel: a review. *Biotechnol. Adv.* 2010, ***28*(4)**, 500-518. doi: 10.1016/j.biotechadv.2010.03.002.

[68] Talha, N.S.; Sulaiman, S. Overview of Catalysts in Biodiesel Production. *ARPN J. Eng. Appl. Sci.* 2016, ***11*(1)**, 439-448.

[69] Ferreira, A.B.; Cardoso, A.L.; da Silva, M.J. Tin-Catalyzed Esterification and Transesterification Reactions: A Review. *ISRN Renew. Energy* 2012, *2012*, 42857. doi:10.5402/2012/142857.

[70] Corma, A.; García, H. Lewis Acids: From Conventional Homogeneous to Green Homogeneous and Heterogeneous Catalysis. *Chem. Rev.* 2003, ***103*(11)**, 4307-436. https://doi.org/10.1021/cr030680z.

[71] Hou, X.; Qi, Y.; Qiao, X.; Wang, G.; Qin, Z.; Wang, J. Lewis acid-catalyzed transesterification and esterification of high free fatty acid oil in subcritical methanol. *Korean J. Chem. Eng.* **2007**, ***24*(2)**, 311-313.

[72] Vasudevan, P.T.; Briggs, M. Biodiesel production—current state of the art and challenges. *J. Ind. Microbiol. Biotechnol.* 2008, ***35*(5)**, 421-430. https://doi.org/10.1007/s10295-008-0312-2.

[73] Atadashi, I.M.; Aroua, M.K.; Aziz, A.R.A.; Sulaiman, N.M.N.N. The effects of catalysts in biodiesel production: A review. *J. Ind. Eng. Chem.* 2013, ***19*(1)**, 14-26. https://doi.org/10.1016/j.jiec.2012.07.009.

[74] Demirbas, A. *Biodiesel: A Realistic Fuel Alternative for Diesel Engines*; Springer-Verlag London Limited, London, Great Britain, 2008.

[75] Pastore, C.; Lopez, A.; Mascolo, G. Efficient conversion of brown grease produced by municipal wastewater treatment plant into biofuel using aluminium chloride hexahydrate under very mild conditions. *Bioresour. Technol.* 2014, ***155***, 91-97. doi: 10.1016/j.biortech.2013.12.106.

[76] Soriano, N.U.; Venditti, R.; Argyropoulos, D.S. Biodiesel synthesis via homogeneous Lewis acid-catalyzed transesterification. *Fuel* 2009, ***88*(3)**, 560-565. doi: 10.1016/j.fuel.2008.10.013.

[77] Barbosa, S.L.; Dabdoub, M.J.; Hurtado, G.R.; Klein, S.I.; Baroni, A.C.M.; Cunha, C. Solvent free esterification reactions using Lewis acids in solidphase catalysis. *Appl. Catal. A* 2006, ***313***, 146-150. doi: 10.1016/j.apcata.2006.07.015.

[78] Casas, A.; Ramos, M.J.; Rodríguez, J.F.; Pérez, Á. Tin compounds as Lewis acid catalysts for esterification and transesterification of acid vegetable oils. *Fuel Process. Technol.* 2013, ***106***, 321-325. https://doi.org/10.1016/j.fuproc.2012.08.015.

[79] Sun, H.-B.; Hua, R.; Yin, Y. $ZrOCl_2 \cdot 8H_2O$: An Efficient, Cheap and Reusable Catalyst for the Esterification of Acrylic Acid and Other Carboxylic Acids with Equimolar Amounts of Alcohols Molecules 2006, 11(4), 263-271. https://doi.org/10.3390/11040263.

[80] Cai, Z.; Wang, Y.; Teng, Y.; Chong, K.; Wang, J.; Zhang, J.; Yang, D. A two-step bio-diesel production process from waste cooking oil via recycling crude glycerol esterification catalyzed by alkali catalyst. *Fuel Processing Technology* 2015, ***137***, doi: 10.1016/j.fuproc.2015.04.017.

[81] Takase, M.; Zhang, M.; Feng, W.; Chen, Y.; Zhao, T.; Cobbina, S.J.; Yang, L., Wu, X. Application of zirconia modified with KOH as heterogeneous solid base catalyst to new non-edible oil for biodiesel. *Energy Convers. Manage.* 2014, ***80***, 117-125. doi: 10.1016/j.enconman.2014.01.034.

[82] Lee, S.L.; Wong, Y.C.; Tan, Y.P.; Yew, S.Y. Transesterification of palm oil to biodiesel by using waste obtuse horn shell-derived CaO catalyst. *Energy Convers. Manage.* 2015, ***93***, 282-288. doi: 10.1016/j.enconman.2014.12.067.

[83] Sirisomboonchai, S.; Abuduwayiti, M.; Guan, G.; Samart, C.; Abliz, S.; Hao, X.; Kusakabe, K.; Abudala, A. Biodiesel production from waste cooking oil using calcined scallop shell as catalyst. *Energy Convers. Manage.* 2015, ***95***, 242-247. doi: 10.1016/j.enconman.2015.02.044.

[84] Amirthavalli, V., Warrier, A.R., Gurunathan, B. Chapter 6 - Various methods of biodiesel production and types of catalysts. In *Biofuels and Bioenergy*, 1st ed.; Gurunathan, B.; Sahadevan, R.; Zakaria, Z.A., Eds.; Elsevier, Amsterdam, Netherlands, 2022, pp. 111-132.

[85] Alaei, S.; Haghighi, M.; Rahmanivahid, B.; Shokrani, R.; Naghavi, H. Conventional vs. hybrid methods for dispersion of MgO over magnetic Mg–Fe mixed oxides nanocatalyst in biofuel production from vegetable oil. Renewable Energy 2020, **154**, 1188-1203. doi: 10.1016/j.renene.2020.03.039.

[86] Panchal, B.L.; Deshmukh, S.A.; Sharma, M.R. Production and kinetic transesterifi-cation of biodiesel from yellow grease with dimethyl carbonate using methanesulfonic acid as a catalyst. *Environmental Progress & Sustainable Energy* 2017, ***36*(3)**, 802-807. https://doi.org/10.1002/ep.12559.

[87] Ortiz-Martínez, V.M.; Salar-García, M.J.; Palacios-Nereo, F.J.; Olivares Carrillo, P.; Quesada-Medina, J.; De los Ríos, A.; Hernández-Fernández, F. In-depth study of the transesterification reaction of Pongamia pinnata oil for biodiesel production using cata-lyst-free supercritical methanol process. *The Journal of Supercritical Fluids* 2016, ***113***, 23-30. doi: 10.1016/j.supflu.2016.03.009.

[88] Goembira, F.; Saka, S. Advanced supercritical Methyl acetate method for biodiesel production from Pongamia pinnata oil. *Renewable Energy* 2015, ***83*(C)**, 1245-1249. doi: 10.1016/j.renene.2015.06.022.

[89] Özçelik, E.; Aydogan, H.; Acaroglu, M. (2015). The Effects of Safflower Biodiesel and its Blends with Diesel Fuel on Engine Performance in a Common-Rail Diesel Engine. In Proceedings of the 20th International Academic Conference, Madrid, Spain, 06 October 2015, vol. 1.

[90] Syamsuddin, Y.; Hameed, B.H. Synthesis of glycerol free-fatty acid methyl esters from Jatropha oil over Ca–La mixed-oxide catalyst. Journal of the Taiwan Institute of Chemical Engineers 2015, **58**, 181-188. https://doi.org/10.1016/j.jtice.2015.06.041.

[91] Gaurav, A.; Ng, F.T.T.; Rempel, G.L. A new green process for biodiesel production from waste oils via catalytic distillation using a solid acid catalyst – Modeling, economic and environmental analysis. *Green Energy & Environment* 2016, ***1*(1)**, 62-74. https://doi.org/10.1016/j.gee.2016.05.003.

[92] Duda, K.; Wierzbicki, S.; Mikulski, M. An experimental analysis of performance and exhaust emissions of a CRDI diesel engine operating on mixtures containing mineral and renewable components. *Combustion Engines* 2019, ***179*(4)**, 27-31. https://doi.org/10.19206/CE-2019-404.

[93] Moffitt, L. ISCC redefines brown grease: Correction. 2021. Available online: https://www.argusmedia.com/en/news/2280379-iscc-redefines-brown-grease-correction (accessed on 15 March 2022).

[94] Kolet, M.; Zerbib, D.; Nakonechny, F.; Nisnevitch, M. Production of Biodiesel from Brown Grease. *Catalysts* 2020, ***10*(10)**, 1189. https://doi.org/10.3390/catal10101189.

[95] Tran, N.N.; Tišma, M.; Budžaki, S.; McMurchie, E.J.; Yung, N.; Morales-Gonzalez, O.M.; Hessel, V. Production of Biodiesel from Recycled Grease Trap Waste: A Review. *Industrial & Engineering Chemistry Research* 2021, ***60*(46)**, 16547-16560. doi: 10.1021/acs.iecr.1c02496.

[96] Hums, M.E.; Amin, H.; Tsao, T.C.; Olson, M.S.; Spatari, S.; Cairncross, R.A. Longitudinal Study of Wastewater Greases and Their Potential for the Production of Biofuels. *Energy and Fuels* 2018, ***32*(2)**, 1831-1842. https://doi.org/10.1021/acs.energyfuels.7b03550.

[97] Zheng, L.; Hou, Y.; Li, W.; Yang, S.; Li, Q.; Yu, Z. Exploring the potential of grease from yellow mealworm beetle (Tenebrio molitor) as a novel biodiesel feedstock. *Applied Energy* 2013, ***101***, 618-621. https://doi.org/10.1016/j.apenergy.2012.06.067.

[98] Spiller, R.; Knoshaug, E.P.; Nagle, N.; Dong, T.; Milbrandt, A.; Clippinger, J.; Peterson, D.; Van Wychen, S.; Panczak, B.; Pienkos, P.T. Upgrading brown grease for the production of biofuel intermediates. *Bioresource Technology Reports* 2020, ***9***, 100344. https://doi.org/10.1016/j.biteb.2019.100344.

[99] Fan, X.; Burton, R.; Austic, G. The Enzymatic Conversion of Brown Grease to Biodiesel in a Solvent-free Medium. *Energy Sources, Part A: Recovery, Utilization, and Environmental Effects* 2013, ***35*(18)**, 1779-1786. doi: 10.1080/15567036.2010.531505.

[100] AutoCentrum. Citroen C5 II 1.6 HDi hatchback 2004-2008. Available online: https://www.autocentrum.pl/dane-techniczne/citroen/c5/ii/hatchback/silnik-diesla-1.6-hdi-109km-2004-2008/ (accessed on 15 March 2022).

[101] Dynomet Roller Dynamometer. Available online: http://www.dynomet.dk/en_rollersetdynoforcars.htm (accessed on 1 march 2022).
[102] Elwico Unit. Exhaust Gas Analyzer and Opacimeter. Available online: https://elwico.com/ (accessed on 15 March 2022).
[103] VitesseAuto. Citroen C5 II 1.6 HDi. Available on line: https://vitesse.auto.pl/chiptuning_lodz.php?auto=Citroen_C5_II_1.6%C2%A0HDi%C2%A080kW (accessed on 15 March 2022)
[104] Labeckas, G.; Slavinskas, S. The effect of rapeseed oil methyl ester on direct injection Diesel engine performance and exhaust emissions. *Energy Conversion and Management* 2006, ***47*(13-14)**, 1954-1967. https://doi.org/10.1016/j.enconman.2005.09.003.
[105] Labeckas, G., Slavinskas, S. The effect of diesel fuel blending with rapeseed oil and rme on engine performance and exhaust emissions. *Journal of KONES* 2005, ***12*(1-2)**, 187-194.
[106] Peterson, C.; Reece, D. Emissions Characteristics of Ethyl and Methyl Esters of Rapeseed Oil Compared with Low Sulfur Diesel Control Fuel in a Chassis Dynamometer Test of a Pickup Truck. *Transactions of the ASAE* 1996, ***39*(3)**, 805-816.

Analysis of longitudinal acceleration during rapid minibus braking on various surfaces

D Kurczyński*, P Łagowski, T Stańczyk, E Szumska, A Zuska, P Grabski, M Jaśkiewicz, R Jurecki

Department of Automotive Vehicles and Transportation, Faculty of Mechatronics and Mechanical Engineering, Kielce University of Technology, Kielce, Poland

* e-mail: kdarek@tu.kielce.pl

Abstract. Vehicle acceleration can be used to control vehicle traffic, evaluate driver behaviour and analyse accident causes, among others. The purpose of the testing was to designate longitudinal acceleration in limit conditions during the test vehicle's braking on various surfaces. The paper presents selected results of testing the motion parameters of a Ford Transit during rapid braking. The testing included measurements of the vehicle's acceleration and speed on test sections with asphalt and concrete surfaces. The test runs were conducted on a dry, wet and icy asphalt surface. The testing included determination of the impact of the road surface's condition on the vehicle's longitudinal acceleration. It was demonstrated that a virtually invisible and thin ice layer covering the road surface substantially affected the acceleration during braking, thereby affecting safety. A clean and wet surface caused a negligible change in the tested vehicle's acceleration. The obtained results can be used in accident cause analysis with consideration of the road surface's condition.

Keywords – vehicle dynamics, hard braking, vehicle

1. Introduction

Vehicle braking manoeuvres in road traffic are very important in terms of the possibility of collision or accident occurrence. They affect safety in a substantial manner. Their unexpected course can lead to the loss of life or health and property damage, including negative environmental effects. This causes the braking process to remain the subject of many studies and analyses.

The course of the braking process depends on many factors. The factors can be classified in three groups. The first group includes vehicle-related factors. These mainly include the vehicle type and braking system design. A driver should be aware of the vehicle's capabilities deriving from its weight, size, centre of gravity, braking system design, technical condition and long-term operation capabilities, tyre type and condition, vehicle control systems, including ASR and ABS, and other assistance systems. This group also includes ergonomic factors aimed at providing the driver with the ability to adapt the workplace and optimise the working environment to ensure comfort.

The second group includes factors related to the environment in which the vehicle is driven. The group includes the surface type and technical condition, whether the surface is dry or wet, icy, covered with snow, sand or other contaminants (e.g. mud, leaves, engine oil, soil, gravel). Roads are used by various users. These include, for example, agricultural machinery or various construction machinery users, thereby causing the roads to be covered with various contaminants. Ambient temperature, atmospheric precipitation and wind speed are also very important parameters. In negative or near-zero degree temperatures, the driver should take into account that the road can be covered with ice.

The third group of factors affecting the vehicle's braking process includes driver-related factors. The driver makes decisions and is responsible for the braking effects. The most important factors affecting vehicle braking deriving from the driver's action include speed, respecting other road traffic regulations, taking into consideration the traffic conditions and the vehicle's capabilities when making decisions. Taking into account personal capabilities, health and well-being on the given day is also important. A good driver is able to conduct self-assessment and not give in to emotions when driving. The driver must be aware that his or her mental and physical capabilities change substantially with age. All three factor groups are interrelated and affect the vehicle's motion and braking, thereby affecting its safety.

Vehicle braking and acceleration are often subjects of scientific publications. Road conditions substantially affect vehicle motion safety. Braking parameters are being tested in various atmospheric conditions. The aim of the testing is to designate the impact of these conditions on the deceleration, braking distance and tyre grip. Braking parameters designated in various conditions provide information on the traction properties of various vehicles on various surfaces and in different conditions and are useful in accident cause analysis. Waluś K.J. tested the acceleration and braking of a passenger vehicle equipped with winter tyres on fresh snow [1]. Subsequent test series were conducted on different days at various weather conditions. It was pointed out that winter conditions feature a considerable dispersion of the obtained results even when using the same vehicle, tyres, test location, driver and measurement instruments. Braking in winter conditions was the subject of numerous studies [2, 3, 4, 5].

Paper [6] presents broad results of testing the emergency braking of buses, tractors and commercial vehicles. The results were compared to the results obtained during passenger vehicle testing. The initial braking speed, delay rise time and mean fully developed deceleration. The parameters were designated in order to be used in road accident reconstruction. The authors of paper [7] tested the braking distance and deceleration of a passenger vehicle on various surfaces: dry, wet and contaminated. Tests were also conducted on mixed surfaces. They included driving one side of the vehicle on dry asphalt and the other on asphalt covered with dry sand, wet sand or wet asphalt covered with wet sand. It was demonstrated that braking on mixed surfaces is very dangerous. Mixed surfaces extend the braking distance and can lead to vehicle stability loss. The braking parameters of an unloaded tractor were tested in paper [8]. The designated mean fully developed deceleration (MFDD) was lower from the values achieved by passenger vehicles by approx. 20-30%. Paper [9] presents the results of braking parameters' measurements conducted for an efficient and damaged braking system as well as efficient and inoperable ABS system. The testing was conducted on a wet skid pad. The average mean fully developed deceleration recorded for an efficient braking system and ABS amounted to 4 m/s^2, while the average braking distance amounted to 24.3 m. In tests conducted on the same surface and in the same conditions, the average mean fully developed deceleration recorded for an efficient braking system and inoperable ABS amounted to 2.33 m/s^2, while the average braking distance amounted to 41.6 m. The lack of an efficient ABS resulted in a substantial extension of the braking distance.

Kordani et al. conducted simulation tests of the impact of weather-dependent friction on the braking process. The testing was conducted for three car models: sedan passenger car, commercial vehicle and bus. The tests were carried out with the use of the Adams/car software [10]. The authors demonstrated that the friction of 0.9, 0.8, 0.7 and 0.6 achieved in the simulation show no substantial differences in braking distance. They can be assigned to weather conditions featuring no atmospheric precipitation. The authors determined that the friction of 0.5, 0.4, 0.28 and 0.18 can be assigned to wet, rainy, snowy and icy weather conditions, respectively.

Pokorski J et al proposed the method of measuring the tyre grip by using the SRT-4 skid resistance tester consisting of a special dynamometric trailer, tractor and measurement instrumentation [11]. The speed characteristics of the tyre grip and grip characteristics in the wheel skid function were designated for various tyre types and in different road conditions. The authors demonstrated differences in the designated characteristics depending on atmospheric conditions and tyre type.

Road traffic conditions change and are often unpredictable for the driver. Road safety can be improved by notifying the driver about the road's slipperiness. This would enable the driver to adapt his or her behaviour (speed, distance between cars) to the road conditions. Jang J. proposed the possibility of road slipperiness detection based on wheel skid and wheel acceleration [12]. He pointed out that digital tachograph sensor data can be used for this purpose.

The paper's authors tested the braking process on different surfaces and with different technical conditions of these surfaces. The testing was conducted, among others, on asphalt covered with a virtually invisible, thin layer of black ice. The tests were carried out on a minibus. It was demonstrated that the surface's technical condition affected the braking parameters.

2. Testing methodology

The testing of motion parameters during intense braking were conducted on test sections (test tracks) at the Kielce University of Technology's Car and Tractor Laboratory and the Jastrząb Autodrome near Radom. The tests were carried out on test sections with asphalt and concrete surfaces. The testing was conducted on dry and wet asphalt and concrete surfaces. Furthermore, the asphalt surface testing was conducted when the surface was covered with a thin, invisible layer of ice (so-called black ice).

The test subject was a sixth-generation Ford Transit intended to transport 9 passengers, with a curb weight of 2,070 kg. The car was equipped with a compression ignition engine, cubic capacity of 2,198 cm^3 and rated power of 92 kW. The vehicle featured traction improvement systems (ABS, ASR, ESP) which were enabled during testing. Such vehicles are often referred to as minibuses and can be used as delivery trucks after adequate interior adaptation. Figure 1 presents the tested vehicle.

Figure 1. Ford Transit during testing.

The vehicle's technical condition, especially the tyres and braking system elements were checked prior to testing. During the tests, the vehicle was loaded with the testing instrumentation and two people: the driver and the person operating the instrumentation (approx. 180 kg). The tested vehicle was equipped with summer Continental ContiVanContact 200, 205/65 R16C tyres with very little wear. The tread depth demonstrated wear of approx. 1 mm when compared to the nominal value.

Rapid (emergency) braking cycles were conducted on the test sections in each test series. The driver was accelerating the vehicle to approx. 50 km/h and then pressed the brake pedal sharply as in the case of an impending collision.

The measurements were used to develop plots for vehicle acceleration and speed during braking. An analysis of the plots allowed for the designation of the initial braking speed V_o and mean fully developed deceleration (MFDD). The mean fully developed deceleration (MFDD) was calculated as the average measured acceleration for the tested vehicle's speed range from 0.8 of the initial braking speed V_o to 0.1 of the initial braking speed V_o.

3. Testing instrumentation

The rapid braking parameters were measured by using specialist measurement instrumentation including the following elements:

IOP Conf. Series: Materials Science and Engineering 1247 (2022) 012014 doi:10.1088/1757-899X/1247/1/012014

- Correvit S-350 Aqua optoelectronic longitudinal speed sensor [13],
- uEEP-12 Datron® data acquisition station with the ARMS® data analysis software [14],
- TAA three-axis linear acceleration sensor [15].

4. Test results and their analysis

The tests conducted during rapid braking allowed for obtaining acceleration over time for the Ford Transit on various surface types and surface conditions. Figure 2 presents acceleration over time during emergency braking tests conducted for three asphalt surfaces: dry, wet and covered with a thin ice layer.

Figure 2.a presents acceleration over time during emergency braking on a dry asphalt surface, while Figure 2.b presents acceleration over time during emergency braking on wet asphalt. No substantial differences in the obtained maximum accelerations can be seen when comparing the acceleration values from both plots. The acceleration presented in Figure 2.b demonstrated greater amplitude of changes in the acceleration recorded during fully developed braking. Two instances of acceleration over time feature greater rapid acceleration changes in the second part of the fully developed braking. The difference in the acceleration over time instances when comparing the wet and dry surfaces is most probably an effect of the operation of the ABS system which provides protection against wheel locking and affects the measured acceleration values. Figure 2.c presents the acceleration over time measured on asphalt covered with a thin ice layer. The acceleration values obtained during fully developed braking are considerably smaller in this instance. These acceleration over times instances demonstrate substantial changes in the recorded acceleration values at the start and, in the case of some plots, end of braking. At the start of braking, the changes are caused by the ABS system that prevents the wheels from locking and uncontrolled vehicle skidding. The amplitudes of changes in the measured acceleration are substantially smaller in the latter part of braking. Some plots feature acceleration increases at the end of braking. In one instance, the recorded acceleration at the end of braking is substantially higher than the value recorded during fully developed braking.

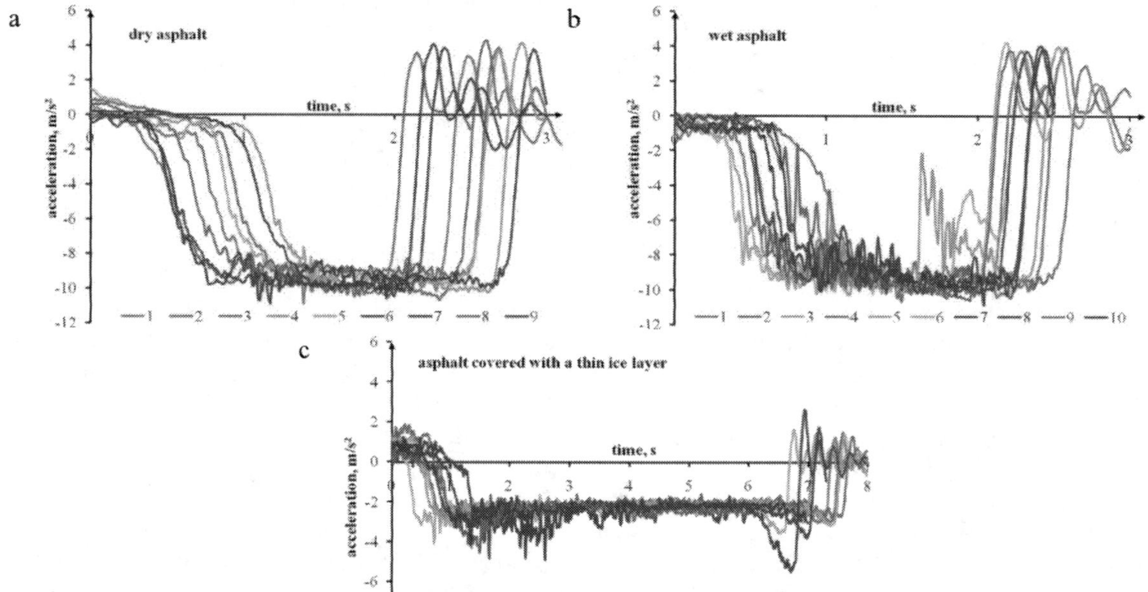

Figure 2. Ford Transit's longitudinal acceleration over time during rapid braking on asphalt surfaces: a) dry, b) wet, c) covered with a thin ice layer.

The initial braking speed V_o and mean fully developed deceleration for particular rapid braking attempts on dry, wet and icy asphalt are presented in Table 1. The table also features mean initial braking speeds V_{om} and average mean fully developed deceleration $MFDD_m$ during rapid braking attempts on asphalt with various surface conditions. The calculations also included the standard deviation for the designated V_o and the MFDD on their average value. The $MFDD_m$ for wet asphalt is slightly lower than the $MFDD_m$ for dry asphalt. On the other hand, the $MFDD_m$ for asphalt covered with a thin ice layer amounted to 2.27 m/s^2 and is substantially lower than the $MFDD_m$ for dry asphalt, which amounted to 9.51 m/s^2. Paper [4] specifies the MFDD on a surface covered with a thick black ice layer of 1.18 to 2.55 m/s^2. On the other hand, paper [16] presents the MFDD on an ice-covered surface for various tyres as well as with the ABS system enabled and disabled. In the case of summer tyres, the MFDD with the ABS system enabled amounted to 1.96–1.28 m/s^2, whereas with the ABS system disabled, the MFDD amounted to 1.57–1.47 m/s^2.

Table 1. Ford Transit's motion parameters during rapid braking on dry, wet and icy asphalt.

asphalt	Test run no.	V_o km/h	V_{om} km/h	SD	MFDD m/s^2	$MFDD_m$ m/s^2	SD
	1	49.1			9.66		
	2	52.2			9.94		
	3	48.56			9.63		
	4	49.69			9.59		
dry	5	52.81	52.10	2.52	9.31	9.51	0.31
	6	54.03			9.72		
	7	52.88			9.54		
	8	56.25			8.87		
	9	53.35			9.32		
	1	52.37			9.05		
	2	52.05			9.38		
	3	51.59			9.32		
	4	51.52			8.03		
wet	5	53.15	50.50	2.21	9.47	9.16	0.45
	6	49.66			9.27		
	7	49.22			9.33		
	8	46.22			9.54		
	9	48.73			9.07		
	1	52.18			2.20		
	2	52.67			2.17		
	3	51.11			2.32		
	4	53.79			2.07		
	5	52.57			2.18		
icy	6	54.62	52.68	1.239	2.23	2.27	0.138
	7	51.84			2.23		
	8	54.21			2.54		
	9	50.94			2.42		
	10	52.82			2.36		

Figure 3 presents acceleration over time during rapid braking on a dry and wet concrete surface. The acceleration measured during braking features substantial change amplitudes for both dry and wet concrete surfaces. Acceleration over time during fully developed braking on wet asphalt has an upward trend during this process.

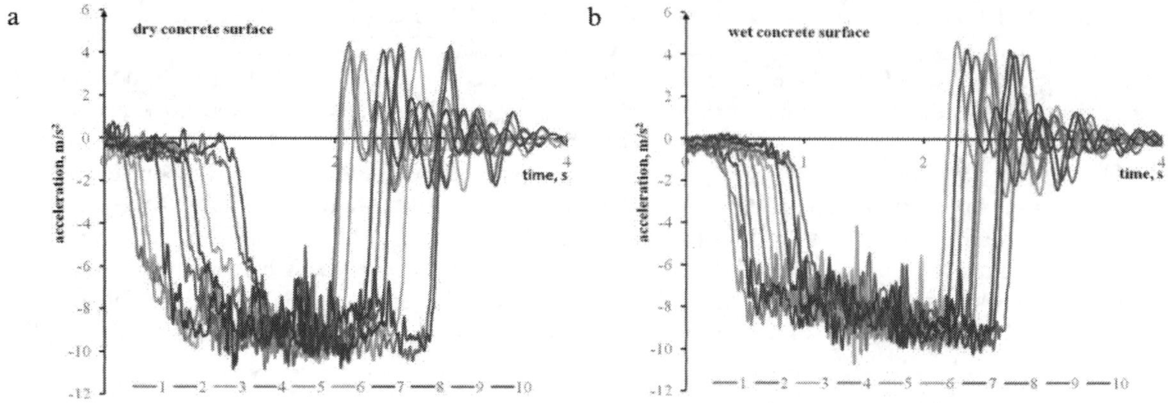

Figure 3. Ford Transit's longitudinal acceleration over time during rapid braking on concrete surfaces: a) dry, b) wet.

Table 2 presents the initial braking speed V_o and mean fully developed deceleration MFDD for particular rapid braking attempts on dry and wet concrete. Other presented values include the mean initial speed V_{om}, $MFDD_m$ as well as standard deviations of the determined values from their average values. The $MFDD_m$ designated for emergency braking amounted to 9.05 m/s^2 on dry concrete and 8.46 m/s^2 on wet concrete.

Table 2. Ford Transit's motion parameters during rapid braking on dry and wet concrete.

concrete	Test run no.	V_o km/h	V_{om} km/h	SD	MFDD m/s^2	$MFDD_m$ m/s^2	SD
dry	1	49.94	51.06	1.75	9.20	9.05	0.19
	2	50.96			8.96		
	3	50.62			9.31		
	4	52.52			9.19		
	5	54.21			9.16		
	6	49.74			9.10		
	7	48.54			8.77		
	8	49.45			9.17		
	9	52.04			8.81		
	10	52.60			8.83		
wet	1	52.43	50.33	1.28	8.20	8.46	0.20
	2	50.62			8.47		
	3	52.01			8.66		
	4	48.84			8.18		
	5	50.72			8.46		
	6	50.28			8.48		
	7	49.74			8.57		
	8	50.33			8.85		
	9	48.15			8.43		
	10	50.16			8.32		

Figure 4 presents a comparison of Ford Transit's average mean fully developed deceleration $MFDD_m$ designated during emergency braking on dry, wet and icy asphalt surfaces as well as dry and wet concrete surfaces. It is clear that the $MFDD_m$ recorded for the ice-covered surface is substantially smaller than in the case of other surfaces.

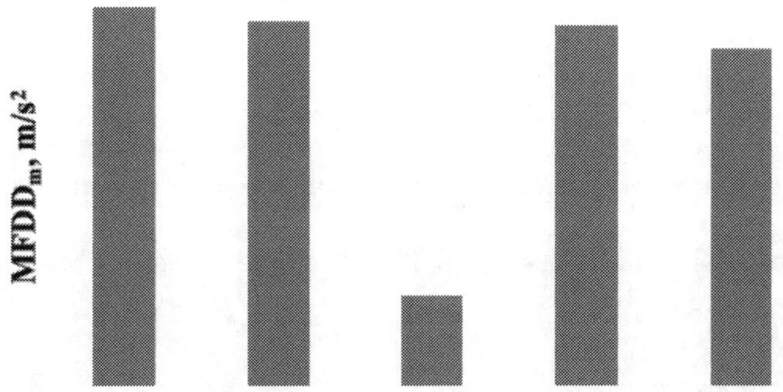

Figure 4. Comparison of Ford Transit's average mean fully developed deceleration $MFDD_m$ designated during emergency braking on various surfaces.

5. Conclusions

Longitudinal acceleration during braking is very important in terms of road traffic safety. It depends on many factors. It is difficult to ensure repeatable braking conditions. The driver's actions are therefore most important for safety. Modern cars are equipped with systems intended to protect the driver and other road users against their own errors.

The parameters of the braking process were tested for various road surfaces and an initial speed of the test car of about 50 km/h. The designated average mean fully developed deceleration $MFDD_m$ for the dry and wet asphalt and concrete surfaces do not differ substantially during emergency braking. The $MFDD_m$ for dry and wet asphalt is slightly higher than the values obtained on concrete. The $MFDD_m$ amounted to 9.51 m.s^2 on dry asphalt and 9.05 m/s^2 for dry concrete. The $MFDD_m$ amounted to 9.16 m/s^2 on wet asphalt and 8.46 m/s^2 on wet concrete. The least favourable $MFDD_m$ value was obtained on asphalt covered with a thin ice layer. It amounted to 2.27 m/s^2. The braking distance is substantially longer for such a $MFDD_m$.

The $MFDD_m$ value determined on the wet asphalt was only 3.7% lower compared to the dry asphalt. The value of $MFDD_m$ on icy asphalt decreased by 76.1% compared to dry asphalt. In contrast, the value of $MFDD_m$ on icy asphalt decreased by 75.2% compared to wet asphalt. An invisible thin layer of ice on the road surface poses a serious threat to road safety.

The acceleration designated as part of the conducted testing can be used to assess the driver's behaviour and in accident cause analysis [17,18]. The acceleration values presented are limit values for the tested vehicle type. A driver who frequently obtains similar acceleration values in normal road traffic may be considered to be driving dangerously.

Acknowledgements
The research was carried out as part of the Innovative system research project supporting the motor vehicle insurance risk assessment dedicated to UBI (Usage Based Insurance) No. POIR.04.01.04 00 0004/19 00 financed by the National Centre for Research and Development.

References

[1] Waluś K J 2016 The intensity of the acceleration and deceleration of a passenger car on a road surface covered with fresh snow. *Procedia Engineering* **136**, pp 187-92.

[2] Ella S, Formagne P-Y, Koutsos V and Blackford J R 2013 Investigation of rubber friction on snow for tyres. *Tribology International* **59** pp 292-301

[3] Koglbauer I, Holzinger J, Eichberger A and Lex C 2018 Autonomous emergency braking systems adapted to snowy road conditions improve drivers' perceived safety and trust. *Traffic Injury Prevention* **19** pp 332-37

[4] Martin D P and Schaefer G F 1996 Tire-road friction in winter conditions for accident reconstruction. *SAE Technical Paper* **60657**

[5] Klein-Paste A and Sinha N K 2010 Comparison between rubber–ice and sand–ice friction and the effect of loose snow contamination. *Tribology International* **43** pp 1145-50

[6] Janczur R and Zawałeń J 2021 Braking parameters of trucks and buses in the reconstruction and analysis of road accidents. XVII Conference Problems of Reconstruction of Road Accidents (Zakopane) pp 105-20

[7] Szumska E, Młodzińska D and Jurecki R 2014 Vehicle braking effectiveness in different road conditions. *Logistyka* **6** pp 10430-39

[8] Waluś K J 2013 Experimental study of intensive deceleration truck unit in sub-zero temperatures (reconnaissance study). *Autobusy - Technika, Eksploatacja, Systemy transportowe* **3** pp 2097-102

[9] Witkiewicz M, Owczarz M and Taryma S 2016 Comparative analysis of the measured value of deceleration of a car with an efficient and damaged brake system. *Autobusy- Technika, Eksploatacja, Systemy transportowe* **8** pp 154-57

[10] Kordani A A, Rahmani O, Nasiri A S A and Boroomandrad S M 2018 Effect of Adverse Weather Conditions on Vehicle Braking Distance of Highways. *Civil Engineering Journal* **4** pp 46-57

[11] Pokorski J, Sar H and Reński A 2019 Influence of exploitation conditions on anti-skid properties of tyres. *Transport* **34** pp 415–24

[12] Jang J 2021 Pavement slipperiness detection using wheel speed and acceleration sensor data. *Transportation Research Interdisciplinary Perspectives* **11** pp 100431

[13] Correvit® S-350 Non-Contact Optical Sensors, Type CS350A... Kistler Patent No. DE 43 13 497 C2 https://www.kistler.com/files/document/000-807e.pdf

[14] µEEP-12 Data Acquisition and Evaluation Typ CMEP1A... Kistler. https://www.techtarget.com.br/site/wp-content/uploads/2016/09/CMEP12A_000-819e.pdf

[15] Acceleration modules SAA (1-axis), DAA (2-axis), TAA (3-axis), Type KCD15760, KCD15951, KCD15911. Kistler https://www.kistler.com/files/document/000-913e.pdf?callee=frontend

[16] Eddie R 1994 Ice, ABS, and Temperature. *SAE Technical Paper* **940724**

[17] Prochowski L, Gidlewsk, M, Dabrowski F, 2018, Analysis of changes in the accident hazard for the elderly in road traffic compared to demographic changes and possibilities of supporting the elderly mobility. *XI International Science-Technical Conference AUTOMOTIVE SAFETY*, 2018 pp. 1–8

[18] Prochowski L, Gidlewski M, Ziubiński M, Dziewiecki K 2021 Kinematics of the motorcar body side deformation process during front-to-side vehicle collision and the emergence of a hazard to car occupants. *Meccanica*, **56(4)** pp. 901–922

IOP Conf. Series: Materials Science and Engineering 1247 (2022) 012015 doi:10.1088/1757-899X/1247/1/012015

Tests of lateral acceleration during a controlled skid - various types of vehicles

D. Frej[1*], M. Jaśkiewicz[1], R. Jurecki[1], D. Kurczyński[1], P. Łagowski[1], T.L. Stańczyk[1], A. Zuska[1], E.M. Szumska[1], P. Grabski[1]

[1] Kielce University of Technology, Aleja Tysiąclecia Państwa Polskiego 7, 25-314 Kielce, Poland
* Correspondence: dfrej@tu.kielce.pl

Abstract: This article presents the values of lateral acceleration during controlled slip for a passenger vehicle and a city bus. Skidding is one of the most dangerous road situations that can occur while driving a vehicle. They can happen to anyone, without exception whether someone has experience or is a young driver. The lateral accelerations during the controlled skidding presented in the article showed higher values for a passenger vehicle. The article refers to drawing attention to the lateral accelerations acting on the vehicle during the slip. It depends on the driving system of the driver whether he will be able to make the correct choice in a dangerous situation. Therefore, improving driving techniques, especially in a skidding situation, should be generally available to drivers.
Keywords: road accidents, controlled skid, automotive safety

1. Introduction

One of the scariest things that can happen while driving is losing control of your vehicle due to skidding. Skidding is the result of a loss of grip between the vehicle tires and the road surface [1-4]. This is usually because the driver accelerates too rapidly, brakes too hard, changes direction abruptly at too high a speed. Front wheel skid occurs when a vehicle's front wheels lose traction. Front-wheel drive vehicles (front-wheel drive) suffer from this problem more often than rear-wheel drive vehicle [5-8].

This is mainly due to the fact that the front wheels drive the vehicle and require more grip than the rear tires [6-7]. Front wheel slippage usually occurs when braking hard, when the front wheels lock up, or when trying to corner too fast for road conditions. Front wheel skid is also known as understeer because the vehicle is going straight no matter how much pressure on the steering wheel [9-11]. Rear wheel skid occurs when the rear wheels lose traction. It happens either during heavy braking, especially in the case of failure of the rear brakes, or more frequently, when the corner is taken too quickly [10-11]. In this respect, vehicles with rear wheels suffer more often from slippage of the rear wheels than vehicles with front-wheel drive because the rear wheels drive the vehicle and thus place additional demands on the tires [12-13].

It should be noted that one of the main causes of vehicles accidents is the speed not being adjusted to the road conditions [14-16]. This means that slippery ground or poor visibility is a serious hazard. Experts argue that sharp turning of the steering wheel will be useless if we skid while driving or we have to avoid an obstacle, eg a pedestrian who suddenly appeared on the road [16-19].

2. Research methodology

The research was carried out on the training and training track located 20 km south of Radom in the Jastrząb commune. The research used a skid-in-run plate (100x20m), equipped with a dynamic plate of the "jerking devices" type. The plate allows the vehicle to skid at speeds above 40 km/h. Figure 1 shows a skid-and-overrun plate with a dynamic plate of the shredder type. When the rear axle of the vehicle travel the plate of the controlled skid station, the hydraulic vehicle axle jerking device will dynamically move the entire plate of the device in the transverse direction to the specified direction of travel. The surface of the plate with a high coefficient of adhesion causes the rear axle of the vehicle to deviate from its previous direction of travel, simultaneously causing it to skid. After reaching the extreme position, the device quickly returns to its initial position. In order to reduce the adhesion of the vehicle to the surface, the skid-run plate was covered with a layer of water throughout the entire duration of the tests.

Figure 1. The skid-on-ramp plate with the dynamic plate of the jerking devices type

The initial speed of the vehicle, for which the horizontal acceleration values were determined, was 40 km/h and 50 km/h. The tests were carried out at an ambient temperature of about 20°C. During a series of tests, the dynamic plate was set to random mode, thanks to which the driver approaching the plate was not able to predict which side the plate would be dislocated. An Audi A6 passenger vehicle and a city bus were used in the research. Each time, a series of 5 measurements was performed. During the tests, the vehicles were loaded with the driver, measuring equipment (about 20 kg) and the person carrying out the measurement. To register the traffic parameters of the tested vehicles, specialized measuring equipment was used, including:

- TAA Datron® triaxial linear acceleration sensor and TANS Datron® linear and angular acceleration
- uEEP-12 Datron® data acquisition stations with ARMS® data analysis software,
- S-350 Aqua Datron® optoelectronic sensor for measuring longitudinal speeds.

The first vehicle to be tested was the Audi A6 passenger vehicle. Its curb mass was 1,920 kg (permissible mass 2,475 kg). The vehicle is equipped with a power unit with a capacity of 2,967 cm^3 and a power of 176 kW. The second research vehicle was a city bus capable of carrying 54 passengers. During the controlled skidding, the vehicles accelerated to the assumed speed, and then ran onto a dynamic plate. Depending on the position of the plate, the rear axle of the vehicle moved to the left or right side. The driver's task was to get the vehicle out of a controlled skid, while keeping the vehicle on a skid plate. Figure 2 shows a city bus during a controlled skid test. On the other hand, in Figure 3 of the Audi A6 passenger vehicles, during a controlled skid test.

IOP Conf. Series: Materials Science and Engineering 1247 (2022) 012015 doi:10.1088/1757-899X/1247/1/012015

Figure 2. City bus during a controlled skid test

Figure 3. Audi A6 passenger vehicle during a controlled skid test

3. Analysis of the results

Figures 4-7 show examples of the transverse velocity of the tested vehicles during a controlled skid. The lateral speed of the Audi A6, during a controlled skid at 50 km/h, recorded the lowest values. The duration of the controlled skidding in the Audi A6 was 5 seconds, while in the city bus it was 9 seconds.

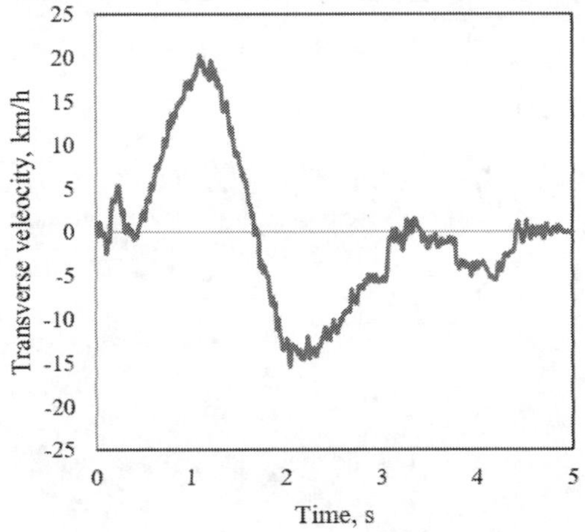

Figure 4. Transverse veleocity of the Audi A6 during a controlled skid from 40 km/h

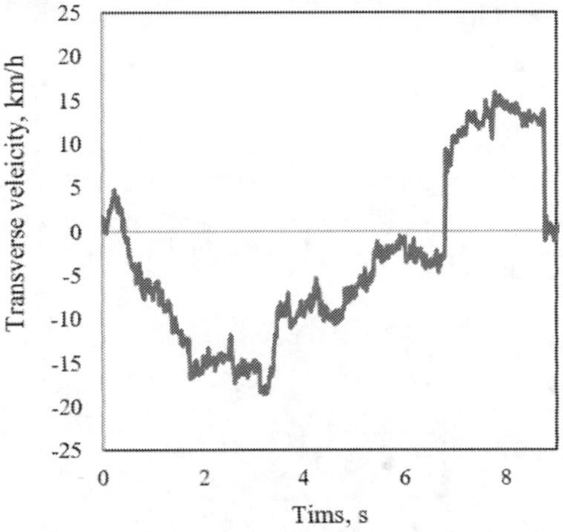

Figure 5. Transverse veleocity of the City bus during a controlled skid from 40 km/h

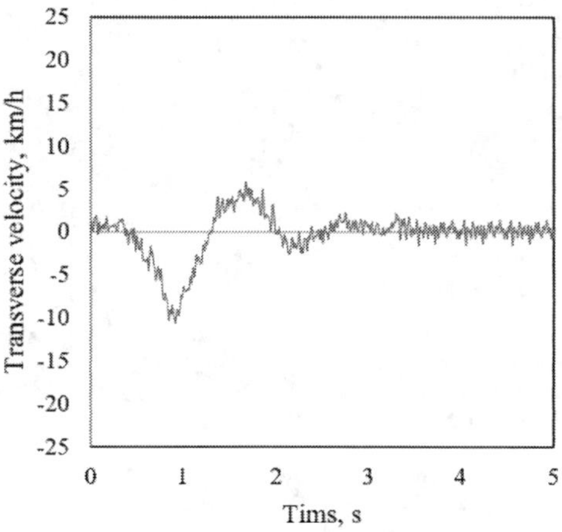

Figure 6. Transverse veleocity of the Audi A6 during a controlled skid from 50 km/h

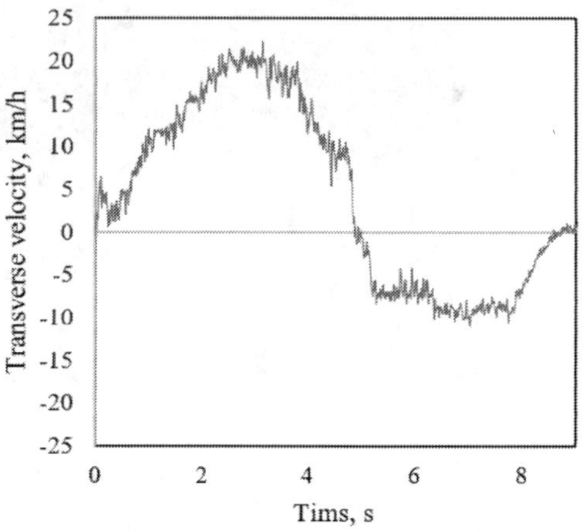

Figure 7. Transverse veleocity of the City bus during a controlled skid from 50 km/h

Figures 8 and 11 show examples of the characteristics of lateral acceleration for the Audi A6. On the other hand, in Figures 16 and 17, examples of the characteristics of lateral acceleration for a city bus.

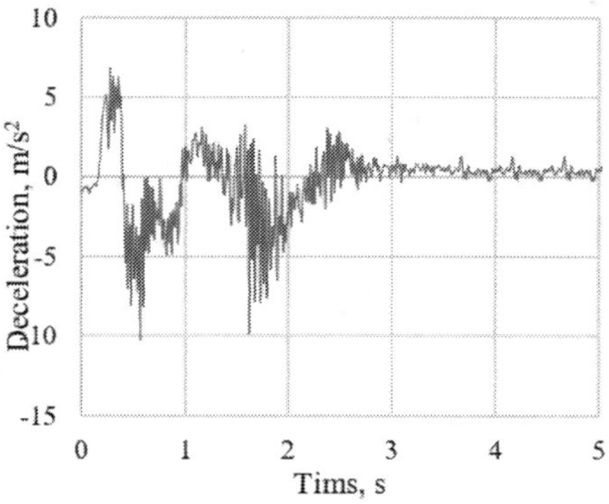

Figure 8. Characteristics of lateral accelerations of the Audi A6 during a controlled skid at a speed of 40 km/h

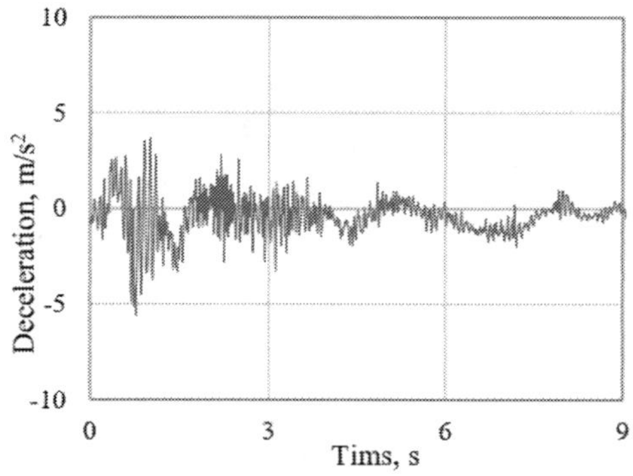

Figure 9. Characteristics of lateral accelerations of the City bus during a controlled skid at a speed of 40 km/h

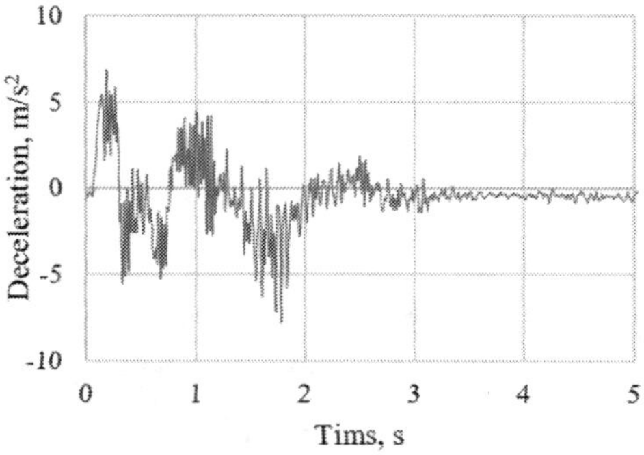

Figure 10. Characteristics of lateral accelerations of the Audi A6 during a controlled skid at a speed of 50 km/h

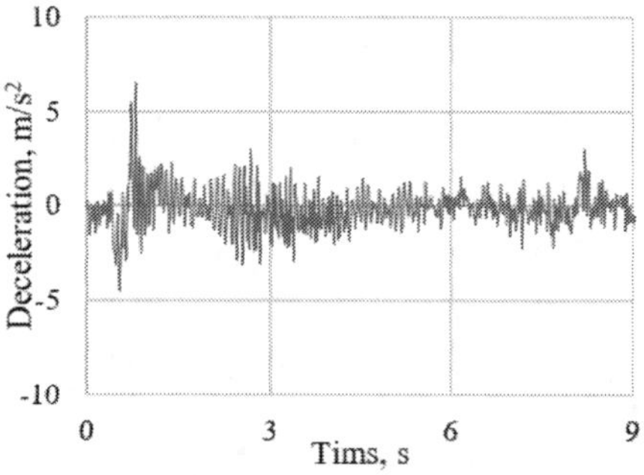

Figure 11. Characteristics of lateral accelerations of the City bus during a controlled skid at a speed of 50 km/h

Table 1 presents the maximum values of deceleration during the controlled skid of the tested vehicles. The highest value of deceleration during a controlled skidding was 11.79 m/s^2 for the Audi A6 vehicles traveling at 40 km/h. In the case of the city bus, the greatest deceleration value was 6.82 m/s^2 during the test at 50 km/h.

Table.1. Deceleration values during a controlled skid from a speed of 40 km/h

lp	Speed 40 km/h Deceleration, m/s^2		Speed 50 km/h Deceleration, m/s^2	
	Audi A6	City bus	Audi A6	City bus
1	11.79	5.61	10.26	6.54
2	11.39	4.98	7.78	5.68
3	10.20	6.18	8.71	5.94
4	9.74	5.39	8.26	6.82
5	10.59	5.51	8.77	6.71
Mean	9.94	5.53	8.76	6.34
SD	1.44	0.39	0.83	0.45

4. Conclusions

Staying alert while driving is critical because being aware of a problem and resolving it in seconds can mean the difference between causing an accident and solving the situation safely. Slippage cannot always be avoided. Every skid is different and involves many factors related to driver activities, road conditions and hazards. The research showed that the Audi A6 passenger vehicle had a much shorter duration of controlled skidding than in the case of the city bus. Skidding can happen to anyone, and the more someone drives, the more likely it is. For our own safety, we should be able to deal with such unexpected behavior of the vehicle, preferably under control conditions. Improving the driving techniques of the driver contributes to increasing the safety of the driver and other road users.

Acknowledgments
The research was carried out as part of the Innovative system research project supporting the motor vehicle insurance risk assessment dedicated to UBI (Usage Based Insurance) No. POIR.04.01.04 00 0004/19 00 financed by the National Centre for Research and Development.

References

[1] Zhao L., Wang H., Yang Y., Sun L., Liu J., **2013**, Dynamic modeling and drift control of high-speed wheel mobile robot, *Control Conference (CCC) 2013 32nd Chinese*, 5827-5832.

[2] Jian Z., Shuang-shuang W., Hua L. Bin L., **2012**, The sliding mode control based on extended state observer for skid steering of 4-wheel-drive electric vehicle, 2012 2nd International Conference on Consumer Electronics, *Communications and Networks (CECNet),* 2195-2200, doi: 10.1109/CECNet.2012.6201445.

[3] Yang Y., Wang X., Zhang Y., **2021**, Study on Mechanism Analysis of Skidding Prediction for Electric Vehicle Based on Time-Delay Effect of Force Transmission. *World Electr. Veh.* 12, 171, https://doi.org/ 10.3390/wevj12040171

[4] Sekour M., Hartani K., Abdelkader M., **2017**, Electric Vehicle Longitudinal Stability Control Based on a New Multimachine Nonlinear Model Predictive Direct Torque Control, *Journal of Advanced Transportation*, 4125384.

[5] Fawzy N., Habib HF., Mokhtari S., **2021**, Performance Evaluation of Electric Vehicle Model under Skid Control Technique. *World Electr. Veh. J*, 12, 83.

[6] Austin L., Morrey D., **2000**, Recent advances in antilock braking systems and traction control systems, *Proc. Inst. Mech.Eng. Part D,* 214, 625–638.

[7] Kang J., Yoo J., Yi K., **2011**, Driving control algorithm for maneuverability, lateral stability, and rollover prevention of 4WD electric vehicles with independently driven front and rear wheels, *IEEE Transactions on Vehicular Technology*, 60, 7, 2987–3001.

[8] Hiraoka T., Nishihara O., Kumamoto H., **2003**, Path Tracking of Four Wheel Steering Vehicle by Using Sliding Mode Control, Transaction of The Institute of System, *Control and Information Engineers*, 16, 10, 520-530.

[9] Fujii K., Fujimoto H., **2007**, Traction Control Based on Slip Ratio Estimation without Detecting Vehicle Speed for Electric Vehicle, *Proceedings of the 2007 Power Conversion Conference*, 688-693.

[10] Martinez MC., Heucke M.,. Wang FY., Gao B., Cao D., **2018**, Driving Style Recognition for Intelligent Vehicle Control and Advanced Driver Assistance: A Survey, in *IEEE Transactions on Intelligent Transportation Systems*, 19, 3, 666-676, doi: 10.1109/TITS.2017.2706978.

[11] Shi B., et al., **2015**, Evaluating Driving Styles by Normalizing Driving Behavior Based on Personalized Driver Modeling, in *IEEE Transactions on Systems, Man, and Cybernetics: Systems*, 45, 12, 1502-1508, doi: 10.1109/TSMC.2015.2417837.

[12] Choi JK.,. Kwon YJ., Kim K., Jeon J., Jang B., **2019**, Driver Behavior Analysis and Warning System for Digital Cockpit Based on Driving Data, *2019 International Conference on Information and Communication Technology Convergence (ICTC)*, 1397-1399, doi: 10.1109/ICTC46691.2019.8939875.

[13] Wang WY., Li IH., Chen MC., Su SF., Hsu SB., **2009**, Dynamic Slip-Ratio Estimation and Control of Antilock Braking Systems Using an Observer-Based Direct Adaptive Fuzzy–Neural Controller, *in IEEE Transactions on Industrial Electronics*, 56, 5, 1746-1756, doi: 10.1109/TIE. 2008.2009439.

[14] Szumska E, Frej D, Grabski P., **2020**, Analysis of the Causes of Vehicle Accidents in Poland in 2009-2019. *LOGI – Scientific Journal on Transport and Logistics*, 11, 2, 76-87. https://doi.org/10.2478/logi-2020-0017

[15] Jurecki R.S., **2020**, Analysis of Road Safety in Poland after Accession to the European Union. *Communications - Scientific Letters of the University of Zilina*, 22, 2, 60-67, https://doi.org/10.26552/com.C.2020.2.60-67

[16] Jaśkiewicz M., Jaskólski J., **2017**, Wypadkowe dane statystyczne na polskich drogach w latach 2001-2016. *Autobusy : technika, eksploatacja, systemy transportowe*, 18, 20–23.

[17] Frej DP, Ludwinek K., **2020**, Analysis of road accidents in 2002–2019 on the example of Poland. *The Archives of Automotive Engineering – Archiwum Motoryzacji*, 89, 3, 5-18. doi:10.14669/AM.VOL89.ART1.

[18] Lozia Z., **2020**, Can anything optimistic be found in the statistics of road accidents in Poland in 1975-2018?, *2020 XII International Science-Technical Conference AUTOMOTIVE SAFETY, 2020*, 1-4, doi: 10.1109/AUTOMOTIVESAFETY47494.2020.9293513.

[19] Prochowski, L., Kochanek, H., Gidlewski, M., Pusty, T., **2018** Analysis of the seasonal and regional variations in the accident hazard in Poland. *WIT Transactions on the Built Environment*, 176, 441–452, doi: 10.2495/UT170381.

[20] Prochowski, L., Gidlewski, M., Dabrowski, F., **2018**, Analysis of changes in the accident hazard for the elderly in road traffic compared to demographic changes and possibilities of supporting the elderly mobility. *XI International Science-Technical Conference AUTOMOTIVE SAFETY*, 2018, pp. 1–8, doi: 10.1109/AUTOSAFE.2018.8373297

Road testing of lateral acceleration in urban bus

E M Szumska*, T L Stańczyk, A Zuska, P Grabski, M Jaśkiewicz, R Jurecki, D Kurczyński, P Łagowski, [1]

[1] Kielce University of Technology, Aleja Tysiąclecia Państwa Polskiego 7, 25-314 Kielce, Poland

* eszumska@tu.kielce.pl

Abstract. Lateral acceleration is one of the parameters used to assess a driver's behaviour and the comfort of urban bus passengers. Because of its purpose, a city bus is a vehicle designed to transport large numbers of people over relatively short distances. The specific driving conditions require constant stopping and accelerating. Lateral acceleration values collected in selected road manoeuvre tests or performance tests conducted in real traffic conditions will determine the lateral vehicle movement dynamics. This paper presents the results of lateral acceleration in a urban bus, obtained during experimental studies in selected road manoeuvres on various surfaces. The surface condition and type do not substantially affect the lateral acceleration when driving on a circular track. During test drives on a circular track, the range of lateral acceleration amounts to 5.56÷7.37 m/s^2. The lateral acceleration obtained during double lane change attempts amounted to 2.84÷4.42 m/s^2. The maximum lateral acceleration values achieved by the urban bus on a dry and wet asphalt surface are similar.

Keywords – lateral acceleration, vehicle, vehicle dynamics

1. Introduction

Urban buses are the main mode of urban passenger transport. A urban bus is a vehicle designed to transport a large number of people over relatively short distances. Buses are mainly driven in urban and suburban traffic conditions. A urban bus is characterized by driving at low speed and stopping often at bus stops. A urban bus driver must drive in accordance with the timetable, try to avoid delays and reach the given stop at the right time. A professional driver spends many hours on driving in various traffic intensity conditions. Dynamic road conditions compel the driver to quickly assess the situation and make specific decisions. Due to their specific occupational features, urban bus drivers spend a lot of time in complex driving conditions that require concentration. The task of a urban bus driver is to ensure a smooth and safe journey for passengers in accordance with the road traffic regulations.

As confirmed by the test results presented in papers [1-4], each driver is characterized by a specific driving behaviour. A driver's behaviour affects many important parameters, such as the passengers' comfort and safety, fuel consumption, the emission of harmful compounds contained in exhaust fumes, or the wear of vehicle components.

A driver's style and behaviour substantially affects the pollutants emitted into the atmosphere. In the case of dynamic and aggressive driving, the vehicle consumes more fuel and generates more exhaust fumes. As confirmed by the test results presented in papers [5-8], the fuel consumption and emission of harmful exhaust fume compounds can be as much as several dozen percent higher in the case of aggressive driving when compared to smooth driving.

A dynamic and aggressive driving style can cause passenger discomfort and lead to dangerous situations. Aside from seats, urban buses include space for standing passengers. Rapidly performed acceleration, braking and turning manoeuvres can lead to situations that endanger the safety of standing or moving passengers [9].

As confirmed by the test results presented in papers [10, 11], the passengers' comfort is a key factor for the selection of a transport mode and its improvement can attract larger numbers of urban transport users. Urban bus comfort studies can be viewed from two perspectives: the first one is related to travel and the other is related to the vehicle itself. Examples of factors related to the passengers' evaluation of comfort with reference to travel can concern subjective feelings during specific manoeuvres, e.g. acceleration, braking or turning. The information comes from the passengers and can be collected using surveys, questionnaires or other instruments by utilizing adequate mobile apps or devices. Examples can be found in papers [12-14]. When considering passenger comfort affected by the vehicle's operation and exploitation, it is necessary to specify such parameters as the vehicle's technical condition, seating ergonomics, noise and temperature in the passenger space. Examples of studies can be found in papers [15-18].

Lateral acceleration is one of the parameters used to assess a driver's behaviour and the comfort of urban bus passengers. This parameter is directly related to the vehicle's lateral motion. A vehicle's lateral acceleration is one of the most important factors used in road and road traffic safety designing. Lateral acceleration is affected by three main factors: the road curve radius, vehicle type and road type. This paper presents the results of lateral acceleration in a urban bus, obtained during experimental tests in selected road manoeuvres. The tests were conducted on suitably prepared measurement sections of a track.

2. Research methodology

The experimental tests included rapid driving manoeuvres on a circular track and double lane changes. The test vehicle was a urban bus with 12 m in length, 2.42 m in width and 3.27 m in height. The vehicle's permissible gross weight amounted to 18,000 kg. During testing, the bus was occupied by the driver and a person operating the measurement instrumentation. The tests were divided into two parts: the first part included driving on a circular track, while the second included double lane change manoeuvres.

The following variables were recorded during the measurements: time, instantaneous speed, instantaneous acceleration. The measurement system installed in the vehicle featured the following:

- Corrsys Datron S-350® Aqua optoelectronic sensor;
- TAA® three-axis accelerometer;
- Datron uEEP12® data acquisition station along with a control tablet and the ARMS® software.

The recording frequency amounted to 100Hz.

The first phase of testing featured the recording of the urban bus parameters when driving on a circular track. The driver's task was to maintain a fixed speed and a fixed steering wheel turn angle. The tests were conducted at a speed subjectively deemed by the driver as safe and not leading to the loss of steerability. The tests were conducted on test tracks with the following surfaces: dry and wet asphalt surfaces as well as dry and wet concrete surfaces. Measurement parameters were recorded after the driver stabilised the speed. Further analyses were carried out on the basis of the following parameters: average absolute values of lateral acceleration from the whole passage and maximum values of lateral acceleration recorded during one passage.

The second phase included double lane change manoeuvres. The test simulates going around an obstacle and returning to the original lane. Table 1 shows the method for establishing the track dimensions.

Table 1. Track dimensions [19]

Section	Dimensions	
	Length	Width
1	15m	1.1B+0,25m
2	L+24m	
3	25m	1.2B+0,25m
4	L+24m	
5	15m	1.1B+0,25m
6	15m	1.1B+0,25m

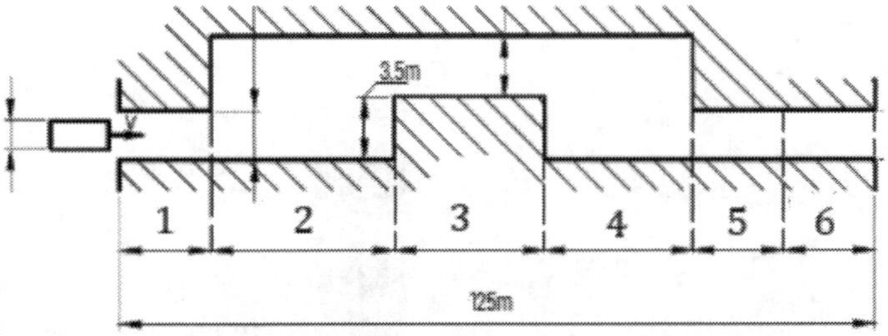

B – vehicle width, L – vehicle length

During the double lane changes, the driver performed the determined manoeuvres at the speed of 50 km/h. The drive speed was selected to ensure safe manoeuvre performance. The tests were conducted on a test track with a dry and wet asphalt surface. Further analyses were conducted for values of lateral acceleration recorded in the 2nd and 4th sections of the track, during obstacle avoidance and return to the previous track. In each trial, the maximum value of lateral acceleration was estimated. These values were used for further statistical analyses.

3. Results

The first phase of testing included driving on a circular track with an asphalt and concrete surface. Figure 1 presents the maximum lateral acceleration values recorded for a urban bus during tests on an asphalt surface.

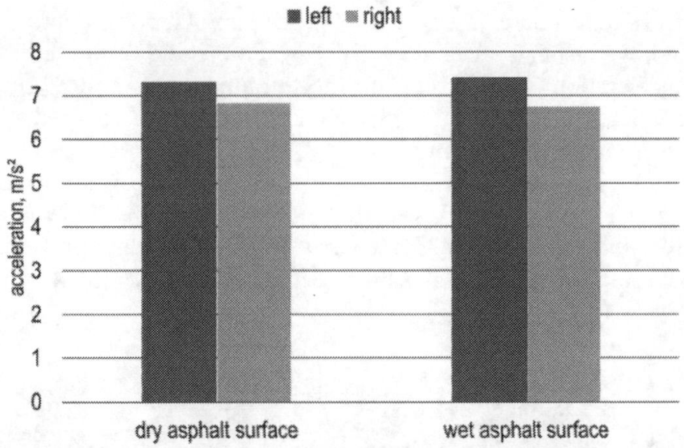

Figure 1. Maximum lateral acceleration values recorded when driving on a circular asphalt track

The average speed recorded when driving on a circular track amounted to approx. 33 km/h. The maximum lateral acceleration when driving to the left amounted to 7.29 m/s^2. The average lateral acceleration recorded when driving to the left on a dry asphalt surface amounted to 5.33 m/s^2. The maximum lateral acceleration when driving to the right amounted to 6.82 m/s^2, while the average acceleration amounted to 5.05 m/s^2. The highest lateral acceleration when driving to the left on a wet asphalt surface amounted to 7.40 m/s^2. The average acceleration in these conditions amounted to 4.94 m/s^2. The maximum lateral acceleration when driving to the right amounted to 6.73 m/s^2, while the average acceleration amounted to 4.57 m/s^2.

IOP Conf. Series: Materials Science and Engineering 1247 (2022) 012016 doi:10.1088/1757-899X/1247/1/012016

Figure 2 presents the maximum lateral acceleration values recorded for a urban bus when driving on a circular concrete track.

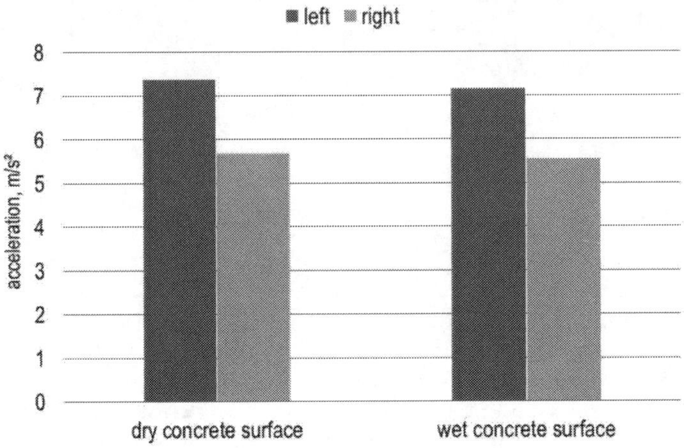

Figure 2. Maximum lateral acceleration values recorded when driving on a circular concrete track

The average speed when driving on a dry concrete surface amounted to 28 km/h. The average lateral acceleration when driving to the left on a dry concrete surface amounted to 5.42 m/s^2. The maximum lateral acceleration when driving to the left amounted to 7.37 m/s^2. The maximum lateral acceleration when driving to the right on a dry concrete surface amounted to 5.68 m/s^2, while the average acceleration amounted to 3.91 m/s^2.
The average lateral acceleration on a wet concrete surface amounted to approx. 24 km/h. The highest lateral acceleration on a wet asphalt surface recorded when driving to the left amounted to 7.15 m/s^2. The average lateral acceleration when driving to the left amounted to 4.69 m/s^2. The average lateral acceleration when driving to the right on a wet concrete surface amounted to 3.39 m/s^2, while the maximum acceleration amounted to 5.56 m/s^2.
The differences in the values of maximum lateral acceleration when traversing a circular track to the left and to the right are a direct result of the driver's behaviour. It was the driver who assessed the driving situation and subjectively adjusted the speed and steering angle.Figure 3 presents the maximum lateral acceleration values achieved by the urban bus on selected surfaces.

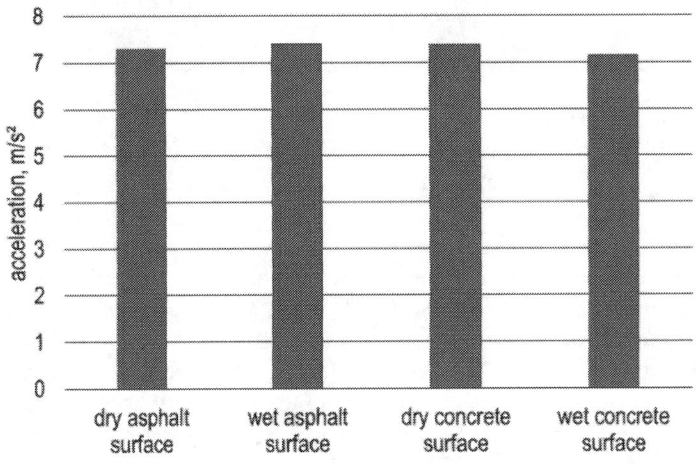

Figure 3. Maximum lateral acceleration values recorded when driving in a circle

As shown in Figure 3, the maximum lateral acceleration values recorded when driving to the left on the analysed surfaces are similar. The highest lateral acceleration recorded on a wet asphalt surface amounted to 7.37 m/s^2. The range of maximum lateral acceleration values when driving in a circle on the analysed surfaces amounted to 7.15÷7.40 m/s^2.
The second part of the testing included double lane change manoeuvres on a dry and a wet asphalt surface. The distribution of the urban bus's lateral acceleration values is presented as box plots in Figure 4.

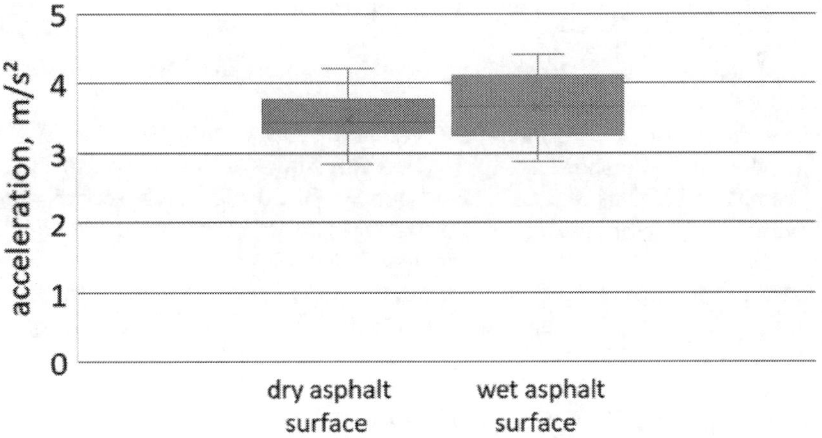

Figure 4. Distribution of lateral acceleration recorded during double lane change attempts

During double lane changes on a dry asphalt surface, the range of lateral acceleration amounted to 2.84÷4.20 m/s^2. The urban bus's average acceleration during the performed manoeuvres amounted to 3.47 m/s^2, whereas the standard deviation amounted to 0.36 m/s^2. The maximum lateral acceleration recorded during double lane changes on a dry surface amounted to 4.20 m/s^2.
The average lateral acceleration achieved by the urban bus on a wet asphalt surface amounted to 3.66 m/s^2. The acceleration's standard deviation during the manoeuvres amounted to 0.43 m/s^2. The lateral acceleration range recorded during double lane changes on a wet asphalt surface amounted to 2.89÷4.42 m/s^2.

4. Conclusion

The paper presents the maximum lateral acceleration values achieved by a urban bus when driving on a circular track and during double lane changes. The testing was conducted on an asphalt and concrete surface. The results of the tests conducted when driving on a circular track demonstrated that the surface condition and type do not substantially affect the lateral acceleration. The highest lateral acceleration recorded during the manoeuvres amounted to 7.37 m/s^2. The lateral acceleration range recorded during double lane changes amounted to 2.84÷4.42 m/s^2. Based on the lateral acceleration values recorded during double lane changes, it can be stated that the maximum lateral acceleration values achieved by the urban bus on a dry and wet asphalt surface are similar.

Acknowledgments
The research was carried out as part of the Innovative system research project supporting the motor vehicle insurance risk assessment dedicated to UBI (Usage Based Insurance) No. POIR.04.01.04 00 0004/19 00 financed by the National Centre for Research and Development.

References

[1] French DJ, West RJ, Elander J, and Wilding JM 1993 Decision-making style, driving style, and self-reported involvement in road traffic accidents *Ergonomics.* **36(6)** 627-644

[2] Puchalski A and Komorksa I 2019 Driving style analysis and driver classification using obd data of a hybrid electric vehicle *Transport Problems.* **15(4)** 83-94

[3] Wang W, Xi J, and Chen H 2014 Modeling and Recognizing Driver Behavior Based on Driving Data: A Survey *Mathematical Problems in Engineering.* **2014** 245641

[4] Lyu N, Wang Y, Wu C, Peng L, and Thomas AF 2022 Using naturalistic driving data to identify driving style based on longitudinal driving operation conditions *Journal of Intelligent and Connected Vehicles.* **5(1)** 17-35

[5] Szumska EM and Jurecki R 2020 The Effect of Aggressive Driving on Vehicle Parameters *Energies.* **13(24)** 6675

[6] Keyvanfar A, Shafaghat A, Muhammad NZ, and Ferwati MS 2018 Driving Behaviour and Sustainable Mobility—Policies and Approaches Revisited *Sustainability.* **10** 1152

[7] Andrzejewski M and Nowak M 2019 The influence of the driving style according to the principles of eco-driving on the emission of toxic substances from a light utility vehicle *AUTOBUSY – Technika, Eksploatacja, Systemy Transportowe.* **24(6)** 33-37

[8] Fuć P, Siedlecki M, Sokolnicka B, and Szymlet N 2017 The influence of the driving style on the exhaust emission from a passenger car with a Euro 5 Diesel engine *Journal of Mechanical and Transport Engineering.* **69(4)** 5-19

[9] Tubis AA, Skupień ET, and Rydlewski M 2021 Method of Assessing Bus Stops Safety Based on Three Groups of Criteria *Sustainability.* **13** 8275

[10] Barabino B, Eboli L, Mazzulla G, Mozzoni S, Murru R, and Pungillo G 2019 An innovative methodology to define the bus comfort level *Transportation Research Procedia.* **41** 461-470

[11] Karekla X and Fang C 2021 Upper body balancing mechanisms and their contribution to increasing bus passenger safety *Safety Science.* **133** 105014

[12] García YD and Aguilar D 2021 Passengers' comfort in horizontal curves on mountain roads: A field study using lateral accelerations *Revista Facultad de Ingeniería Universidad de Antioquia.* **98** 94-103

[13] Eboli L, Mazzulla G, and Pungillo G 2016 Measuring Bus Comfort Levels by using Acceleration Instantaneous Values *Transportation Research Procedia.* **18** 27-34

[14] Zhang Z, Ma T, Ji N, Hu Z, and Zhu W 2019 An assessment of the relationship between driving skills and driving behaviors among Chinese bus drivers *Advances in Mechanical Engineering.* **11(1)** 1-10

[15] Demić M, Lukić J, and Milić Ž 2002 Some aspects of the investigation of random vibration influence on ride comfort *Journal of Sound and Vibration.* **253(1)** 109-128

[16] Wu G, Fan G, and Guo J 2021 Ride comfort evaluation for road vehicle based on rigid-flexible coupling multibody dynamics *Theoretical and Applied Mechanics Letters.* **3(1)** 013004

[17] Kernytskyy I, Yakovenko Y, Horbay, O, Ryviuk M, Humenyuk R, Sholudko Y, Voichyshyn Y, Mazur Ł, Osiński P, Rusakov K, and Koda E 2021 Development of Comfort and Safety Performance of Passenger Seats in Large City Buses *Energies.* **14** 7471

[18] Prochowski, L, Gidlewski M, Ziubiński M and Dziewiecki K 2021 Kinematics of the motorcar body side deformation process during front-to-side vehicle collision and the emergence of a hazard to car occupants. *Meccanica* **56(4)** 901–922.

[19] Pieniążek W, Wolak S 2017 Supporting of forensic expert's opinion by automobile road experimental tests. Some examples *AUTOBUSY – Technika, Eksploatacja, Systemy Transportowe.* **7-8** 202-207

Experimental testing of longitudinal acceleration in urban buses

E M Szumska*, T L Stańczyk, A Zuska, P Grabski, M Jaśkiewicz, R Jurecki, D Kurczyński, P Łagowski [1]

[1] Kielce University of Technology, Aleja Tysiąclecia Państwa Polskiego 7, 25-314 Kielce, Poland

* eszumska@tu.kielce.pl

Abstract. A vehicle's longitudinal acceleration is one the parameters that characterize a vehicle's motion dynamics and used, among others, to assess the comfort of public bus passengers. Rapid acceleration and braking performed by urban bus drivers can be deemed by passengers as uncomfortable and dangerous in some situations, especially for standing passengers. This paper presents an analysis of a public bus's longitudinal acceleration recorded during extreme acceleration and braking attempts. The experimental tests were conducted on dry and wet asphalt and concrete surfaces. The test results allow for the statement that lateral acceleration depends largely on the road's surface and its type. The maximum longitudinal acceleration values during rapid acceleration, depending on the surface, are within the narrow range of 2.18-2.81 m/s^2. During braking, a public bus's minimum longitudinal acceleration was within the range of -5.58÷-8.54 m/s^2.

Keywords – lateral acceleration, vehicle, vehicle dynamics

1. Introduction

Vehicle experimental testing features bench testing and road testing. Due to the nature of the test subject, such testing includes the testing of specific elements, assemblies or complete vehicles. Complete vehicles are tested in road conditions [1,2], whereas specific elements or entire assemblies are tested on test benches [3-5]; however, there are test benches intended for conducting complete tests [6]. The manner in which tests are conducted depends mainly on the tests' objectives. Road testing is intended to test specific vehicle assemblies [7-9].

A complete vehicle's experimental testing can be conducted in terms of safety, technical properties, performance or dynamic properties. Depending on the conditions, vehicle testing can be conducted on public roads (in real traffic conditions), on isolated road sections that enable road testing in specific and repeatable conditions, e.g. on a test track or yard, and on various simulators (e.g. driving simulator, chassis dynamometer).

A vehicle's dynamic properties depend on many factors. They can differ substantially depending on the operating conditions and manner of exploitation, the drive system type and the driver's behaviour. One of the parameters that can be used to determine a car's dynamic properties is longitudinal acceleration. This parameter is directly related to the vehicle's longitudinal motion, i.e. with acceleration and braking performed by the driver. Longitudinal acceleration can be used to analyse and assess many aspects related to a vehicle's correct operation and safety. It is also one of the parameters used in the dynamics and safety testing and analyses conducted for public buses.

A public bus is a vehicle designed to transport a large number of people over relatively short distances, characterized by frequent acceleration and braking. Aside from seats, public buses include space for standing passengers, wide and comfortable aisles between seats and wide entry doors. The dynamic properties of public buses differ substantially from the properties of passenger cars, commercial trucks and even long-distance buses.

The acceleration profile characteristics of public buses can be used to assess the drivers' behaviour. In the case of public buses, the driver's behaviour and driving style are important as they directly affect the passengers' safety. The results presented in papers [10-12] confirmed that each driver views the environment in a unique way and reacts subjectively to the changing traffic conditions. Longitudinal acceleration exceeding 2.0 m/s^2 is deemed by passengers as very uncomfortable [13]. Papers [14,15] present longitudinal acceleration recorded at various speeds and in various traffic conditions, deemed as uncomfortable by sitting, standing and moving passengers. As demonstrated in papers [15-17], longitudinal acceleration exceeding 2.0 m/s^2 is dangerous for standing and moving passengers. Balance is at risk in possible-accident situations and the passengers may not be able to cling to the handrail. Some publications mention the limit value of 1.5 m/s^2, above which situations dangerous for standing passengers can take place in road traffic, e.g. on high-risk intersections [18]. The publication [19] demonstrates that this acceleration value is deemed as the equilibrium for a standing passenger that uses a handrail.

The use of adequate acceleration during possible accident reenactment can be an important aspect related to acceleration measurements during various manoeuvres [].

The purpose of this paper is to demonstrate the extreme longitudinal acceleration ranges for public buses recorded during rapid braking from 50 km/h and rapid acceleration from a standing start to 50 km/h. The experimental testing was conducted on specifically isolated test sections. The presented test results include longitudinal acceleration obtained on dry and wet asphalt and concrete surfaces.

2. Research methodology

The experimental testing included rapid acceleration and braking manoeuvres. The tests were conducted on test tracks with the following surfaces: dry and wet asphalt surfaces as well as dry and wet concrete surfaces. During the acceleration attempts, the driver was tasked with reaching the speed of 50 km/h from a standing start as soon as possible. During the braking attempts, the driver was tasked with stopping the vehicle from a starting speed of 50 km/h as soon as possible. For acceleration tests, the analyses considered the maximum acceleration value recorded during the road test. For braking, the minimum acceleration value recorded during the road test was considered for further analysis. Then, the maximum and minimum acceleration values recorded during the road tests were further analyzed.

The test vehicle was a public bus measuring 12 m in length, 2.42 m in width and 3.27 m in height. The vehicle's permissible gross weight amounted to 18,000 kg. During testing, the bus was occupied by the driver and a person operating the measurement instrumentation. The measurement system installed in the vehicle featured the following:

- Corrsys Datron S-350® Aqua optoelectronic sensor;
- TAA® three-axis accelerometer;
- Datron uEEP12® data acquisition station along with a control tablet and the ARMS® software.

The following variables were recorded during the measurements: time, instantaneous speed, instantaneous acceleration. The recording frequency amounted to 100Hz. The effects of mass and tires on longitudinal acceleration values were not studied.

3. Results

The longitudinal acceleration values collected during the rapid acceleration and braking attempts were subject to statistical analysis. Figure 1 presents the distribution of the longitudinal acceleration values recorded for a public bus during braking attempts on an asphalt surface.

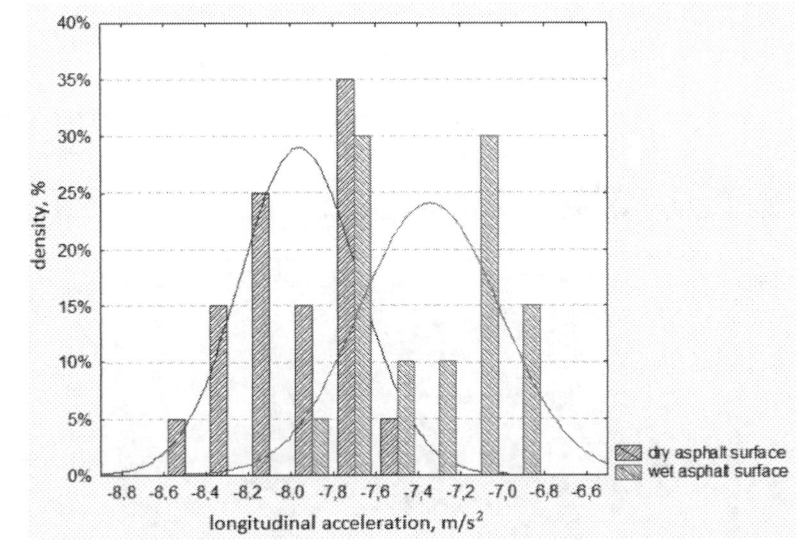

Figure 1. Distribution of minimum longitudinal acceleration during braking attempts on an asphalt surface

The negative longitudinal acceleration obtained during braking attempts was in the range of -8.54÷-7.96 m/s^2. The average acceleration recorded during the braking attempts amounted to -7.96 m/s^2, whereas the standard deviation amounted to 0.27 m/s^2. The average negative acceleration recorded for the public bus on a wet asphalt surface was approx. 8% lower than that recorded on a dry surface and amounted to -7.35 m/s^2. The acceleration's standard deviation during the braking attempts amounted to 0.32 m/s^2. The acceleration recorded during the braking attempts on a wet asphalt surface was in the range of 2.89÷4.42 m/s^2. The standard deviation of acceleration obtained during braking attempts performed for various surfaces was small and amounted to 0.27 m/s^2 and 0.32 m/s^2, respectively, thereby pointing to a high concentration of the obtained results and the measurements' correctness.
Figure 2 presents the distribution of acceleration values recorded during braking attempts on a concrete surface.

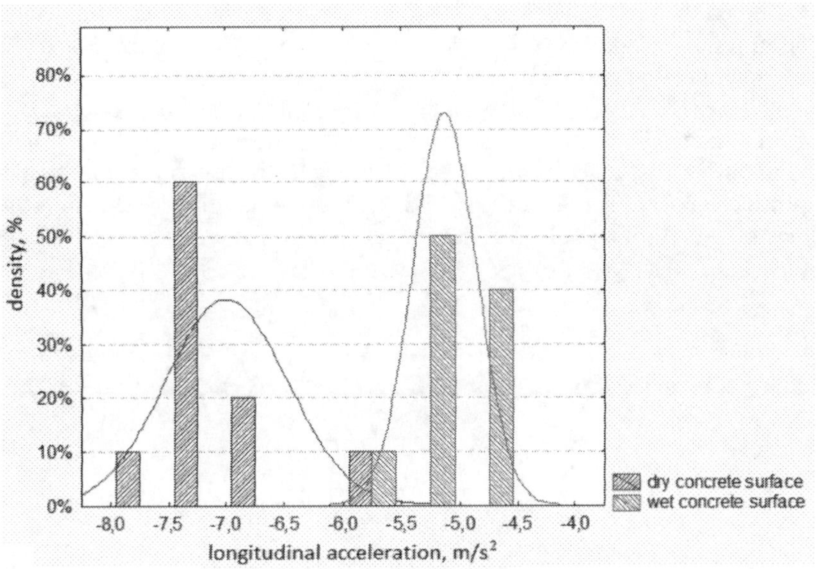

Figure 2. Maximum lateral acceleration values recorded when driving on a circular asphalt track

During the public bus's braking attempts on a dry concrete surface, the delay was in the range of -7.58÷-7.01 m/s^2 and the designated average negative acceleration amounted to -7.01 m/s^2, whereas the delay's standard deviation amounted to 0.49 m/s^2. On a wet concrete surface, the average negative acceleration was smaller by over 26% when compared to the dry surface and amounted to -5.15 m/s^2 with a standard deviation of 0.26 m/s^2. In this case, the acceleration recorded during the braking attempts was in the range of -5.58÷-5.15 m/s^2.
Figure 3 presents the distribution of the minimum longitudinal acceleration values recorded for a public bus during braking attempts on an asphalt and a concrete surface.

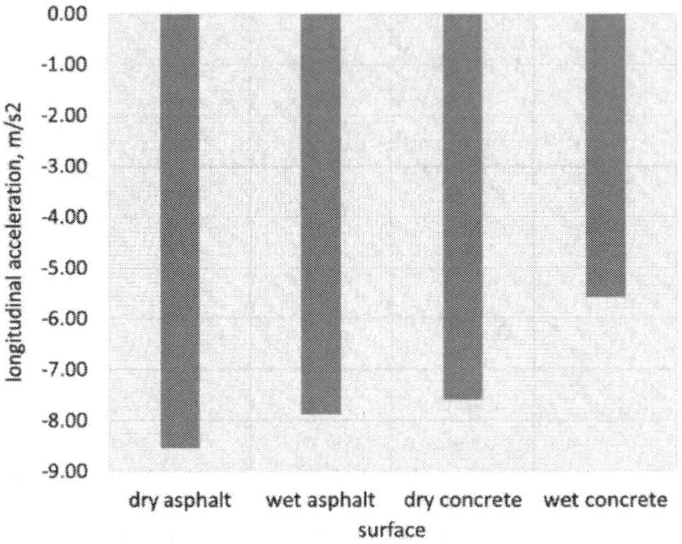

Figure 3. Minimum longitudinal acceleration recorded during braking attempts

The lower of the longitudinal acceleration values (-8.54 m/s^2) during rapid braking attempts was recorded on a dry asphalt surface. On a wet asphalt surface, this value was 8% higher than the minimum acceleration recorded on the dry asphalt surface. Higher longitudinal acceleration values were recorded during braking attempts on a concrete surface. The lowest acceleration was recorded during braking on a wet surface and amounted to 5.58 m/s^2.
The acceleration intensity is a relatively important parameter that can be taken into account when dealing with the comfort and safety of public buses. This parameter determines the possibility of dangerous situations occurring during passenger transport, but also affects the driver's ability to perform defensive manoeuvres (rapid acceleration) in road traffic [20-23]. Figure 4 presents the distribution of longitudinal acceleration values recorded for a public bus during acceleration attempts on an asphalt surface..

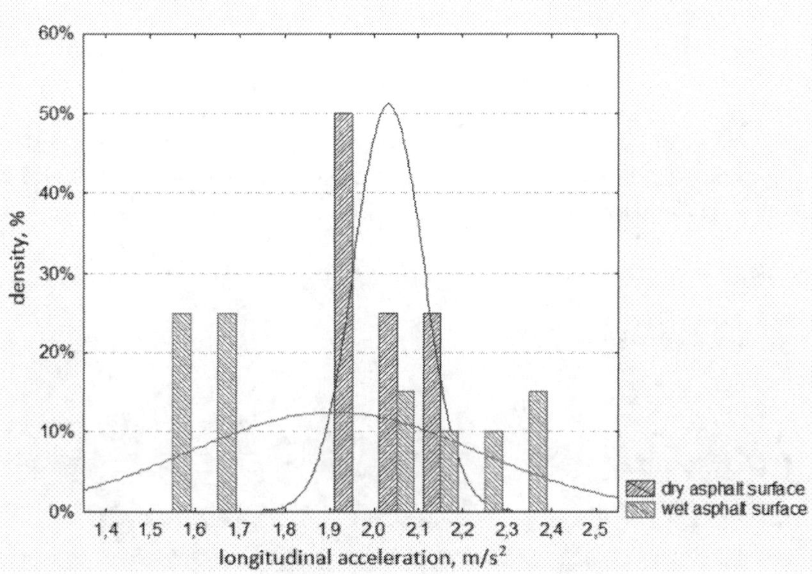

Figure 4. Distribution of longitudinal acceleration during acceleration attempts on an asphalt surface

During attempts to rapidly accelerate with a public bus from 0 km/h to 50 km/h on a dry asphalt surface, the acceleration was in the range of 1.92÷2.18 m/s^2. The average acceleration during the attempts performed on a dry asphalt surface amounted to 2.03 m/s^2, whereas the standard deviation amounted to 0.08 m/s^2. The acceleration recorded on a dry asphalt surface was slightly higher and was in the range of 1.52÷2.33 m/s^2. The acceleration's standard deviation during the attempts amounted to 0.81 m/s^2. The average acceleration on a wet asphalt surface was lower by 5.9% than the value obtained on a dry surface and amounted to 1.91 m/s^2.
Figure 5 presents the distribution of longitudinal acceleration values recorded for a public bus during acceleration attempts on a concrete surface.

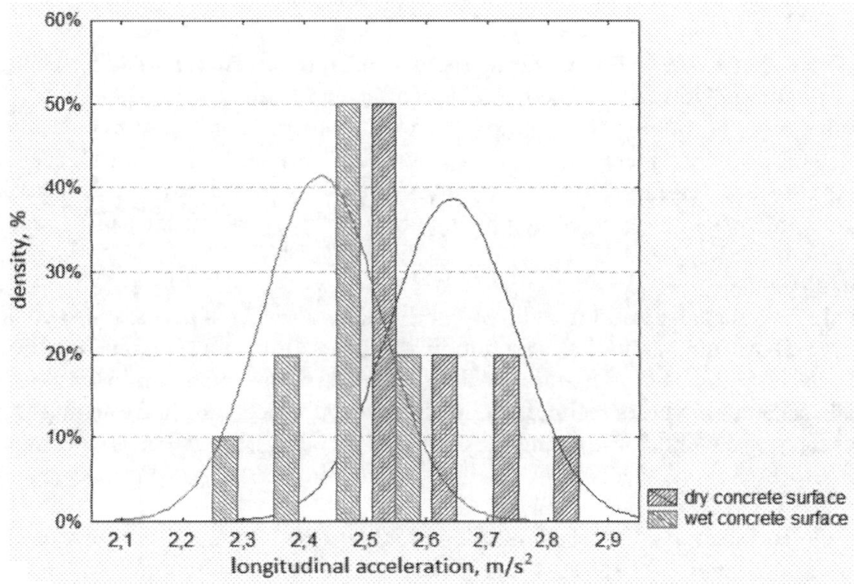

Figure 5. Distribution of longitudinal acceleration during acceleration attempts on a concrete surface

The public bus's acceleration values recorded during rapid acceleration on a dry concrete surface were higher than the ones presented earlier and were in the range of 2.51÷2.81 m/s^2. The average acceleration during the attempts performed on a dry concrete surface amounted to 2.64 m/s^2, whereas the standard deviation amounted to 0.10 m/s^2. The acceleration recorded on a wet concrete surface was in the range of 2.21÷2.55 m/s^2. The acceleration's standard deviation during the attempts performed on a wet concrete surface amounted to 0.09 m/s^2. The average acceleration was lower by 7.9% when compared to the dry surface and amounted to 2.43 m/s^2.
Figure 6 presents the maximum acceleration values recorded for a public bus during rapid acceleration on the various surfaces aforementioned.

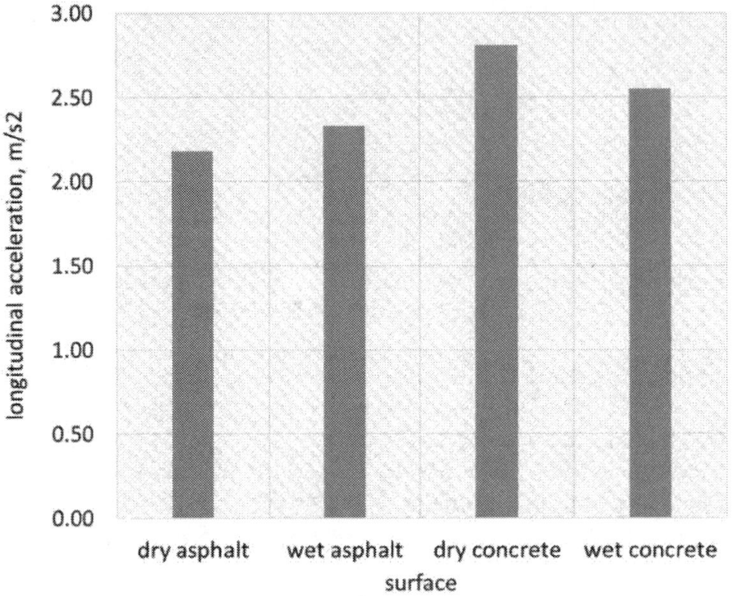

Figure 6. Maximum acceleration values recorded during acceleration attempts

The highest acceleration values obtained during rapid acceleration were recorded on a concrete surface. The highest acceleration here was recorded on a dry surface and amounted to 2.81 m/s^2. The maximum acceleration value recorded during the attempts on a wet concrete surface was lower by 9% than the maximum acceleration value recorded on a dry concrete surface. The lowest acceleration values obtained during the acceleration attempts were recorded on a wet asphalt surface. The highest acceleration recorded during the acceleration attempts on a wet asphalt surface amounted to 2.33 m/s^2.

4. Conclusion

In order to obtain reliable and real longitudinal acceleration values, it is necessary to conduct adequate tests using dedicated instrumentation. The longitudinal acceleration values obtained in the experimental testing provide reliable information about a vehicle's dynamic properties. In the case of public buses, the longitudinal acceleration values recorded during braking and acceleration are an important parameter that determines the passengers' comfort and the ability to perform effective defensive manoeuvres. The results of the performed rapid acceleration attempts demonstrated that the acceleration values depend on the surface's type and condition.
Depending on the surface type, the maximum longitudinal acceleration values recorded during rapid acceleration are in the range of 2.18÷2.81 m/s^2, whereas during rapid braking – 5.58÷-8.54 m/s^2. The obtained acceleration values differing for various roads and measurement conditions demonstrate the need to conduct further related research. As is shown, the measured values are substantially affected by the surface type, and surface types are rather diverse. As demonstrated in the Report on the technical

condition of roads in 2020 in Poland [24], nearly 60% of roads are in good technical conditions but nearly 14% of roads are in a bad condition. This is related to the substantial differences in the friction coefficient between the best and worst road sections, which broadens the perspective for further research.
As can be seen from the presented road research, the type and condition of surface has a significant impact on the values of longitudinal accelerations. Therefore, it is worth to extend the research with the analysis of the influence of vehicle load and different types of tires.

Acknowledgments
The research was carried out as part of the Innovative system research project supporting the motor vehicle insurance risk assessment dedicated to UBI (Usage Based Insurance) No. POIR.04.01.04 00 0004/19 00 financed by the National Centre for Research and Development.

References
[1] Jurecki R and Stańczyk TL 2021 Analysis of Vehicle Moving Parameters in Various Road Conditions *Communications - Scientific Letters of the University of Zilina.* **23(3)** F58-F70
[2] De Gennaro M, Paffumi E, Martini G, Manfredi U, Vianelli S, Ortenzi F, and Genovese A 2015 Experimental Test Campaign on a Battery Electric Vehicle *SAE International Journal of Alternative Powertrains.* **4(1)** 100-114
[3] Bielaczyc P, Szczotka A, Pajdowski P, and Woodburn J 2013 Development of automotive emissions testing equipment and test methods in response to legislative, technical and commercial requirements *Silniki Spalinowe.* **52(1)** 28-41
[4] Kurczyński D 2021 ffect of the RME Biodiesel on the Diesel Engine Fuel Consumption and Emission *Communications - Scientific Letters of the University of Zilina.* **23(4)** B308-B316
[5] Jurecki R, Jaśkiewicz M, and Wdowski T 2014 Testing the influence of car load and pressure in tyres on the value of damping of shock absorbers specified with the use of the Eusama method *Diagnostyka.* **15(3)** 45-50
[6] Šarkan B and Stopka O 2018 Quantification of road vehicle performance parameters under laboratory conditions *Adv. Sci. Technol. Res. J.* **12(3)** 16–23
[7] Frej D, Zuska A, and Szumska EM 2021 Laboratory studies of the influence of the working position of the passenger vehicle air suspension on the vibration comfort of children transported in the child restraint system *Open Engineering.* **11(1)** 470-482
[8] Stańczyk TL, Jurecki RS, Jaśkiewicz M, Zuska A, Szumska E, Grabski P, Frej D, and Łagowski P 2020 Braking performance tests for different types of vehicles Proceedings of 2020 XII International Science-Technical Conference AUTOMOTIVE SAFETY. 1-6
[9] Kim J 2009 Identification of lateral tyre force dynamics using an extended Kalman filter from experimental road test data *Control Engineering Practice.* **17(3)** 357-367
[10] Bokare PS and Maurya AK 2017 Acceleration-Deceleration Behaviour of Various Vehicle Types *Transportation Research Procedia.* **25** 4733-4749
[11] Eboli L, Mazzulla G, and Pungillo G 2016 Combining speed and acceleration to define car users' safe or unsafe driving behaviour *Transportation Research Part C: Emerging Technologies.* **68** 113-125
[12] Vaiana R, Iuele T, Astarita V, Caruso MV, Tassitani A, Zaffino C, and Giofrè VP 2014 Driving Behavior and Traffic Safety: An Acceleration-Based Safety Evaluation Procedure for Smartphones *Modern Applied Science.* **8(1)** 88-96
[13] He Y, Yan X, Wu C, Chu D, and Peng L 2013 Effects of Driver's Unsafe Acceleration Behaviors on Passengers' Comfort for Coach Buses Proceedings of Second International Conference on Transportation Information and Safety ICTIS. 1649 - 1655
[14] Barabino B, Eboli L, Mazzulla G, Mozzoni S, Murru R, and Pungillo G 2019 An innovative methodology to define the bus comfort level *Transportation Research Procedia.* **41** 461-470

[15] Krašna S, Keller A, Linder A, Silvano AP, Xu JC, Thomson R, and Klug C 2021 Human Response to Longitudinal Perturbations of Standing Passengers on Public Transport During Regular Operation *Front. Bioeng. Biotechnol.* **21** 680883
[16] Karekla X 2020 Passenger Safety: The impact of smoother bus acceleration on passenger experience aboard London buses Proceedings of ETC European Transport Conference. 1-14
[17] Karekla X, Gkiotsalitis K, and Tyler N 2020 The impact of a passenger-safety-driven acceleration limit on the operation of a bus service *Accid Anal Prev.* **148** 105790
[18] Olejnik K 2021 Analysis of dangers in the operation of city buses at the intersections *Open Engineering.* **11(1)** 414-418
[19] Palacio A, Tamburro G, O'Neill D, and Simms CK 2009 Non-collision injuries in urban buses—Strategies for prevention, *Accid Anal Prev.* **41(1)** 1-9
[20] Jurecki R 2009 An analysis of collision avoidance manoeuvres in emergency traffic situations Non-collision injuries in urban buses—Strategies for prevention, *Archives of Automotive Engineering - Archiwum Motoryzacji.* **72(2)** 73-93
[21] Jurecki R, Jaskiewicz M, and Zuska A 2013 A variety of the behaviour of drivers at risk accident situations, *Scientific Journals Of The Maritime University Of Szczecin-Zeszyty Naukowe Akademii Morskiej W Szczecinie.* **35(107)** 38-46
[22] Hryciów Z, Jackowski J, Żmuda M 2020 The Influence of Non-Pneumatic Tyre Structure on its Operational Properties. *International Journal of Automotive and Mechanical Engineering,* **17(3)** pp. 8168–8178
[23] Gidlewski M, Jackowski J, Jemioł L, Żardecki D 2021 Sensitivity of a vehicle lane change control system to disturbances and measurement signal errors – Modeling and numerical investigations. *Mechanical Systems and Signal Processing* **147** 107081
[24] Radzikowski M, Foryś G, and Bukowski L 2020 Report on the technical condition of the surface of the national road network at the end of 2020, GDDKiA 2020

Effects of motorbike front suspension component operation on torsional vibration frequencies

A Dębowski [1]

[1] Faculty of Mechanical Engineering, Military University of Technology, Gen. Sylwester Kaliski's 2 St., 00-908 Warsaw, Poland, e-mail: andrzej.debowski@wat.edu.pl

Abstract. Riding a motorbike requires much more experience than other types of vehicles. However, not only insufficient skills may contribute to dangerous driving situations, but also vibrations generated in the steering system of a motorbike. This paper presents a method of vibration analysis on the example of a motorbike steering system using a transmittance model and shows how to extract the relevant frequencies for analysis. The effects of selected stiffness and damping parameters are analysed and an assessment is made of how the frequencies for wobble vibration will change over the life of the steering components. The results were presented using Bode diagrams.

Kew words: suspension, motorbike, vehicle dynamics, vibration comfort

1. Introduction

The observed trend of decreasing the number of fatalities in accidents of different vehicle groups is not present among motorbikes [1]. On the example of Poland it may be stated that motorbikes constitute 4.8% of all registered motor vehicles. However, motorcyclists are involved in 6.8% of accidents of which 9.7% are fatal accidents. The most common cause of accidents is still not adjusting the speed to the road conditions (54%) [2]. Unfortunately, due to the way statistics are collected, it is usually impossible to determine which number of accidents was caused by typical reasons (not adapting speed to the conditions, improper overtaking, failing to keep a safe distance, failing to yield right of way), and which could have been preceded by peculiar dynamic processes (including vibrations) leading to loss of control over the vehicle. In addition, this is extremely difficult to establish. Therefore, in addition to the development of active and passive protection systems in motorbikes, it is important that the motorbike is properly constructed and thus naturally stable and resistant to the generation and propagation of vibrations.

The human body has certain properties by which vibrations can be damped, and the sitting position of the rider has a significant influence here. As part of the vibrations are damped by the lower body, only a small part of the vibrations is affected by the human head. It should be noted, however, that a person in this position is 10 times more sensitive to vibrations with a frequency of 5 Hz than 100 Hz [3], and vibrations with frequencies of up to 12 Hz affect all organs in the human body [4]. Vibration can cause muscle fatigue and increase susceptibility to spinal injuries [5]. One method of reducing vibration is described in [6], [7], but nothing can replace a well-designed motorbike structure.

Unlike a car, a motorbike steering system does not allow the introduction of the gear ratio found in a car steering gearbox or other assist mechanisms. This makes the effects of the roadway on the wheel, and thus the rider's hands, much more noticeable. Sometimes they can be so great that they overcome the resistance force exerted by the driver and cause a loss of stability and control of the vehicle. For this

reason, proper front suspension design and geometry are particularly important for proper motorbike handling.
Over time, the performance of vehicle components deteriorates. For this reason, research is being conducted to assess to what extent this occurs. As indicated in the work [5], decreasing or increasing the value of the stiffness coefficient by 1/3 of the initial value is of little significance in this respect. The same work also analysed the effect of changing the value of the shock absorber damping coefficient, where the initial value of this parameter was reduced and increased by half. In this case, a rather obvious conclusion was drawn that as the value of the damping coefficient increases, the amplitude of the acceleration decreases. An obvious conclusion, as this is the purpose of using dampers in suspension. However, while in the case of coil springs there are fewer factors causing a change in parameters, for shock absorbers the parameters will depend, among other things, on the operating conditions.
In the case of coil springs used in motorbike suspensions, they can shorten over time, reducing their resting length and therefore their preload. During suspension operation, the wire from which the coil spring is made can wear away, resulting in a reduction in spring stiffness [8]. In some types of suspension, the spring preload can be adjusted, but above a certain value set by the manufacturer, replacement is necessary. For example, for the Yamaha R6 motorbike, the resting length of the spring is 247.0 mm, while the limiting length, is 242.1 mm [9].
The damping force of shock absorbers varies with the speed at which the piston moves, and this in turn causes the temperature to rise. Of course, like engine oils, the oils used to fill shock absorbers can operate over a wide temperature range, nevertheless the viscosity of the oil will change. It can also be different even for the same type of shock absorber, as shown in [10]. The sealing rings used in the suspension (see Fig. 2) also decrease in elasticity as a result of ageing processes, and small cracks can form on their surface. The sealing rings are particularly stressed during heavy or emergency braking, when the rear wheel can detach from the ground and transfer the entire weight of the motorbike and rider to the front wheel, resulting in a loss of shock absorber oil. The effect of shock absorber oil loss was investigated in [11]. Although a shock absorber with as much as 40% oil loss was tested, it was found that this causes a distortion of the performance and speed graphs compared to a new shock absorber. The instantaneous values of damping forces and damping power were also reduced. Therefore, motorbike manufacturers recommend relatively frequent inspection of the front suspension (every 7000 km or 6 months) [9].
If it is not possible to eliminate the vibrations, it becomes necessary to use damping systems. The need for increased torsional vibration damping in the motorbike steering system has already been recognised in older motorbike designs. Numerous patents have been issued with various means of reducing vibration, many of which include viscous dampers. Correct vibration damper design should include frequency analysis, but very few models have been developed for frequency analysis of motorbike steering system vibrations. It is still difficult to find works in which a detailed analysis of the influence of individual parameters describing the dynamics of the steering system has been carried out. In addition, the studies discussed earlier only concern car suspensions. The few works based on an analytical model include [12], where an analysis of the dynamics of a simplified motorbike model and an evaluation of its behaviour based on eigenvalues was undertaken. The small number of publications in recent years related to motorbike vibration can be attributed to some kind of shift away from analytical methods to engineering software and MBS and finite element methods [13], but these do not provide a description useful to the controller.
Therefore, the following section will present a model of a motorbike steering system in the form of an operator transmittance. This is not only a useful form used for controller development, such as in [14], but can also find application in mechanical system design, as will be presented in this paper.

2. Equations and mathematical expressions

The simulation studies used the motorbike steering model described in detail in the paper [15], which describes, among other things, its verification process. The following is the most important information necessary to understand the analyses carried out in the following section. The physical model corresponds to the motorbike steering system mounted in a special drum test rig presented in [15], which was used to verify the mathematical model.

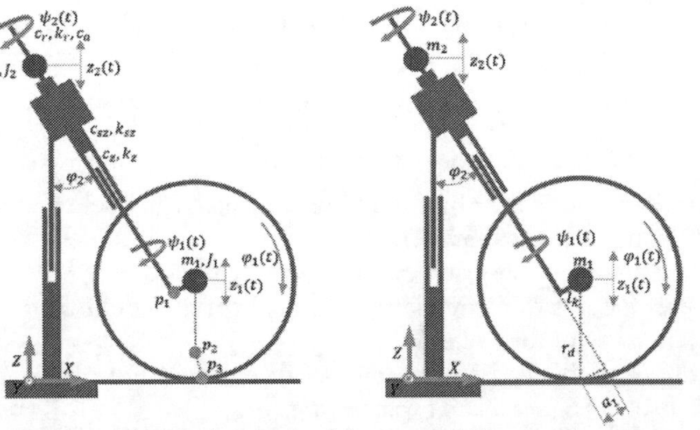

Figure 1. A physical model of the motorbike steering system

ψ, φ – steering and steering wheel angles and rotations, z – vertical displacements, J – moment of inertia, m – mass, c – damping coefficient, k – stiffness coefficient, p_1, p_2, p_3 – end point of suspension, wheel rim and wheel contact with the road, X, Y, Z – principal axes of the coordinate system, r_d – dynamical radius, l_k – offset of the running wheel axle from the steering axis of the head tube, w_{po} – tyre profile height, a_1 – actual overtaking distance

This model has five degrees of freedom, so that the operation of the suspension and its torsional compliance are taken into account. Despite the simplifying assumptions discussed in [16], it represents the steering dynamics well and is sufficient to demonstrate the method of analysis presented in this paper. The procedure described in the following section allows the use of any other model, as the procedure will not change. It is easy to see the usefulness of the method described hereafter, especially at the initial design stage, where estimates and preliminary calculations of the structure are carried out. As it was written earlier, it can also be applied to other types of vehicles, as only the mathematical model itself is changed, which is only the first step in the procedure.

Despite the use of a relatively simple physical model, the resulting system of equations has a complicated form, which is mainly due to the inclusion of the head angle of the frame. However, it reproduces the real system sufficiently well, as confirmed by the graph below (see Figure 2).

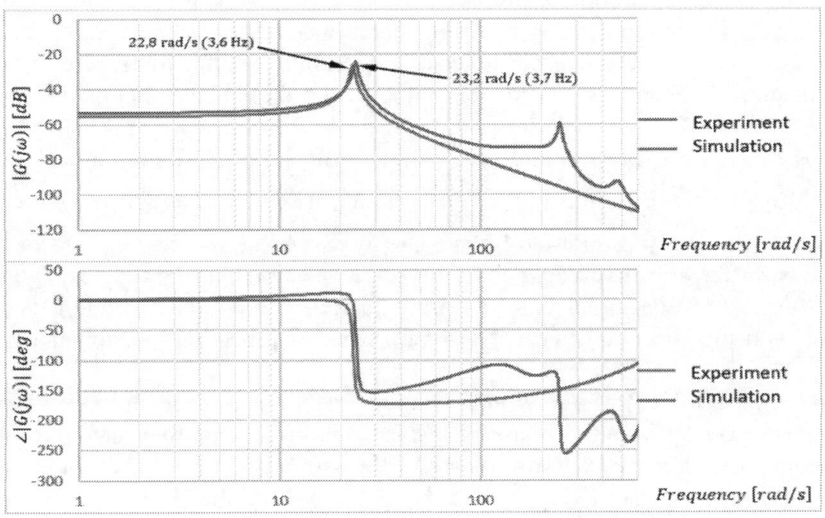

Figure 2. Comparison of the frequency characteristics of the angular acceleration of the contact wheel $\ddot{\psi}_1$ experiment and simulation model [15]

Ten different transmittances can be written for such a system [17], but four of them are important, namely: the transmittance for the vertical displacements of the masses m_1 and m_2 under vertical excitation with force F_Z ($G_1(s) = \frac{Z_1(s)}{F_Z(s)}$, $G_2(s) = \frac{Z_2(s)}{F_Z(s)}$) and the transmittance for angular displacements of the masses m_1 and m_2 under torsional excitation M_S ($G_3(s) = \frac{\Psi_1(s)}{M_s(s)}$, $G_4(s) = \frac{\Psi_2(s)}{M_s(s)}$).
These transmittances can be obtained by following the steps listed below and are as follows:
1. Development of the mathematical model;
2. Linearisation of the equations describing the system dynamics;
3. Write the equations in operator form and determine the transmittances;
4. Perform stability analysis of the system;
5. Decomposition of transmittance into summation form;
6. Frequency analysis of each member of transmittance.

On the other hand, analyses based on the transmittances obtained in this way were performed using the LabVIEW software and an application developed with its assistance for the purposes of this paper. It is worth noting that, in contrast to dual-mass models describing automotive suspension, the operator transmittances obtained here are of higher order. It should be noted, however, that in contrast to the above-mentioned models, the model used here has a feature that has a key effect on its dynamics - the angle of inclination of the head tube.
The decomposition of the transmittance into simple fractions is a relatively simple task in the case where one-time calculations are necessary. However, in the case of more extensive analyses, it is necessary to repeat the same calculations several times, which at the same time prompts the use of appropriate tools to facilitate the calculation process. However, it is rare for there to be a ready-made tool with advanced capabilities dedicated to very specific tasks. Although the analytical determination of the summation form of the operator transmittance does not cause major difficulties, it is not so with performing such calculations in a software manner and a number of difficulties can be encountered, as mentioned in [15].

3. Simulation results

Using the developed program, the Bode diagrams given below were obtained together with the significant values read from the diagrams, which were tabulated. In addition to the characteristic values for the graphs in question (frequency, maximum amplitude value, natural vibration period, relative damping coefficient), the result of the Hurwitz stability analysis for the highest sub-designator is also included. The most important logarithmic amplitude and phase diagrams are presented here, as there would be additionally 18 diagrams analogous to the ones presented below. However, it was decided to omit them, because of the lack of changes of the analysed values at the change of the selected parameter, and also some duplication of diagrams would occur due to the applied decomposition into simple fractions.
The presented graphs concern the transmittance describing vertical and torsional vibrations occurring on the steering wheel ($G_2(s) = \frac{Z_2(s)}{F_Z(s)}$ oraz $G_4(s) = \frac{\Psi_2(s)}{M_s(s)}$), which arise at the handlebars of the motorbike and from the road wheel and are transmitted to them. The focus was on handlebar vibrations because, although these vibrations affect the unsprung mass, their negative effects can be reduced by introducing additional damping directly on the handlebar (the aforementioned torsional vibration damper). The excitation signal for both torsional and vertical vibrations was a force pulse mapped by the Dirac function $\delta(t)$.
It can be seen from fig. 3, tab. 1 that the torsional damping coefficient in the suspension, C_{sz}, that it relates to high frequency steering wheel resonances. This also indicates that the rings used, in addition to their sealing function, also damp vibrations. Which at the same time means that their replacement may not only be necessary in the case of visible oil leaks, but also when the elastomer from which they are made begins to lose its original properties. Therefore, the solution proposed in the work [18], compensates for insufficient damping of sealing rings with an increase in the mass of a motorbike, and in this case, it may be enough to design them properly, without deterioration of other properties. It is also worth noting that the value of the C_{sz} parameter in the context of modelling the dynamics of a motorbike, apart from the vibration damping mentioned, does not affect the stability of the steering

system and if the research does not focus on the vibrations of the system, this value does not need to be precisely determined.

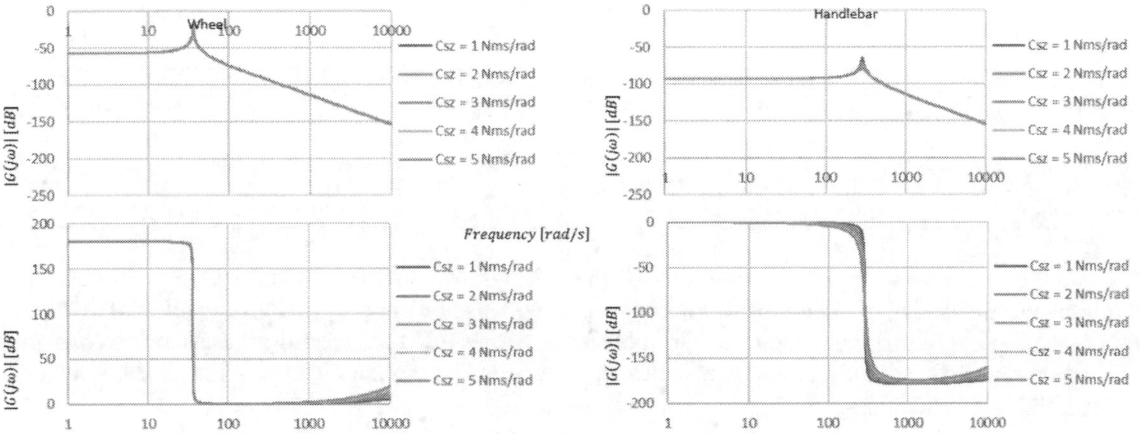

Figure 3. Amplitude and phase characteristics of the transmittance of the steering angle position when the parameter values are changed C_{sz}

Table 1. The parameter values read from the graphs in Figure 3.

Changing parameter	from mas m_1					from mas m_2				
	Frequency	Max. amplitude	Stability	Oscillation period	Relative suppression	Frequency	Max. amplitude	Stability	Oscillation period	Relative suppression
	[rad/s]	[dB]	[-]	[s]	[-]	[rad/s]	[dB]	[-]	[s]	[-]
1 Nms/rad	36,381594	-18,840356	0,000315	0,027431	0,005746	284,734004	-64,704775	0,000136	0,003504	0,019382
2 Nms/rad	36,381676	-19,049099	0,000324	0,027431	0,005903	284,575139	-70,635241	0,000270	0,003504	0,038529
3 Nms/rad	36,381814	-19,252067	0,000333	0,027431	0,006061	284,311175	-74,126801	0,000404	0,003504	0,057676
4 Nms/rad	36,382008	-19,449395	0,000341	0,027431	0,006219	283,941821	-76,605729	0,000538	0,003504	0,076824
5 Nms/rad	36,382258	-19,641216	0,000350	0,027431	0,006376	282,308070	-78,523875	0,000673	0,003505	0,095972

It is important to note that the value of the damping factor affects both low and high frequency vibrations. However, for low frequencies these vibrations are completely damped, while for higher frequencies they are only slightly damped. This is also manifested by the value of the coefficient of the amplification constant, which changes the value of the logarithmic modulus of the transmittance, which influences the damping of the system.

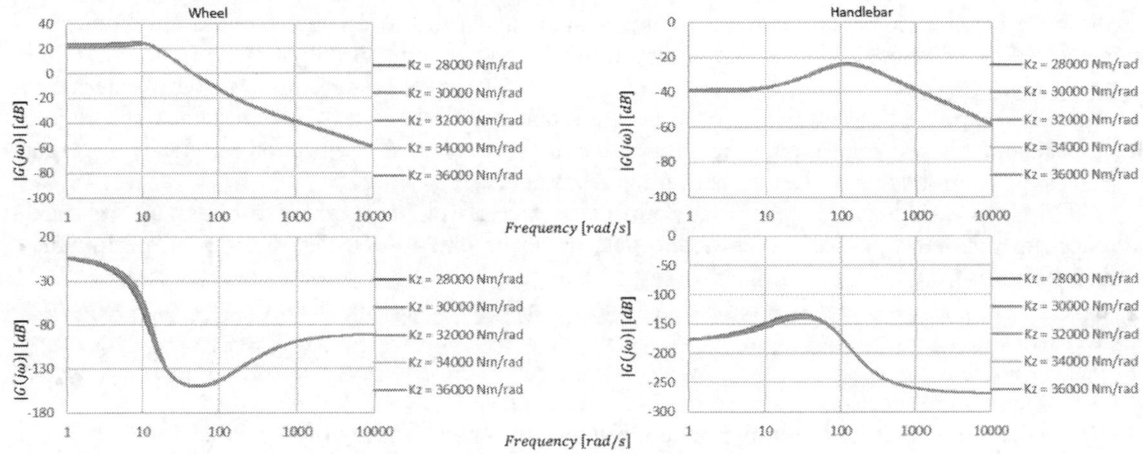

Figure 4. Amplitude and phase characteristics of the transmittance of the vertical position of the steering control with a change in the parameter value K_z

Table 2. The values of the parameters read from the graphs in Figure 4.

Changing parameter	**from mas m_1**					**from mas m_2**				
	Frequency	**Max. amplitude**	**Stability**	**Oscillation period**	**Relative suppression**	**Frequency**	**Max. amplitude**	**Stability**	**Oscillation period**	**Relative suppression**
	[rad/s]	**[dB]**	**[-]**	**[s]**	**[-]**	**[rad/s]**	**[dB]**	**[-]**	**[s]**	**[-]**
28000 Nm/rad	8,527749	24,542363	0,082981	0,082895	0,500518	117,681195	-23,596452	0,012459	0,008215	0,758326
30000 Nm/rad	9,171745	24,154614	0,076914	0,080482	0,477835	117,681195	-23,810908	0,012331	0,008168	0,754847
32000 Nm/rad	9,740655	23,806457	0,071618	0,078306	0,457297	117,681195	-24,020465	0,012207	0,008122	0,751427
34000 Nm/rad	10,306919	23,491283	0,066956	0,076331	0,438592	117,681195	-24,225315	0,012085	0,008078	0,748064
36000 Nm/rad	10,783719	23,204015	0,062823	0,074528	0,421468	117,681195	-24,425637	0,011967	0,008034	0,744756

Of course, in addition to the sheer flexibility of the O-rings, the stiffness of the steering column and which suspension design of the motorbike is considered will also be important. Some benefits of inverted suspension are also apparent, as due to the increased stiffness of the steering, the effect of constant amplification is reduced and the resonant vibration frequency of the handlebars is increased, which is good for the rider's health.

4. Summary and conclusions

Motorbike dynamics is a complex issue both in terms of research and mastery of safe driving skills by the driver. The former is evidenced by the still small number of publications compared to car research, and the latter by road accident statistics. To a large extent, the effects of gyroscopic moments and the countersteering technique, as well as the vibrations generated in the structure of the motorbike, make it difficult for novice drivers to drive a motorbike. These can sometimes be caused by interactions transmitted from the road to the other components of the motorbike, while in other cases they can be self-excited. There are also more types of vibration that are specific to motorbikes (e.g. high-side, chatter). In addition, a number of factors can make it difficult to stabilise the motorbike after their onset, and torsional vibration dampers are not fitted to every motorbike just as steering assist systems are rare. This paper presents the results of analysis for selected model parameters using the method of analysis of motorbike steering vibrations with the use of a tool developed with the help of LabVIEW.

It is important to note that some of the analysed parameters are influenced by the user himself through the way of driving the motorbike or during periodic replacement of components. While the use of a torsional vibration damper only has a positive effect on the behaviour of the steering system, the selection of damping settings of telescopic shock absorbers requires experience as it is associated with low frequency vibrations. However, for the values of the coefficients of longitudinal elasticity, similar results were obtained as in the work [5], namely their insignificant influence on the amplitudes and frequencies of vibrations. In the case of sealing rings, it can be seen that they play an important role in damping high-frequency vibrations, and therefore it is worthwhile to determine to what extent their elastic-damping properties deteriorate even before replacement dictated by loss of shock absorber oil due to leakage. It is also important to note that there may be a change in vibration frequency over time, as well as the natural damping capacity resulting from the design.

In future work, it is planned to carry out analyses in relation to weight distribution on the vehicle and steering geometry, as well as studies to determine the extent to which suspension performance deteriorates over time.

5. Literature

[1] European Commission, Traffic Safety Basic Facts on Motorcycles & Mopeds, European Commission, Directorate General for Transport, June 2017.

[2] Komenda Główna Policji, Biuro Ruchu Drogowego: Wypadki drogowe w Polsce w 2016 roku, Warszawa 2017.

[3] Mansfield N. J. Human response to vibration, CRC Press, London, 2005.

[4]Merkisz J, Pielecha I. Układy elektryczne pojazdów hybrydowych. Wydawnictwo Politechniki Poznańskiej, Poznań 2015; 12(19): 139-154.

[5] Dukalski P, Będkowski B, Parczewski K, Wnęk H, Urbaś A, Augustynek K. Dynamics of the vehicle rear suspension system with electric motors mounted in wheels. Eksploatacja i Niezawodnosc – Maintenance and Reliability 2019; 21 (1): 125–136, http://dx.doi.org/10.17531/ein.2019.1.14.
[6] Agostoni S, Cheli F, Leo E, Pezzola M.E. Methodology to design a vibration absorption footplate for motorcycle application: From phenomena investigation to prototype performance evaluation, Mechanical Systems and Signal Processing, Volume 30, July 2012, Pages 296-305, https://doi.org/10.1016/j.ymssp.2011.11.017.
[7] Agostoni S, Cheli F, Leo E, Pezzola M.E. An innovative multi dof TMD system for motorcycle handlebars designed to reduce structural vibration and human exposure, Mechanical Systems and Signal Processing, Volume 31, August 2012, Pages 298-315, doi:10.1016/j.ymssp.2011.11.018.
[8] Dmowski R. Diagnozowanie podzespołów i zespołów motocykli, Podręcznik do kształcenia w zawodzie mechanik motocyklowy, Wydawnictwo Komunikacji i Łączności, Warszawa 2019.
[9] Yamaha Motor Corporation, Service Manual, YZFR6V(C), 2C0-28197-10, First edition, U.S.A., December 2005.
[10] Konieczny Ł. "Analysis of simplifications applied in vibration damping modelling for a passive car shock absorber." Shock and Vibration 2016 (2016).
[11] Konieczny Ł, Śleziak B. Wpływ wybranych parametrów na charakterystyki tłumienia amortyzatorów hydraulicznych, zeszyty naukowe politechniki śląskiej, Seria: TRANSPORT z. 64, Nr kol. 1803, 2008, 145-150.
[12] Zanarini A, Brugnoni E. Frequency analysis of a motorbike under motion conditions. In: ISMA2012-USD2012 Conferences, Volume: Multi-Body Dynamics and Control. 2012.
[13] Meng Z, Li M. Vibration Forecast and Vibration Attenuation Research of the Finished Motorcycle based on Virtual Analysis. Applied Mechanics and Materials. Vol. 705. Trans Tech Publications Ltd, 2015.
[14] Szczerba P, Rzucidło P, Szczerba Z, Drupka G. Vision system supporting the pilot in variable light conditions. Eksploatacja i Niezawodnosc – Maintenance and Reliability 2019; 21 (1): 60–67, http://dx.doi.org/10.17531/ein.2019.1.8.
[15] Dębowski A.: *Torsional vibration and its suppression in motorcycle steering system*, Military University of Technology, Warsaw 2021. ISBN 978-83-7938-322-1
[16] Dębowski A.: *Analysis of the Effect of Mass Parameters on Motorcycle Vibration and Stability*, MDPI, Energies 2021, 14(16), 5090; https://www.mdpi.com/1996-1073/14/16/5090.
[17] Dębowski A. Automatyka. Podstawy teorii, Wydawnictwo Naukowo-Techniczne, Łódź 2007.
[18] Baad S. M, Patil R. J, Qaimi M. G. Hand Arm Vibration Alleviation of Motorcycle Handlebar using Particle Damper, Engineering and Manufacturing, 2017, Vol.7, No.1, pp. 26-40.

IOP Conf. Series: Materials Science and Engineering 1247 (2022) 012019 doi:10.1088/1757-899X/1247/1/012019

Modelling of dynamics of the suspension of tandem axles of a multi-axle vehicle provided with a novel load-equalising system

R Pilatowicz[1], W Luty[1]

[1]Łukasiewicz Research Network – Automotive Industry Institute, Warsaw, Poland

Abstract: The paper presents the problem of using multi-axle suspension systems in vehicles, concerning unbalanced distribution of axle loads while moving in off-road conditions. An innovative load balancing system for tandem axles, which limits the occurring overloads, is presented. A mathematical model is proposed to represent the suspension of a front "quarter" of a four-axle vehicle provided with a novel system to equalise the loads on two closely spaced (tandem) axles. The "quarter" of the vehicle is described as a set of rigid bodies linked with each other by spring and damping elements. The model has three degrees of freedom and defines vehicle's responses to excitations (inputs) generated by road irregularities. The mathematical models known from the literature describe typical suspension systems and specific types of load balancing systems for two- and three-axle vehicles. This makes it impossible to apply them directly to simulation test of a vehicle with two front axles. A new model dedicated to the proposed load equalizing system, which allows simulation tests for the design and verification of the new system's operation, has been prepared.

Keywords – vehicle, vehicle dynamics, suspension

1. Introduction

One of the crucial problems that arise from using in vehicles multi-axle suspension systems is the necessity to equalise the loads on all the vehicle axles when moving on uneven surfaces typical for unpaved roads or roadless tracts [1,2]. The loads on vehicle axles must be kept uniform to prevent axle overloads, which might result in damage to components of vehicle suspension or power transmission systems. Moreover, the correct distribution of loads on vehicle axles is important for the performance of vehicle braking and steering systems, which is critical for driving safety. The multi-axle vehicles' suspension systems, where each of a pair of closely spaced axles, referred to as "tandem axles" or "twin axles", is suspended separately from each other, do not ensure uniform distribution of axle loads, especially when the vehicle is negotiating separate and repeating ground bumps or pits. As an example, when the wheels of one of the axles run onto a high bump, the other axle may be suspended in mid-air. In such a situation, one of the tandem axles carries the entire static load normally carried by both of them [2, 3]. To overcome this problem, engineering solutions are used whose common feature is the fact that the suspensions of each of the tandem axles are coupled together. Examples of the solutions of this kind, having the form of mechanical, hydraulic, or pneumatic load equalizing systems, can be found in the literature (Fig. 1).

a)

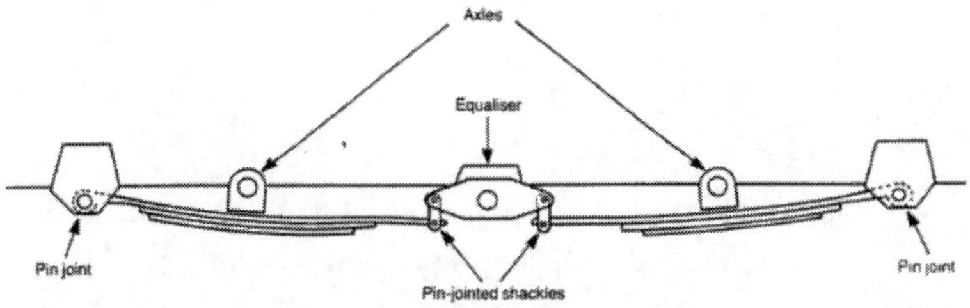

b)

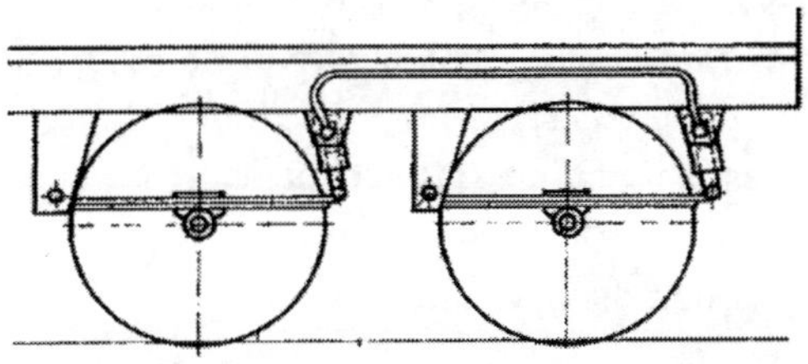

c)

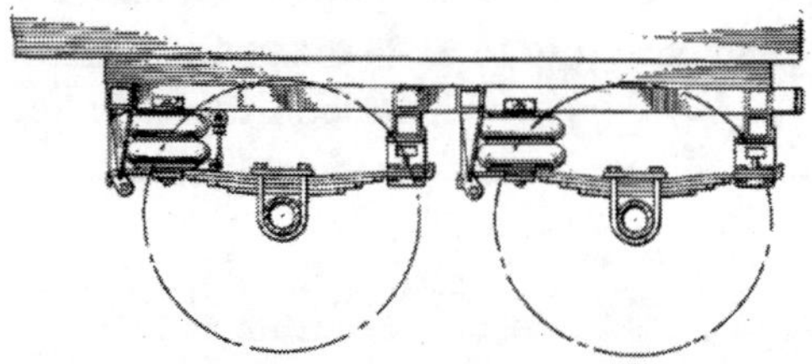

Fig. 1. Schematics of example systems to equalise the loads on tandem axles of multi-axle vehicles: a) lever-type system; b) hydraulic system; c) pneumatic system [4, 5, 6]

Based on results of analyses of the good and bad points of the solutions used, a new concept of the system to equalise the loads on tandem axles has been proposed. A schematic of such a system has been shown in Fig. 2. The conventional arrangement of the front suspension system of a four-axle vehicle with rigid drive axles and leaf springs on each wheel has been developed by adding four hydraulic cylinders fixed to the vehicle frame over the axles next to the leaf springs and four other hydraulic cylinders pin-jointed in place of spring shackles. All the cylinders are of the single-acting type. The cylinders situated on one vehicle side are interconnected in such a way that the cylinder situated over the first axle is hydraulically linked with the one pin-connected to the end of the leaf spring of the second axle and the cylinder situated over the second axle is hydraulically linked with the one pin-connected to the end of the leaf spring of the first axle. There is some clearance between each axle and the piston rod of the cylinder situated over it. In the rest position, there is no overpressure in the hydraulic system, the pistons in the cylinders situated over the axles are fully extended, and the pistons in the cylinders situated at the leaf spring ends are fully retracted. In the schematic in Fig. 2, the system has been presented in its working position, without showing the damping elements.

IOP Conf. Series: Materials Science and Engineering 1247 (2022) 012019 doi:10.1088/1757-899X/1247/1/012019

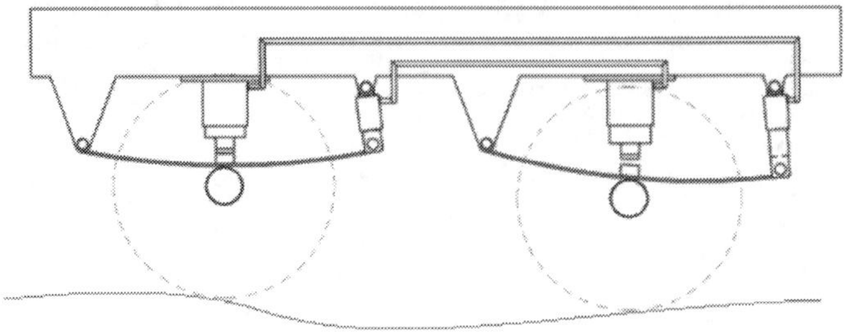

Fig. 2. A concept of the multi-axle vehicle's suspension
with a hydraulic load equalising system [7]

The most important good points of the solution proposed include:

- extended range of system operation in comparison with the designs known hitherto;
- applicability to all the vehicles of this kind, whether newly manufactured or already being in service (after minor modifications in the design of the latter);
- ensured correct operation of the suspension in case of a failure of the equalising system;
- simple construction;
- elimination of permanent presence of high pressure in the hydraulic system.

The essence of the problem addressed in the scientific work undertaken lies in the modelling of the system proposed. During the tests of vehicle dynamics, simplified "quarter-vehicle" or "half-vehicle" models, described in numerous publication [8, 9, 10, 11, 12, 13, 14] are often used. Ready-to-use equations of the dynamics are used, substituting only parameters of spring and damping elements of the suspension and values of sprung and unsprung masses. In spite of the fact that the first load balancing systems for axles suspended on leaf springs appeared at the beginning of the 20th century, the use of these systems has not become widespread until now and no universal mathematical models for this type of systems have been developed. The simulation models described in the literature refer to two-axle vehicles in which the front and rear axle shock absorbers were replaced by double-acting hydraulic cylinders [15] and three-axle vehicles in which the suspension of all axles was connected hydraulically [16]. These models refer to suspension systems that use two cylinders per axle (one per wheel). Fig. 3 shows examples of models related to specific design solutions.

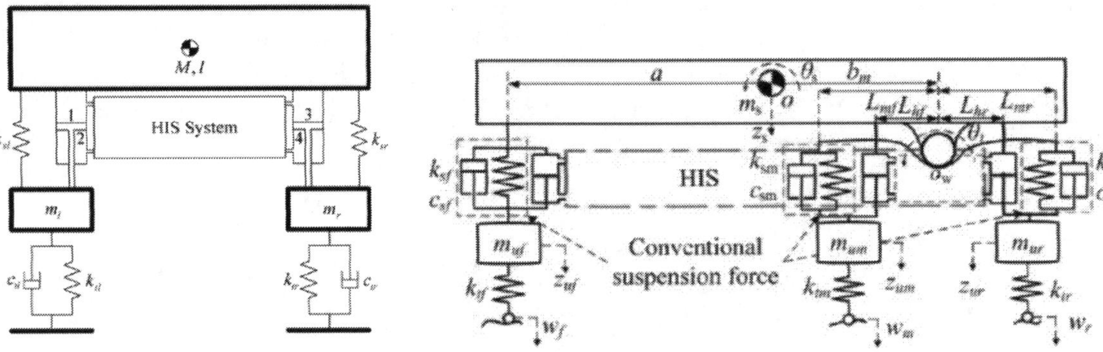

Fig. 3. „Half-vehicle" models with hydraulically interconnected suspension of:
a) a two-axle vehicle, b) a three-axle vehicle [15, 16]

2. Model of the suspension with a load equalising system

To obtain effective operation of the system installed in a vehicle, many tests must be carried out both in laboratory conditions and in real driving tests. First, however, engineering computations and simulation tests of the system functioning are necessary for the prototype to be built. With this goal in view, a model of dynamics of the suspension of tandem axles using a load-equalising system was prepared. The model was developed to represent the suspension system of front tandem axles of a four-axle motor truck (Fig. 4), taken as an example.

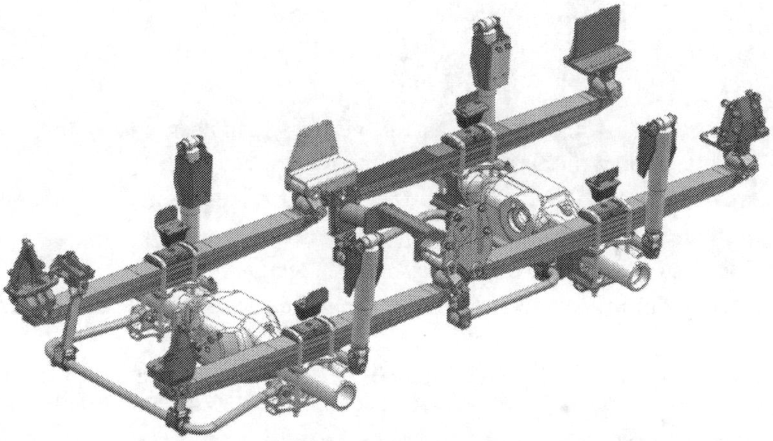

Fig. 4. Visualisation of the suspension system of tandem axles of an off-road truck [17]

When building the model, the following simplifying assumptions were adopted:

- the sprung mass is a rigid solid;
- the vehicle frame remains in the horizontal position;
- the system has three degrees of freedom;
- only vertical movements are considered;
- the vehicle body has a longitudinal symmetry plane;
- the damping and stiffness characteristics are linear;
- the ground surface is undeformable;
- the impact of antiroll bars and rubber-metal elements is ignored;
- the tangent tyre-ground interactions are ignored;
- the gravitational components of the forces acting in the suspension system are ignored;
- the vehicle is driven along a rectilinear path with a constant velocity.

The physical model of the equalising system proposed has been presented in Fig. 5. Its components with definite physical properties include:

- unsprung masses m_1 [kg] and m_2 [kg] as the parts of the vehicle axle mass taken per a single wheel of the first and second axle, respectively;
- sprung mass m_3 [kg] as the part of the vehicle body mass acting upon the two wheels of the tandem axle that are situated on the same vehicle side;
- displacements z_1 [m] and z_2 [m] as the vertical displacements of the first and second axle, respectively, relative to the axles' positions of static equilibrium;
- displacement z_3 [m] as the vertical displacement of the vehicle body relative to its position of static equilibrium;

- clearance values L_1 [m] and L_2 [m] as the distances between the top surface of the first and second axle, respectively, and the end surface of the piston rod of the hydraulic cylinder situated over the corresponding axle;
- damping values c_1 [N·s/m] and c_2 [N·s/m] as the values of the viscous damping of the first and second axle's shock absorbers, respectively;
- stiffness values k_{11} [N/m] and k_{12} [N/m] as the component stiffness values of the first axle's leaf spring;
- stiffness values k_{21} [N/m] and k_{22} [N/m] as the component stiffness values of the second axle's leaf spring;
- dry friction T_1 [N] and T_2 [N] between leaves of the springs of the first and second axle, respectively;
- displacements h_{11} [m] and h_{21} [m] as the vertical displacements of pistons of the hydraulic cylinders situated over the first and second axle, respectively;
- displacements h_{12} [m] and h_{22} [m] as the vertical displacements of pistons of the hydraulic cylinders situated at leaf springs' ends;
- areas s_1 [m^2] and s_2 [m^2] of hydraulic cylinders' pistons;
- leaf spring length l [m] as the distance between the points of fastening the spring to the vehicle frame;
- dimensions a [m] and b [m] as the distances between the axle and individual points of fastening the spring to the vehicle frame;
- damping value c_k [N·s/m] as the value of tyre damping;
- stiffness value k_k [N/m] as the value of tyre stiffness;
- coordinate $q_{(t)}$ [m] representing the kinematic input generated by the road profile.

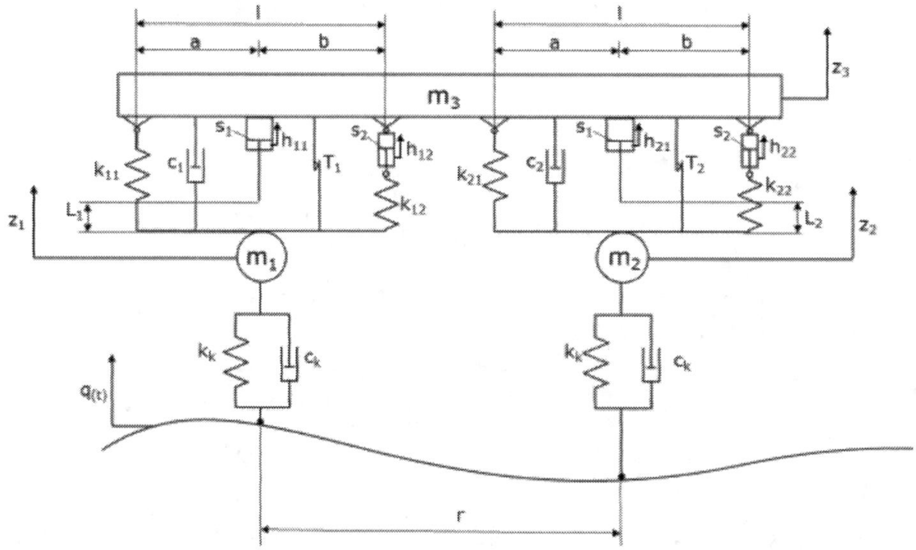

Fig. 5. Physical model of a front "quarter" of a four-axle vehicle

For the model cleared of its constraints, a system of forces acting on individual masses of the model (Fig. 6) was identified, which included:

- elasticity forces F_{S11} [N] and F_{S21} [N] as the elasticity force components acting on the points of fastening the front ends of the first and second axle's springs, respectively;

- elasticity forces F_{S12} [N] and F_{S22} [N] as the elasticity force components acting on the piston rods of the hydraulic cylinders at the rear ends of the first and second axle's springs, respectively;
- damping forces F_{t1} [N] and F_{t2} [N] as the forces of viscous damping of the first and second axle's shock absorbers, respectively;
- forces F_{p1} [N] and F_{p2} [N], exerted by the hydraulic cylinders situated over the first and second axle, respectively;
- forces F_{T1} [N] and F_{T2} [N], generated by dry friction between leaves of the springs of the first and second axle, respectively;
- elasticity forces F_{sk1} [N] and F_{sk2} [N] generated in tyres of the first and second axle, respectively;
- damping forces F_{tk1} [N] and F_{tk2} [N] generated in tyres of the first and second axle, respectively.

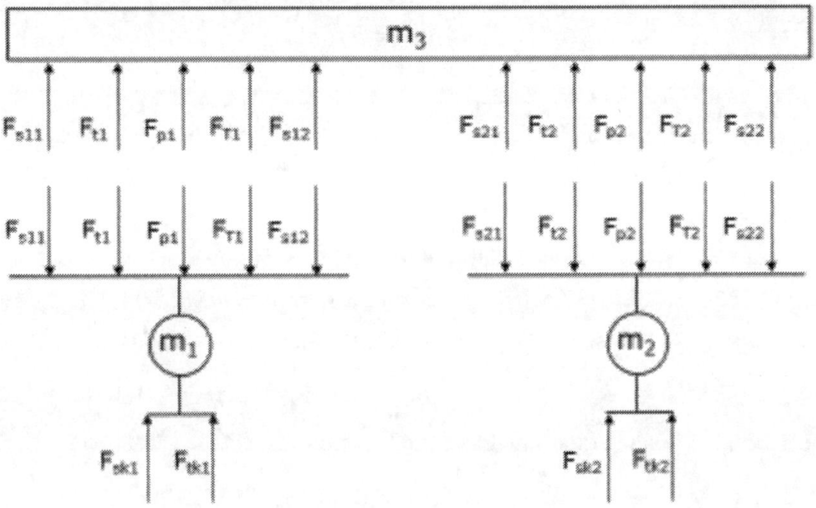

Fig. 6. System of forces acting in the equivalent model of suspension dynamics

Based on the system of forces acting in the model of suspension dynamics, equations of the dynamics of individual masses in the model have been formulated as follows, using the d'Alembert's method:

$$m_1 \cdot \ddot{z}_1 = F_{sk1} + F_{tk1} - F_{s11} - F_{t1} - F_{p1} - F_{T1} - F_{s12} \quad (1)$$

$$m_2 \cdot \ddot{z}_2 = F_{sk2} + F_{tk2} - F_{s21} - F_{t2} - F_{p2} - F_{T2} - F_{s22} \quad (2)$$

$$m_3 \cdot \ddot{z}_3 = F_{s11} + F_{t1} + F_{p1} + F_{T1} + F_{s12} + F_{s21} + F_{t2} + F_{p2} + F_{T2} + F_{s22} \quad (3)$$

2.1. Leaf spring elasticity forces

In the model, the leaf spring has been represented as a system of two coil springs combined with a rigid horizontal beam, which can move in vertical direction only. For the first axle, the equations have the form as shown below:

$$F_{S11} = u_{11} \cdot k_{11} \quad (4)$$

$$F_{S12} = u_{12} \cdot k_{12} \quad (5)$$

$$k_{11} = k_1 \cdot \frac{b}{l} \quad (6)$$

IOP Conf. Series: Materials Science and Engineering 1247 (2022) 012019 doi:10.1088/1757-899X/1247/1/012019

$$k_{12} = k_1 \cdot \frac{a}{l} \tag{7}$$

$$F_{S1} = F_{S11} + F_{S12} \tag{8}$$

where: k_1 is stiffness of the leaf spring of the first axle and u_{11} and u_{12} are equivalent dynamic deflections [m] of the first and second coil spring of the first axle, determined from equations:

$$u_{11} = u_1 \tag{9}$$

$$u_{12} = z_1 - z_3 + h_{12} \tag{10}$$

where: u_1 is deflection [m] of the first axle's suspension, determined from an equation:

$$u_1 = z_1 - z_3 \tag{11}$$

and, simultaneously:

$$h_{12} = h_{21} \cdot i_h \tag{12}$$

where: i_h is hydraulic ratio:

$$i_h = \frac{S_1}{S_2} \tag{13}$$

$$for \quad u_2 \leq L_2 \quad h_{21} = 0 \tag{14}$$

$$for \quad u_2 > L_2 \quad h_{21} = u_2 - L_2 \tag{15}$$

and: u_2 is deflection [m] of the second axle's suspension, determined from an equation:

$$u_2 = z_2 - z_3 \tag{16}$$

For the second axle, the equations have the form:

$$F_{S21} = u_{21} \cdot k_{21} \tag{17}$$

$$F_{S22} = u_{22} \cdot k_{22} \tag{18}$$

$$k_{21} = k_2 \cdot \frac{b}{l} \tag{19}$$

$$k_{22} = k_2 \cdot \frac{a}{l} \tag{20}$$

$$F_{S2} = F_{S21} + F_{S22} \tag{21}$$

where: k_2 is stiffness of the leaf spring of the second axle and u_{21} and u_{22} are equivalent dynamic deflections [m] of the first and second coil spring of the second axle, determined from equations:

$$u_{21} = u_2 \tag{22}$$

$$u_{22} = z_2 - z_3 + h_{22} \tag{23}$$

and, simultaneously:

$$h_{22} = h_{11} \cdot i_h \tag{24}$$

$$for \quad u_1 \leq L_1 \quad h_{11} = 0 \tag{25}$$

$$for \quad u_1 > L_1 \quad h_{11} = u_1 - L_1 \tag{26}$$

IOP Conf. Series: Materials Science and Engineering 1247 (2022) 012019 doi:10.1088/1757-899X/1247/1/012019

2.2. Damping forces in the viscous damper

$$F_{t1} = \dot{u}_1 \cdot c_1 \tag{27}$$

$$\dot{u}_1 = \dot{z}_1 - \dot{z}_3 \tag{28}$$

$$F_{t2} = \dot{u}_2 \cdot c_2 \tag{29}$$

$$\dot{u}_2 = \dot{z}_2 - \dot{z}_3 \tag{30}$$

where: $\dot{u}_1$ and $\dot{u}_2$ are suspension deflection velocities [m/s] of the first and second axle, respectively; $\dot{z}_1$ and $\dot{z}_2$ are displacement velocities [m/s] of the first and second axle, respectively; and $\dot{z}_3$ is vehicle body displacement velocity [m/s].

2.3. Forces exerted by the hydraulic cylinders situated over the axles

The interaction between the leaf spring and the hydraulic cylinder is only possible when the clearance between the axle and the piston rod end is reduced to zero, i.e. when these parts come into contact with each other, which is described as follows:

$$for\ \ u_1 \leq L_1\ \ F_{p1} = 0 \tag{31}$$

$$for\ \ u_1 > L_1\ \ F_{p1} = F_{s22} \cdot i_h \tag{32}$$

and

$$for\ \ u_2 \leq L_2\ \ F_{p2} = 0 \tag{33}$$

$$for\ \ u_2 > L_2\ \ F_{p2} = F_{s12} \cdot i_h \tag{34}$$

2.4. Forces generated by dry friction between spring leaves

$$F_{T1} = sig(\dot{u}_1) \cdot T \tag{35}$$

$$F_{T2} = sig(\dot{u}_2) \cdot T \tag{36}$$

2.5. Forces coming from vehicle tyres

$$F_{sk1} = (q_1 - z_1) \cdot k_k \tag{37}$$

$$F_{sk2} = (q_2 - z_2) \cdot k_k \tag{38}$$

$$F_{tk1} = (\dot{q}_1 - \dot{z}_1) \cdot c_k \tag{39}$$

$$F_{tk2} = (\dot{q}_2 - \dot{z}_2) \cdot c_k \tag{40}$$

Unlike the well- known models, in which the leaf spring is replaced by a single spring element, the prepared model uses two spring elements located at the ends of a horizontal beam, the combined stiffness of which corresponds to the leaf spring stiffness. The value of the force, exerted by the hydraulic cylinder situated over the axle, depends on the deflection of the spring and the load value of the second axle. The mathematical model includes a variable for the clearance between the hydraulic cylinder and the axle to ensure that the system only acts when it is needed, i.e. when a specified spring deflection limit is exceeded. Once the model has been implemented into a computational environment such as MATLAB Simulink, it will be possible to carry out a number of simulation tests for different loading scenarios.

IOP Conf. Series: Materials Science and Engineering 1247 (2022) 012019 doi:10.1088/1757-899X/1247/1/012019

The results of the simulation tests will be used to select the parameters of the load equalizing system. In the next stage, experimental studies are planned to verify the model.

3. Conclusion

A load equalising system is required to ensure a balanced load distribution on tandem axles of multi-axle vehicles operating in off-road conditions. These systems are not commonly used, and the available mathematical models are not universal and their application is limited to specific construction solutions. The development of a new concept of the load equalising system required the preparation of a mathematical model, taking into account the characteristic features of the system. The equivalent model described herein and representing a "quarter" of a four-axle vehicle provided with the innovative system to equalise the loads on tandem vehicle axles may be a tool for the carrying out of simulation tests aimed at the preparation of data for prototype designing and construction purposes. It may also be used to check whether the system functioning ensures the obtaining of the effects expected, i.e. an adequate reduction in the dynamic loads on vehicle axles during drives in off-road conditions. At the next stage of the work, the model developed must be verified by comparing the simulation results with results of experimental testing of the vehicle when driven in real off-road conditions.

References

[1] Davis L and Bunker J 2009 Dynamic load sharing for heavy vehicles - a new metrics *Road & Transport Research* **18** 23

[2] Kamiński E and Pokorski J 1983 *Dynamika zawieszeń i układów napędowych pojazdów samochodowych (Dynamics of suspension systems and powertrains of motor vehicles)* (Warszawa: Wydawnictwo Komunikacji i Łączności) pp 157-168

[3] Kamiński E 1983 *Dynamika pojazdów i teoria zawieszeń (Dynamics of vehicles and theory of suspension systems)* (Warszawa: Wydawnictwo Politechniki Warszawskiej) pp 139-154

[4] Simmons I C P and Mitchell C G B 1990 *The equalization of multi-axle bogies fitted to commercial vehicles* Research report 277 (Crowthorne: Transport and Road Research Laboratory Department of Transport)

[5] Schaeff F 1973 *Axle assembly* GB1459877A

[6] Harbers H C and Hamlet B C 1958 *Vehicle-suspension means employing leaf and air spring assemblies in combination* US3063732A

[7] Luty W, Piłatowicz R and Wróbel T 2021 *Układ równoważenia obciążeń sąsiednich osi jezdnych pojazdu (System to equalise loads on tandem vehicle axles)* P.438617

[8] Lozia Z 1998 *Analiza ruchu samochodu dwuosiowego na tle modelowania jego dynamiki (Analysis of motion of a two-axle automotive vehicle against the background of modelling of its dynamics)* (Warszawa: Oficyna Wydawnicza Politechniki Warszawskiej) pp 139-154

[9] Pikosz H and Ślaski G 2010 Charakterystyki elementów sprężystych i tłumiących zawieszenia samochodu osobowego oraz zastępcze charakterystyki ich modeli (Passenger car suspension spring and damper characteristics and substitute characteristics of their models) *The 7th Scientific and Technical Conference LOGITRANS* 2115

[10] Lozia Z 2012 Modele symulacyjne ruchu i dynamiki dwóch pojazdów uprzywilejowanych (Vehicle dynamics simulation models of two emergency vehicles) *Czasopismo Techniczne. Mechanika* **109** 19

[11] Ferdek U and Łuczko J 2011 Analiza ćwiartkowego modelu zawieszenia pojazdu z tłumikiem hydraulicznym (The analysis of the quarter model of the vehicle suspension with the hydraulic damper) *Symulacja w Badaniach i Rozwoju* **2** 67

[12] Lozia Z and Zdanowicz P 2017 Wyznaczenie optymalnej wartości współczynnika asymetrii amortyzatora pasywnego zawieszenia samochodu z wykorzystaniem modelu „ćwiartki samochodu

(Calculation of shock absorber asymmetry coefficient optimal value in car passive suspension using "quarter-car" model) *Prace Naukowe Politechniki Warszawskiej. Transport* **119** 249

[13] Szyca M, Maciejewski I and Chamier-Gliszczyński N 2014 Analiza dynamiki pojazdu samochodowego z wykorzystaniem modelu symulacyjnego oraz badań terenowych (The analysis of a vehicle dynamics using simulation model) *Logistyka* **3** 6164

[14] Vaishnav A, Sarvaiya M and Dhabaliya P 2016 Mathematical Modelling and Comparison of Two Degree of Freedom Suspension System of Quarter Car *Imperial Journal of Interdisciplinary Research* **2** 128

[15] Smith W A and Zhang N 2010 Hydraulically interconnected vehicle suspension: optimization and sensitivity analysis *Proc. IMechE* **224** D 1335

[16] Ding F, Han X, Luo Z and Zhang N 2012 Modelling and characteristic analysis of tri-axle trucks with hydraulically interconnected suspensions *Vehicle System Dynamics* **50(12)** 1877

[17] Stańco M, Iluk A and Sawicki M 2019 Static and Dynamic Tests of Suspension System Heavy Off-road Vehicle *Proceedings of the 14th International Scientific Conference: Computer Aided Engineering* 752

IOP Conf. Series: Materials Science and Engineering 1247 (2022) 012020 doi:10.1088/1757-899X/1247/1/012020

Vibration damping performance of the electrohydraulic self-leveling unit of the tracked platform of robotic bricklaying combine

R. Dindorf [1*], P. Woś [1]

[1]Kielce University of Technology, al. Tysiąclecia Państwa Polskiego 7, 25-314 Kielce, Poland

*E-mail: dindorf@tu.kielce.pl

Abstract. The study proposes the use of a semi-active hydro-pneumatic spring damper unit for damping pressure pulsations in the servohydraulic leveling system of an electrohydraulic self-leveling unit of a tracked platform of a robotic bricklaying combine (RBC). The semi-active hydro-pneumatic spring damper unit consisted of a gas spring in the hydro-pneumatic accumulator and a controlled hydraulic damping valve. A model of a hydro-pneumatic accumulator with a connection tube was presented, using lumped hydraulic elements in the form of resistance, inductance, and capacitance (RLC). This model was transformed into a form of frequency impedance, from which the amplitude and phase characteristics were determined. The complex impedance of the throttle valve was determined, in which the real part is resistance and the imaginary part is reactance. Finally, the damping performance characteristics of a semi-active hydro-pneumatic spring damper unit were analyzed.

Keywords: vibration, vehicle, automotive safety

1. Introduction

The implementation of advanced technology in construction using robots significantly accelerates and leads to an improvement in the quality of the bricklaying and plastering work. Recently, productivity improvement has not been a major problem in the construction sector, but there is a lack of a sufficient supply of skilled construction workers. Currently, the construction sector is approaching a decisive period of rapid automation and robotization of traditional and innovative technological processes. Robotics is considered a potential solution to improve the efficiency of the construction industry. There have been many reports in literature on the implementation of masonry robots. One study [1] presented research on the possibility of automating masonry. A masonry robot capable of autonomously building walls using cinder blocks is designed. Pritschow et al. demonstrated an automated masonry construction process on a building site using a mobile robot [2]. Madsen analyzed the benefits and weaknesses of SAM100 (Semi Automated Mason) cooperating with a mason worker, which smooths out excess mortar [3]. SAM100 is more effective for large box-shaped structures, such as warehouses. These structures were primarily composed of bricks and other masonry materials. An alternative method is panelization or prefabrication of brick panels in a plant environment [4]. In [5], the author presented a comprehensive methodological process for implementing automation and robotics in housing construction.
In previous studies, the authors presented a hydraulic control system for the self-leveling system of a tracked platform for a robotic bricklaying combine (RBC) [6–7]. In cooperation with the industrial

partner Strabag Poland, an innovative research project aimed at designing, manufacturing, and implementing the first mobile RBS in Poland. Strabag is a large international company that focuses on building, construction, civil engineering, and transportation infrastructures. The RBC project was designed for masonry technology classified as a "mobile masonry robot". The main applications of mobile RBS are the facades and partition walls of offices and residential buildings, as well as industrial halls. The mobile RBC enables bricklaying of walls with dimensions limited to the working space of the robot. RBS optimizes the time, cost, and efficiency of masonry work and reduces the amount of waste. The mobile RBC is characterized by a new and innovative design solution, the task of which is to robotize the time-consuming and heavy bricklaying work traditionally performed by hand by masons. Figure 1 shows a 3D model of an RBC. This RBC is built with a bricklaying robot, a tracked platform (built-up tracked undercarriage with hydraulic unit), the hydraulic self-leveling unit, the brick warehouse, the brick feeder, a control cabinet, and the control panel. Due to the large mass of the bricklaying robot, the brick warehouse with feeders, and the control cabinet, a stable, robust and vibration-resistant leveling system is required. Four hydraulically extendable legs were used to stabilize the position of the tracked platform during automatic bricklaying. The main task of this hydraulic self-leveling unit is to raise the tracked platform to the desired height and precisely level it. The use of an electrohydraulic self-leveling unit is justified in the following cases: level control is needed; level control needs to work frequently; level control needs to react quickly; robust components are required; hydraulic supply is available. The operating conditions of the heavy bricklaying robot with variable load mass caused mechanical vibrations of the structural connections and hydraulic elements of the hydraulic leveling unit. Excess mechanical vibration affects the accuracy of RBC operation. These vibrations excite pressure pulsations in the hydraulic actuators of the self-leveling units. A semi-active hydro-pneumatic spring damper unit was used to reduce the pulsation loads in hydraulic actuators. It is desirable to use a semi-active hydro-pneumatic spring damper unit in which the damping of pressure pulsations is controlled in response to constantly changing excitations of the hydraulic actuator. The semi-active hydro-pneumatic spring damper unit consists of a gas spring in the hydro-pneumatic accumulator and a controlled hydraulic damping valve. The air spring absorbs the excited pressure pulsations, and the hydraulic damper absorber dissipates vibration energy.

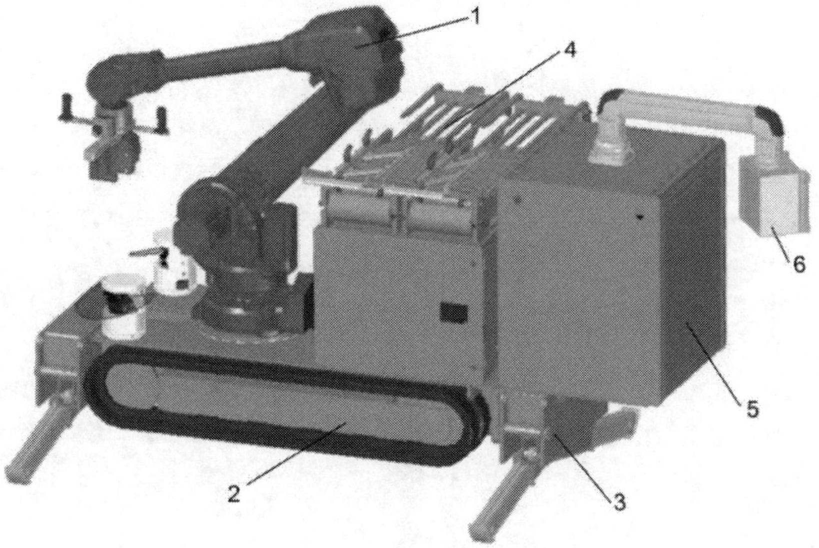

Figure 1. 3D model of an RBC: 1 – bricklaying robot, 2 – tracked platform, 3 – hydraulic self-leveling unit, 4 – brick warehouse and brick feeder, 5 – control cabinet, 6 – control panel.

IOP Conf. Series: Materials Science and Engineering 1247 (2022) 012020 doi:10.1088/1757-899X/1247/1/012020

Two electrohydraulic self-leveling units were used to lift and level the tracked platform of an RBC, in which four support legs mounted on a cross-layout are hydraulically extended. The electrohydraulic self-leveling unit is shown in Figure 2.

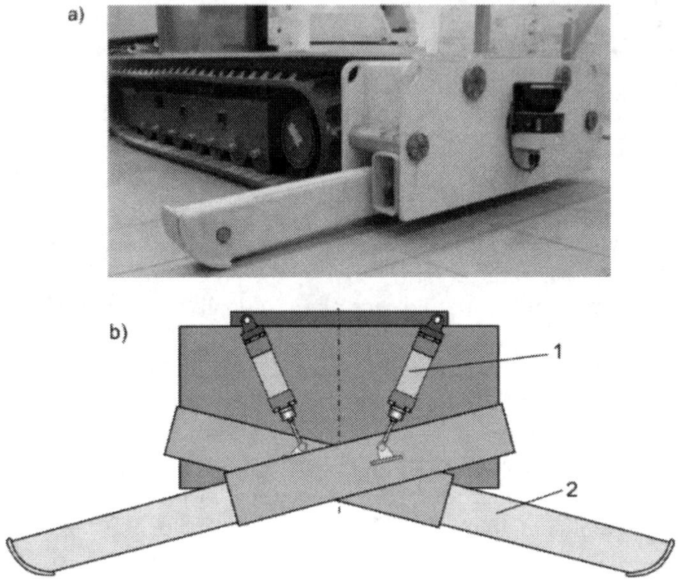

Figure 2. Electrohydraulic self-leveling unit of an RBC: a) view of the extendable one support leg, b) cross-layout of the hydraulically extended support legs: 1 – hydraulic actuator, 2 – support leg.

Figure 3 shows the servohydraulic leveling system with a semi-active hydro-pneumatic spring damper unit used in further analytical considerations.

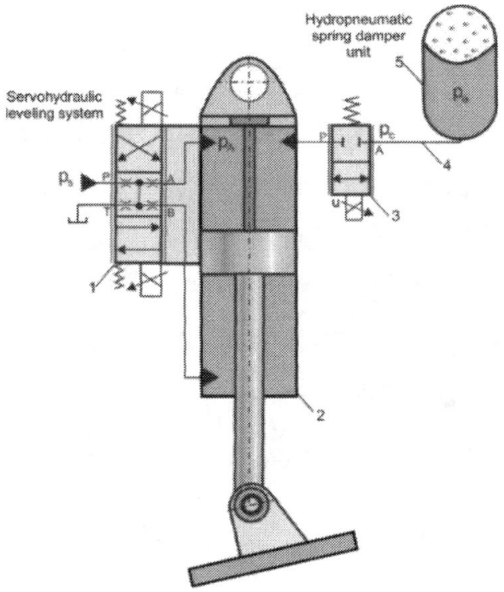

Figure 3. Schematic diagram of the servohydraulic leveling system with a semi-active hydro-pneumatic spring damper unit: 1 – proportional directional control valve, 2 – hydraulic leveling actuator, 3 – proportional throttle valve, 4 – connecting tube, 5 – hydro-pneumatic accumulator.

IOP Conf. Series: Materials Science and Engineering 1247 (2022) 012020 doi:10.1088/1757-899X/1247/1/012020

The leveling of an RBC occurs in two stages: maximum extension of the supporting legs and then lifting and leveling of the tracked platform. In the first stage, the cylinders extend the supporting legs without external load until they come into contact with the ground. In the second stage, the actuators pressed the supporting legs until the tracked platform was raised to the required height. This is followed by leveling in which the servohydraulic leveling systems are controlled to the horizontal position of the track platform. After leveling, the tracked platform is locked mechanically. During robotic bricklaying, the tracked platform was dynamically loaded. These loads generate vibrations transmitted to the bricklaying robot and the servohydraulic leveling system. Therefore, a semi-active hydro-pneumatic spring damper was introduced to maintain the stable position on the track platform exposed to vibration during robotic bricklaying. A semi-active hydro-pneumatic spring damper unit was introduced to maintain a stable position of the dynamically loaded tracked platform during robotic bricklaying. An electrohydraulic self-leveling unit comprises two support legs and two servohydraulic leveling systems with a semi-active hydro-pneumatic spring damper unit. The servohydraulic leveling system consists of a hydraulic leveling actuator, a 4/3 proportional directional control valve, directly operated with electrical position feedback, and integrated control electronics (ICE). The semi-active hydro-pneumatic spring damper unit consists of a 2/2 proportional throttle valve, a connecting tube, and a hydro-pneumatic accumulator. When a controlled throttle valve is placed between the hydraulic actuator and the accumulator, adjustable damping is achieved, which occurs in semi-active shock absorbers. The throttle valve continuously controls the flow resistance (throttling) using the voltage of the proportional solenoid coil. As the coil voltage increases/decreases, the valve spool shifts to increase/decrease flow resistance.

2. Transfer function of the hydro-pneumatic accumulator including the connecting tube

The hydro-pneumatic accumulator with the connecting tube can be modelled by lumped circuit elements in the form of resistance, inductance, and capacitance (RLC),

$$L_h \frac{dq_c}{dt} + R_h q_c = p_c - p_a, \tag{1}$$

$$q_c = C_a \frac{dp_a}{dt} \Rightarrow \frac{dp_a}{dt} = \frac{1}{C_a} q_c , \tag{2}$$

where q_v is the flow rate in the connecting tube, p_a is the pressure of the gas (nitrogen) in the accumulator, p_c is the pressure in the connecting tube,
L_h is the hydraulic inductance of the connecting tube,

$$L_h = \frac{\rho a}{\pi r^2}, \tag{3}$$

where r is the inner diameter of the connecting tube, ρ is the density of the oil, a is the velocity of the sound

$$a = \sqrt{\frac{K}{\rho}}, \tag{4}$$

where K is the bulk modulus of oil,
R_h is the hydraulic resistance of the connecting tube from the Hagen–Poiseuille equation,

$$R_h = \frac{8 l \eta}{\pi r^4}, \tag{5}$$

where l is the length of the connecting pipe, η is the dynamic viscosity of oil,

C_a is the gas capacitance of the hydraulic accumulator [4],

$$C_a = \frac{dV}{dp} = \frac{V_0}{n\,p_0}\left(1 - \frac{\Delta V}{V_0}\right)^{n+1} \approx \frac{V_0}{n\,p_0}, \tag{6}$$

where p_0, V_0 is the pressure and volume of the gas in the initial state and n is the polytropic index of change in the gas state.
After differentiation of Eq.(1) and substituting Eq.(2), it is obtained;

$$C_a L_h \frac{d^2 q_c}{dt^2} + C_a R_h \frac{dq_c}{dt} + q_c = C_a \frac{dp_c}{dt}. \tag{7}$$

After the Laplace conversion, Eq.(7) was transformed into a transfer function,

$$G_a(s) = \frac{q_c(s)}{p_c(s)} = \frac{C_a\,s}{C_a L_h s^2 + C_a R_h s + 1} = \frac{C_a\,s}{\frac{1}{\omega_n^2} s^2 + \frac{2\zeta}{\omega_n} s + 1}, \tag{8}$$

where ω_n is the natural frequency,

$$\omega_n = \frac{1}{\sqrt{C_a L_h}}, \tag{9}$$

ζ is the damping coefficient,

$$\zeta = \frac{1}{2} C_a R_h \omega_n. \tag{10}$$

The transfer function $G_v(j\omega)$ in the frequency domain of the accumulator, including the connecting tube, has the form;

$$G_a(j\omega) = \frac{q_c(j\omega)}{p_c(j\omega)} = \frac{C_a\,j\,\omega}{1 - \frac{\omega^2}{\omega_n^2} + \frac{2\zeta}{\omega_n} j\omega} = \frac{1}{\alpha\left[\beta + j\left(\frac{\omega}{\omega_n} - \frac{\omega_n}{\omega}\right)\right]}, \tag{11}$$

where α and β are the constant coefficients,

$$\alpha = \frac{1}{C_a \omega_n}, \tag{12}$$

$$\beta = 2\zeta. \tag{13}$$

3. Impedance of the hydro-pneumatic accumulator with the connecting tube

Fluctuations of a pressure and flow rate in the hydro-pneumatic accumulator including the connecting tube can be modelled in the frequency domain using the impedance $Z(j\omega)$, which, analogously to the electrical impedance in an AC circuit (ratio voltage to current), is determined by the ratio of pressure Δp and flow rate q_v;

$$Z(j\omega) = \frac{\Delta p(j\omega)}{q_v(j\omega)} \quad \text{in} \quad \frac{P_a s}{m^5}. \tag{14}$$

The impedance $Z_a(j\omega)$ of the hydro-pneumatic accumulator with the connecting tube is reciprocal to (11);

$$Z_a(j\omega) = \frac{p_c(j\omega)}{q_c(j\omega)} = \alpha\left[\beta + j\left(\frac{\omega}{\omega_n} - \frac{\omega_n}{\omega}\right)\right]. \tag{15}$$

From Eq.(15) results the impedance modulus;

$$|Z_a(j\omega)| = \alpha\sqrt{\beta^2 + \left(\frac{\omega}{\omega_n} - \frac{\omega_n}{\omega}\right)^2}, \tag{16}$$

and the impedance argument;

$$\arg\left[Z_a(j\omega)\right] = \arctan\frac{\left(\frac{\omega}{\omega_n} - \frac{\omega_n}{\omega}\right)}{\beta}. \tag{17}$$

Eq.(16) was presented in a dimensionless form;

$$|Z_A(j\omega)| = \frac{|Z_a(j\omega)|}{\alpha} = \sqrt{\beta^2 + \left(\frac{\omega}{\omega_n} - \frac{\omega_n}{\omega}\right)^2}. \tag{18}$$

The characteristics of the impedance amplitude and impedance phase of the hydro-pneumatic accumulator for the basic parameters introduced: $l = 0.01$ m, $r = 0.0025$ m, $p_0 = 1$ MPa, $V_0 = 0.001$ m^3 are shown in Figures 4 and 5.

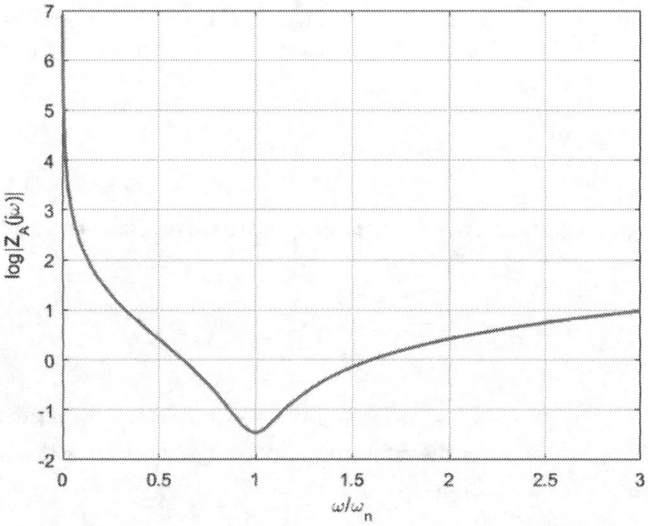

Figure 4. Characteristics of the impedance amplitude.

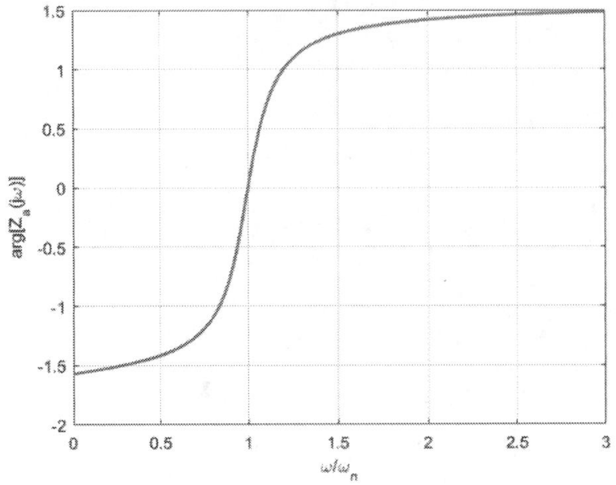

Figure 5. Characteristics of the impedance phase.

4. Damping performance of a semi-active hydro-pneumatic spring damper unit

The best mathematical way to write the impedance of the valve is a complex form, the real part is the 'resistance' R and the imaginary part is the 'reactance' of an inductive element $Z_L = j\omega L$;

$$Z(j\omega) = R + Z_L = R + j\omega L\,, \tag{19}$$

or of a capacitance element $Z_C = 1/j\omega C$;

$$Z(j\omega) = R + Z_C = R + \frac{1}{j\omega C}\,, \tag{20}$$

where ω is the angular frequency, R is the resistance, L is the inductance, and C is the capacitance. The phenomenon of flow through the valve describes not only the resistance R, but also the inductive effect L observed as a result of the high flow velocity. The capacitive effect can be explained by the volume of fluid contained in the valve chamber between the connection port and the orifice, the flow rate through the throttle control valve;

$$q_v(z) = \frac{C_d\, A(z)}{\sqrt{\rho}} \sqrt{\Delta p}\,, \tag{21}$$

where C_d is the discharge coefficient, A is the area of throttle valve, Δp is the pressure drop, z is the shift of a spool valve;

$$z = \frac{z_{\max}}{u_{\max}} u = K_z u\,, \tag{22}$$

where K_z is the gain factor and z_{max} is the maximum deflection of the spool from the center position, $z_{max} = max(|z_{smin}|, |z_{smax}|)$, z_{smin} is the minimum spool stroke, z_{smax} is the maximum spool stroke, the spool path is limited to $z_{smin} \leq z \leq z_{smax}$.
Eq.(18) linearization for small perturbations around the operating n-point;

$$q_v(z_n) = \frac{\sqrt{2}}{2} \frac{C_d\, A(z_n)}{\sqrt{\rho}\,\sqrt{\Delta p_n}} \Delta p = \frac{1}{R(z_n)} \Delta p\,, \tag{23}$$

where R is the flow resistance through the throttle control valve;

$$R(z_n) = \frac{2}{\sqrt{2}} \frac{\sqrt{\rho}}{C_d \, A(z_n)} \sqrt{\Delta p_n} \,. \tag{24}$$

The dynamic valve model is presented as the first-order differential equation given by

$$q_v = \frac{1}{R} \Delta p + C \frac{d\Delta p}{dt}, \qquad \Rightarrow \qquad q_v(s) = \frac{1}{R} \Delta p(s) + C \, s \, \Delta p(s), \tag{25}$$

where q_v is the flow rate in the control valve, Δp is the pressure drop, C is the capacitance of the valve, and R is the flow resistance.
The impedance of the throttle valve is characterized by a PT1 transfer function;

$$Z_v(s) = \frac{\Delta p(s)}{q_v(s)} = \frac{R}{1 + R\,C\,s} = \frac{R}{1 + T\,s}, \tag{26}$$

where T is the constant time $T = RC$;

$$Z_v(j\omega) = \frac{\Delta p(j\omega)}{q_v(j\omega)} = \frac{R}{1 + j\omega T} \,. \tag{27}$$

It is possible to remove the effects of the impedance of the capacitive component of the valve impedance by shifting the point at which the impedance is measured inside the valve so that the effective volume is zero. The impedance measured in this way is closer to the simple resistance model, and then (27) has the form $Z_v = R$.
The impedance function results from the flow continuity equation in a semi-active hydro-pneumatic spring damper unit,

$$q_c = \frac{1}{Z_a} p_c = q_v = \frac{1}{Z_v} \Delta p = \frac{1}{Z_v} (p_A - p_c) \tag{28}$$

$$p_A = \left(\frac{Z_v}{Z_a} + 1 \right) p_c \qquad \Rightarrow \qquad Z_p = \frac{p_A}{p_c} = 1 + \frac{Z_v}{Z_a} \tag{29}$$

where p_0 is the upstream pressure set by the control valve and p_c is the downstream pressure.
The impedance function was calculated from (29);

$$Z_p(j\omega) = 1 + \frac{R}{\alpha\beta + \alpha j \left(\frac{\omega}{\omega_n} - \frac{\omega_n}{\omega} \right)} \tag{30}$$

The impedance modulus was determined as follows;

$$|Z_p(j\omega)| = \sqrt{1 + \frac{R^2 + 2R\,\alpha\beta}{\alpha^2\beta^2 + \alpha^2 \left(\frac{\omega}{\omega_n} - \frac{\omega_n}{\omega} \right)^2}} \tag{31}$$

The semi-active hydropneumatic spring damper unit can be evaluated by the damping performance defined as follows;

$$D = 20 \log |Z_p(j\omega)| \quad dB \tag{32}$$

Based on (32), the characteristics of the damping performance of the semi-active hydro-pneumatic spring damper unit were determined. Figure 6 shows the characteristics of the damping performance for different input voltages u of the controlled throttle valve at the upstream pressure $p_A = 1$ MPa. Figure 7 shows the characteristics of the damping performance for different upstream pressures p_A at the input voltage $u = 5$ V of the throttle valve.

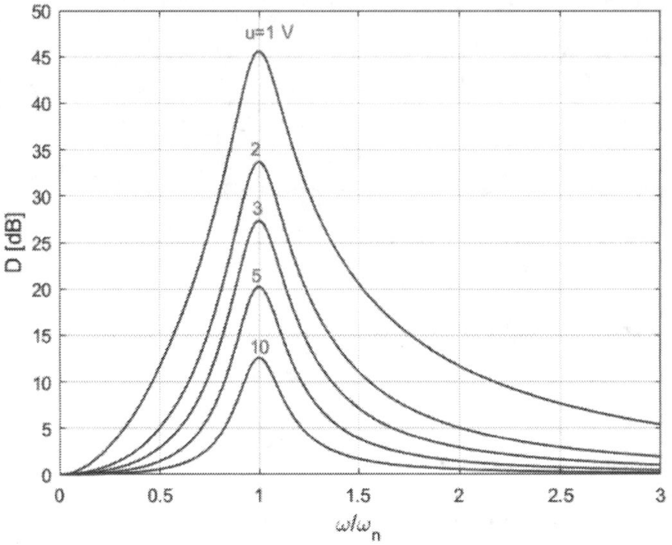

Figure 6. Characteristics of damping performance for different input voltages u of the controlled throttle valve at upstream pressure $p_A = 1$ MPa.

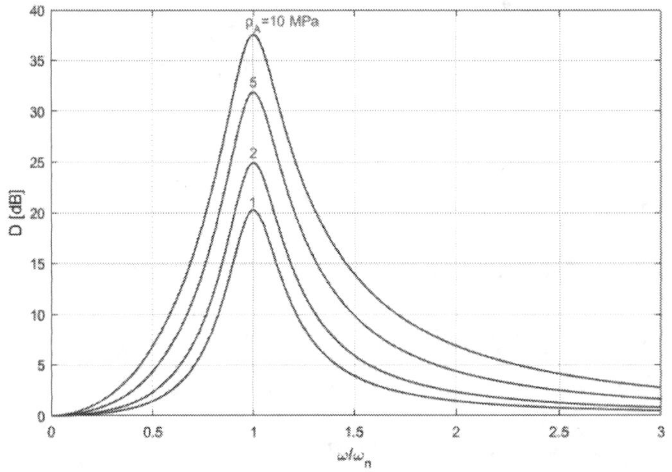

Figure 7. Characteristics of damping performance for different upstream pressures p_A at input voltage $u = 5$ V of the throttle valve.

The effectiveness of the damping system was evaluated using the peak pressure overshoot ratio defined as follows;

$$\delta_p = \frac{\Delta p}{p_{pp}} = \frac{p_{pp} - p_{sp}}{p_{pp}} \tag{33}$$

where Δp is the overshoot of the peak pressure, p_{pp} is the peak pressure, and p_{sp} is the set pressure. Figure 8 compares the simulation and the experimental shock pressure ratio for different shock forces.

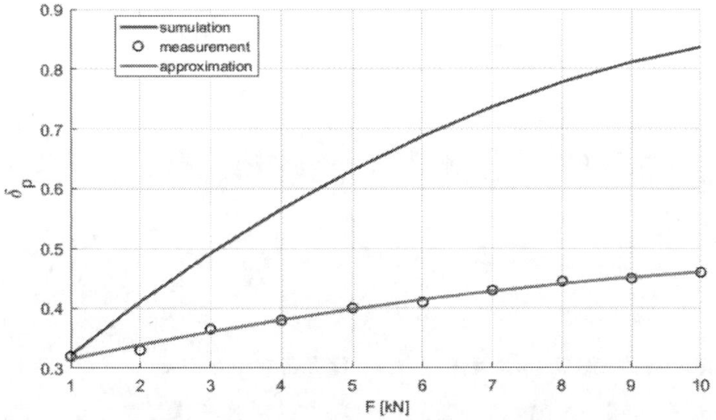

Figure 8. Comparison of the peak pressure overshoot ratio for various shock forces.

5. Conclusions

This study focuses on the frequency analysis of the damping performance of a semi-active hydro-pneumatic spring damper unit in the servo-hydraulic leveling control of a semi-active electrohydraulic leveling unit of a tracked platform of an RBC. Based on the impedance characteristics, the parameters of the hydro-pneumatic accumulator and connected tube were selected to minimize pressure pulsation for a given flow rate. By controlling the flow resistance of the throttle valve, the damping performance characteristics of the semi-active hydro-pneumatic spring damper unit were selected for the vibrations of the servo-hydraulic leveling system excited by external loads of the robotic bricklaying combine. Moreover, this work requires further research on the control system of a semi-active hydro-pneumatic spring damper unit.

Acknowledgments

This study, part of a research project POIR.04.01.02-00-0045/18-00 ‘Development and demonstration of a robotic bricklaying and plastering system for use in the construction industry’, was financially supported by the National Centre for Research and Development in Poland within the framework of the Smart Growth Operational Programme 2014-2020.

References

[1] Dakhli Z and Lafhaj Z. 2017 Robotic mechanical design for bricklaying automation. *Cogent Engineering* vol.4 pp.1-22.
[2] Pritschow G, Dalacker M and Kurz J. 1995 Gaenssle, M. Technological aspects in the development of a mobile bricklaying robot. In Proc. 12the ISARC I. Sym. on Auto. and Rob. in Const. Warsow, Poland, 30.05-01.06.1995 pp.1–8.
[3] Madsen A J. 2019 The SAM100: Analyzing labor productivity. Construction Management Department, California Polytechnic State University, San Luis Obispo USA.
[4] Rihani R A and Bernold L E. 1996 Methods of control for robotic brick masonry. *Automation in Construction* 4 pp.281-292.
[5] Pan W. Methodological development for exploring the potential to implement on-site robotics and automation in the context of public housing construction in Hong Kong. Ph.D. Thes. Lehrstuhl für Baurealisierung und Baurobotik, Technische Universität München, München 2020, Germany.
[6] Wos P, Dindorf R and Takosoglu J. 2020 The electro-hydraulic lifting and leveling system for the bricklaying robot. *In Book: Lecture Notes in Mechanical Engineering* Springer no 13 pp 215-227.
[7] Wos P, Dindorf R and Takosoglu J. 2021 Bricklaying robot lifting and leveling system. *Communications - Scientific Letters of the University of Žilina* vol 23 no 4 pp B257-B264.

IOP Conf. Series: Materials Science and Engineering 1247 (2022) 012021 doi:10.1088/1757-899X/1247/1/012021

The bicycle model of a 4WS car lateral dynamics for lane change controller

A Dębowski[1], J J Faryński[2,*] and D P Żardecki[1]

[1] Military University of Technology, Faculty of Mechanical Engineering, 00-908 Warsaw, Gen. Sylwestra Kaliskiego 2, Poland.
[2] Military University of Technology, Doctoral School, 00-908 Warsaw, Gen. Sylwestra Kaliskiego 2, Poland.

[*] jakub.farynski@wat.edu.pl

Abstract. The article shows the validity of four-wheel steering (4WS) bicycle model. At the beginning it is presented in which tasks, connected with 4WS vehicles, the bicycle model is applied. Then the equations in local coordinate system and their forms in global coordinate system are presented. The research tested the response of the model to a lane change manoeuvre, using transfer functions and a "bang-bang" control signal. The transfer functions was connected with static ratio characteristic to construct full four-wheel steering model. A reference signal generator and vehicle model were constructed in analytical form and checked in Matlab&Simulink using sample vehicle's data from another literature item. One example of simulations was shown. The results presented led to interesting conclusions and indicated directions for further research. This research will include the issue of regulators derivation. This will allow for the construction of a fully functioning control system. The effects of this activities will be tested again in simulation environment.

Keywords: road safety, infrastructure, automotive safety

1. Introduction

It was decided to take up the above topic of the article due to the authors' work on the topic of automatic control the lane change manoeuvre by a car with four steered wheels (4WS). The bicycle model of a 4WS car lateral dynamics is treated as the basic model for the synthesis of algorithms of the lane change controller. The idea behind this system was shown in the Figure 1.

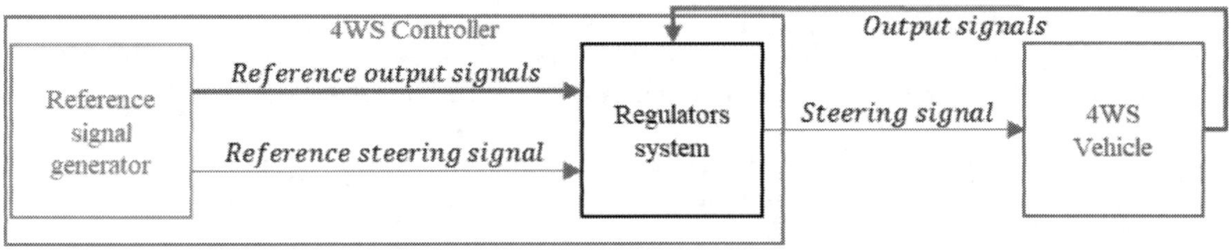

Figure 1. The idea of 4WS control system.

The proposed control system consists of a reference signal generator and regulators [5]. The reference signal generator is based on the lane change process reference model. It generates a reference steering signal and corresponding reference output signals. According to the theory of control systems and the practice of drivers [7], the reference steering signal has a form of "bang-bang" type signal. This signal is then corrected by regulators which work on the signal errors between reference and measured (in real car) output signals. The regulators correct the reference control signal so that, after correction, the actual control signal acting on the car ensures the minimization of errors, and finally a good implementation of lane change process. The algorithms of the regulators are also based on a reference model describing the lane change process. The reference model is therefore crucial for the design of the lane change controller. As shown in the literature analyzes, the bicycle model turned out to be the correct reference model used in many solutions of other authors.

The bicycle model is one of the models that forms the basis for considering the control concept in cars (2WS as well as 4WS), which can be seen from several publications in which this model appears together with other solutions. In papers [1], [9], [10] and [21] the bicycle model is used together with the MPC (Model Predictive Control) technique for the process of lane change. A wider spectrum has even devoted several review papers to the bicycle model: [5], [6], [8], [16] and [17]. It has also been used in, among other things: various computer simulations [2], real-world studies [3], in the study of variable vehicle dynamics [4], vehicle performance [12], active steering [15], comparison of 2WS and 4WS vehicle control [18] and control system design [20].

2. The "bicycle model" in ODE form

The bicycle model is a plane model with three degrees of freedom. It assumes that the vehicle's centre of mass is located on the plane of the road. The vehicle itself is symmetrical with respect to the longitudinal vertical plane. The axles are reduced to individual wheels. This model assumes that the left and right wheels of the vehicle generate the same lateral forces and depend only on the drift angle of the entire vehicle axis (Figure 2).

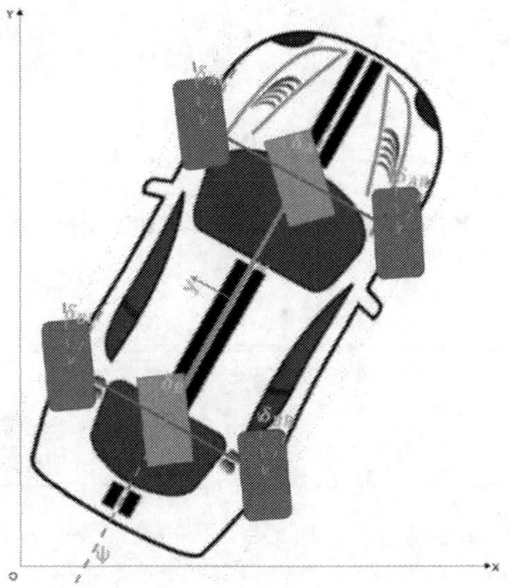

Figure 2. Idea of bicycle model.

Notation:

- (X, Y) – global coordinate system,
- (x, y) – local coordinate system,
- m – vehicle mass,
- J – mass moment of inertia (J_{zz} to p. C),
- $L_A,\ L_B$ – distances AC and BC between the center of mass to the front / rear axis,

- K_A, K_B – front and rear tyre cornering stiffness (yaw coeficcients to p. A and B),
- δ_A, δ_B – front and rear steering angles,
- Ω – angular velocity of the vehicle,
- V – longitudinal vehicle speed in the local coordinate system,
- U – lateral vehicle speed in the local coordinate system.

The considered issues concern the control of motion of a 4WS vehicle travelling at a steady speed on a straight, level road with good adhesion, so that the well-known "bicycle model" can be used to describe the dynamics. The vehicle is treated as a moving single mass planar object on elastic torsion pneumatics, symmetric about its longitudinal axis. The road reaction forces are perpendicular to the wheel rim side planes and depend directly (functional dependence) on the drift angles, so the angles between the wheel rim side planes and the wheel tyre line marks [11]. The road reaction forces are related to the points defined by the vertical projections of the wheel axis centres. Using the assumptions, one obtains ordinary differential equations (ODE) form:

$$m\dot{U}(t) + \frac{K_A+K_B}{V}U(t) + \frac{mV^2+K_AL_A-K_BL_B}{V}\Omega(t) = K_A\delta_A(t) + K_B\delta_B(t) \tag{1}$$

$$J\dot{\Omega}(t) + \frac{K_AL_A{}^2+K_BL_B{}^2}{V}\Omega(t) + \frac{K_AL_A-K_BL_B}{V}U(t) = K_AL_A\delta_A(t) - K_BL_B\delta_B(t) \tag{2}$$

$$X(t) = \int_0^t \dot{X}(\tau)d\tau = \int_0^t \big(V\cos(\psi(\tau)) - U(\tau)\sin(\psi(\tau))\big)d\tau \tag{3}$$

$$Y(t) = \int_0^t \dot{Y}(\tau)d\tau = \int_0^t (V\sin(\psi(\tau)) + U(\tau)\cos(\psi(\tau)))d\tau \tag{4}$$

$$\psi(t) = \int_0^t \Omega(\tau)d\tau \tag{5}$$

When $\psi(t)$ is small (for example during the lane change process) the linearisation of non-linear equations is possible and in this case equations (3), (4) obtain the linear form:

$$X(t) = \int_0^t \dot{X}(\tau)d\tau = Vt \tag{6}$$

$$Y(t) = \int_0^t \dot{Y}(\tau)d\tau = \int_0^t \big(V\psi(\tau) + U(\tau)\big)d\tau \tag{7}$$

3. Transmittance formulation of the model

The analysis of the 4WS bicycle model will be based on the research of the 4WS vehicle lane change process through a suddenly appearing obstacle [13] and [14]. The driver or controller during this manoeuvre generates a "bang-bang" signal (a jerk of the steering wheel, a hold, a jerk in the other direction, a hold and a return to the zero position) (Figure 3).

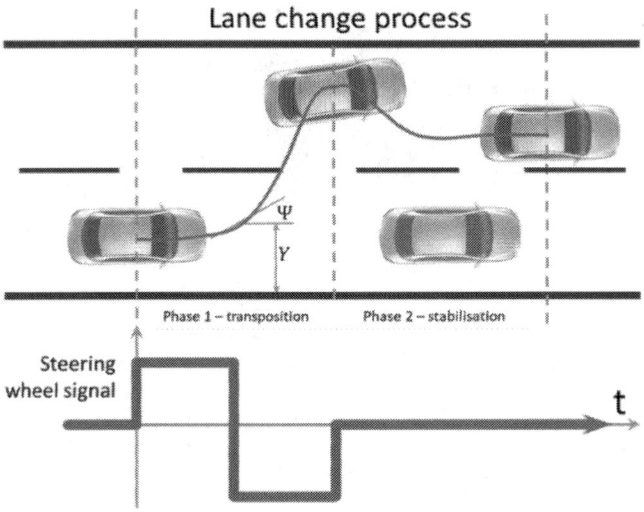

Figure 3. The concept of time decomposition of lane change control in 2WS vehicle [6].

After the equations have been linearised, they can be subjected to the Laplace transform and then the model was determined as a transfer function forms (Figure 4):

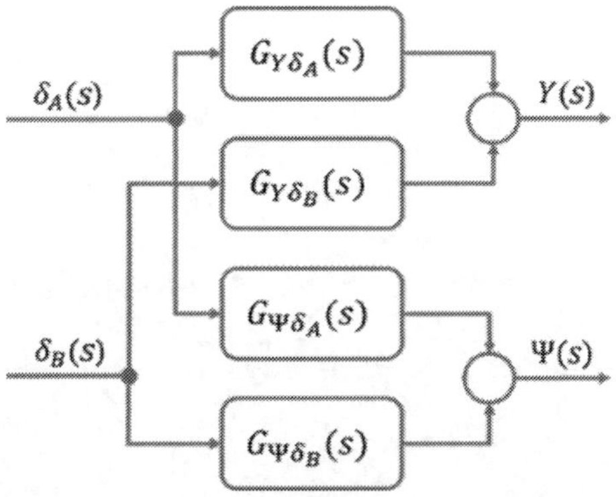

Figure 4. Model in transfer functions version.

$$G_{Y\delta_A}(s) = \frac{K_{Y\delta_A}\left(T^2_{Y\delta_A}s^2+2\xi_{Y\delta_A}T_{Y\delta_A}s+1\right)}{s^2(T_0^2s^2+2\xi_0T_0s+1)} \qquad G_{Y\delta_B}(s) = \frac{K_{Y\delta_B}(T^2_{Y\delta_B}s^2+2\xi_{Y\delta_B}T_{Y\delta_B}s-1)}{s^2(T_0^2s^2+2\xi_0T_0s+1)} \quad (8, 9)$$

$$G_{\psi\delta_A}(s) = \frac{K_{\psi\delta_A}(T_{\psi\delta_A}s+1)}{s(T_0^2s^2+2\xi_0T_0s+1)} \qquad G_{\psi\delta_B}(s) = \frac{K_{\psi\delta_B}\left(T_{\psi\delta_B}s+1\right)}{s(T_0^2s^2+2\xi_0T_0s+1)} \quad (10, 11)$$

where:

$$K_{Y\delta_A} = K_{Y\delta_B} = VK_0 \quad K_{\psi\delta_A} = K_0 \quad K_{\psi\delta_B} = -K_0 \quad K_0 = \frac{K_AK_B(L_A+L_B)V}{K_AK_B(L_A+L_B)^2-mV^2(K_AL_A-K_BL_B)} \quad (12\text{-}15)$$

$$T_0 = V\sqrt{\frac{mJ}{K_AK_B(L_A+L_B)^2-mV^2(K_AL_A-K_BL_B)}} \qquad \xi_0 = \frac{m(K_AL_A{}^2+K_BL_B{}^2)+J(K_A+K_B)}{2\sqrt{mJ(K_AK_B(L_A+L_B)^2-mV^2(K_AL_A-K_BL_B))}} \quad (16, 17)$$

$$T_{\psi\delta_A} = \frac{mL_AV}{K_B(L_A+L_B)} \quad T_{\psi\delta_B} = \frac{mL_BV}{K_A(L_A+L_B)} \quad T_{Y\delta_A} = \sqrt{\frac{J}{K_B(L_A+L_B)}} \quad T_{Y\delta_B} = \sqrt{\frac{J}{K_A(L_A+L_B)}} \quad (18\text{-}21)$$

$$\xi_{Y\delta_A} = \frac{L_B}{2V}\sqrt{\frac{K_B(L_A+L_B)}{J}} \qquad \xi_{Y\delta_B} = \frac{L_A}{2V}\sqrt{\frac{K_A(L_A+L_B)}{J}} \quad (22, 23)$$

The above representation of the transfer functions show a very strong connection between the transfer functions parameters and the mechanical parameters. The transfer functions show the essence of the effects of steering angles on lateral and angular displacements.

In classic 4WS controls, the rear wheels are controlled by transforming the front wheel control accordingly (Figure 5).

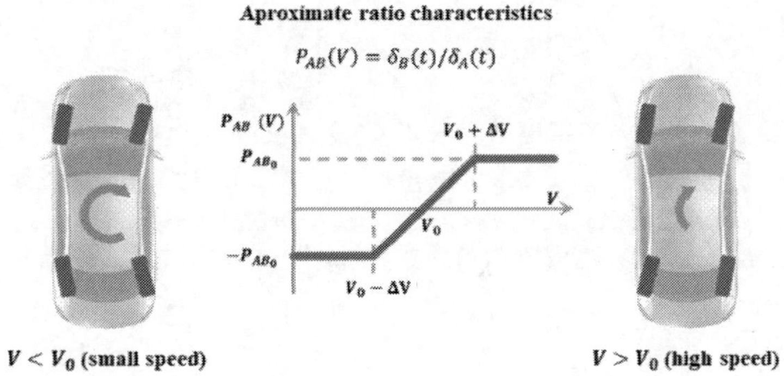

Figure 5. Overall idea of steering front and rear wheels in 4WS vehicles.
Notation: P_{AB} – gear ratio between front and rear wheel twist angles (δ_A, δ_B)
V – vehicle speed, V_0 – characteristic speed when δ_B=0.

With the assumed ratio characteristics, the transfer functions model appears as follows (Figure 6):

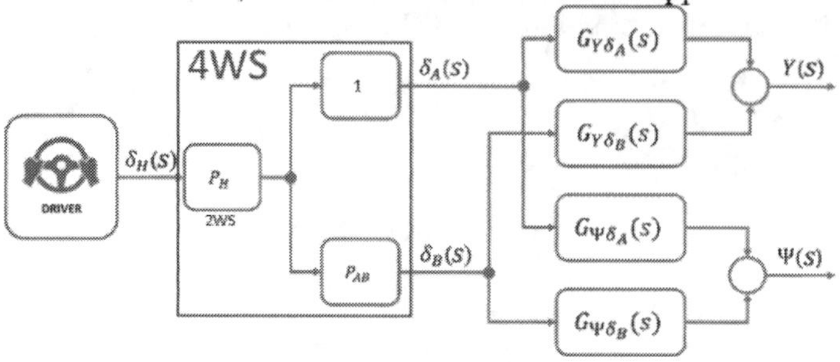

Figure 6. Model including rear wheel ratio.

$$Y(s) = G_{Y\delta}(s)\delta(s) \quad \text{where} \quad G_{Y\delta}(s) = G_{Y\delta_A}(s) + P_{AB}(V)G_{Y\delta_B}(s) \tag{24, 25}$$

$$\psi(s) = G_{\psi\delta}(s)\delta(s) \quad \text{where} \quad G_{\psi\delta}(s) = G_{\psi\delta_A}(s) + P_{AB}(V)G_{\psi\delta_B}(s) \tag{26, 27}$$

$$\text{from: } G_{Y\delta}(s) = \frac{K_{Y\delta}(T_Y^2 s^2 + 2\xi_{Y\delta}T_{Y\delta}s + 1)}{s^2(T_0^2 s^2 + 2\xi_0 T_0 s + 1)} \qquad G_{\psi\delta}(s) = \frac{K_{\psi\delta}(T_{\psi\delta}s + 1)}{s(T_0^2 s^2 + 2\xi_0 T_0 s + 1)} \tag{28, 29}$$

where:

$$K_{Y\delta} = (1 - P_{AB}(V))VK_0 \qquad T_{Y\delta} = \sqrt{\frac{J\left(\frac{1}{K_B} + \frac{P_{AB}(V)}{K_A}\right)}{(L_A + L_B)(1 - P_{AB}(V))}} \qquad \xi_{Y\delta} = \frac{L_B + P_{AB}(V)L_A}{2V(1 - P_{AB}(V))\sqrt{\frac{J\left(\frac{1}{K_B} + \frac{P_{AB}(V)}{K_A}\right)}{(L_A + L_B)(1 - P_{AB}(V))}}} \tag{30-32}$$

$$K_{\psi\delta} = (1 - P_{AB}(V))K_0 \qquad T_{\psi\delta} = \frac{mV\left(\frac{L_A}{K_B} - P_{AB}(V)\frac{L_B}{K_A}\right)}{(L_A + L_B)(1 - P_{AB}(V))} \tag{33, 34}$$

4. Theoretical and simulation based on analyses

The next step is determine and analyse the limes of the transfer functions. In particular, the "bang-bang" input waveforms (Figure 7) should be considered, as they approximate the real waveforms during lane change process.

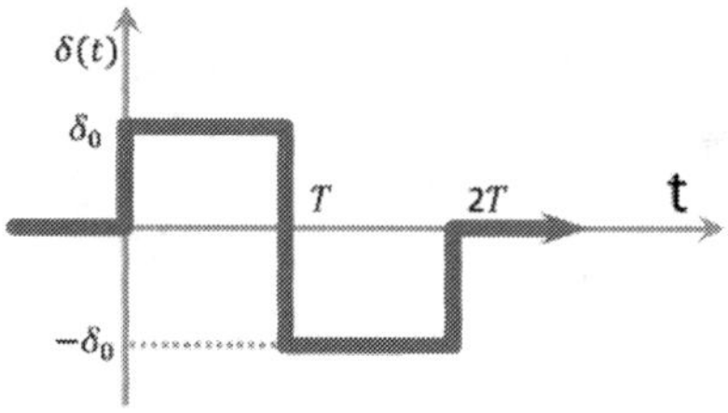

Figure 7. "Bang-bang" signal.

$$\delta(t) = \delta_0(1(t) - 2 \cdot 1(t - T) + 1(t - 2T) \qquad \delta(s) = \delta_0 \frac{(1 - e^{-sT})^2}{s} \tag{35, 36}$$

$$\delta_A(t) = \delta(t), \quad when \quad \delta_B(t) = P_{AB}(V)\delta(t) \qquad P_{AB}(V) > 0 \tag{37-39}$$

$$\lim_{t\to\infty}(\psi(t)) = \lim_{s\to 0}(s\psi(s)) = \lim_{s\to 0}\left(s(G_{\psi\delta_A}(s)\delta_A(s) + G_{\psi\delta_B}(s)\delta_B(s)\right) = 0 \tag{40}$$

$$\lim_{t\to\infty}(Y(t)) = \lim_{s\to 0}(sY(s)) = \lim_{s\to 0}\left(s\left(G_{Y\delta_A}(s)\delta_A(s) + G_{Y\delta_B}(s)\delta_B(s)\right)\right) = (1 - P_{AB}(V))\delta_0 T^2 V K_0 \tag{41}$$

It was decided to use also the model in a reduced version (members affecting only the transients are omitted), as this enables an analytical synthesis of the control of classical 4WS vehicles.

$$G_{Y\delta_R}(s) = \frac{K_{Y\delta}}{s^2} \qquad G_{\psi\delta_R}(s) = \frac{K_{\psi\delta}}{s} \tag{42, 43}$$

The same results of the boundary analysis are also obtained assuming a reduction of the initial transfer functions model. The limit defined by equation (39) is the steady-state displacement value Y_0, also the magnitude of the lane change. It should be noted here that theoretically, according to formula (40), the steady-state yaw angle is zero, so the vehicle moves on a track parallel to the original track. However, equations (40) and (41) are not sufficient to determine the signal parameters δ_0 and T. Therefore, an analogous study was performed here by updating the state after step excitation at time $t = T$ using (43). In this case, the angular displacement $\psi(T) = \psi_0$ reaches the maximum value ψ_0 defining the assumed linearity of the model. Based on these analyses, formulas for the values of T i δ_0 can be calculated (Figure 8):

$$T = \frac{Y_0}{V\psi_0} \qquad \delta_0 = \frac{V\psi_0^2}{K_{\psi\delta}Y_0} \tag{44, 45}$$

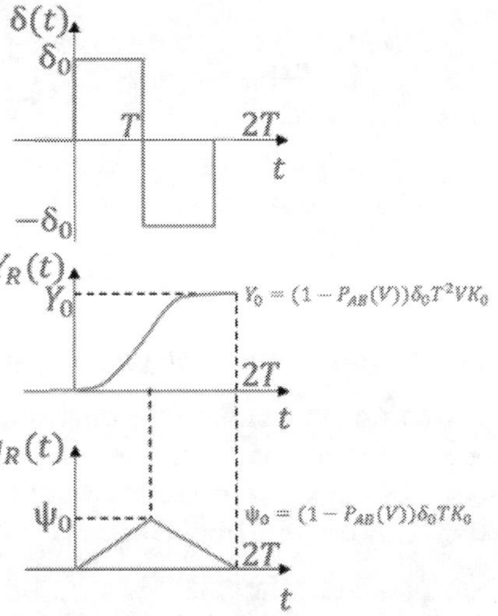

Figure 8. Reference signals in the control system.

The form of the individual elements for the control process must be defined at the outset in the planned control concept. This mainly concerns the form of the reference signal generator and the vehicle, which will generate the real signals on the basis of the reference signals. In the feedback loop, so-called failures will arise as a result of the interaction of the vehicle's individual parameters with the prevailing traffic situation. In order to check the operation of the analysed reference signal generator and 4WS vehicle, it was implemented into the Matlab&Simulink program. As a result of not having an real 4WS vehicle, this paper uses vehicle's data from the article [22] (Table 1).

Table 1. Vehicle's data.

Parameter	Unit	Value
m	$[kg]$	1740
J	$[kg \cdot m^2]$	3048
L_A	$[m]$	1,035
L_B	$[m]$	1,655
K_A	$[N/rad]$	35000
K_B	$[N/rad]$	37500
V_0	$[m/s]$	15
ΔV	$[m/s]$	5
$P_{AB\,0}$	$[-]$	0,1
Y_0	$[m]$	2
ψ_0	$[rad]$	0,11

The tests were carried out for several speed variants V with process time $t = 10\,s$ to check the functionality of the system. The graphs show reference (green color) and real (blue color) values. Example simulation result is presented in Figure 9.

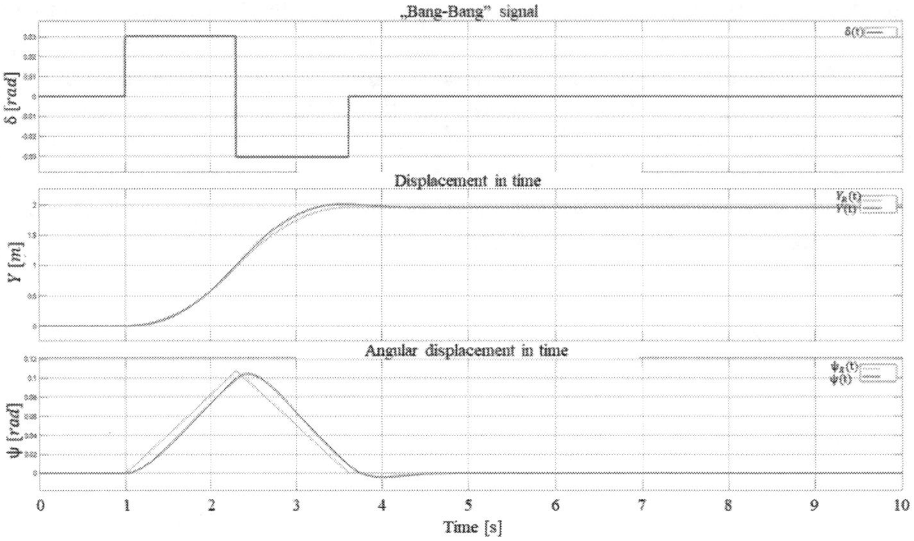

Figure 9. The result of simulation for $V = 50\ km/h$.

As a result of the simulations carried out, it can be seen what role the extended and reduced transfer functions have. Simulations were repeated for many different sets of data (not presented here). They had shown the sensitivity of the model, but in every cases the errors between the results of reduced and extended forms of transfer functions model were not high.

5. Conclusions

This study shows how important and significant role the bicycle model plays in the overall control concept of the 4WS cars. The linearisation of the model and its subsequent presentation in the form of operator calculus allowed the derivation of transfer functions, which greatly facilitated the process of analysing the model in terms of lane changes. Appropriate use of the "bang-bang" signal and a reduced model with the application of limes let to make reference signal generator. The bicycle model in reduced form can be used for analytical synthesis of the regulation system, which correct the reference steering signal The proposed model in extended as well as in reduced form seems to be good proposition for synthesis and analyses the lane change control system algorithms for 4WS vehicles. The simulations, which were carried out, had shown that constructed model works as theoretical. This simulations show very well the effect of transition parameters on model. Graphs of extended transfer functions are not as ideal as graphs of reduced transfer functions. It is possible to construct fully regulation system. The next step for further research is to derivate regulators to proposed system on Figure 1. Everything will be checked again in Matlab&Simulink system for verification sensitivity of the model for different vehicles' data.

6. References

[1] Ahmadi B., *Design a Reliable Model Predictive Control for Path Following with Application to the Autonomous Vehicle and Considering Different Vehicle Models,* Waterloo: University of Waterloo, 2021, pp. 13-43.

[2] Bilen Ö. B., *Path Control Algorithms for Autonomous Steering and Braking of Heavy Vehicles,* Göteborg: Chalmers University of Technology, 2010, pp. 10-52.

[3] Bin Ramil A. S., *Study on Influences of Rear-Wheel Turn Angle to Performance of Four Wheel Steering (4WS) Vehicle,* Tronoh: Universiti Teknologi PETRONAS, 2010, pp. 26-41.

[4] Fahimi F., "Full drive-by-wire dynamic control for four-wheel-steer all-wheel-drive vehicles," *Vehicle System Dynamics,* vol. 51, no. 3, pp. 360-376, 2013.

[5] Faryński J. J., Żardecki D. P., Dębowski A., "Study of the problem and the idea of operation of four-wheel-steering cars," *The Archives of Automotive Engineering - Archiwum Motoryzacji,* vol. 94, no. 4, pp. 39-59, 2021.

[6] Faryński J. J., Dębowski A., Żardecki D. P., "Wstępne studium problematyki sterowania w samochodach 4WS," *Biuletyn Wojskowej Akademii Technicznej,* vol. 69, no. 4, pp. 95-113, 2020.

[7] Frej D., Grabski P., Szumska E., "The Importance of Alternative Drive Vehicles in Road Transport in Poland and the European Union," *LOGI - Scienetific Journal on Transport and Logistics,* vol. 12, no. 1, pp. 67-77, 2021.

[8] Furukawa Y., Yuhara N., Sano S., Takeda H., Matsushita Y., "A Review of Four-Wheel Steering Studies from the Viewpoint of Vehicle Dynamics and Control," *Vehicle System Dynamics: International Journal of Vehicle Mechanics and Mobility,* vol. 18, no. 1-3, pp. 151-186, 1989.

[9] Gidlewski M., Jemioł L., Żardecki D. P., "Badania symulacyjne dynamiki procesu nagłego omijania przeszkody przez samochód," *The Archives of Automotive Engineering – Archiwum Motoryzacji,* vol. 73, no. 3, pp. 31-46, 2016.

[10] Gidlewski M., Jankowski K., Muszyński A., Żardecki D. P., "Vehicle Lane Change Automation with Active Steering - Theoretical Studies and Numerical Investigations," *Society of Atomotive Engineers International,* vol. 1555, pp. 1-11, 2017.

[11] Hryciów Z., Jackowski J., Żmuda M., "The Influence of Non-Pneumatic Tyre Structure on its Operational Properties," *International Journal of Automotive and Mechanical Engineering,* vol. 17, no. 3, pp. 8168-8178, 2020.

[12] Owen E. D., *The Benefits of Four-Wheel Drive for a High-Performance FSAE Electric Racecar,* Cambridge: Massachusetts Institute of Technology, 2018, pp. 21-64.

[13] Prochowski L., Pusty T., Gidlewski M., Jemioł L., "Experimental studies of the car-trailer system when passing by a suddenly appearing obstacle in aspect of active safety of autonomous vehicles," in *IOP Conference Series: Materials Science and Engineering*, 2018.

[14] Prochowski L., Ziubiński M., Szwajkowski P., Mirosław G., Pusty T., Stańczyk T. L., "Impact of control system model parameters on the obstacle avoidance by an autonomous car-trailer unit: Research results," *Energies,* vol. 14, no. 10, pp. 1-31, 2021.

[15] Scalzi S., *Integrated Control of Active Steering and Electronic Differentials in Four Wheel Drive and Steering Vehicles,* Rzym: University of Rome Tor Vergata, 2009, pp. 44-118.

[16] Spentzas K., Alkhazali I., Demic M., "Dynamics of four-wheel-steering vehicles," *Springer-Verlag,* vol. 66, pp. 260-266, 2001.

[17] Spentzas K., Alkhazali I., Demic M., "Kinematics of four-wheel-steering vehicles," *Springer-Verlag,* vol. 66, pp. 211-216, 2001.

[18] Tourajizadeh H., Sarvari M., Ordoo A. S., "Modeling and Optimal Control of 4 Wheel Steering Vehicle Using LQR and its Comparison with 2 Wheel Steering Vehicle," *International Journal of Robotics,* vol. 9, no. 1, pp. 20-32, 2020.

[19] Weigel-Milleret K., *Wpływ regulacji sił napędowych na stabilność ruchu wąskich pojazdów czterokołowych,* Kraków: Politechnika Krakowska, 2019, pp. 17, 23.

[20] Xia Y., Lin M., Zhang J., Fu M., Li C., Li S., Yang Y., "Trajectory planning and tracking for four-wheel steering vehicle based on differential flatness and active disturbance rejection controller," *International Journal of Adaptive Control and Signal Processing,* vol. 35, no. 11, pp. 2214-2244, 2021.

[21] Yakub F., Abu A., Sarip A., Mori Y., "Study of Model Predictive Control for Path-Following Autonomous Ground Vehicle Control under Crosswind Effect," *Journal of Dynamics Systems, Measurement and Control,* vol. 133, pp. 1-6, 2011.

[22] Yin G.-D., Chen N., Wang J.-X., Wu L.-Y., "A Study on μ-Synthesis Control for Four-Wheel Steering System to Enhance Vehicle Lateral Stability," *Journal of Dynamics Systems, Measurement and Control,* vol. 133, pp. 1-6, 2011.

The issue of empirical testing of Highly Automated Vehicles

Dariusz Więckowski *

[1]Institute of Vehicles and Construction Machinery Engineering, Faculty of Automotive and Construction Machinery Engineering, Warsaw University of Technology, ul. Narbutta 84, 02-524 Warszawa, Poland

* wiec.darek@gmail.com

Abstract. The study concerns the issue of empirical testing of Highly Automated Vehicles (HAVs). The driver assistance systems used in the HAVs have been listed. The principles of testing conventional vehicles have been presented and reference has been made to specific tests described in normative documents. The dissimilarity of testing the HAVs has been presented against this background.

Keywords: experimental testing of motor vehicles, experimental testing of Highly Automated Vehicles

1. Introduction

The measures taken to minimise the number of hazardous situations include the introduction of driver assistance systems, which can eliminate or reduce the risk of a hazard or collision by monitoring of, warning against, or autonomic responding to, the threats having emerged. The active safety of vehicles can be improved by e.g. using Advanced Driver Assistance Systems (ADAS) [1]. Such systems are becoming increasingly often standard production equipment of new motor vehicles. When the human driver is replaced with an automated system, the impact of driver reaction time and his/her motor skill (time of motion), which vary from person to person and depend on the current psycho-physical driver's condition, will be eliminated; instead, the time values being constant and characteristic for the automated system will be taken into account. The use of ADAS makes a direct step to replace the human driver with a machine and thus to build the autonomous vehicle capable of controlling itself and counteracting the loss of its controllability thanks to its capability of recognising the environment and performing collision-free motion without intervention from outside [1]. Simultaneously, the monotony of passive ride in the vehicle lengthens the driver reaction time in case of a sudden hazard. Another third-level problem is the possibility that some messages sent by the vehicle may escape driver's attention. Such a case resulted in a fatal accident with a vehicle of this type [2]. The development and increasing commercial use of highly automated motor vehicles is unavoidable. The works on such vehicles are being carried out by numerous companies, see e.g. [3, 4]. Therefore, the issue of testing the vehicles of this type has become particularly important, especially due to the fact that their specificity of driving considerably differs from that of conventional ones.

2. The specificity of highly automated vehicles

To enable the full-scale existence of autonomous vehicles, the available automatic driver assistance systems should be continuously developed and brought to compatibility with each other for their cooperation and reciprocal supplementing to be made possible. The Highly Automated Vehicles (HAVs) must support the following basic functions:

- to raise driver's consciousness of the situation at least by e.g. displaying information about changes in the current speed limits or an unintentional lane change;
- to provide early warning of the possibility of a hazard by emitting acoustic signals, light signals in the form of tell-tale lamps or pictorial messages displayed on LCD screens, or haptic signals coming from driver's seat or steering wheel;
- to intervene in order to prevent an accident or to reduce the accident effects at the instant when the driver reacts too late or fails to react at all; the system should be able to stop or slow down the vehicle when it recognizes that the driver does not take any action to avoid an imminent hazard or a collision.

The adequate operation of vehicles' active safety systems is only possible when they incorporate many sensors, which may be divided, in the most general terms, into two groups, i.e. active and passive ones [5, 6]. The more so, for the HAVs to function correctly, they must be provided with a number of such devices, e.g. [7, 8, 9]: radars [10, 11], lidars [12, 13, 14], supersonic sensors [15, 16, 17, 18, 19], satellite navigation [20, 21, 22, 23, 24], a variety of cameras [25, 26, 27, 28, 29, 30, 31, 32, 33, 34, 35], rate and directional gyroscopes [36], and accelerometers [37, 38].

Based on the sensors used, specific driver assistance systems could be built and applied, such as:

Dedicated Short Range Communication (DSRC) [39, 40];
Satellite navigation (GPS, GLONASS, Beidou, Galileo) [20];
Inertial Measurement Unit (IMU) – determining the position and orientation of an object in space by measuring its acceleration and angular velocity [41, 42];
Blind Spot Warning [43, 44, 45];
Cross Traffic Warning [46, 47, 48];
Lane Departure Warning [49, 50];
Lane Keep Assist [51, 52];
Cruise Control [53, 54];
Adaptive Cruise Control (ACC) [55, 56];
Cooperative Adaptive Cruise Control (CACC) [57];
Adaptive Cruise Control with Look-ahead Anticipation [58];
Forward Collision Warning (FCW) [59];
Autonomous Emergency Braking (AEB) [60];
Pedestrian Detection [61, 62];
Collision imminent steering [63, 64];
Do-not-pass warning [65, 66, 67, 68];
Left Turn Assist [69, 79, 71];
Intersection Movement Assist [72, 73, 74];
Emergency Electronic Brake Light (EEBL) [75, 76, 77, 78];
Traffic Sign Recognition (TSR) [79];
Traffic Jam Chauffer (TJS) [80, 81, 82];
Park Assist [83, 84];
Platooning – integrated escorting [85, 86, 87];
Raster map, vector map [88, 89].

Three separate but interconnected real-time operation phases may be distinguished in the operation of the HAVs' systems (similarly as it is in the industrial automatic control systems) [90, 91]:

1) starting phase, i.e. transition from the steady state of standstill to the operation state of the system;

2) working phase, i.e. the phase when the system begins its "normal" operation (the control algorithms reproduce the preset values);
3) termination phase, i.e. controlled transition from the normal operation to the standstill.

The entire process runs dynamically. If the quantities that describe the dynamic state of the system change insignificantly, then the control algorithms do not undergo structural changes. On the other hand, if the process is very dynamic (where big changes take place in a short time and/or various decisions must be made in real time), the object controlled requires the control algorithm structure to be changed. This means that the changes should be effected in such a case by physical switching over between the physical information transmission channels. Preferably, the transitions between individual states should be as smooth as possible. However, this requires an automatized control algorithm with a non-linear structure to be prepared for such a system. In consequence, the functioning of the system as a whole in real operation conditions must be verified by testing it in testing-ground and road conditions.

3. Vehicle testing

In the case of conventional vehicles operating within the man-vehicle-environment (MVE) system, the most important system component is the human vehicle driver, who is the source of the inputs that are strongly coupled with the basic movement of the vehicle and whose actions are decisive for the triggering or accepting of individual situations related to the vehicle movement.

Extensive research on the running characteristics of motor vehicles has made it possible to define and adopt the recommended test types and the test results evaluation criteria. Among such tests, the steerability and stability tests cover a set of problems related to the properties that determine not only the ease of steering a vehicle but also have a direct impact on active safety issues. The said properties are described by a set of characteristics of the vehicle as a dynamic system, representing its reactions (responses) to various external and internal forces and making it possible to predict the behaviour of the driver-vehicle system during various manoeuvres. The recognition and then appropriate shaping of vehicle's running properties helps the driver to maintain vehicle's stability while driving, especially in critical situations. The standard tests include:

steady-state tests on a circular track [92, 93]; step input applied to the steering wheel [94]; braking in a turn [95]; quick lane change [96, 97] (double lane-change manoeuvre test representing an attempt to avoid an obstacle); single-period sinusoidal input applied to the steering wheel [98, 99]; continuous sinusoidal input applied to the steering wheel [95, 99]; impulse and random steering wheel input [95, 99, 100]; single lane-change attempt [101].

Due to the dissimilarity of the HAVs and the specificity of their controlling, a different approach to the road testing of such vehicles is necessary. Therefore, other tests are proposed:

test conditions according to Standard ISO 17361 [102]; test conditions according to Standard SAE-J2808-201701 [103]; test conditions according to Standard SAE J3045-201507 [104]; test conditions according to Standard Regulation UN ECE 131 [105, 106]; division of test protocols Euro NCAP (New Car Assessment Programme) [107]; test conditions according to the Euro NCAP AEB VRU scenario [108]; test conditions according to the Euro NCAP AEB C2C scenario [16]; test conditions according to the Euro NCAP LSS scenario [105]; test conditions according to Standard Regulation UN ECE 130 [109]; Lane Departure Warning (LDW) [110]; Lane Keep Assist (LKA) (road edge: dashed line, solid line) [110]; Emergency Lane Keep (ELK) (road edge, oncoming, overtake) [110]. Examples of the tests have been shown in Figures 1 to 4.

IOP Conf. Series: Materials Science and Engineering 1247 (2022) 012022 doi:10.1088/1757-899X/1247/1/012022

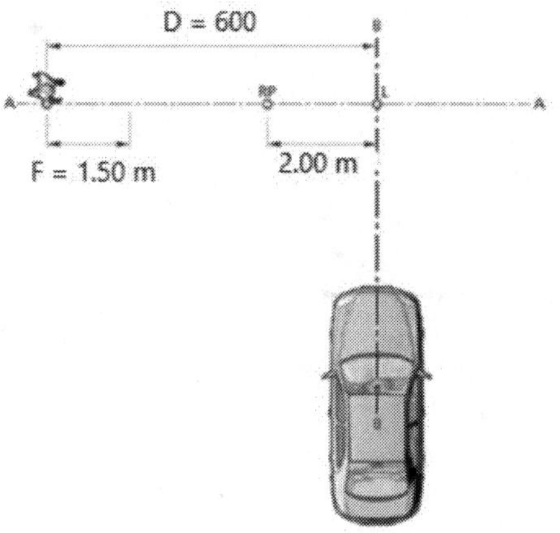

Fig. 1. AEB Pedestrian, Case CPFA-50. A simulation where the moving vehicle is on a collision path with an adult pedestrian running across the path from the driver's side. The collision with the pedestrian, in the case that the system fails to intervene, would occur in the front part of the vehicle in 50% of its width. During the test, the vehicle brakes shall not be applied manually [108].

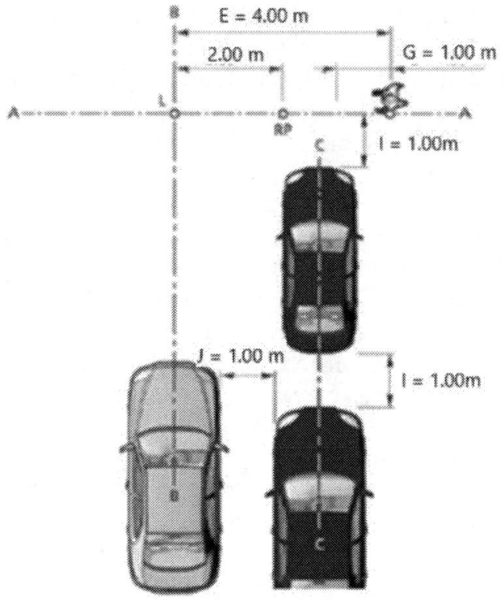

Fig. 2. AEB Pedestrian, Case CPNC-50. A simulation where the moving vehicle is on a collision path with a child running across the path from the passenger's side. The collision with the pedestrian, in the case that the system fails to intervene, would occur in the front part of the vehicle in 50% of its width. During the test, the vehicle brakes shall not be applied manually [108].

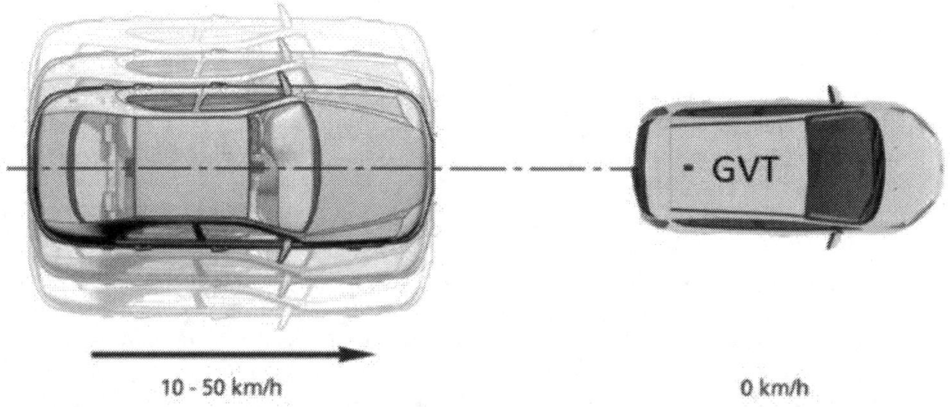

Fig. 3. AEB City. Only the AEB (Autonomous Emergency Braking) system is tested. Here, a single AEB CCRs (Car-to-Car Rear stationary) scenario has been taken into account, where a vehicle moving forwards hits its front part on the rear of a stationary vehicle. The test is carried out for different values of the offset between the symmetry axes of the vehicles involved. The impacting vehicle may only come to a halt when its test speed does not exceed 20 km/h [108].

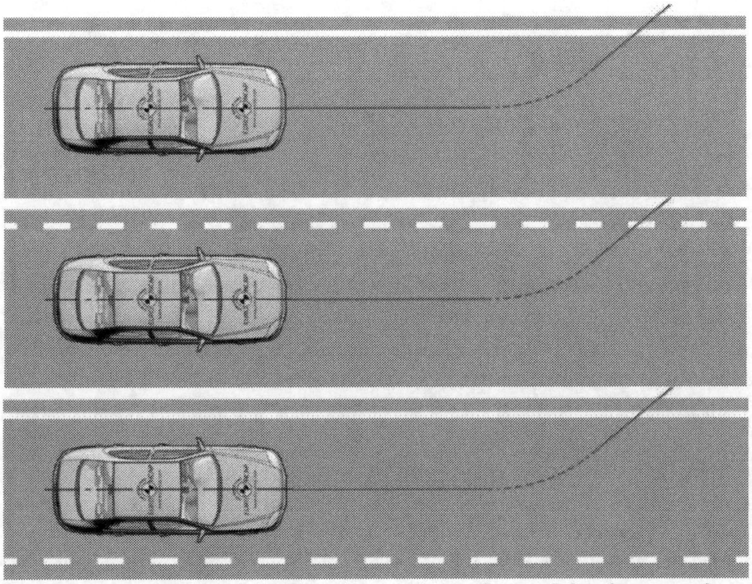

Fig. 4. Lane Departure Warning (LDW). The test is performed for three road marking types: a solid line on one side, a broken line on one side, and complete lane marking [110].

4. Summary

Autonomous vehicles make one of the main goals of development of the automotive industry. With the technology development, updated or new versions of legal instruments are introduced in order to regulate and standardize the modes of operation and methods of testing of advanced driver assistance systems and autonomous vehicles. Such actions are, *inter alia*, to make it possible for the autonomous vehicles to move on public roads as well as to standardize the requirements such systems must meet. This also forces the testing of such systems before they are launched for commercial use. The testing of the highly automated vehicles (autonomous from a level higher than 3) is based, first of all, on examining such systems in simulation conditions and in real operation conditions on testing grounds and in urban traffic. The providing of adequate infrastructure for the tests is critical because this has an impact not only on the test results but also on the safety of the tests. During the tests in real conditions, continuous communication with the autonomous vehicle should be ensured by a DSRC (Dedicated Short Range Communication) system, which not only is responsible for uninterrupted supervision of the test but also makes it possible to transmit simulated information about e.g. traffic intensity or the operation of traffic lights to the vehicle under test.

References

[1] https://www.techemergence.com/self-driving-car-timeline-themselves-top-11-automakers/, access 04.02.2022.
[2] https://www.bloomberg.com/news/features/2018-05-07/who-s-winning-the-self-driving-car-race, access 04.02.2022.
[3] https://www.dimensionengineering.com/info/accelerometers, access 04.02.2022
[4] Pusty T., Kowieski H,, et el., Increasing the research potential of the Vehicle Testing Laboratory in the area of advanced safety systems, Task report no: DDS-109-BLP, Automotive Industry Institute Warsaw Poland, 2018.
[5] Majcher P.: Autonomiczny robot mobilny zbierający piłki do tenisa stołowego An autonomous mobile robot that collects table tennis balls. Praca magisterska, Politechnika Warszawska, rok akademicki 2011/2012 Master thesis, Warsaw

University of Technology Poland, academic year 2011/2012. https://pl.wikipedia.org/wiki/Teledetekcja.
[6] https://www.optyczne.pl/98-s%C5%82ownik-CCD.html, access 04.02.2022
[7] Mueck M., Ingolf K.: Networking Vehicles to Everything. Evolving Automotive Solutions.
[8] https://www.sensorsmag.com/components/three-sensor-types-drive-autonomous-vehicles, access 04.02.2022.
[9] Stańczk T.L., Więckowski D., The idea of choice of steering wheel input parameters in relation to basic conditions of manoeuvre execution. Asian Journal of Science and Technology. Vol. 09, Issue, 12, ISSN: 0976-3376 pp. 9059-9072, December, 2018.
[10] Dubik A.: 1000 słów o laserach i promieniowaniu laserowym 1000 words about lasers and laser radiation. Warszawa: Wydawnictwo Ministerstwa Obrony Narodowej Warsaw: Publishing House of the Ministry of National Defense, Poland, 1989, pp. 154-155.
[11] https://leddartech.com/?utm_term=autonomous%20vehicle%20perception&utm_campaign=LeddarVision&utm_source=adwords&utm_medium=ppc&hsa_acc=1395223153&hsa_cam=12342818838&hsa_grp=117258683029&hsa_ad=498578393700&hsa_src=g&hsa_tgt=kwd-353136743711&hsa_kw=autonomous%20vehicle%20perception&hsa_mt=b&hsa_net=adwords&hsa_ver=3&gclid=EAIaIQobCh MIl871-rHm9QIViwCiAx3lnwVLEAAYAiAAEgK3QfD_BwE, access 04.02.2022.
[12] https://www.htpow.com/blog/bid-103.html, access 04.02.2022.
[13] http://laboratoria.net/artykul/17603.html, access 04.02.2022.
[14] https://leddartech.com/technology-fundamentals/, access 04.02.2022.
[15] https://en.wikipedia.org/wiki/Leddar, access 04.02.2022.
[16] https://geoforum.pl/gnss/beidou, access 04.02.2022.
[17] https://scholar.google.pl/scholar?q=ultrasonic+sensors+in+autonomous+vehicles&hl=pl&as_sdt=0&as_vis=1&oi=scholar, access 05.02.2022.
[18] https://www.balluff.com/local/pl/products/sensors/ultrasonic-sensors/.
[19] https://www.dimensionengineering.com/info/accelerometers, access 20.11.2018.
[20] https://geoforum.pl/gnss/beidou, access 26.11.2018.
[21] http://www.technologiagps.org.pl/dzialanie-gps.html, 04.02.2022.
[22] https://www.multigps.pl/gps, access 04.02.2022.
[23] https://www.gps.gov/, access 04.02.2022.
[24] http://www.astronautix.com/g/glonass.html, access 04.02.2022.
[25] http://www.kadrzasada.pl/matryce-ccd-i-cmos/, access 08.10.2018.
[26] https://vesta.astro.amu.edu.pl/Education/ccd2/node8.html, access 04.02.2022.
[27] http://www.fotoporadnik.pl/matryca-cmos-1.html, access 08.10.2018.
[28] http://www.avicon.pl/czy-kolor-zawsze-jest-potrzebny.html, access 11.10.2018.
[29] https://dsp.stackexchange.com/questions/24268/what-is-the-difference-between-monocular-and-stereo-image, access 04.02.2022.
[30] https://robotics.stackexchange.com/questions/6880/monocular-vs-stereo-computer-vision- robustness-for-object-detection, access 04.02.2022.
[31] https://www.dpreview.com/articles/7939734865/samsung-announces-nx300-3d-capable-20mp-mirrorless-camera-with-wvga-touchscreen, access 04.02.2022.
[32] http://www.avicon.pl/kamera-basler-tof-wchodzi-do-seryjnej-produkcji.html, access 15.10.2018.
[33] https://www.mobility.siemens.pl/main_imo/6711.htm, access 04.02.2022.
[34] https://whatis.techtarget.com/definition/dedicated-short-range-communication-DSRC, access 04.02.2022.

[35] https://www.coulton.com/beginners_guide_to_ultrasonic_level_transmitters.html, access 04.02.2022.
[36] https://www.invensense.com/solutions/automotive/, access 04.02.2022.
[37] Szypytko J.,Hyla P.: Wybrane metody sterowania pojazdami automatycznymi w terminalach kontenerowych, Akademia Górniczo-Hutnicza w Krakowie Wydział Inżynierii Mechanicznej i Robotyki AGH University of Science and Technology in Kraków Poland Faculty of Mechanical Engineering and Robotics, file:///C:/Users/48602/AppData/Local/Temp/156-1.pdf, access 01.02.2022 .
[38] Stasiak-Cieślak B., Szczepański T., Ślęzak M., Skarbek-Żabkin A., Malawko P., Pojazdy autonomiczne jako ułatwienie mobilności kierowców z niepełnosprawnością Autonomous vehicles to facilitate the mobility of drivers with disabilities, Niepełnosprawność – zagadnienia, problemy, rozwiązania Disability - issues, problems, solutions. Nr II/2018(27), Poland, pp. 7 – 15.
[39] https://www.google.com/search?client=firefox-b-&q=dedicated+short+range+communication+%28dsrc%29, access 02.02.2022.
[40] https://whatis.techtarget.com/definition/dedicated-short-range-communication-DSRC, access 04.02.2022.
[41] http://agropilot.pl/dodatki/agropilot-imu, access 15.10.2018.
[42] https://elektronikab2b.pl/technika/12098-zyroskopy-i-akcelerometry-mems-w-elektronice-uzytkowej#.Wsb4b9SLTs2, access 15.10.2018.
[43] https://www.caranddriver.com/car-accessories/a37283347/aftermarket-blind-spot-monitor/, access 04.02.2022.
[44] https://support.volvocars.com/pl/cars/Pages/owners-manual.aspx?mc=Y413&my=2014&sw=13w46&article=1d73d6b3c4f5bb36c0a801e8012e9503#, access 04.02.2022.
[45] https://owner.ford.com/how-tos/vehicle-features/safety/blind-spot-information-system-with-cross-traffic-alert.html, access 04.02.2022.
[46] http://www.bklasapl.net/mazda_2-1574/system_monitoruj_261_cego_ruch_poprzeczny_przy_cofaniu_rcta_-1704.html, access 04.02.2022.
[47] https://mycardoeswhat.org/safety-features/rear-cross-traffic-alert/, access 04.02.2022.
[48] https://www.honda.co.nz/technology/safety/cross-traffic-warning/, access 04.02.2022.
[49] https://www.google.com/search?client=firefox-b- d&q=lane+departure +warning+system, access 02.02.2022.
[50] https://www.nissan-global.com/EN/TECHNOLOGY/OVERVIEW/ldw.html access 03.02.2022.
[51] https://www.acura.com/mdx/modals/lane-keeping-assist-system , access 04.02.2022.
[52] http://www.toyota-global.com/innovation/safety_technology/safety_technology/technology_file/active/lka.ht ml , access 04.02.2022.
[53] https://www.kia.com/dm/discover-kia/ask/what-is-cruise-control.html, access 04.02.2022.
[54] https://ctms.engin.umich.edu/CTMS/index.php?example=CruiseControl&section=SystemModeling, access 02.02.2022.
[55] https://www.researchgate.net/figure/Adaptive-Cruise-Control-ACC-fault-tolerant-paradigm_fig1_325702711, access 04.02.2022.
[56] https://www.acura.com/tlx/modals/adaptive-cruise-control-with-low-speed-follow, access 02.04.2022.
[57] https://arxiv.org/ftp/arxiv/papers/1809/1809.02867.pdf, access 02.02.2022.
[58] Energis Md Abdus Samad Kamal 1,* , Kotaro Hashikura1, Tomohisa Hayakawa2 , Kou Yamada1, and Jun-ichi Imura, Adaptive Cruise Control with Look-ahead Anticipation for Driving on Freeways.

[59] https://mycardoeswhat.org/deeper-learning/forward-collision-warning/, access 04.02.2022.
[60] https://blog.checkmybus.pl/newsy/innowacyjny-system-detekcji-pieszych-stworzony-przez-daimler-064, access 19.10.2018.
[61] https://www.nissan-global.com/EN/TECHNOLOGY/OVERVIEW/autonomous_emergency_steering_system.html, access 04.02.2022.
[62] https://www-esv.nhtsa.dot.gov/Proceedings/22/files/22ESV-000111.pdf, access 04.02.2022.
[63] https://www.nissan-lobal.com/EN/TECHNOLOGY/OVERVIEW/autonomous _emergency_steering_system.html access 04.02.2022.
[64] https://www.bosch-mobility-solutions.com/en/solutions/assistance-systems/evasive-steering-support/, access 04.02.2022.
[65] https://local.iteris.com/cvria/html/applications/app16.html#tab-7, access 04.02.2022
[66] http://slideplayer.com/slide/2351070/, access 04.02.2022.
[67] https://togethermagazyn.pl/v2v-technologia-bezpieczenstwa, access 25.10.2018.
[68] https://www.press.bmwgroup.com/global/article/detail/T0108975EN/the-left-turn-assistant-%E2%80%93-turning-left-safely-by-looking-out-for-oncoming-traffic?language=en, access 04.02.2022.
[69] http://sites.ieee.org/connected-vehicles/2015/09/10/continental-presents-left-turn-assist-based-v2x-frankfurt-motor-show/, access 04.02.2022.
[70] https://www.press.bmwgroup.com/poland/article/detail/T0109833PL/asystent-skr%C4%99tu-w-lewo-%E2%80%93-bezpieczne-skr%C4%99canie-w-lewo-dzi%C4%99ki-wykrywaniu-nadje%C5%BCd%C5%BCaj%C4%85cych-pojazd%C3%B3w, access 25.10.2018.
[71] http://savari.net/wp-content/uploads/2017/01/Savari-V2VApps-DataSheet-FINALOct2016.pdf, access 04.02.2022.
[72] https://www.continental-automotive.com/en-gl/Passenger-Cars/Chassis-Safety/Software-Functions/Active-Safety/Intersection-Movement-Assist, access 04.02.2022.
[73] https://local.iteris.com/cvria/html/applications/app36.html, access 04.02.2022
[74] https://www.digitaltrends.com/cars/ford-electronic-brake-light-uses-wireless-tech-to-shorten-stopping-distances, access 04.02.2022.
[75] https://www.continental-automotive.com/en-gl/Passenger-Cars/Chassis-Safety/Software-Functions/Active-Safety/Emergency-Brake-Assist/Electronic-Emergency-Brake-Light, access 04.02.2022.
[76] https://local.iteris.com/cvria/html/applications/app23.html, access 04.02.2022.
[77] https://www.contracthireandleasing.com/car-leasing-news/ford-tests-new-brake-light-that-talks-to-the-car-behind, access 04.02.2022.
[78] https://www.carwow.co.uk/guides/glossary/how-does-road-sign-recognition-work, access 04.02.2022.
[79] Onaszkiewicz P.: Rozpoznawanie znaków drogowych w sekwencjach wideo Recognition of road signs in video sequences, Politechnika Warszawska, Wydział Elektroniki i Technik Informacyjnych, Warszawa, Warsaw University of Technology, Faculty of Electronics and Information Technology, Warsaw, Poland 2015.
[80] http://www.volkswagen.co.uk/technology/proximity-sensing/traffic-jam-assist#overlay-close, access 04.02.2022.
[81] https://www.audi-technology-portal.de/en/electrics-electronics/driver-assistant-systems/audi-a4-traffic-jam-assist-en, access 04.02.2022.
[82] https://www.milescontinental.co.nz/news/features/traffic-jam-assist/, access 05.02.2022.

[83] http://antyweb.pl/asystent-parkowania-jak-to-dziala-czy-rzeczywiscie-wyrecza-kierowce/, access 30.10.2018.
[84] https://www.freightlink.pl/knowledge/artyku%C5%82y/truck-platooning-jako-szansa-na-zmian%C4%99-%C5%9Brodowiska-transportu, access 05.02.2022.
[85] https://www.eutruckplatooning.com/About/default.aspx, access 05.02.2022.
[86] https://truckfocus.pl/nowosci/28836/iveco-wzielo-udzial-w-pierwszym-na-swiecie-truck-platooning-challenge, access 30.10.2018.
[87] EUROPEAN NEW CAR ASSESSMENT PROGRAMME (Euro NCAP) TEST PROTOCOL – AEB VRU systems, Version 2.0.2, November 2017.
[88] Łabędzki P.: Metody generowania modelu sceny dla autonomicznego robota kroczącego z wielomodowym systemem wizyjnym Methods of generating a scene model for an autonomous walking robot with a multi-mode vision system, Politechnika Poznańska, Poznań Poznań University Of Technology, Poznan, Poland 2014.
[89] T. Bailey, H. Durrant-Whyte, Simultaneous Localizationand Mapping: Part I. IEEE Robotics & Automation Magazine, s. 99-108, June 2006.
[90] Dąbrowski A., Automatyka – podstawy teorii Automation - the basics of theory, Scientific and Technical Publishing House, Warsaw, Poland 2012 Edition II.
[91] Dąbrowski A., Automatyka – technika regulacji Automation - regulation technique. Scientific and Technical Publishing House, Warsaw, Poland 2013.
[92] ISO 4138; Passenger Cars - Steady-state circular driving behaviour - Open-loop test procedure. 1996.
[93] ISO 14792 Road vehicles – Heavy commercial vehicles and buses – Steady state circular test.
[94] ISO 7401 Road Vehicles - Lateral Transient Response Test Methods 1988.
[95] ISO 7975 Road Vehicles - Braking in a Turn - Open Loop Test Procedure 2006.
[96] ISO 3888-1 Passenger cars - Test track for a lane-change manoeuvre. Part 1: Double lane-change 1999.
[97] ISO TR 3888 Road Vehicles - Test Procedure for a Severe Lane - Manouvre. 1975.
[98] ISO TR 8725 Road Vehicles - Transient open-loop response test method with one period of sinusoidal input. 1988.
[99] ISO 7401 Road Vehicles - Lateral Transient Response Test Methods - Open-Loop Test methods. 2003.
[100] ISO TR 8726 Road Vehicles - Transient open-loop response test method with pseudo-random steering input. 1988.
[101] ISO 14791 Road vehicles – Heavy commercial vehicles combinations and articulated buses – Lateral stability test method.
[102] International Standard: Intelligent transport systems — Lane departure warning systems — Performance requirements and test procedures. ISO 17361:2017.
[103] Surface Vehicle Information Report: Lane Departure Warning Systems: Information for the Human Interface, SAE J2808, January 2017.
[104] Surface Vehicle Recommended Practice: Track and Bus Lane Departure Warning System Test Procedure, SAE J3045, July 2015.
[105] LSS Test Protocol v2.0.2, January 2019, access https://s3.amazonaws.com/cdn.ancap.com.au/app/public/assets/fe81d348f6676aae0bf2de95b7be66a2c6c74a80/original.pdf?
[106] Regulation No 131 of the Economic Commission for Europe of the United Nations (UN/ECE) —Uniform provisions concerning the approval of motor vehicles with regard to the Advanced Emergency Braking Systems (AEBS) (AEBS).
[107] NCAP Euro (The European New Car Assessment Programme) https://www.euroncap.com/en, access 01.02.2022.

[108] EUROPEAN NEW CAR ASSESSMENT PROGRAMME (Euro NCAP) TEST PROTOCOL– AEB systems, Version 2.0.1, November 2017, https://cdn.euroncap.com/media/32278/euro-ncap-aeb-c2c-test-protocol-v201.pdf, access 01.02.2022.
[109] COMMISSION REGULATION (EU) No 130/2012 concerning type-approval requirements for motor vehicles with regard to vehicle access and manoeuvrability and implementing Regulation (EC) No 661/2009 of the European Parliament and of the Council concerning type-approval requirements for the general safety of motor vehicles, their trailers and systems, components and separate technical units intended therefor.
[110] SAS Test Protocol v2.0, January 2018, https://www.euroncap.com/en/for-engineers/protocols/safety-assist/, access 01.02.2022.

Experimental testing of selected energy absorbers in the context of their use in rear underrun protective devices

J Jackowski[1], P Posuniak[2*], Z Rożnowicz[2]

[1] Military University of Technology (MUT), ul. gen. Sylwestra Kaliskiego 2, 01-476 Warszawa
[2] Łukasiewicz Research Network – Automotive Industry Institute (ŁUKASIEWICZ – PIMOT), ul. Jagiellońska 55, 03-301 Warszawa

* pawel.posuniak@pimot.lukasiewicz.gov.pl

Abstract. The energy-absorbing structures have been being developed for many years by various research centres all over the world. The article presents experimental test results obtained for selected structures designed for the absorbing of kinetic energy in order to reduce the loads on occupants of a motorcar when hitting the rear of a motor truck. The differences between selected energy-absorbing structures have been analysed, based on impact resistance indicators having been determined. The main objective of this research work was to verify the applicability of selected energy absorbers to the rear underrun protective devices as the truck parts accountable for vehicles' passive safety.

Keywords: safety, vehicle accidents, crash test

1. Introduction

The searching for effective impact energy absorbers makes a challenge for various industries, e.g. automotive industry, aviation, or armed forces. The elements accountable for the absorption of kinetic energy should meet definite requirements and be characterised by appropriate parameters [1, 2, 3], depending on their intended use. In aviation, for example, the parameter of critical importance is the ratio of the energy absorbed to the absorber mass. The materials intended for military applications should be characterised by good effectiveness at high impact velocity values. In the automotive industry, the above characteristics are certainly very important, but a critical matter is the absorber size, especially when the energy-absorbing elements are to be applied to small L-category vehicles. Another issue that cannot be considered unimportant is the absorber cost, because such an element installed in a vehicle should not noticeably increase the vehicle price. Moreover, the absorber installed in a motor vehicle must ensure comparable effectiveness in varying operational conditions (especially temperature and humidity), which additionally raises the difficulty of selecting the most effective absorber design.

An analysis of the commercially available engineering solutions of energy-absorbing structures shows that in the automotive industry, three-dimensional structures are commonly used, having the form of hollow profiles with diversified cross-sections and additionally embossed areas, made of steel or aluminium (less frequently, composite material can also be met). As another solution, structures of the multi-cell type are used, having the form of a pack of many thin-walled pipes with different cross-sections. The large number of pipes connected together make a structure in the form of e.g. a honeycomb, a good point of which is low mass and relatively low production cost. As a similar solution, materials referred to as "foams" are also used [4].

In consideration of their intended use, motor trucks are designed for their cargo space volume and load capacity to be as big as possible. The questions of road traffic safety are dominated by the issue of active safety [5]. The elements related to the passive safety are limited to the vehicle cabin (driver's safety) and to the front, side, and rear underrun protective devices (to ensure safety for the other road users). The basic function of the rear underrun protective devices (RUPDs) is to prevent passenger cars from running under the rear of a motor truck [6]. Despite the fact that the RUPD is to ensure safety for the occupants of a passenger car when hitting the rear of a motor truck, the impact energy absorption issue is generally ignored (see e.g. the UN ECE Regulations) or marginalised (the US or Canadian regulations) [7, 8, 9, 10]. In principle, such a device is only to be a rigid barrier for the car, while this may cause in some cases an increase in the loads acting on the car [11].

In consideration of the growth in social consciousness as well as the authorities' pursuits of improvement in the road traffic safety (the "Vision Zero" strategy etc.), it is indispensable to develop modern engineering designs for motor trucks as well, the more so that the rear-end collisions make a significant percentage in all the road accidents [12, 13].

2. Research methods

To determine the energy-absorption characteristics of various engineering designs, dimensions, and materials adopted to build energy absorbers, some absorber samples characterised in Table 1 were prepared. The sample denoted as HC1.71_x2 was made of aluminium sheet and had a honeycomb structure. The absorber denoted as HC1.71box_x2, unlike the one above, was additionally enclosed with steel sheet 1.2 mm thick. The absorbers denoted as ST-KMC, ST-OAG and ST-RNO were structures based on rectangular hollow steel profiles. The sample denoted as AL-TMC was a rectangular hollow profile, too, but made of aluminium. It was assumed that the sample length could not exceed 0.2 m because of the limited space available for the mounting of a large-size RUPD.

The dynamic tests of the samples were carried out on a test stand at the Łukasiewicz Research Network – Automotive Industry Institute (Łukasiewicz – PIMOT). The test stand has been shown in Figure 1. The absorber sample under test was installed on an impact trolley with a mass of 325 kg, which struck an undeformable impact barrier with a velocity of 6.67 m/s. During the experiment (impact test), the trolley deceleration was measured. Additionally, the course of each test was filmed by means of a high-speed camera.

Table 1: Description of test samples

	Height [m]	Width [m]	Length [m]	Mass [kg]
HC1.71box_x2	0.125	0.245	0.15	1.56
HC1.71x2	0.125	0.245	0.15	0.34
ST-KMC	0.060	0.115	0.135	1.20
ST-OAG	0.073	0.070	0.187	1.14
ST-RNO	0.100	0.072	0.158	1.34
AL-TMC	0.100	0.060	0.132	0.66

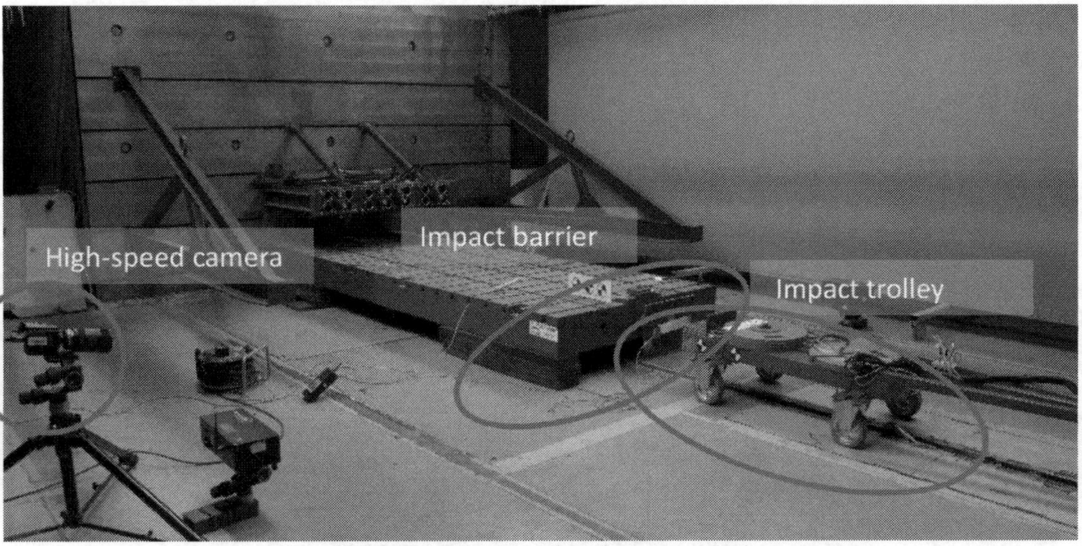

Figure 1: Energy absorber test stand

3. Test results

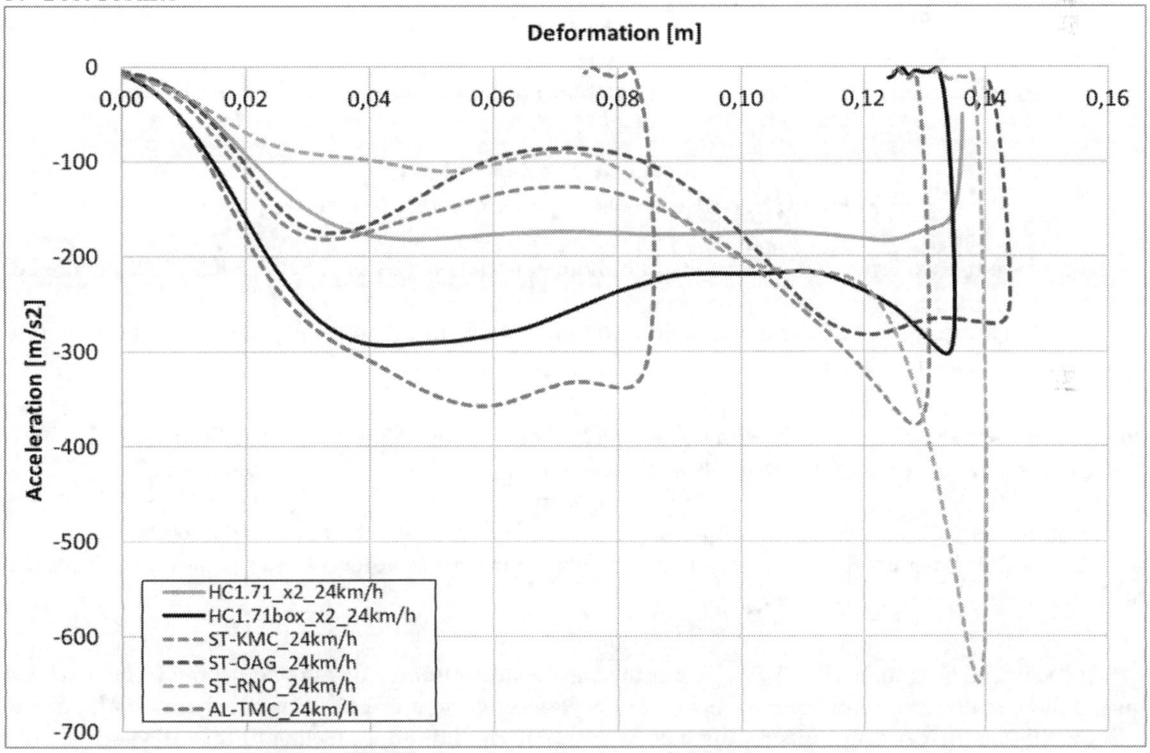

Figure 2. Trolley acceleration vs absorber deformation curves

The experimental test results have been shown in Figures 2-7 and in Table 2. The curves in Figure 2 represent impact trolley accelerations as functions of the absorber deformation. It can be noticed that for all the absorber materials chosen, apart from that of the AL-TMC sample, the maximum deformation values remained at a similar level. The extreme acceleration values for the ST-OAG and HC1.71box_x2 samples were −28 g and −30 g, respectively; for the ST-KMC and AL-TMC samples, they varied between −35 g and −37 g. In the test with the HC1.71_x2 sample, the trolley acceleration curve did not reach −20 g; in contrast, the highest deceleration values were reached for the ST-RNO sample (the acceleration was almost −65 g).

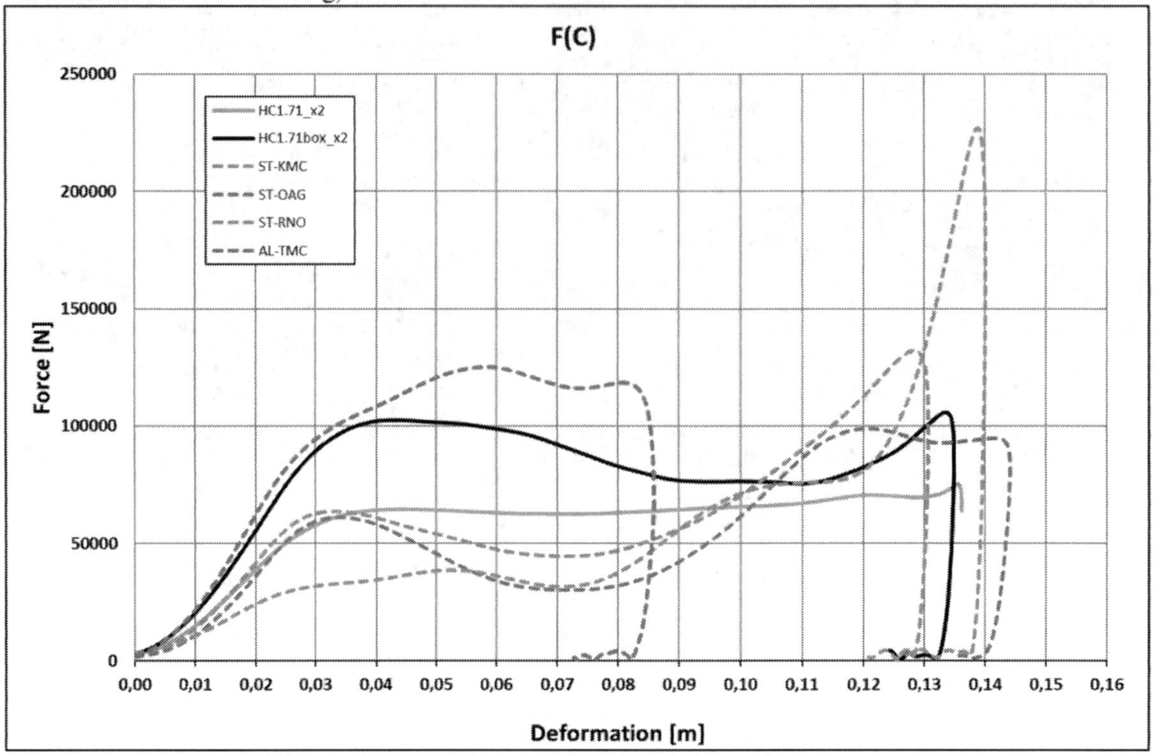

Figure 3. Impact force vs absorber deformation curves

To analyse the experimental test results, the indicators denoted as EA, SEA, MCF, PCF, and CLE were used.

The EA indicator is defined as the total deformation energy absorbed during the plastic deformation. It is represented by a formula:

$$EA = \int_0^x F(x)dx \tag{1}$$

where: F(x) is the instantaneous crushing force and x is the deformation value (see Figure 3).

SEA is the ratio of EA to the mass of the structure under test, according to the formula:

$$SEA = EA/m \tag{2}$$

where: m is the total mass of the structure under test (in our case, it is the mass of the test sample).

MCF is the mean crushing force; for the known deformation value, it was determined from the formula:

$$MCF = \frac{1}{x}\int_0^x F(x)dx \tag{3}$$

The peak crushing force (PCF) was determined as the maximum value of the crushing force recorded during the test. It has a significant influence on the safety of the impacting car occupants [14]. For the loads acting on the car occupants during a collision to be minimised, the value of this indicator should also be as low as possible. By comparing the PCF and MCF values with each other, we may determine the CLE indicator value [3], defining the uniformity of the load acting on the structure:

$$CLE = MCF/PCF \tag{4}$$

The results of the above calculations have been presented in Table 2 and in the graphs below.

Table 2. Results of dynamic strength tests.

	Absorbed Energy [J]	PCF [kN]	SEA [kJ/kg]	MCF [kJ/m]	CLE	Deceleration [m/s^2]
HC1.71box_x2	7 763	106	4.97	52	0.49	302
HC1.71x2	7 755	75	22.81	52	0.69	215
ST-KMC	7 713	132	6.43	57	0.43	377
ST-OAG	7 701	99	6.76	41	0.42	283
ST-RNO	7 696	227	5.74	49	0.21	649
AL-TMC	7 696	125	11.66	58	0.47	358

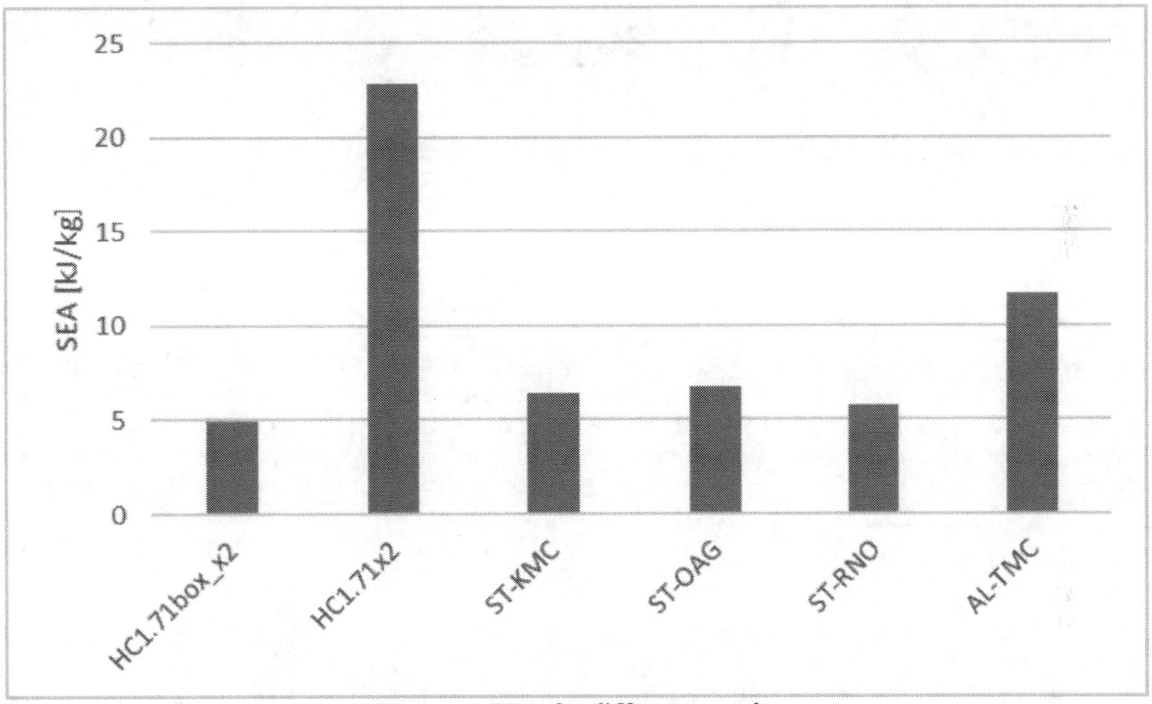

Figure 4. SEA for different samples

Figure 4 shows the SEA values determined for individual samples. It can be noticed that the highest SEA value (almost 23 [kJ/kg]) was achieved for the HC1.71x2 sample. Noteworthy is also the result obtained for the HC1.71box_x2 sample, which was made of the same material as the one used for the sample mentioned above, but was additionally enclosed with steel sheet. For this absorber, the SEA value was reduced to only somewhat more than one-sixth of that for the former sample.

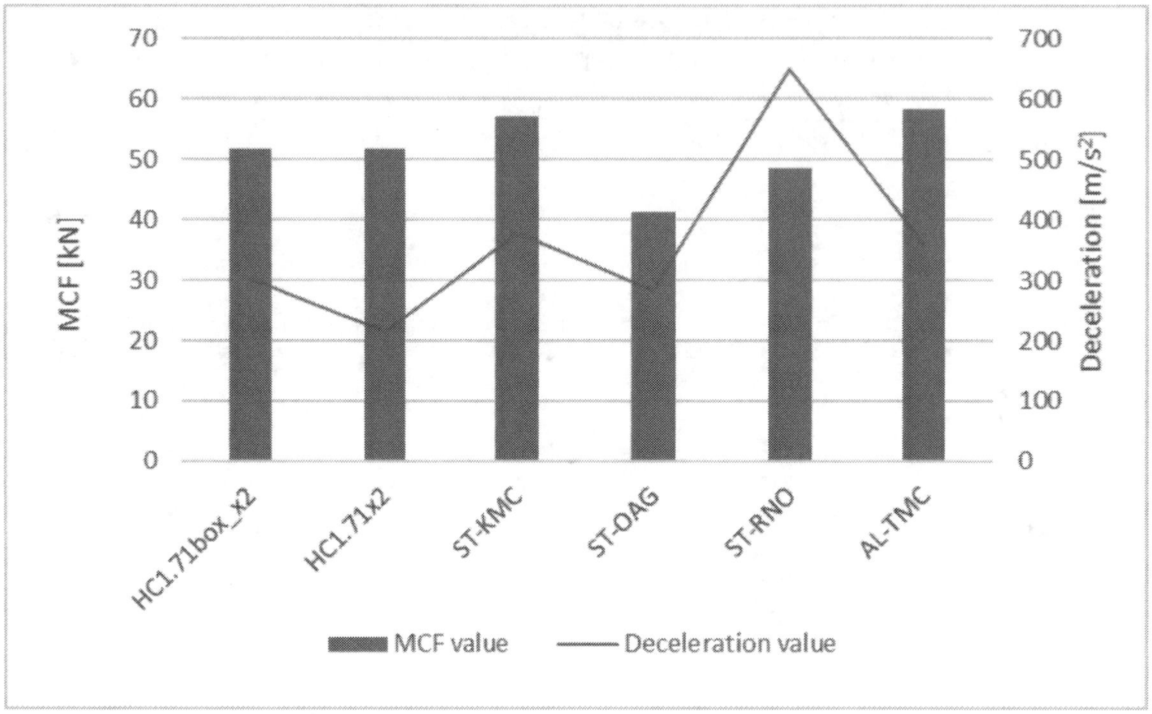

Figure 5. MCF and deceleration for different samples

According to an analysis of the MCF calculation results, the lowest values of this indicator were obtained for two samples, i.e. ST-OAG and ST-RNO. For the other four samples, the MCF values varied between 50 kN and 60 kN.

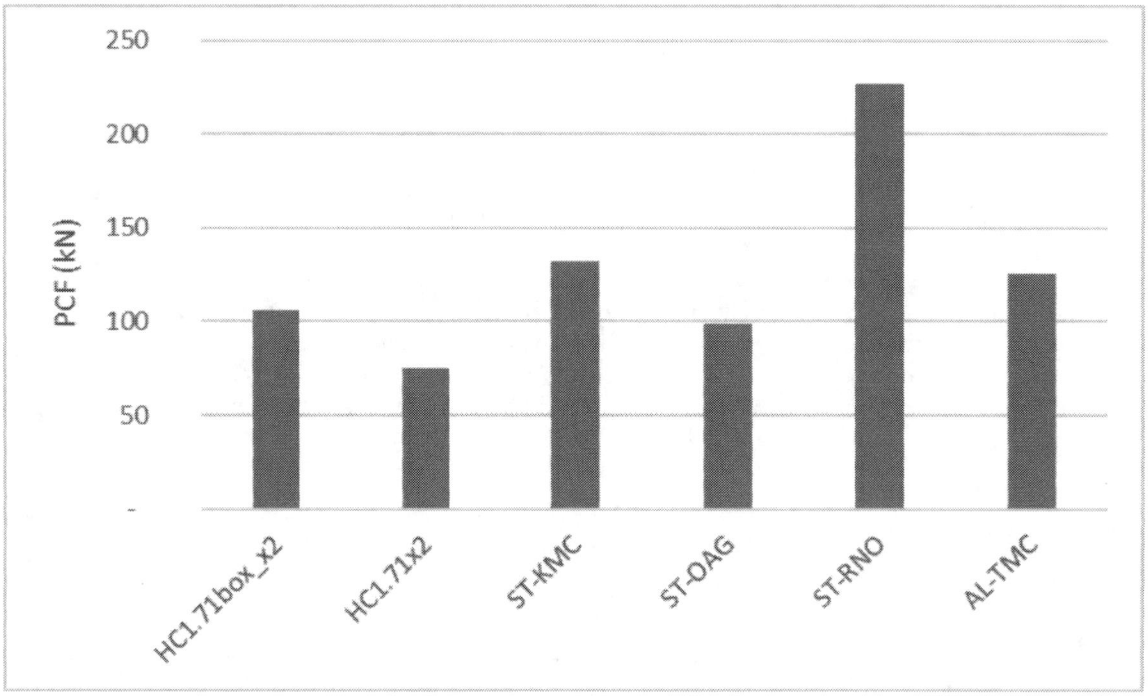

Figure 6. PCF for different samples

Results of PCF calculations for the samples under test have been presented in Figure 6. For most of the samples, the values of this indicator were at a similar level, except the standing-out value for the ST-RNO sample.

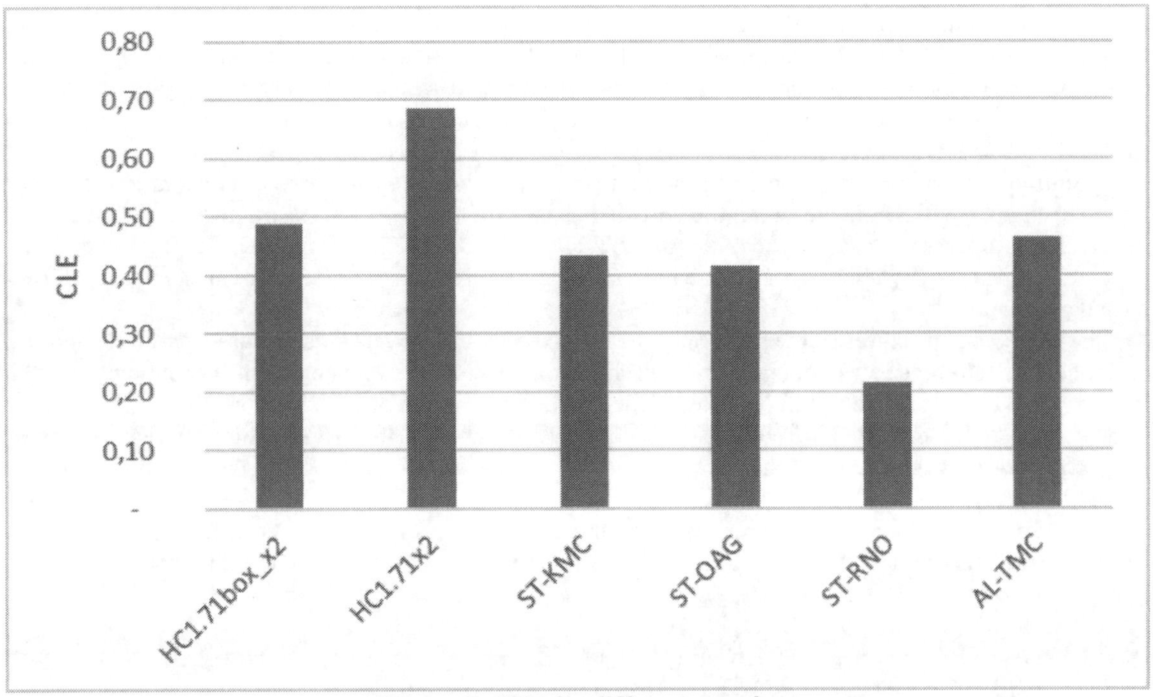

Figure 7. CLE for different samples

The CLE values shown in Figure 7 were in most cases at a similar level of about 0.45. Exceptions were the highest and the lowest values of the structure load uniformity indicator, amounting to 0.69 and 0.21 for the HC1.71x2 and ST-RNO samples, respectively.

4. Conclusion

An analysis of the impact trolley deceleration (load) values recorded as a function of the energy absorber deformation has shown that the lowest overloading values took place in the case of the sample made entirely of aluminium with a honeycomb structure and the highest values occurred, especially in the final deformation phase, for the steel profile denoted as ST-RNO. This indicates that the latter sample underwent ultimate deformation. It can also be noticed that for most of the absorber designs, the deformation values were close to each other. Moreover, significant exceedances of the 100 kN value were found to occur (for the ST-KMC, ST-RNO, and AL-TMC samples). Therefore, the reasonability of using such materials as energy absorbers in RUPDs should be reconsidered because their stiffness is close to the stiffness of structural components of the conventional RUPD [15], which in consequence may reduce the absorber's effectiveness, especially if a reduction of the loads acting on car occupants during a collision is concerned.

A simple qualitative analysis of the test results, e.g. of the force or acceleration vs displacement curves, does not offer an unequivocal answer to the question which of the samples under test will be the most suitable for being used as an energy absorber in an RUPD. To indicate the most effective absorber, the features that should be considered the most important should be identified.

Based on the SEA indicator, the honeycomb structure (HC1.71_x2) should be considered the most appropriate. However, if the absorber is to be used in a motor truck, where the absorber mass makes merely about one-thousandth of the total vehicle mass, then the absorber mass may be deemed unimportant. Hence, the SEA indicator is actually inappropriate for evaluating the properties of the material sample under test from the point of view of using this material for an energy absorber.

If the test results are evaluated on the grounds of the MCF value, then two structures, i.e. ST-KMC and AL-TMC, for which the MCF was 57 kJ/m and 58 kJ/m, respectively, may be indicated as the most desirable. For the next two samples (HC1.71_x2 and HC1.71box_x2), this value was 52 kJ/m, i.e. by about 9% lower than the values mentioned above.

If the PCF value (detrimental to the safety of the impacting car occupants) is taken into account in the analysis, then the ST-RNO absorber should be considered the least effective in reducing the loads acting on the occupants of the car hitting the rear of the motor truck. A definitely better solution was the HC1.71_x2 structure, for which the PCF value was equal to about one-third of the value determined for the ST-RNO sample.

Attention should however be paid to the fact that if the same PCF parameter is taken into account together with the MCF indicator, then the absorbers based on the honeycomb structure show the energy absorption properties close to those of the ST-KMC and AL-TMC samples, but at lower peak deceleration values (by about 15-20%), which may be decisive for the safety of car occupants during a road accident.

Therefore, the structures of the multi-cell type based on the honeycomb structure may be recognised as capable of finding successful application in the automotive industry as elements accountable for the passive safety of motor vehicles. Moreover, their structure enables them to absorb the impact energy almost over their whole length without rapid changes (peaks) in the deceleration curves. A similar property was only showed by the structure made of the aluminium profile, but at much higher deceleration values.

The research results presented are treated by the authors as preliminary. The work on the selection of the energy-absorbing structures suitable for rear underrun protective devices will be continued.

References

[1] Aktay L, Johnson A F and Kröplin B, 2008, Numerical modelling of honeycomb core crush behaviour, *Engineering Fracture Mechanics*, **75**, 2616-30

[2] Sun G, Li G, Hou S, Zhou S, Li W and Li Q, 2010, Crashworthiness design for functionally graded foam-filled thin-walled structures, *Materials Science and Engineering: A*, **527**, 1911-19

[3] Zhang Y, Ge P, Lu M and Lai X, 2018, Crashworthiness study for multi-cell composite filling structures, *International Journal of Crashworthiness*, 32-46

[4] Yusof N S B, Sapuan S M, Salutan M T H, Jawaid M and Maleque M A, 2017, Design and materials development of automotive crash box: a review, *Ciência & Tecnologia dos Materiais*, **29**, 129-144

[5] Prochowski L, Żuchowski A, 2016, *Samochody ciężarowe i autobusy (Goods vehicles and buses)*, **4**, Warszawa, Wydawnictwa Komunikacji i Łączności; ISBN: 978-83-206-2047-4

[6] Proposal for Draft Amendments to UN ECE Regulation No 58, Justification for amendments proposed in document GRSG/2013/27, 2013

[7] Standard No 223: Rear Impact guards; Federal Motor Vehicle Safety Standards USA; Available online: https://www.federalregister.gov/d/2014-16018 (accessed Dec. 06, 2021)

[8] Standard No 224: Rear impact protection; Federal Motor Vehicle Safety Standards USA; Available online: https://www.federalregister.gov/d/2014-16018 (accessed Dec. 06, 2021)

[9] Standard No 223: Rear Impact guards; Motor Vehicle Safety Regulations Government of Canada; Available online: https://tc.canada.ca/en/corporate-services/acts-regulations/motor-vehicle-safety-regulations-crc-c-1038 (accessed Dec. 06, 2021)

[10] National Standard of the People's Republic of China GB 11567-2017: Motor vehicles and trailers – Lateral protection and rear underrun protection requirements; AQSIQ; Republic of China 2017

[11] Cappello F, Ingrassia T and Nigrelli V, 2008, Design of a new high energy rear underrun protective device, *WIT Transactions on the Built Environment*, **97**, 325–335

[12] *Wypadki drogowe w Polsce w 2020 roku (Road accidents in Poland in 2017)*; Komenda Główna Policji Biuro Ruchu Drogowego (National Police Headquarters – Road Traffic Office), Poland 2021

[13] Fatality Facts 2018 Large trucks; IIHS, Available online: https://www.iihs.org/topics/fatality-statistics/detail/large-trucks (ac-cessed Dec. 06, 2021)
[14] Rogala M, Gajewski J and Ferdynus M, 2020, The Effect of Geometrical Non-Linearity on the Crashworthiness of Thin-Walled Conical Energy-Absorbers, Materials
[15] Regulation No 58 of the Economic Commission for Europe of the United Nations – Revision 3, including the amendments which entered into force on 14 September 2017 s; UN ECE Transport Vehicle Regulations

Assessment of the vehicle body side stiffness influence on the process of hazard generating in front-to-side vehicles' collision

M Ziubiński[1] and L Prochowski[1,2]

[1] Military University of Technology (WAT), Faculty of Mechanical Engineering, Institute of Vehicles and Transportation, ul. gen. Sylwestra Kaliskiego 2, 00-908 Warsaw, Poland
[2] Łukasiewicz Research Network – Automotive Industry Institute (Łukasiewicz-PIMOT), ul. Jagiellońska 55, 03-301 Warsaw, Poland

* mateusz.ziubinski@wat.edu.pl

Abstract. A vehicles' collision is a complex energy process, determined by many factors. In the front-to-side collision, the hazards for he occupants of the struck vehicle results from the inertial loads they are subjected to together with the vehicle and from excessive deformation of the bodywork, often involving the space occupied by people. Action is needed to reduce this hazard. One of such actions is to increase the stiffness of the side part of the vehicle body in order to limit the range of its deformation. However, this is an activity that requires a compromise approach, because as stiffness increases, greater lateral forces on the struck vehicle will occur, which will result in inertial loads. The aim of the study is to assess the impact of changes in the stiffness of the side part of the bodywork on the course of the hazard emergence process during a front-to-side vehicles' collision. Model tests in the PC-Crash program have been prepared and implemented. Their results allowed to establish the relationship between the increased side-stiffness and the characteristic values of physical quantities, which are the measure of the hazard and its effects in a vehicles' collision.

Keywords: front-to-side vehicles' collision, motorcar bodywork stiffness, passive safety, hazards for occupants during vehicle lateral impact, vehicle movement

1. Introduction

A vehicles' collision can be treated as a complex energy process, the course of which is determined by the properties of motorcars, road surface and the state of vehicles traffic. It was assumed that at the initial time of the collision, the cars were moving on mutually intersecting trajectories. Their relative position at the time of contact phase beginning determine which areas of the bodywork are deformed. It is known that the energy-absorbing zones of the vehicle body (front, side, rear) have various properties, significantly depending on the type and shape of the car. This complexity of the boundary conditions indicates the extensity of the research area.

In a front-to-side vehicles' collision, one of the cars hits by its front-end into the side of the another vehicle. As a result of this incident, the vehicles and the occupants are subjected to significant dynamic loads, creating the risk of the injury. The course of the front-to-side collision is more complex than the frontal collision [1]. In such a collision, the vehicles move relative to each other not only in the direction of the impact, but also in the transverse direction [2]. Significant hazard in the event of a side impact occur in case of the vehicle body deformation into the passenger compartment and of the dynamics of

IOP Conf. Series: Materials Science and Engineering 1247 (2022) 012024 doi:10.1088/1757-899X/1247/1/012024

the struck vehicle lateral displacement. Deformation often involves the space where the occupants are present [3]. It was established in [4] that the greatest injury risk in a front-to-side collision is in the case of an impact into the central area on the body-side between the wheel axles. Then we have the greatest depth of deformation. The dependence of the injury risk on the deformation depth of the vehicle body side was confirmed in [5,6]. Thus, in addition to inertial loads resulting from significant accelerations during a collision, the hazard in front-to-side vehicles' collision also results from the impact of an occupant by the deforming vehicle body. Hence, the occupants injury risk on the car being struck from the side is 2 to 8 times greater than the risk for the occupants in striking vehicle [7-9].
There are many activities to improve the passive safety of the side part of vehicle's body. Examples of such actions we have in [10-16]. The main activities are focused on improving of the doors and the middle pillar construction, the result is an increase in the stiffness of the side part of the vehicle's body (side-stiffness). The aim is to reduce the collision deformation depth. The results of these actions, however, are very limited, which results from the restricted space for the arrangement of the reinforcement elements on the vehicle's body side, which has a small thickness to not reduce the passenger compartment space. Hence, the processes occurring during the front-to-side collision, leading to occur of the hazard for the occupants, require further research in order to find a compromise between increasing the side-stiffness and reducing its adverse effects, e.g. as a result of limiting the energy dissipated on deformation.
Figure 1 shows the front-to-side vehicles' collision in two phases - before the beginning of the collision contact phase (Figure 1a) and during the collision (Figure 1b). The figure show the vectors of the linear and angular velocity of vehicle A (striking) and B (being struck) as well as significant external forces acting on the vehicle B during the collision: the impact force F_{DB} in the bodyworks contact zone and the tangential force in the tire-road contact area Y_1 and Y_3 (for the wheels of the left side of the vehicle), resulting from the lateral sliding of the vehicle during the collision. The shaded area in Figure 1 indicates the area of the vehicle body that is deformed. There are occupants in this space.

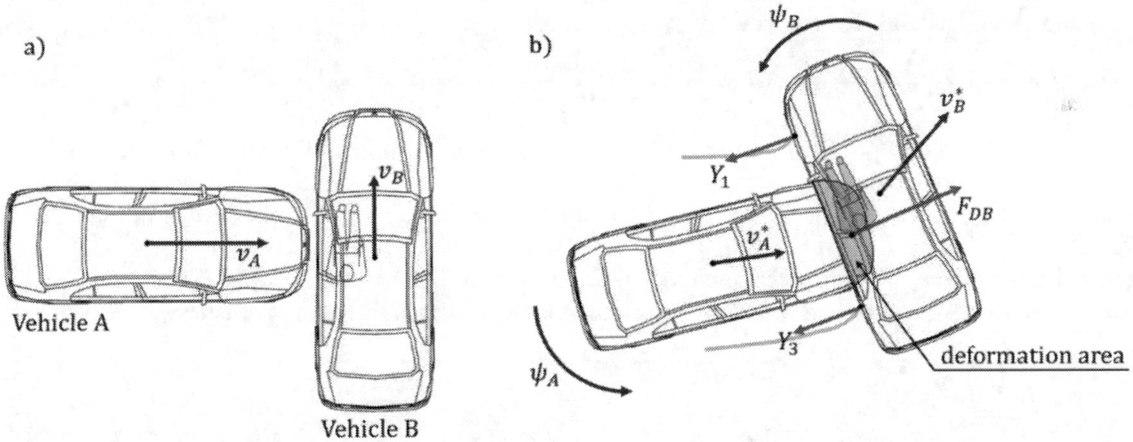

Figure 1. The state of vehicles' movement before front-to-side collision (a) and result of a side impact of the vehicle B (b).

Based on Figure 1, the energy balance of the front-to-side vehicles' collision was written:

$$E_{KA} + E_{KB} = E_{KA}{}^* + E_{KB}{}^* + W_{DAB} + W_{TAB} + W_{FAB} + E_R \qquad (1)$$

where:
E_{KA}, E_{KB} – kinetic energy of the impacting vehicle (A) and the vehicle struck on its side (B) at the instant of the beginning of the contact phase of vehicle collision (impact energy);
$E_{KA}{}^*$, $E_{KB}{}^*$ – kinetic energy at the end of the vehicle collision process;

$W_{DAB} = W_{DA} + W_{DB}$ – work of deformation of the bodies of vehicles A and B; as a result of impact force F_D act;
W_{TAB} – work of the friction forces in the zone of contact between the vehicle bodies;;
$W_{FAB} = W_{FA} + W_{FB}$ – work of the sliding resistance forces of vehicles A and B in the impact force act direction - the longitudinal displacement of vehicle A and the lateral displacement of vehicle B; hence in a front-to-side collision we have $W_{FB} \gg W_{FA}$;
E_R – the remaining energy part, related to other processes that take place during the vehicle collision.
In result of the front-to-side vehicle collision, the initial energy $(E_{KA} + E_{KB})$ is divided into the energy received by vehicle A $(E_{KA}{}^* + W_{DA})$, energy received by vehicle B $(E_{KB}{}^* + W_{DB})$, and dissipated energy $(W_{TAB} + W_{FAB} + E_R)$.
The greater the kinetic energy $E_{KB}{}^*$ is received by vehicle B as a result of the collision, the higher risk for the occupants will be. The change of kinetic energy result in inertial loads to which the vehicle and occupants are subjected. Part of the impact energy is dissipated into the work of deformation of the body side W_{DB}. This energy dissipation makes it possible to limit the value of $E_{KB}{}^*$. However, the value of this work is directly dependent on the side-stiffness of the vehicle body and deformation occurring during the collision, which is a source of hazard for occupants. This raises the need to find a compromise in the considered area of automotive design improvement. By increasing the side-stiffness, deformation during side impact can be limited. However, the effect of such action is an impact force increase which causes the lateral movement of the vehicle, and also a possible reduction of the energy dissipated on deformation. The aim of the study is to assess the impact of the side-stiffness change on the course of the hazard emergence process during a front-to-side vehicles' collision. The measures and effects of the collision will be considered. The influence of the impact velocity on the hazard level generated during the collision will also be shown. The change of the body stiffness is one of the possible aspects of the design changes leading to the vehicle passive safety improvement. The risk will be assessed on the basis of physical quantities describing the loads to which vehicles and occupants are subjected during a road accident. Acceleration of the driver's seat, the depth of vehicle's body deformation and changes in the collision energy balance components (eq. (1)) are considered.

2. Collision model and its validation

The PC-Crash 13.0 [17] program was used for the calculations. A model of the front-to-side vehicles' collision in a spatial approach was prepared. The physical model of the collision has been shown in Figure 2. The global coordinate system $\{O_E\}$ and the local systems related to the centres of mass of vehicles $\{C_A\}$, $\{C_B\}$ and the ti
re-road contact area $\{\boldsymbol{O_T}\}$ have been marked. A separate coordinate system $\{\boldsymbol{O_{Tn}}\}$ is introduced for each wheel. The kinematic excitation is the initial velocity of the vehicles movement $\boldsymbol{v_{0s}}$.

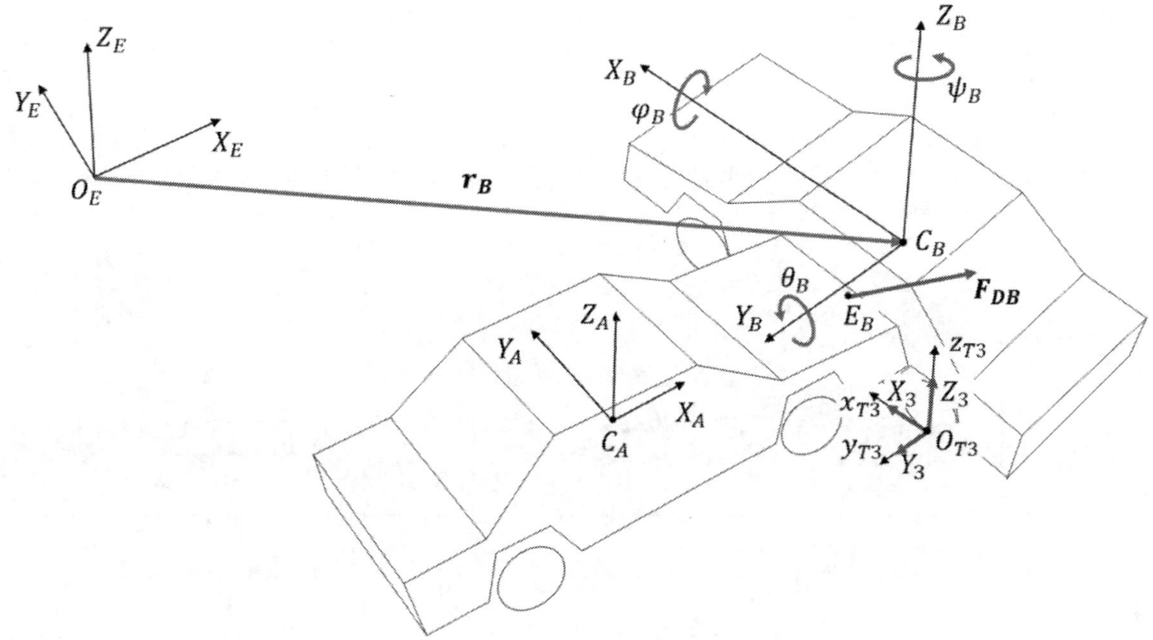

Figure 2. Physical model of the front-to-side vehicles' collision.

The sth vehicle (s=[A, B]) can be described by vectorial equations [18]:

$$m_s(\ddot{\boldsymbol{r}}_s + \boldsymbol{\Omega}_s \times \dot{\boldsymbol{r}}_s) = \sum_i^n \boldsymbol{F}_{si} \quad (2)$$

$$T_s\dot{\boldsymbol{\Omega}}_s + \boldsymbol{\Omega}_s \times T_s\boldsymbol{\Omega}_s = \sum_j^k \boldsymbol{M}_{sl} \quad (3)$$

where:
m_s — vehicle mass;
$\boldsymbol{r_s}$ — vector from the origin of the global coordinate system to the centre of mass C_s in the global coordinate system $\{O\}$; $\boldsymbol{r_s} = [x_s \quad y_s \quad z_s]^T$ (cf. Figure 2)
$\boldsymbol{F_{si}}$, $\boldsymbol{M_{sj}}$ — generalized external forces and moments acting on vehicle s;
T_s — tensor of inertia of vehicle s relative to the vehicle centre of mass in the local coordinate system $\{C_s\}$;
$\boldsymbol{\Omega_s}$ — vector of the yaw velocity of the body of vehicle s in the local coordinate system $\{C_s\}$; $\boldsymbol{\Omega_s} = [\dot{\varphi}_s \quad \dot{\theta}_s \quad \dot{\psi}_s]^T$.
In Figure 2 has been shown the external forces which act on vehicle B during the collision. Those are:
- impact force $\boldsymbol{F_{DB}}$ acting on the body B, calculated according to the force-based impact model [19];
- tangential reaction forces X_n, Y_n in the tire-road contact area, calculated in the non-linear model TM-Easy [20-22];
- normal components of reaction forces Z_n from the road surface.
The results of crash tests carried out by the Research Network Łukasiewicz-Automotive Industry Institute (Łukasiewicz-PIMOT) [23] were used to validate and parameterize the model. In these tests, a front-to-side collision of passenger cars of the same model was carried out (Honda Accord, model year 1998). The vehicles had 195/60 R15 tires. The results of a crash test were used, in which the vehicle A hit at a velocity of 55 kph into the centre area between the axes of the stationary vehicle B. At the beginning of the collision contact, the longitudinal axes of the motorcars were perpendicular. The data describing the geometry and mass-inertia properties of vehicles were taken from the above-mentioned experimental research to modelling.
The lacking data for modelling was obtained in the process of model parameterization using the trivial method and on the basis of the analysis of the other experimental studies carried out results, for example, by NHTSA [24] and also [25-27]. On their basis, the parameters describing the energy-consuming properties of the front-end part of the vehicle body, necessary to determine the impact force $\boldsymbol{F_{Ds}}$, were

established. The data collection process was also based on the effects of the authors' previous research on the issue under consideration. They were the basis, for example, to estimation of the energy-consuming properties of the side part of the vehicle body [28]. The exemplary results of the model validation have been shown in Figure 3. The course of the vector component of the velocity of the center of mass of the striking (A) and being struck (B) vehicle as well as the course of the process of increasing the summarize deformation of the vehicles' bodies in the experimental (dashed line in the figure) and model tests (solid line) are shown.

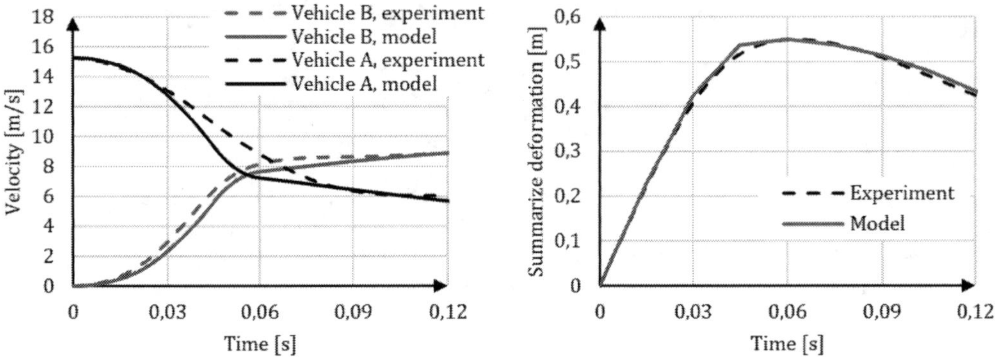

Figure 3. The course of changes in the velocity vector component of the centre of masses of vehicle A and B and the summarize deformation of vehicles' bodies in experimental and model tests

3. Model tests and results analysis

Model tests were prepared and carried out to assess the influence of selected factors on the course and effects of the front-to-side vehicles' collision. The analysed factors, their base values and the range of changes in relation to the base value are summarized in Table 1. In total, 42 variants have been considered.

Table 1. Factors investigated in model tests and the range of changes in their values

Factor	Symbol, unit	Base value	Changes range
Side-stiffness of the vehicle B body	k_B [kN/m]	650	100 – 130 %
Impact velocity	v_{0A} [km/h]	40	40 – 65 kph

The research results are presented in the form of kinematic and dynamic quantities describing the course and effects of a front-to-side vehicles' collision. The following physical quantities and their characteristic values were taken into account:

- resultant acceleration of the driver's seat a_W;
- vehicle's body deformation depth c_B;
- work of deformation W_{DB};
- kinetic energy of vehicle B, obtained as a result of a collision $E_{KB}{}^*$.

The work of deformation W_{DB} of vehicle B body at the instant t_j was calculated from the equation:

$$W_{DB}(t_j) = W_{DB}(t_{j-1}) + 0.5 \cdot \left[F_{DB}(t_j) + F_{DB}(t_{j-1})\right] \cdot \left[c_B(t_j) - c_B(t_{j-1})\right] \quad (4)$$

where c_B is the deformation of the side part of the vehicle body, determined on the basis of the dependencies proposed in [25]. The change in kinetic energy of vehicle B was determined from the dependence:

$$E_{KB}{}^*(t_j) = 0.5 \cdot m_B \cdot v_B^2(t_j) + 0.5 \cdot \left[T_B \cdot \Omega_B^2(t_j)\right] - E_{KB}(t = 0) \quad (5)$$

$v_B = \sqrt{\dot{x}_{CB}^2 + \dot{y}_{CB}^2 + \dot{z}_{CB}^2}$ – resultant velocity of the vehicle B.

IOP Conf. Series: Materials Science and Engineering 1247 (2022) 012024 doi:10.1088/1757-899X/1247/1/012024

The side-stiffness analyzed in this work determines the value of the F_{DB} impact force (see Figure 1) resulting from the deformation c_B [18], from where we have:

$$F_{DB}(t_j) = k_B c_B(t_j) \tag{6}$$

Figures 4 and 5 show changes in physical quantities depending on the side-stiffness and the vehicle A impact velocity into B. The values in the figures are the percentage change of a given physical quantities in relation to its values obtained in the calculations for the base value of the analyzed factor (Table 2). The results showing the influence of the side-stiffness of vehicle body have been presented for the impact velocity equal to 55 kph, and the results showing the influence of the impact velocity - for the basic value of the side-stiffness (100%).

The side-stiffness is a factor describing the structural properties of the vehicle body in terms of energy absorption and determining dynamic deformation process during a collision. The essence of activities in the field of improving the passive safety of the bodywork structure is the search for the possibility of limiting this deformation depth into the passenger compartment. Hence, in this study, the impact of increasing this side-stiffness on the course and effects of the front-to-side vehicles' collision is analysed. Increasing the stiffness allows to limit the range of deformation c_B (Figure 4). For the stiffness increased by 30%, the deformation depth is limited by nearly 6%. This is a beneficial and expected effect of the analysed change in the design properties of the vehicle bodywork. However, as the side-stiffness increases, the acceleration value of the driver's seat is higher (Figure 4). This is a negative effect in the context of hazard for the driver, but the range of changes in these loads is not large, as it does not exceed 3%. This is due to the' collision process, in which both vehicles are subjected to deformation under the action of impact force. It creates a system of two deformed objects with different stiffnesses, which can be treated as a series connection of two springs. In this case, changes to one stiffness do not proportionally affect the equivalent stiffness of the system. On the other hand, this resultant stiffness will determine the achieved impact force, which has a direct impact on the acceleration of the vehicles. The stiffness of the bodywork is directly related to the process of its deformation and the associated energy dissipation in the form of deformation work W_{DB}. Its value decreases by nearly 2% (Figure 5), which results in an increase in kinetic energy $E_{KB}{}^{*}$. However, this energy increase by about 1.4% with a change of side-stiffness by 30%. The course of the observed changes in individual factors values is approximately linear. The obtained results show that with the increase of the body stiffness, the deformation depth is limited, and the percentage change in this parameter is greater than the changes in parameters indicating the negative effects of modifying the properties of the vehicle structure. Thus, by increasing the stiffness of the side of the vehicle bodywork, the occupants injury risk can be reduced.

The velocity impact of the vehicle A into vehicle B has a logical influence on the process of hazard arising in a front-to-side collision. With the increase in this velocity, we have greater acceleration of the driver's seat and the depth of deformation (Figure 4), as well as significant changes in the values of the analysed components of the collision energy balance (Figure 5).

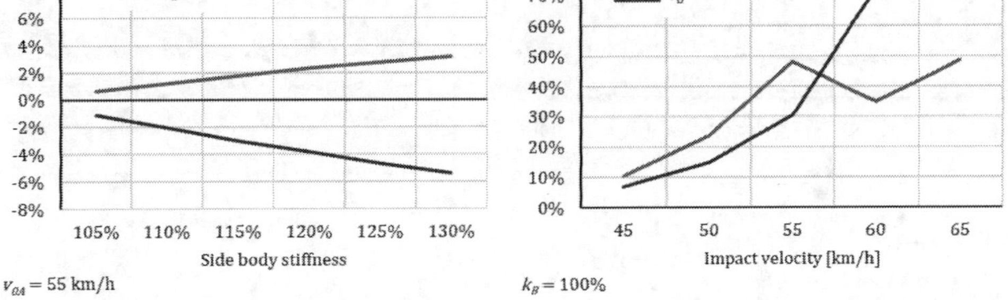

Figure 4. Relative change of the driver's seat resultant acceleration a_W and the depth of deformation of the side of the vehicle body c_B depending on the side-stiffness (left) and the impact velocity (right)

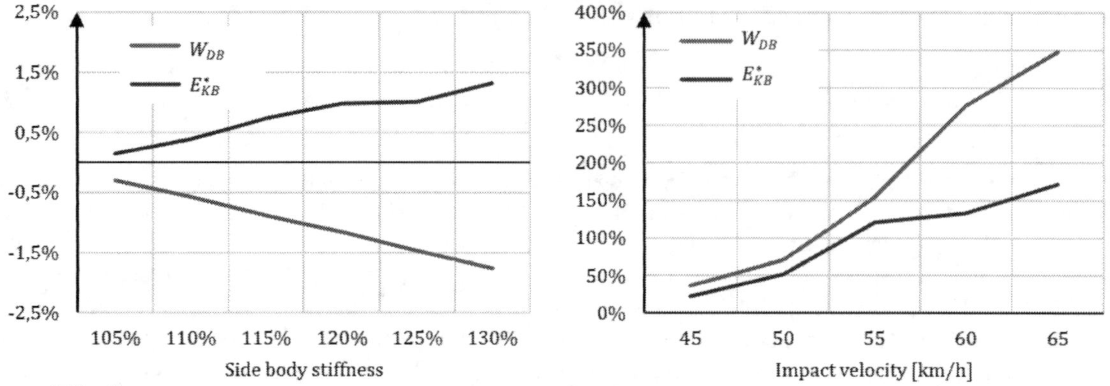

v_{0A} = 55 km/h k_B = 100%

Figure 5. Relative change of the work of deformation W_{DB} and kinetic energy change $E_{KB}{}^*$ depending on the side-stiffness (left) and the impact velocity (right)

On the basis of the model tests results, it was verified how the change in the side-stiffness will translate into the values of the analysed values and parameters depending on the impact velocity. The results of such a combination have been shown in Figure 6. The different colours of the lines refer to the calculations results of individual values obtained for the increased value of the side-stiffness (110, 120 and 130%), compared to the value of these parameters at the base stiffness (100%). This reveals the non-linear nature of the relationship between the effects of the increased side-stiffness for different impact velocities. There is particular advantages of increasing side-stiffness for impact speeds above 60 kph. Then, the greatest limitation of the deformation depth c_B is achieved, with slight changes in the acceleration of the driver's seat a_W, which is the source of inertial loads for the occupant. There is no clear limitation of the value of the deformation work W_{DB} here, and the observed increase in the value of the kinetic energy $E_{KB}{}^*$ has no effect on the acceleration a_W. This is due to the phenomena where for higher impact velocities there is a significant angular movement of the vehicle body, which increases the kinetic energy value, but is not a source of acceleration in the driver's seat, because it is near of the vehicle centre of mass. On the other hand, the negative effect of increasing the side-stiffness at low impact velocities (<45 kph) is visible. There is an increase in the acceleration value a_W, with a disproportionate limitation of the deformation range c_B and unfavourable changes in the values of the energy balance components.

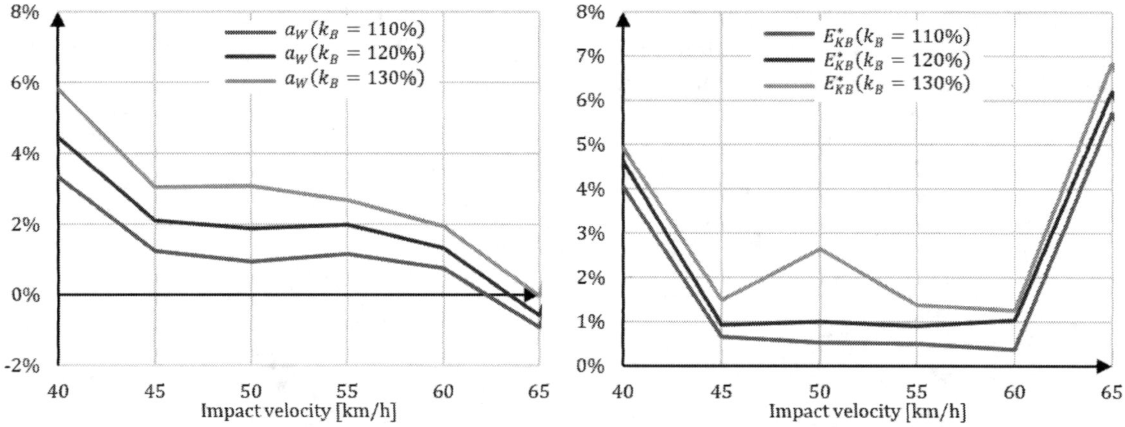

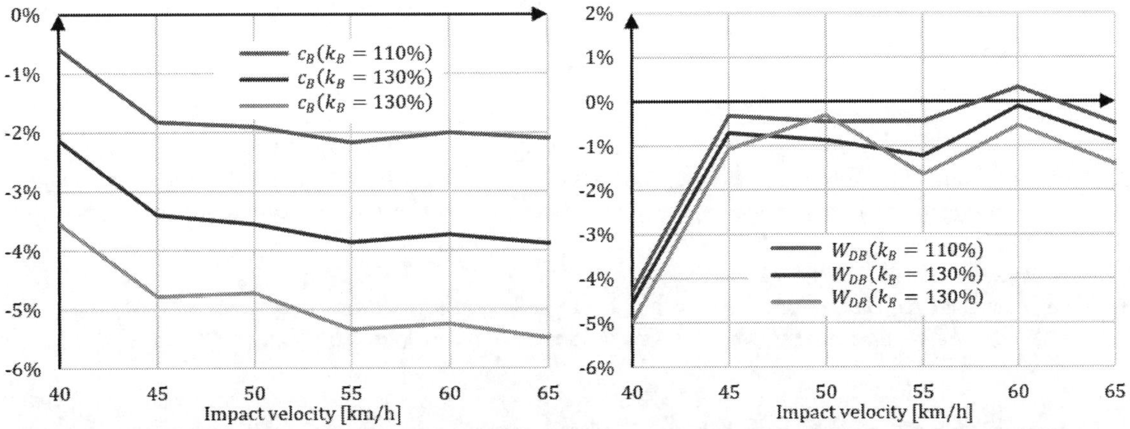

Figure 6. Relative change of the driver's seat resultant acceleration a_W, depth of deformation of the side of the vehicle body c_B, work of deformation W_{DB} and kinetic energy change $E_{KB}{}^*$ depending on the impact velocity for different side-stiffnesses of vehicle B body

4. Summary and conclusions

The hazard arising during the front-to-side vehicles' collision result from inertial loads and from excessive deformation of the side part of the vehicle body, the depth of which usually includes the space containing the occupants. One of the possible actions to improve the passive safety of the motorcar is to increase the stiffness of the side part of the vehicle body. However, this action requires looking for a compromise solution, because by increasing the side-stiffness, a beneficial limitation of its deformation during a side impact can be accomplish, but the effect of such action is an increase in the lateral impact force as well as a possible reduction of the energy absorbed on this deformation. For this reason, the aim of the study was to assess the impact of changes in the side-stiffness on the course of the hazard arising process during a front-to-side vehicles' collision. The aim was achieved through model tests in the front-to-side collision model in the PC-Crash program. The measures and effects of the collision were considered. Based on the results of the tests was concluded:

- Increasing the stiffness of the side part of the vehicle body by 30% allows for a favourable reduction of its deformation depth by up to 6% (Figure 4);
- However, this causes a slight increase in the acceleration of the driver's seat (Figure 4), which is a source of hazard for the driver;
- The effect of increasing of the side-stiffness is not proportional for different values of the impact velocity (see Figure 6), and the greatest benefits in the form of limiting bodywork deformation in the absence of significant changes in inertial loads (acceleration of the driver's seat) are available for high impact velocities (> 60 kph).

The results of the considerations in this work constitute the basis for further actions. It is planned to use proprietary tools for modelling the hazard arising process for the occupants of a struck vehicle during front-to-side collision, enabling the assessment of the injury risk depending on inertial loads and the deformation process of the side part of vehicle body.

Founding

This work was co-financed by Military University of Technology under research project UGB 22-755/2020.

Acknowledgments

The authors would like to gratitude to professor Mirosław Gidlewski for sharing the results of experimental research, enabling the validation of the prepared model for model tests in this work.

References

[1] Gidlewski, M., Prochowski, L., Jemioł, L., & Żardecki, D. (2019). *The process of front-to-side collision of motor vehicles in terms of energy balance*. Nonlinear Dynamics, 97(3), 1877-1893.

[2] Prochowski, L., Ziubiński, M., & Gidlewski, M. (2018, April). *Experimental and analytic determining of changes in motor cars' positions in relation to each other during a crash test carried out to the FMVSS 214 procedure*. In 2018 XI International Science-Technical Conference Automotive Safety (pp. 1-5). IEEE.

[3] Arbelaez, R. A., Baker, B. C., & Nolan, J. M. (2005, June). *Delta Vs for IIHS side impact crash tests and their relationship to real world crash severity*. In 19th ESV Conference, Paper (pp. 05-0049).

[4] Viano, D. C., & Parenteau, C. S. (2010). *Severe injury to near-and far-seated occupants in side impacts by crash severity and belt use*. Traffic injury prevention, 11(1), 69-78.

[5] Huang, M. (2002). *Vehicle crash mechanics*. CRC press.

[6] Vangi, D. (2020). *Vehicle Collision Dynamics: Analysis and Reconstruction*. Butterworth-Heinemann.

[7] Summers S. M., Prasad A., Hollowell W. T., *NTHSA's research program for vehicle aggressivity and fleet compatibility*, National Highway Traffic Safety Administration, USA, Paper #249

[8] Gabler H. C. (2003), *The Evolution of Side Crash Compatibility Between Cars, Light Trucks and Vans*, SAE International

[9] Gabler H. C., Hollowell G. W., *The Aggressivity of Light Trucks and Vans in Traffic Crashes*, SAE Technical Paper Series, Airbag Technology, SP-1333

[10] Huh, H., Lim, J. H., Song, J. H., Lee, K. S., Lee, Y. W., & Han, S. S. (2003). *Crashworthiness assessment of side impact of an auto-body with 60TRIP steel for side members*. International Journal of Automotive Technology, 4(3), 149-156

[11] Njuguna, J. (2011). *The application of energy-absorbing structures on side impact protection systems*. International Journal of Computer Applications in Technology, 40(4), 280-287.

[12] Ikpe, A. E., Owunna, I. B., & Satope, P. (2017). *Design optimization of a B-pillar for crashworthiness of vehicle side impact*. Journal of Mechanical Engineering and Sciences, 11(2), 2693-2710.

[13] Marklund, P. O., & Nilsson, L. (2001). *Optimization of a car body component subjected to side impact*. Structural and Multidisciplinary Optimization, 21(5), 383-392.

[14] Teng, T. L., Chang, K. C., & Nguyen, T. H. (2008). *Crashworthiness evaluation of side-door beam of vehicle*. Technische Mechanik-European Journal of Engineering Mechanics, 28(3-4), 268-278.

[15] Ikpe, A. E., Owunna, I. B., & Satope, P. (2017). *Design optimization of a B-pillar for crashworthiness of vehicle side impact*. Journal of Mechanical Engineering and Sciences, 11(2), 2693-2710.

[16] Jurecki, R.S. (2020. *Analysis of road safety in Poland after accession to the European Union*. Communications - Scientific Letters of the University of Zilina, 22(2), pp. 60–67.

[17] Steffan, H., & Moser, A. (1996). *The collision and trajectory models of PC-CRASH* (No. 960886). SAE Technical Paper.

[18] Datentechnik, S. (2008). *PC-CRASH-A Simulation Program for Vehicle Accidents*. Linz, Austria.

[19] Steffan, H. (2009). *Accident reconstruction methods*. Vehicle system dynamics, 47(8), 1049-1073.

[20] Rill G. (1994), *Simulation von Kraftfahrzeugen*, Vieweg & Sohn Verlag GmbH, Braunschweig/Wiesbaden.

[21] W. Hirschberg, G. Rill & H. Weinfurter (2007) *Tire model TMeasy*, Vehicle System Dynamics, 45:sup1, pp. 101-119, DOI: 10.1080/00423110701776284

[22] G Rill (2013), *TMeasy–A Handling Tire Model based on a three-dimensional slip approach*, Proceedings of the XXIII International Symposium on Vehicle Dynamics…

[23] Research Network Łukasiewicz-Automotive Industry Institute (Łukasiewicz-PIMOT), Warsaw, Poland, https://pimot.lukasiewicz.gov.pl/

[24] National Highway Traffic Safety Administration Vehicle Crash Test Database, https://www.nhtsa.gov/research-data/databases-and-software

[25] Jackowski, J. & Żmuda, M. (2016), *Using different study objects during researches at low speed Crashes*. Transport Means - Proceedings of the International Conference, 2016-October, pp. 402–407.

[26] Gidlewski, M., Jackowski, J. & Posuniak, P. (2022) *Review and Analysis of Technical Designs of Rear Underrun Protective Devices (RUPDs) in Terms of Regulatory Compliance*, Sensors, vol. 22(7), 2645.

[27] Jaśkiewicz, M., Frej, D. & Poliak, M. (2022) *Simulation of a dummy crash test in Adams*, Communications - Scientific Letters of the University of Zilina, vol. 24, pp. B20–B28

[28] Prochowski, L., Ziubiński, M., & Pusty, T. (2018, September). *Experimental and analytic determining of the characteristics of deformation and side stiffness of a motor car body based on results of side-impact crash tests*. In IOP Conference series: materials science and engineering (Vol. 421, No. 3, p. 032025). IOP Publishing.

IOP Conf. Series: Materials Science and Engineering 1247 (2022) 012025 doi:10.1088/1757-899X/1247/1/012025

The influence of seatback reclination on body kinematics during low-speed frontal impact

A Górniak[1], W Górniak[1], J Matla[1], M Zawiślak[1]

[1] Wroclaw University of Science and Technology, Mechanical Faculty, Division of Automotive Engineering, Na Grobli 13, 50-421 Wrocław, Poland

* aleksander.gorniak@pwr.edu.pl

Abstract. Safety tests are generally performed in accordance with strict procedures that involve the use of a carefully positioned dummy. Safety tests are rarely conducted with the occupant being in a non-standard position. In real life driving, occupants, especially passengers, choose a more comfortable reclined position, and in such a case the body dynamics are different from those investigated in the standard safety test. Furthermore, with the rapid development of autonomous vehicles, which provide the possibility of the position of occupants being unrestricted, the investigation of "out of position" body kinetics is becoming more important. The presented study aimed to evaluate body dynamics with various seatback reclinations. Body dynamics were verified by simulating frontal impact on a sled system with the use of a standard 50 percentile Hybrid III dummy. Points of interest located on the dummies head, neck, pelvis, and legs were traced, which allowed its trajectory to be evaluated. Additionally, the maximal extrusion and the time of motion were evaluated. It was found that the maximal extrusion in the longitude direction is the same for semi and fully reclined seats. Furthermore, a reclined seat causes head rotation, which can result in neck injuries.

Keywords – dummy, accident, frontal impact

1. Introduction

The development of personal transport in the form of fully automated vehicles will mean that a minimal contribution from the driver is required. The driver will also become a passenger and will be able to travel in a more comfortable position (e.g. with a reclined seat). Therefore, there will be new configurations and seating positions that will differ from the solutions that are currently commonly used [1]. Initially, however, autonomous cars will use the current passenger and driver restraint systems. Most of the available results of the passive safety test were achieved in accordance with strictly defined standards in order to ensure the repeatability of the tests. Tests concerning the position of the passenger with the seat fully reclined are omitted in standard procedures.

It was found that a partially reclined seating position was the most comfortable for travelers, and also the most frequently chosen option [2]. However, different seat positions cause the unsatisfactory operation of the restraining system, which is designed for a specific position of a driver and passengers. In consequence, safety tests are performed in a range of defined operation regimes of the restrain system. The reclined seatback position has a significant impact on the performance of a seatbelt and airbag. The angle of the seat has an effect on pelvic movement in a crash, and may therefore increase the likelihood of serious injury [3, 4].

The most frequent cause of injury is the improper performance of a seatbelt caused by reclined seatback – i.e. submarining [5]. Submarining causes severe injuries of the lumbar spine and lumbar internal organs, which are often the reason for the many fatalities and permanent disabilities of passengers (including permanent lower body paralysis). Worldwide literature shows that submarining tests are mainly carried out in a numerical environment, and models that predict the risk of injury to reclined passengers are limited by a lack of sufficient reference data [3, 6]. The occurrence of the submarining phenomenon is influenced by body posture, gender, or BMI (body mass index) [7-9]. Apart from the seatback and belt position, the extrusion of the knee, the rotation of the torso and the difference of knee and head extrusion can be used for predicting the effect of submarining [10, 11]. Many researchers have investigated the severity of crashes for reclined seats, but there are very few papers that describe the body dynamics when the seat is reclined in various positions. Altered body dynamics can lead to the ineffective operation of the seatbelt, and in consequence, poor airbag performance [12]. Therefore, overall body dynamics should be considered. For example, submarining generates higher head impact energy due to the greater distance of the head to the airbag. Furthermore, it has been proven that the distance between the seat belt anchor and the hips corresponds to a reduction in head injuries and an increase of sternal deflection [13].

The study aimed to understand the influence of the sitting position on a reclined seat on the body dynamics during the sled test simulation of a frontal impact. For this reason, the trajectory of the head, neck, pelvis and legs is analysed with respect to various seatback angles. Additionally, the times after which individual parts of the body reacted to the impact pulse were also checked.

2. Materials and methods

The entire experiment was performed in the Laboratory of Vehicle Dynamics and Safety, which is located in the Research Complex GEO-3EM ENERGY ECOLOGY EDUCATION. The investigation consisted of verifying body dynamics during an out of position test. In this case, variable seat reclination was considered. The reclination ranged from the standard seatback angle, which is recognised as being comfortable – i.e., 110°, through 130°, and finally to a fully reclined seatback of 145°. The biomechanics of a crash were evaluated with the aid of a standard 50 percentile Hybrid III male ATD dummy. Each test was recorded using the Phantom VEO 410L high-speed camera. Every test was recorded at 5000 FPS with a maximum resolution of 800x1280. TEMA Automotive software was used for record analysis. High-speed record analysis consisted of tracking the points of interest located on the head, neck, hips and legs. The test station prepared for the experiments, with named points of interest, is shown in figure 1. The trajectory of the designated points on the dummy's body was emphasized in the analysis. The reference point was fixed at the centre of the seat, and in consequence all trajectories were found for a movable reference point. Additionally, the dummy was equipped with an accelerometer in its head. The sample rate was set to 15kHz. The acceleration signals presented in the study are filtered with a CFC 1000 filter. Furthermore, in the body dynamics investigation, a backward head motion was expected, especially during the test with a reclined seat. For this reason, only the X direction signal was taken into account, which is due to the fact that the resultant signal of the triaxial measurement would reverse the negative acceleration measured in the centre of the head.

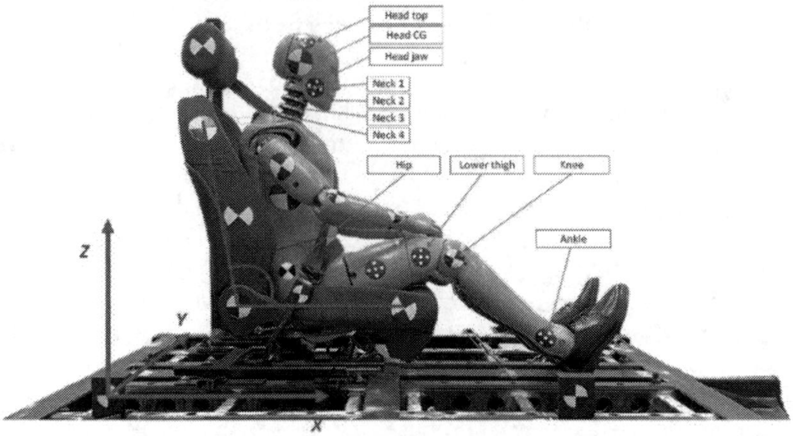

Figure 1. The test station and coordinates system used for the experiments.

The crash pulse was unchanged through the studies. The average acceleration was 5 g and the trolley velocity was Δ7 m/s. The average crash pulse is shown in figure 2, with the shaded area representing the standard deviation.

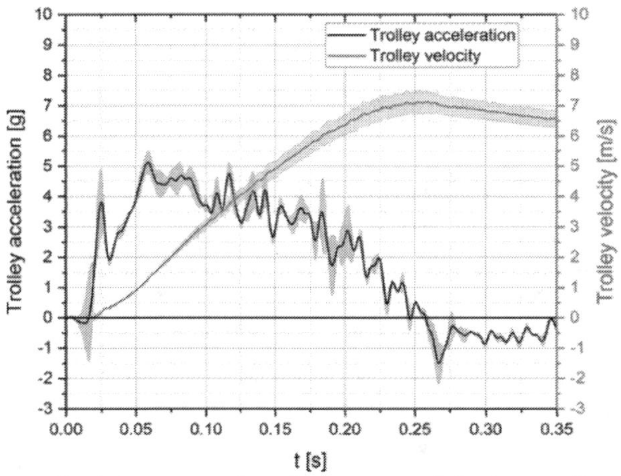

Figure 2. Average crash pulse generated during each test; the shaded area constitutes the standard deviation.

3. Results

The result of the high-speed analysis shows that the body dynamics differ for different reclinations of the seat. As could be expected, the higher the reclination of the seat, the longer the distance the dummy needs to travel, which in turn makes the time of the motion longer. The path of the points and the time of the maximal extrusion is shown in figure 3. As can be seen, the paths of the markers are similar for the seatback angles of 110° and 130°. Moreover, despite the long-distance that the dummy needed to travel, the time did not differ significantly, which suggests a higher velocity of motion. On the other hand, for the seatback angle of 145°, the path was entirely different from the previous cases, and the duration of the motion was the longest. Additionally, the fully reclined seat caused a greater force to be applied on the dummies heals, which in turn prevented the forward motion of the ankles. In all repetitions of the test at fully reclined seat positions, the dummies shins were not thrusted forward.

IOP Conf. Series: Materials Science and Engineering 1247 (2022) 012025 doi:10.1088/1757-899X/1247/1/012025

Seat back angle 110° Seat back angle 130° Seat back angle 145°

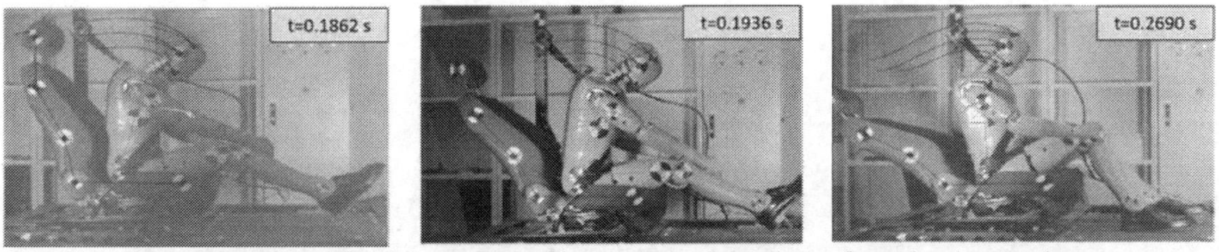

Figure 3. The path of the points of interests and the time of maximal extrusion.

The comparison of the trajectory of points of interest located at the head and the neck are shown in figure 4. It appears that there was virtually no difference between the extrusion in the X direction for the seatback angles of 130° and 145°. However, when considering the overall trajectory of the points, it can be seen that the fully reclined seat provoked the longest path. Additionally, there was a significantly higher Z extrusion when compared to the previous tests.
At the seatback reclination of 110° and 130°, the head motion was rapidly reversed, which is manifested by a sharp peak in the trajectory lines. It is anticipated that such a tendency could have a beneficial influence on neck extension when compared with the test with a fully reclined seat. This, however, is impossible to prove in these studies due to the fact that the dummy used for this research was not equipped with appropriate measurement apparatus. In this study, only the trajectory of the neck points could be evaluated, and as expected, the trajectory of these points resembled the path of the head markers.

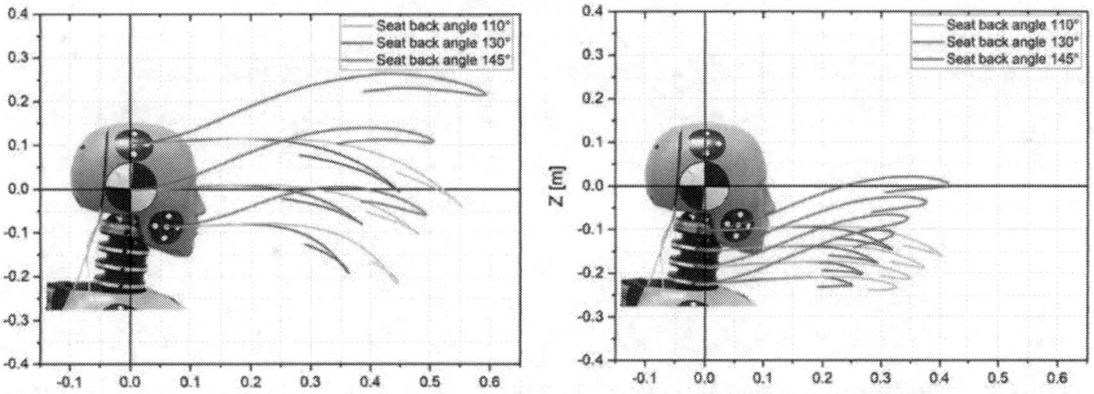

Figure 4. The trajectory of the points of interest located at the head and the neck of the dummy.

The crash pulse was unchanged throughout the test, but the seatback position caused different acceleration signals (figure 5). The standard seat back position provoked an intermediate value of head acceleration. For the seatback angle of 110°, the acceleration was 15g, whereas for the angle of 130° the acceleration was lower – approximately 11g. For the fully reclined seat, the acceleration reached 20g.

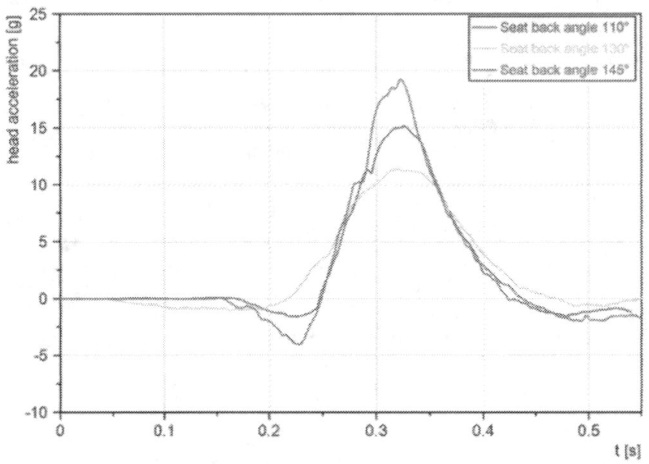

Figure 5. Head acceleration. ISO/TS 13499 description: S1HEADCG00H3ACXA

The difference in acceleration signals can be explained on the basis of the trajectory of the points, which were determined from the observer's perspective. At the angle of 110°, the head moved forward immediately after a crash pulse, and there was a minimal negative acceleration in such a case. The head rotation was relatively small (figure 6a), which caused an increase of the acceleration over time. However, when the seat was further reclined, the head experienced backward motion, which is manifested by the relatively long negative acceleration on the graph. Subsequently, the head was reversed on a wider path, which in turn led to a wide and low acceleration signal (figure 6b). Finally, when the seat was fully reclined, the head was moved backwards at the initial part of the crash pulse, which is manifested by the negative value of acceleration. After that, the head suddenly moved in the opposite direction, which caused the occurrence of a high peak of acceleration (figure 6c).

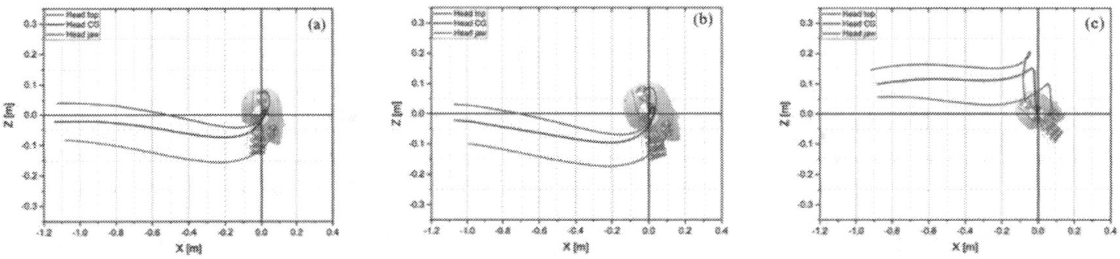

Figure 6. Trajectory of the points of interests on the head, as seen from the observer's perspective: (a) seatback angle of 110°; (b) seatback angle of 130°; (c) seatback angle of 145°.

The extrusion of the hips in the X and Z directions is shown in figure 7. As expected, displacement of the hips was the shortest for the seat back angle of 110°. In this case, the maximum displacement in the X direction reached approximately 6 cm. The seatback angle of 130° provoked the longest X extrusion, and a similar Z extrusion to the seatback angle of 110°. The intermediate X displacement was found for the seatback angle of 145°.The Z extrusion was initially negative in every case, which means that the hips were pushed downwards against the seat cushion. Afterwards, the direction of motion was reversed. The reversed motion is a consequence of the hips being stopped by the seat belt. The motion of the upper body, and also the position of legs, forced the hip to move upwards. As can be seen in figure 6, the seat back angle of 145° provoked the most significant Z extrusion of the hip. The submarining effect was not observed in any of the tests.

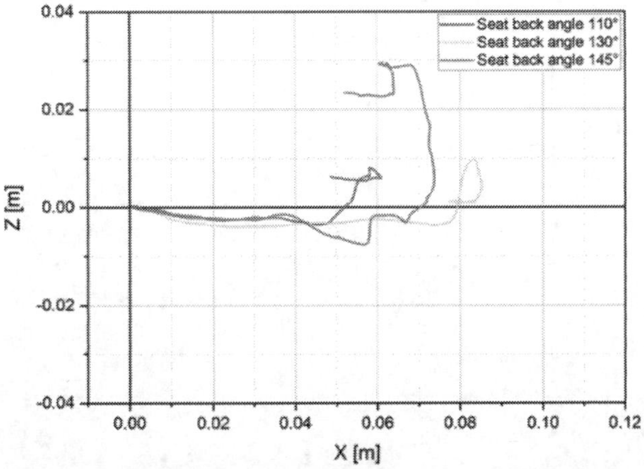

Figure 7. The trajectory of the points of interest located on the hips.

The hip motion corresponds to the extrusion of the legs – in particular the X extrusion. The extrusion on the points located on the legs is shown in figure 8. Here, the seat back angle of 110° provoked the largest Z extrusion. In the case of both the 110° and 130° angles, the Z motion was similar, with the point being displaced being 7-8 cm. However, the Z displacement for the fully reclined seat was significantly lower when compared to the seatback positions with a smaller angle. This is due to the fact that the legs were not straightened in all of the repetitions of the test . The impact energy forced the heals to be pushed against the floor, and therefore the knees were not straightened and all the X motions for the knees were due to the whole body being in a forward motion.
The ankle motion for the seatback angles of 110° and 130° were similar. In both cases, the legs were fully extended, in turn causing the overall trajectory to be similar. However, for the fully reclined seat, the foot was not lifted, and the X motion originated from the hip forward movement.

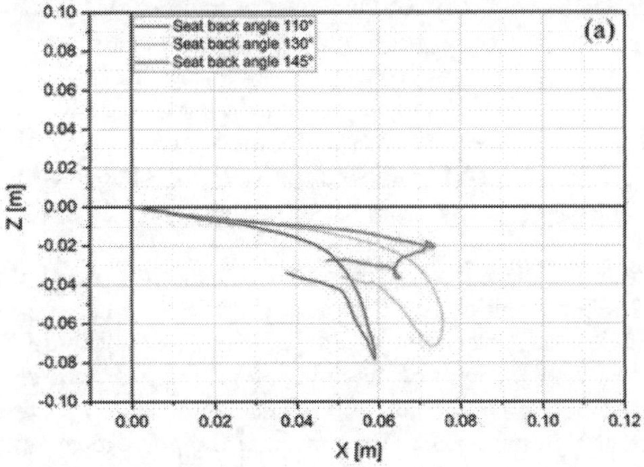

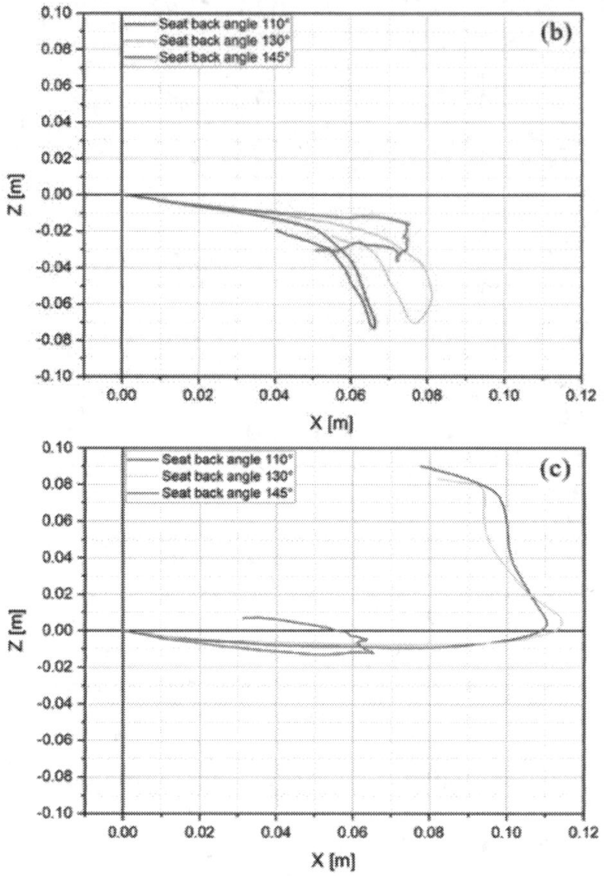

Figure 8. Trajectory of the points of interest located on the leg: (a) lower thigh; (b) knee; (c) ankle.

The comparison of the motion of all the points of interest in the X and Z directions is shown in figure 9. It appears that extrusion in the X direction is virtually the same for when the seat is reclined by 130° and 145°. The extrusion of the points of interest at the standard seatback position (angle of 110°) was generally lower. Furthermore, and as expected, the most significant differences exist for the points that are distant from the hips.
The extrusion in the Z direction represents a different tendency. In general, the displacement measured for the tests with the seatback angles of 110° and 130° represents a similar tendency. However, considering the data spread, it can be concluded that those extrusions are different. For the test with the seatback angle of 145°, the displacement of the point of interest was entirely different. Unique in this case is the displacement of the points located on the neck. It appears that those points were moved in the Z-direction more than the dummies head (in particular the point located on the jaw). This indicates significant neck tension and possible whiplash injuries. The Z extrusion of the points of interest located on the legs was the lowest for the fully reclined seat, because for this test the legs were not straightened.

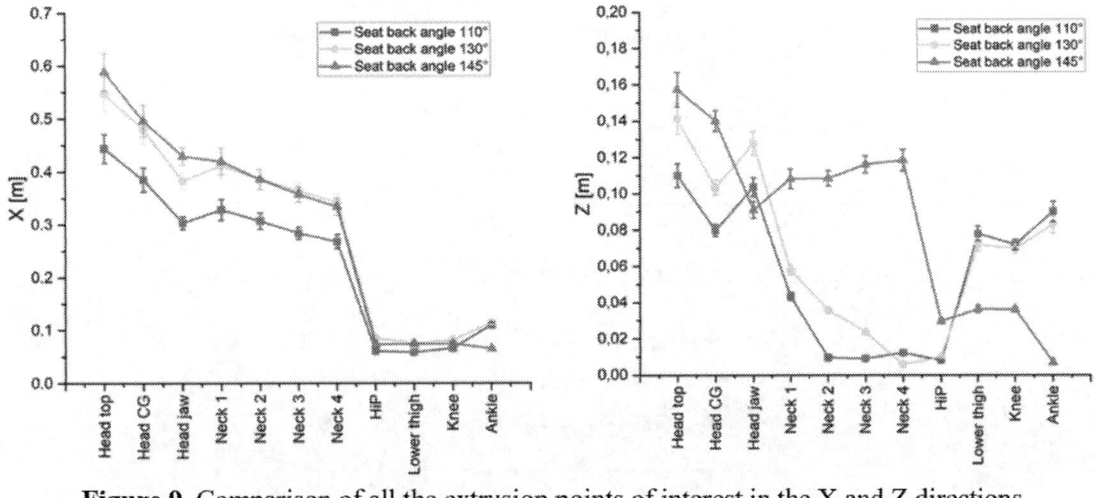

Figure 9. Comparison of all the extrusion points of interest in the X and Z directions.

The seatback position not only had a significant impact on the trajectory of different parts of the dummies body, but also on the time that the specific point of interest began the motion. What is under consideration here is the time difference between the trolley motion and the reaction of the body. This time difference is shown in figure 10. It should be noted that the figure only contains the instance of motion of the points of interest, with no information regarding the direction of its trajectory. At the seatback angle of 110°, the difference in the time between the trolley starting and the dummy beginning its motion is the lowest when compared to all the other tests. In this test, the hips and the neck reacted at almost the same time. However, the reaction of the head was the longest, which is due to the neck flexibility and inertia forces. The similar difference in the time for the neck and the hips suggests that the lab belt and the chest belt caught the body in a similar instance, therefore minimizing the severity of the impact.

The seatback angle of 130° represents a greater delay of the body's motion with regards to the trolley's motion. In this case, similarly as for the seat back angle of 110°, the hips were the first to react to the pulse of the collision. The response time was comparable to the other tests, regardless of the angle of the seat. The delay of the subsequent points of interest was significantly higher. The neck and the lower portion of the head reacted after a longer time when compared with the top of the head. This is due to the backwards movement of the head, which exists for dummies sitting on a reclined seat. This indicates the occurrence of head roll, which may negatively affect the neck. In the case of the fully reclined seat, the reaction time was the longest – i.e. each point responded to the trolley's motion after the longest time when compared to the previous tests. The point at the top of the head started its motion before the rest of the points located at the head.

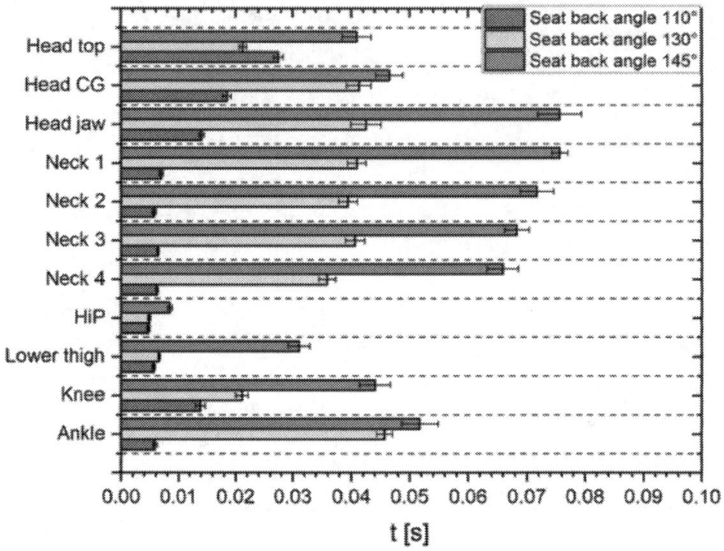

Figure 10. Comparison of the time delay of body parts with regards to the troley's motion.

4. Discussion

Reclined body dynamics in out of position testing is rarely investigated. Investigations regarding this matter are mostly of a numerical nature, where the greatest emphasis is given to the submarining effect and lumbar spine injuries.

The research presented in this paper found that the seat position significantly alters body kinematics. In consequence, injuries to occupants are also different for the same crash pulse. As was determined in [14], a greater load exists when the occupant is in a more reclined position. Moreover, the load is especially greater on the lower spine, and the likelihood of submarining increases. Additionally, during a frontal crash, the torso pitch is a result of the rotation of the pelvis [15]. Body kinematics that would cause minimal injuries occur when occupants move at a constant acceleration. This is usually achieved by active restrained systems [16, 17]. A reclined body posture unfavourably aligns body regions such as the pelvis, lumbar, spine and ribcage to existing restrain systems, in turn increasing the severity of the crash [18].

Papers describing body dynamics during vehicle crashes mainly focus on the standard position of the seat. Furthermore, the performance of standard ATD dummies is often supplemented with PMHS (post-mortem human surrogates). Research has shown that dummy models and human models give different results [19, 20]. The low-speed sled and low acceleration sled test with a volunteer, which focuses on body dynamics and the influence of pre-crash muscle tension, was described in [21] and [22]. However, again only one seatback position was considered, and therefore the results of only one seat position can be compared. Understandably, due to much lower acceleration than in our study, and the participation of live volunteers (representing much greater biofidely), the extrusion presented in our research is greater. However, taking into consideration the general displacement of all the points of interest that are present in both studies, it can be concluded that the results are comparable. Furthermore, it was found that the body dynamics of live occupants vary if the occupants are braced or relaxed before impact. Generally, a greater displacement of the occupants occurs when they are relaxed [23-25]. When comparing displacement data of the human body with that of an ATD dummy, it can be concluded that the dummy has a similar performance to a braced live occupant. The kinetic and kinematic response of ATDs does not entirely correspond to the performance of PMHSs, with discrepancies being visible for foot plate, seat back, steering column, and lap belt forces [26, 27].

The body kinetics during a crash test performed with a velocity of 40 km/h were investigated by Ash et al. [28]. The authors described the corridors of a body displaced in three dimensional environments. The position of the body, which was determined by the authors, is similar to that in [29]. However, the obtained results are 2 times smaller due to test speed being more than 2 times lower. In addition, the general shape of the corridors was determined for the standard seat inclination of 110°. However, the results of the trajectory of the points for the tests of inclined seats extend beyond the corridors described in this paper.

5. Conclusion

The conducted research was carried out to evaluate the body kinematics for a passenger sitting on a seat that is reclined at different angles. It was found that there is a difference of body kinematics when sitting on a seat that is reclined. Forward motion (X direction) for the standard position of the seat provoked the lowest extrusion. For the intermediate seatback angle, as well as for the fully reclined seat, there was no difference in extrusion in the forward movement. An opposite tendency existed in the Z extrusion, where the fully reclined seat provoked the most significant motion of all the points of interest in this direction. It was also found that the duration of the total movement was different for all the seat positions. As expected, the hips were the first to react to the trolley's motion. In the case of the standard seat position, the neck reacted almost instantly. Due to neck flexibility, the inertia forces caused the points at the head to react after a longer time. The head stayed in place when the torso and the neck moved. The points at the head then began to move. In all cases, rotation of the head was observed. The greater angle of the reclined seat, the greater the head rotation. For the fully reclined seat, the rotation was significant enough to cause neck injuries. Therefore, investigation of neck and whiplash injuries during a frontal crash can be seen to be justified. Furthermore, the greater the seatback angle, the greater the difference in time between the hip motion and the motion of the subsequent points of interest.

Acknowledgments

The research was carried out using the R&D infrastructure of the project Complex GEO-3EM ENERGY ECOLOGY EDUCATION.

References

[1] Koppel S, Octavio J, Bohman K, Logan D, Raphael W, Jimenez L and Lopez-Valdes F 2019 Seating configuration and position preferences in fully automated vehicles *Traffic Inj Prev.* **20** S103-9

[2] Dissanaike S, Kaufman R, Mack C, Mock C and Bulger E 2008 The effect of reclined seats on mortality in motor vehicle collisions *J Trauma.* **64** 614-9

[3] Richardson R, Donlon J, Chastain K, Shaw G, Forman J, Sochor S, Jayathirtha M, Kopp K, Overby B, Gepner B and Kerrigan J 2019 *26th Int. Technical Conf. on the Enhanced Safety of Vehicles* (Eindhoven: Netherlands, 10-13 June/ESV) 19-0243

[4] Tang L, Zheng J and Hu J 2020 A numerical investigation of factors affecting lumbar spine injuries in frontal crashes *Accid Anal Prev* **136** 105400.

[5] Couturier S, Faure J, Satué R, Huguet J and Hordonneau J 2007 *20th Int. Conf. on the Enhanced Safety of Vehicles* (Lyon: France, 18-21 June/ESV) 07-0481

[6] Gepner B, Draper D, Mroz K, Richardson R, Ostling M, Pipkorn B, Forman J and Kerrigan J 2019 *Int. Research Council on the Biomechanics of Injury* (Florence: Italy, 11-13 September/IRCOBI) IRC-19-44, p 293-307

[7] Rawska K, Gepner B, Moreau D and Kerrigan J 2020 Submarining sensitivity across varied seat configurations in autonomous driving system environment *Traffic Inj Prev.* **21** S1-6

[8] Rawska K, Gepner B, Kulkarni S, Chastain K, Zhu J, Richardson R, Perez-Rapela D, Forman J and Kerrigan J 2019 Submarining sensitivity across varied anthropometry in an autonomous driving system environment *Traffic Inj Prev.* **20** S123-7

[9] Reed M, Ebert S and Jones M 2019 Posture and belt fit in reclined passenger seats. *Traffic Inj Prev.* **20** S38-S42

[10] Rouhana S, Jedrzejczak E and McCleary J 1990 Assessing submarining and abdominal injury risk in the Hybrid III family of dummies: part II—development of the small female frangible abdomen *SAE Transactions* **99** 1760-88

[11] Hu J, Klinich K, Reed M, Kokkolaras M and Rupp J 2012 Development and validation of a modified Hybrid-III six-year-old dummy model for simulating submarining in motor-vehicle crashes *Med Eng Phys.* **34** 541-51

[12] Östh J, Bohman K and Jakobsson L 2020 *Int. Research Council on the Biomechanics of Injury* (Munich: Germany, 8-10 September/IRCOBI) IRC-20-50, p 358-80

[13] Whyte T, Kent N, Keay L, Coxon K and Brown J 2020 Frontal crash seat belt restraint effectiveness and comfort accessories used by older occupants *Traffic Inj Prev.* **21** 60-5

[14] Boyle K, Reed M, Zaseck L and Hu J 2019 *Int. Research Council on the Biomechanics of Injury* (Florence: Italy, 11-13 September/IRCOBI) IRC-19-43, p 282-92

[15] Richardson R, Jayathirtha M, Chastain K, Donlon J, Forman J, Gepner B, Ostling M, Mroz K, Shaw G, Pipkorn B and Kerrigan J 2020 Thoracolumbar spine kinematics and injuries in frontal impacts with reclined occupants *Traffic Inj Prev.* **21** S66-S71

[16] Cheng Z, Pellettiere J, Crandall J and Pilkey W 2009 Optimal occupant kinematics and crash pulse for automobile frontal impact *Shock and Vibration* **16** 61-73

[17] Ilie S and Tabacu S 2007 Study of the occupant's kinematics during the frontal impact. *Ann Oradea Univ Fascicle Manag Technol Eng* **VI** 542-51

[18] Richardson R, Donlon J-P, Jayathirtha M, Forman J, Shaw G, Gepner B, Kerrigan J, Ostling M, Mroz K and Pipkorn B 2020 Kinematic and injury response of reclined PMHS in frontal impacts. *Stapp Car Crash J* **64** 83-153

[19] Forman J, Lin H, Gepner B and Wu T 2019 Occupant safety in automated vehicles-effect of seatback recline on occupant restraint *Int. J. Automot. Eng* **10** 139-43

[20] Pipkorn B, Lopez-Valdes F, Juste-Lorente O, Insausti R, Lundgren C and Sunnevang C 2016 Assessment of an innovative seat belt with independent control of the shoulder and lap portions using THOR tests, the THUMS model, and PMHS tests *Traffic Inj Prev* **17** 124-30

[21] Ito D, Ejima S, Kitajima S, Katoh R, Ito H, Sakane M, Nishino T, Nakayama K, Ato T and Kimura T 2013 *Int. Research Council on the Biomechanics of Injury* (Gothenburg: Sweden, 11-13 September/IRCOBI) IRC-13-100, p 831-41

[22] Ejima S, Zama Y, Ono K, Kaneoka K, Shiina I and Asada H 2009 *21st Int. Technical Conf. on the ESV* (Stuttgart: Germany, 1-18 June/ESV) 09-0913

[23] Beeman S, Kemper A, Madigan M and Duma S 2011 Effects of bracing on human kinematics in low-speed frontal sled tests *Ann Biomed Eng* **39** 2998-3010

[24] Beeman S, Kemper A, Madigan M, Franck C and Loftus S 2012 Occupant kinematics in low-speed frontal sled tests: Human volunteers, Hybrid III ATD, and PMHS *Accid Anal Prev* **47** 128-39

[25] Dehner C, Schick S, Hell W, Richter P, Kraus M and Kramer M 2013 In-vivo kinematics of the cervical spine in frontal sled tests *Glob J Health Sci* **5** 115-26

[26] Beeman S, Kemper A, Madigan M and Duma S 2013 Kinetic and kinematic responses of post mortem human surrogates and the Hybrid III ATD in high-speed frontal sled tests *Accid Anal Prev* **55** 34-47

[27] Lopez-Valdes F, Lau A, Lamp J, Riley P, Lessley D, Damon A, Kindig M and Kent R 2010 Analysis of spinal motion and loads during frontal impacts. Comparison between PMHS and ATD *Ann Adv Automot Med.* **54** 61-78

[28] Ash J, Lessley D, Forman J, Zhang Q, Shaw C and Crandall J 2012 *Int. Research Council on the Biomechanics of Injury* (Dublin: Ireland, 12-14 September/IRCOBI) IR-12-21, p 142-54

[29] Horsch J and Hering W 1989 A kinematic analysis of lap-belt submarining for test dummies *SAE transactions* **98** 1847-54

IOP Conf. Series: Materials Science and Engineering 1247 (2022) 012026 doi:10.1088/1757-899X/1247/1/012026

Frontal collision simulation in laboratory conditions

D Frej[1]*, M Jaśkiewicz[1], P Kubiak[2], A Zuska[1], D Więckowski[3]

1 Department of Automotive Engineering and Transport, Kielce University of Technology, Avenue Tysiaclecia Państwa Polskiego 7, 25-314 Kielce, Poland
2 Department of Transport, Warsaw University of Technology, Plac Politechniki 1, 00-661 Warsaw, Poland
3 ŁUKASIEWICZ R&D NETWORK - Industrial Automotive Institute Warszawa, Poland
* dfrej@tu.kielce.pl

Abstract: The article presents the displacement of individual body parts of volunteers during a controlled crash test at low speed. The crash test stand used is equipped with a vehicle seat with standard seat belts, which moves along the ramp rails. The stand enables both front, side and rear crash tests. The stand enables crash tests from 5 km/h to 20 km/h. The aim of the research is to compare the displacements of individual parts of the volunteers' body, taking into account the division into sex. A study carried out on 130 volunteers (80 men and 50 women), at a collision speed of 15 km/h, showed slight differences in the trajectory of the volunteers' head movement. Before the study, the volunteers were measured and weighed, and then assigned to the appropriate population percentile. Volunteers were classified into a given percentile group on the basis of the mean of 15 anthropometric dimensions of individual body parts. The obtained results are the basis for building a physical model of a dummy.

Keywords: automotive safety, crash test, anthropometric dummy

1. Introduction

Research issues concerning the protection of the driver and passengers of passenger vehicles against the consequences of road collisions has been constantly developed since the creation of passenger vehicles. In an accident at a speed of 50 km/h, a person mass over 2 tons. It is compared this kind of impact with a fall from 3-4 stories. The consequences of such a collision are, in fact, internal injuries, broken bones and serious injuries, which can be fatal. Since the creation of the first Carl Benz vehicle in 1886, the problem of safety has been considered by many scientists around the world [1-6]. Undoubtedly, in order to prevent the tragic consequences of events during vehicle accidents, vehicle manufacturers around the world are working on more and more modern safety measures. Current passenger protection systems play a key role in preventing the serious consequences of road accidents. Newly built vehicles are becoming safer and the safety requirements are becoming more and more stringent. Despite the better and better road infrastructure and more perfect safety systems installed in vehicles, accidents still happen. In the European Union alone, there are over a million accidents each year, with both injured and killed [6-10].

The complicated nature of the loads acting on the driver and passengers during a road accident contributes to the development of safety systems, in particular the model test of the introduced or improved passive safety systems of the vehicle. There are two types of model analyzes [11, 12, 13]:

- Computer simulation studies, with the use of specialized simulation programs using virtual models reflecting real models of antropometric dummies,
- Crash-test experimental studies that use material models and antropometric dummies that reflect human shapes and behavior during collisions.

Experimental research is characterized by high testing costs. Moreover, this type of research reduces the frequency of tests. It is especially visible during crash tests, where a large amount of data should be collected during one test and a repetition cannot be expected because it involves additional costs [14, 15, 16]. Therefore, nowadays simulation tests are becoming more and more popular. It is characterized by low costs of building simulation models and great ease of carrying out research on virtual models. This type of research does not put such pressure on the research team, as each attempt can be repeated without additional costs and long preparations. Despite the great advantages, simulation studies also have disadvantages. Their main importance is accuracy. The simulation model must undergo validation, which is based on comparing the results of simulation studies with experimental studies [17,18,19].

The crash test dummy models used consist of assemblies and elements resembling the human body. An ideal crash test dummy should have a simple structure that does not require frequent calibration and, at the same time, be highly durable and repeatable results [20, 21]. Currently, in the case of frontal collisions, with the most advanced Hybrid III and THOR dummy. The Hybrid III dummies have a high rate of compatibility with the human body. Research on the behavior of dummies during crash tests led to the evaluation of head, face, chest, abdomen, lower and upper limb injuries, which are used by all countries around the world [21, 22].

2. Experimental research

Experimental tests with the use of volunteers were carried out on a test stand designed for low-speed crash tests and for measuring the force in seat belts during a collision. An experimental crash test was carried out in the laboratory of motor vehicles and tractors at the Kielce University of Technology. The stand consists of a 10m long track along which a vehicle seat moves. At the end of the track, where are shock absorbers that simulate a collision with a stationary obstacle. The speed of the collision depends on the length of the road and the gradient of the tracks. The vehicle seat is on rollers that allow it to move freely along the tracks of the test stand. The measurement takes place when the electro valve is released, which keeps the chair with the dummy or the volunteer at a given height. Measurements on the experimental stand were recorded with a high-speed Phantom camera with a speed of 2500 frames/second.

Figure 1. Test stand for crash tests at low speed

People aged 18 to 45 participated in the experimental studies. A total of 130 people took part in the crash tests - including 80 men and 50 women. Before the measurement test, all people were measured and weighed, and then allocated in the given percentile of the population. Volunteers were classified into a given percentile group on the basis of the mean of 15 anthropometric dimensions of individual body parts. Examples of anthropometric dimensions for a 5-percentile woman and a 50-percentile male are presented in Table 1.

Table 1. Volunteer anthropometric dimensions

Dimensions	Woman 5th	Man 50th
body height, cm	165	183
mass, kg	62	91
knee height, cm	42	50
shoulder width, cm	41	45
upper limb length, cm	64	85
length elbow axis of the handle, cm	32	43
hand length (forearm), cm	17,5	18,5
hand length, cm	10,5	11
crotch height (lower limb length), cm	88	96
thigh length (buttock knee), cm	59,5	59
medial height of the foot, cm	7,8	9
arm length, cm	33	37
head circumference, cm	56	57,5
head height, cm	21,5	21,5
chest circumference, cm	96	90

The results of the experiment using volunteers were obtained with the help of the TEMA program [23]. This program allows you to analyze the movement of objects recorded with the camera. The program included the trajectory of the volunteers' head movement in individual percentile groups. The head displacements of volunteers in relation to the X axis are presented in Figures 2 to 7. The head displacement in the case of women and men representing the 5th percentile in the first phase of impact (0.05 s) ranges from 0.45 m to 0.57 m. In the second phase of impact (0.125 s), it ranges from 0.55 m to 0.62 m. In the case of people representing the 50th percentile of the male population in the first phase of the impact, the displacement of the head in the direction of the X axis ranges from 0.57 m to 0 , 75 m, in the case of women from 0.59 m to 0.72 m. The displacement of the head in relation to the X axis in the second phase of impact for volunteers representing the 50th percentile of the male population ranges from 0.55 m to 0.62 m, and for volunteers representing the 50th percentile of the female population from 0.55 m to 0.60 m. The head displacement for women and men representing the 95 percentile in the first phase of the impact ranges from 0.65 m to 0.82 m, and in the second - in border from 0.70 m to 0.77 m. In the case of in the case of male volunteers, it ranges from 0.65 m to 0.85 m in the first phase of the impact, and in the range from 0.71 m to 0.77 m in the second phase.

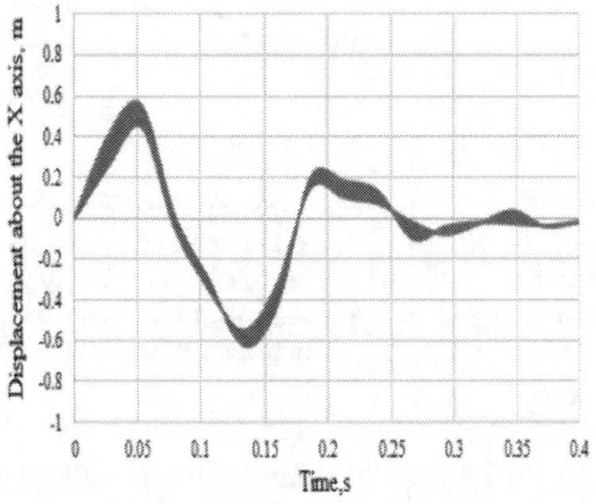

Figure 2. Displacement of the head along the X axis of volunteers representing the 5th percentile of the male population

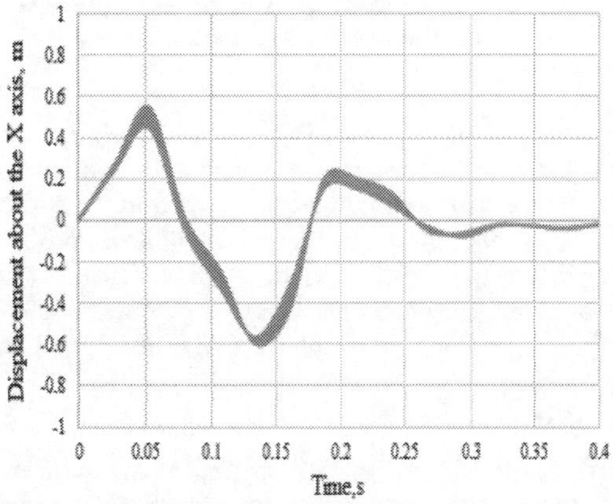

Figure 3. Displacement of the head along the X axis of volunteers representing the 5th percentile of the female population

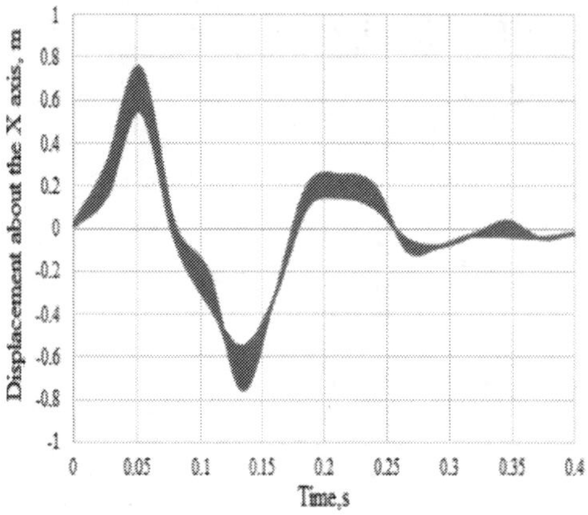

Figure 4. Displacement of the head along the X axis of volunteers representing the 50th percentile of the male population

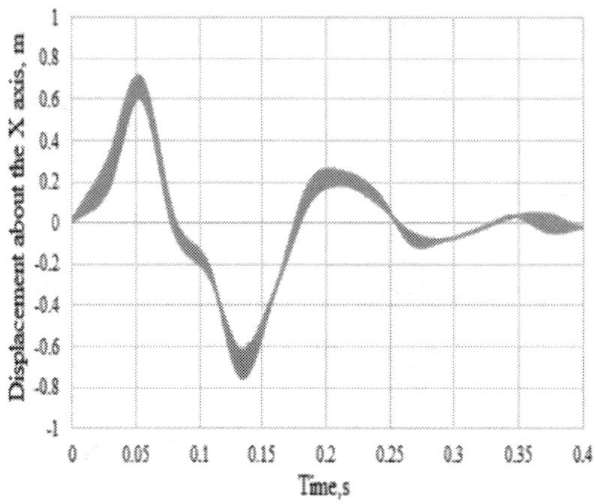

Figure 5. Displacement of the head along the X axis of volunteers representing the 50th percentile of the female population

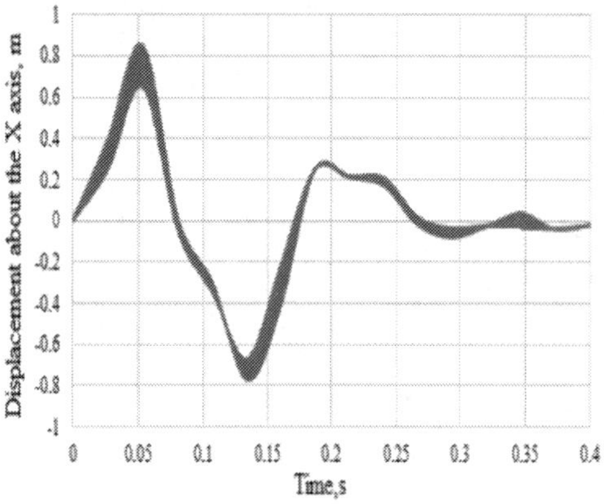

Figure 6. Displacement of the head along the X axis of volunteers representing the 95th percentile of the male population

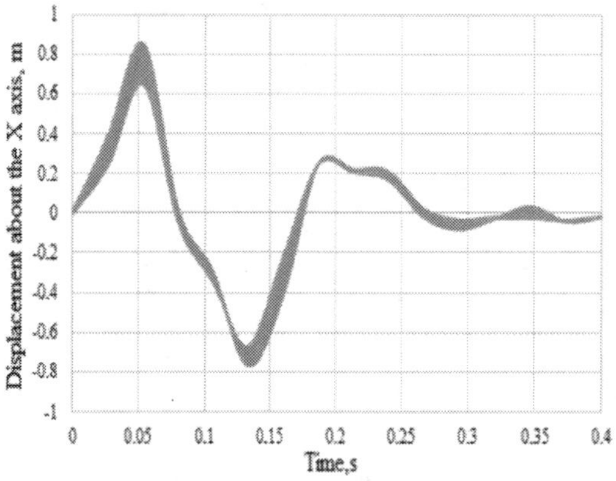

Figure 7. Displacement of the head along the X axis of volunteers representing the 95th percentile of the female population

The head displacements of volunteers in relation to the Y axis are presented in Figures 8 to 13. The head displacement in the case of women and men representing the 5th percentile in relation to the Y axis in the first phase of the impact ranges from 0.31 m to 0.4 m, and in the second phase in the range from 0.52 m to 0.62 m. In the case of volunteers representing the 50 percentile of the population, the head displacement in relation to the Y axis for men in the first phase of the impact ranges from 0.32 m to 0.45 m, and in the second phase from 0 , 55 m to 0.70 m. For women, the displacement of the head relative to the Y axis in the first phase ranges from 0.37 m to 0.50 m, and in the second phase from 0.56 m to 0.70 m.

IOP Conf. Series: Materials Science and Engineering 1247 (2022) 012026 doi:10.1088/1757-899X/1247/1/012026

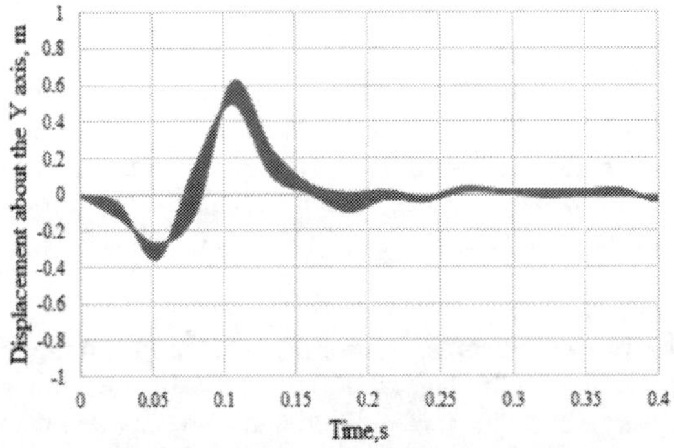

Figure 8. Displacement of the head along the Y axis of volunteers representing the 5th percentile of the male population

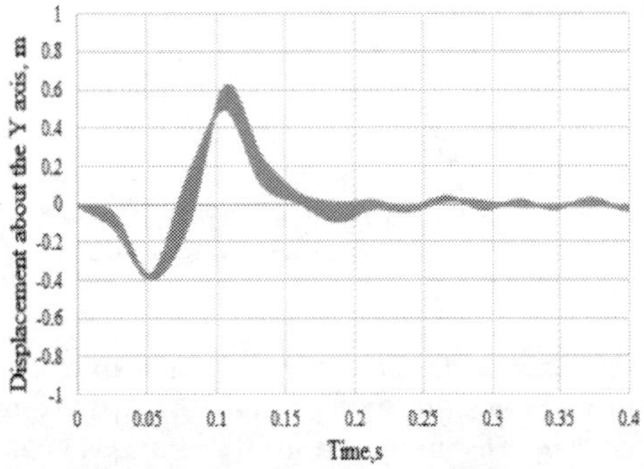

Figure 9. Displacement of the head along the Y axis of volunteers representing the 5th percentile of the female population

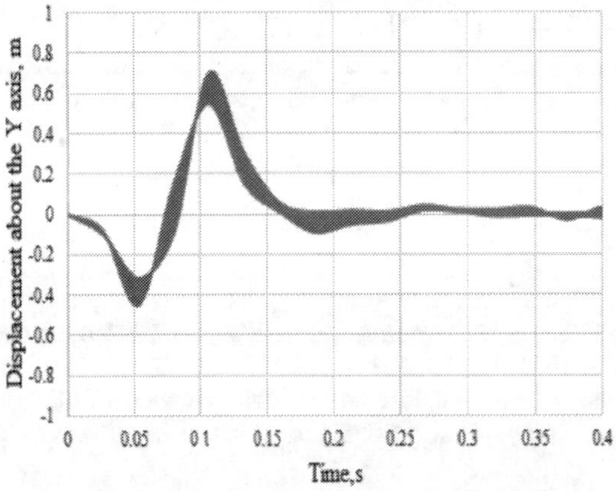

Figure 10. Displacement of the head along the Y axis of volunteers representing the 50th percentile of the male population

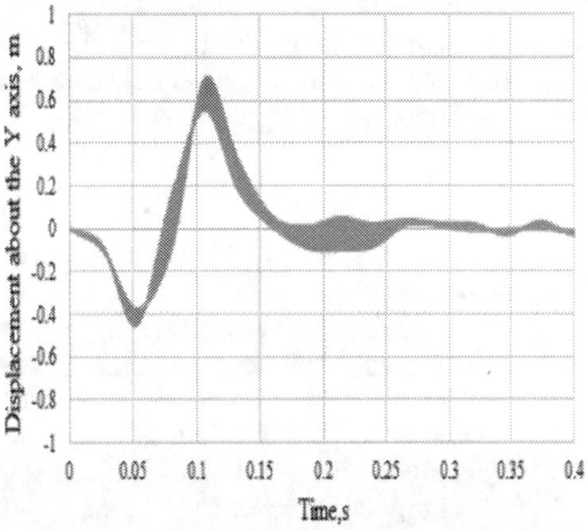

Figure 11. Displacement of the head along the Y axis of volunteers representing the 50th percentile of the female population

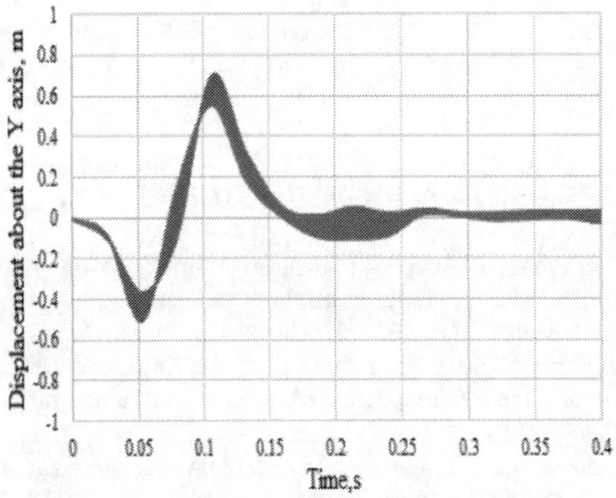

Figure 12. Displacement of the head along the Y axis of volunteers representing the 95th percentile of the male population

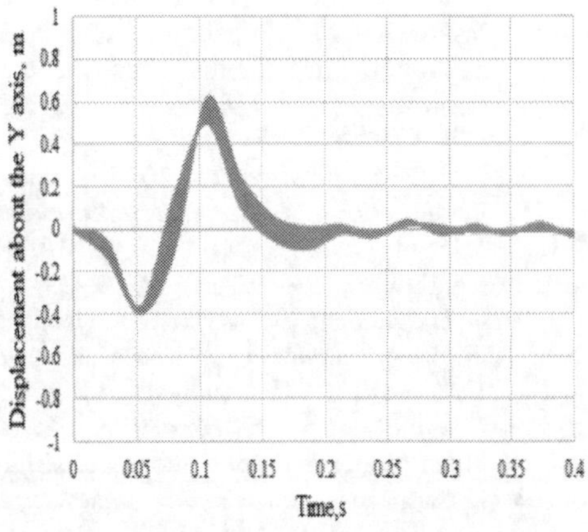

Figure 13. Displacement of the head along the Y axis of volunteers representing the 95th percentile of the female population

3. Conclusions

More than 70 years have passed since the first mannequin was created (Sierra Sam Dummy), and there is still a need to design new joints and the structure of the individual body parts of the anthropometric dummy to be similar to the human body. Modern crash test dummies are modeled on the human body and use more than 200 sensors to record data. Their measuring points are distributed over the entire surface of the manikin. In the case of low speed crash tests, there is no need to use such extensive models of anthropometric dummies to correctly identify the behavior of individual parts of the human body [23-24].

In the case of crash tests aimed at increasing the level of passive safety of passenger vehicles, simulation programs are of great importance. Computer simulation in the simplest way is to recreate the phenomena that take place in the real world using mathematical models. They are defined and also operated using computer programs. In the case of crash tests, experimental tests are associated with high costs. On the other hand, simulation tests are much cheaper than experimental tests because they do not require a test stand. Contemporary simulation programs such as Dytran, Madymo or ADAMS are geared towards crash test simulations and data validation using experimental tests.

On the basis of the volunteers' head displacement in relation to the X and Y axes, it can be noticed that there are differences of 5% between men and women. Therefore, it should be noted that anthropometric dummies should not be divided into sex during a collision with a low speed of 5 to 20 km/h.

Bibliografia

[1] Moravčík E., Jaśkiewicz M., **2018**, Boosting car safety in the EU, *2018 XI International Science-Technical Conference Automotive Safety*, 1-5, doi: 10.1109/AUTOSAFE.2018. 8373307.

[2] Sabbella RR., Arunachalam M., **2019**, Functional Safety Development of Motor Control Unit for Electric Vehicles," *2019 IEEE Transportation Electrification Conference (ITEC-India)*, 1-6, doi: 10.1109/ITEC-India48457.2019.ITECINDIA2019-270.

[3] Szumska E., Frej D., Grabski P., **2020**, Analysis of the Causes of Vehicle Accidents in Poland in 2009-2019. LOGI – Scientific Journal on Transport and Logistics, Vol.11 (Issue 2), pp. 76-87. https://doi.org/10.2478/logi-2020-0017

[4] Frej D., Zuska A., Cadge K., **2019**, Analysis of vertical vibrations affecting a child transported in a child seat during a car passing over the release speed bump, *The Archives of Automotive Engineering –Archiwum Motoryzacji*. 86, 4, 111-125. doi:10.14669/AM. VOL86.ART8.

[5] Jaśkiewicz M., Więckowski D., Witaszek K., **2013**, Przegląd i analiza manekinów stosowanych do testów zderzeniowych, *Zeszyty Naukowe. Mechanika/Politechnika Opolska*, 101, 47–48.

[6] Jaśkiewicz M., Więckowski D., Poliak M., Frej D., **2020**, Shoulder joint design of the anthropometric dummy used for crash tests," *2020 XII International Science-Technical Conference AUTOMOTIVE SAFETY*, 1-5, doi: 10.1109/AUTOMOTIVESAFETY 47494.2020.9293535.

[7] Frej DP., Ludwinek K., **2020,** Analysis of road accidents in 2002–2019 on the example of Poland. *The Archives of Automotive Engineering – Archiwum Motoryzacji*, 89, 3, 5-18. https://doi.org/10.14669/AM.VOL89.ART1

[8] Koralewski G., Wrona R., **2019**, Zagrożenia i wypadki drogowe spowodowane nie ustępowaniem pierwszeństwa. (Road hazards and accidents caused by failure to give way), *Autobusy : technika, eksploatacja, systemy transportowe*, 20, 1(2), 84–87.

[9] Jurecki RS., **2020**, Analysis of Road Safety in Poland after Accession to the European Union. *Communications - Scientific Letters of the University of Zilina*, 22(2), 60-67. https://doi.org/10.26552/com.C.2020.2.60-67

[10] Ludwinek K. , Jurecki R., Jaskiewicz M., Szumska E., Sulowicz M., **2018**, A test stand for the experimental analysis of physical quantities during crash test at low speeds, *2018 XI International Science-Technical Conference Automotive Safety*, 1-7, doi: 10.1109/AUTOSAFE.2018.8373306.

[11] Liu Z., Zhang X., Chen H., Shi Y., **2010,** Optimization of Calibration Parameters for Hybrid III 5th Female Dummy Neck Calibration Based on Taguchi Method, *2010 Third International Conference on Information and Computing, Wuxi*, 97-100, doi: 10.1109/ICIC.2010.295.

[12] Noureddine A., Eskandarian A., Digges K., **2002**, Computer modeling and validation of a hybrid III dummy for crashworthiness simulation, *Pergamon, athematical and computer modeling*, 885-893

[13] Amiri S., Naserkhaki S., Parnianpour M., **2018**, Modeling and validation of a detailed FE viscoelastic lumbar spine model for vehicle occupant dummies, *Computers in Biology and Medicine*, 99, 191-200, https://doi.org/10.1016/j.compbiomed. 2018.06.013.

[14] Stańczyk TL., Zuska A., **2014**, Review of anthropodynamic dummies used to evaluate the effect of vibrations on sitting human (vehicle driver), *The Archives of Automotive Engineering – Archiwum Motoryzacji*. 65(3), 65-74.

[15] Kent R., Bass C., Woods W., Salzar R., Lee S., Melvin J., **2006**, The Role of Muscle Tensing on the Force-deflection Response of the Thorax and a Reassessment of Frontal Impact Thoracic Biofidelity Corridors, *Journal of Automobile Engineering, Proceedings of the Institution of Mechanical Engineers*, 220, 853-868.

[16] Stanczyk TL. et al., **2020**, Braking performance tests for different types of vehicles, *2020 XII International Science-Technical Conference AUTOMOTIVE SAFETY, 2020*, 1-6, doi: 10.1109/AUTOMOTIVESAFETY47494.2020.9293532.

[17] Wang W., Zhang D., Ji J., Tian H., Zhao H., **2016**, Seat belt protection of train driver during secondary impact, *2016 IEEE International Conference on Intelligent Transportation Engineering (ICITE),* 85-88, doi: 10.1109/ICITE.2016.7581313.

[18] Shameem S., Prasad GRK., Ch VT., Bharath Kumar K., Yaswanth M., Ch M., **2018**, Design Simulation and Analysis of Crash Sensor for Air Bag System, *2018 3rd International Conference on Inventive Computation Technologies (ICICT)*, 718-723, doi: 10.1109/ICICT43934.2018.9034425.

[19] Zhang Y., Ju C., Yue G., Chen X., Sun H., **2012**, Simulation Analysis of the Rear Seat Female in Front Impact, *2012 Third International Conference on Digital Manufacturing & Automation*, 767-770, doi: 10.1109/ICDMA.2012.181.

[20] Ruhai G., Manjiang H., Dong X., **2010**, Simulation Study of Bus Occupant Restraint System in Emergency Brake, *2010 International Conference on Optoelectronics and Image Processing*, 392-395, doi: 10.1109/ICOIP.2010.29.

[21] Lai X., Ma C., Hu J., Zhou Q, **2012**, Impact direction effect on serious-to-fatal injuries among drivers in near-side collisions according to impact location: focus on thoracic injuries. *Accid Anal Prev*, 48, 442-450. doi: 10.1016/j.aap.2012.02.023

[22] Thuong LNP., **2018**, Vehicle Frontal Impact to Pole Barrier Simulation Using Computer Finite Element Model, *2018 4th International Conference on Green Technology and Sustainable Development (GTSD)*, 273-277, doi: 10.1109/GTSD. 2018.8595702.

[23] TEMA, (in:) https://www.imagesystems.se/tema/, (access in 23.01.2022)

IOP Conf. Series: Materials Science and Engineering 1247 (2022) 012027 doi:10.1088/1757-899X/1247/1/012027

Methods for determining the physical quantities describing the vehicle movement and the driver during a road accident

M Gidlewski [2*], T Pusty [1], L Jemioł [3] and H Kochanek [3]

[2] Military University of Technology, gen. Sylwestra Kaliskiego 2, 00-908 Warszawa, Poland Jagiellońska 55, 03-301 Warszawa, Poland
[1] Łukasiewicz Research Network – Automotive Industry Institute, Jagiellońska 55, 03-301 Warszawa, Poland
[3] Kazimierz Pulaski University of Technology and Humanities in Radom, Jagiellońska 55, 03-301 Warszawa, Poland

* miroslaw.gidlewski@wat.edu.pl

Abstract. The paper compares the displacement and speed courses of cars, motorbikes and motorcyclists during a road accident obtained with the use of two measurement methods. In the first method, the course of the impact recorded by a fast camera recording 1000 frames per second was used ("frame by frame" film analysis). In the second method, the courses of accelerations and angular velocities of the objects were used (numerical integration of the recorded courses). The measurements were carried out during crash tests involving a frontal-side collision of two cars or a motorcycle with a car. The results has been shown in the time interval 0-0.5 s. During this period, the following phases of the experiment are observed: vehicle collision and the process of their deformation, the separation of vehicles, the start of independent movement of both vehicles after the collision. The measurement results obtained with the use of both methods are burdened with errors resulting, among others, from limited data sampling frequency, offset errors, calibration errors and sensor noise, duration of the analyzed waveforms. The use of two of the above-mentioned measurement methods simultaneously allows to minimize measurement errors.

Key words: crash test, automotive safety, vehicle testing

1. Introduction

Obtaining physical quantities describing the movement of vehicles and people is of interest to many entities researching: passive safety systems [1, 2, 3], active safety systems e.g. during the modelling of autonomous lane change algorithms [4, 5, 6], dynamics of movement [7], compliance of technological solutions with the applicable standard documents, e.g. LDWS (Lane Departure Warning System), experts analyzing the records of UDS (Unfall Daten Speicher), ADR (Accident Data Recorder) black boxes [8, 9], or road accidents experts [10, 11]. The obtained data is subject to uncertainty due to the limited frequency of data sampling, errors of offset, calibration and noise of sensors, duration of measurements. Therefore, it seems advisable to conduct research focused on determining the uncertainty range of the obtained results of certain physical quantities describing the motion parameters with various methods. The paper compares the course of displacement and velocity of objects during the collision and immediately after the collision, determined on the basis of post-frame analysis of films and

IOP Conf. Series: Materials Science and Engineering 1247 (2022) 012027 doi:10.1088/1757-899X/1247/1/012027

numerical integration of acceleration and angular velocity waveforms recorded during several frontal-lateral collisions of vehicles carried out in the Łukasiewicz Research Network - The Automotive Industry Institute in Warsaw.

2. The aim and scope of this work

2.1. Test objects

The aim of this work is compares the trajectory courses of cars, motorbike and motorcyclist during a road accident obtained with the use of two measurement methods based on:

- analysis the course of the impact recorded by a fast camera recording 1000 frames per second ("frame by frame" video analysis);
- analysis the courses of linear accelerations and angular velocities of the objects (numerical integration of the recorded courses - calculations).

Post-impact motion was analyzed. The one frame video in first contact has been presented in Figure 1. Photo a) represent two colliding vehicles under test, photo b) represent motorcyclist hits against a motorcar.

a) b)

Figure 1. The one frame video in first contact; a) a photo of two colliding vehicles under test, b) a photo of motorcyclist hits against a motorcar

2.2. Analysis of the motorcar collision's

The data acquisition system of the acceleration and angular velocity measurements was located in the trunk of the vehicle in a crash test case. At the stage of recording the measurement results, a 10 kHz filter was used and during the calculations, a 50 Hz filter was used, so that the measurement results were reduced to a frequency band that can be reproduced in the process of modelling and solving equations. The selection of the filter cut-off frequency (tests with 100 Hz and 25 Hz filters were carried out) was a compromise between the expected accuracy of calculations, the possibility of interpreting their results and the possibility of obtaining conclusions relevant and useful conclusions during the modelling and reconstruction of this type of road accidents. Recording cameras the position of the collision vehicles with a frequency of 1000 frames per second were installed above the place of collision and has been presented in the Figure 2a and 2b. Figure 3 represents the silhouettes of the cars at the time of the maximum deformation.

IOP Conf. Series: Materials Science and Engineering 1247 (2022) 012027 doi:10.1088/1757-899X/1247/1/012027

a) b)

Figure 2. Recording cameras

Figure 3. The silhouettes of the cars at the time of the maximum deformation

2.3. Analysis of the motorcyclist collision

The data acquisition system of the acceleration and angular velocity measurements was placed in the dummy's head and torso. In addition, the motorcycle and the car have two three-component linear acceleration sensors, the motorcycle - near the centre of mass and on the edge of the seat, and the car near the centre of mass and in the trunk. Accelerations and angular velocity were recorded at a frequency of 10 kHz. In the direction perpendicular to the plane where the motorcyclist was moving, mounted a camera which recorded video at a frequency of 1000 frames per second.

a) b)

c)

d)

Figure 4. The one frame video under test; a) beginning of the process of a motorcycle impact against the motor car side, b) motorcyclist's helmet came into contact with the edge of the car roof, c) restitution phase, d) motorcycle lifted to the maximum height

3. Calculations and results

During calculations, the following coordinate systems were used:

- local coordinate systems, rigidly related to the object. The $O_ix_iy_iz_i$ coordinate system has the origin of O_i at the center of the mass of the object "i", the axis of O_ix_i is parallel to the longitudinal axis of the object;
- local horizontal coordinate systems related to the object "i";
- $O_GX_GY_GZ_G$ global coordinate system related to the road.

The relations between the coordinate systems may be described as follows:

$$\begin{bmatrix} a_{xPi} \\ a_{yPi} \\ a_{zPi} \end{bmatrix} = \begin{bmatrix} \cos\Theta_i & \sin\Theta_i \sin\Phi_i & \sin\Theta_i \cos\Phi_i \\ 0 & \cos\Phi_i & -\sin\Phi_i \\ -\sin\Theta_i & \cos\Theta_i \sin\Phi_i & \cos\Theta_i \cos\Phi_i \end{bmatrix} \begin{bmatrix} a_{xi} \\ a_{yi} \\ a_{zi} \end{bmatrix} \quad (1)$$

$$\begin{bmatrix} \ddot{X}_{Gi} \\ \ddot{Y}_{Gi} \\ \ddot{Z}_{Gi} \end{bmatrix} = \begin{bmatrix} \cos\Psi_i & -\sin\Psi_i & 0 \\ \sin\Psi_i & \cos\Psi_i & 0 \\ 0 & 0 & 1 \end{bmatrix} \begin{bmatrix} a_{xPi} \\ a_{yPi} \\ a_{zPi} \end{bmatrix} \quad (2)$$

$$\begin{bmatrix} \dot{\Phi}_i \\ \dot{\Theta}_i \\ \dot{\Psi}_i \end{bmatrix} = \begin{bmatrix} 1 & \sin\Phi_i \ \mathrm{tg}\Theta_i & \cos\Phi_i \ \mathrm{tg}\Theta_i \\ 0 & \cos\Phi_i & -\sin\Phi_i \\ 0 & \dfrac{\sin\Phi_i}{\cos\Theta_i} & \dfrac{\cos\Phi_i}{\cos\Theta_i} \end{bmatrix} \cdot \begin{bmatrix} P_i \\ Q_i \\ R_i \end{bmatrix} \quad (3)$$

where:

Ψ_i, Θ_i, Φ_i – quasi-Euler angles, defining the orientation of the local system $\{C_S\}$ relative to the global system $\{0\}$;

P_i, Q_i, R_i - components of the angular velocity vector of the body in local coordinate systems.

The trajectory of the car A and B in the global coordinate system has been presented in Figure 3 and the angle of rotation of the cars in relation to the vertical axis has been presented in Figure 4. Continuous lines represents calculations results, pointed line represents results of the "frame by frame" analysis of films from high-speed cameras (high speed cameras).

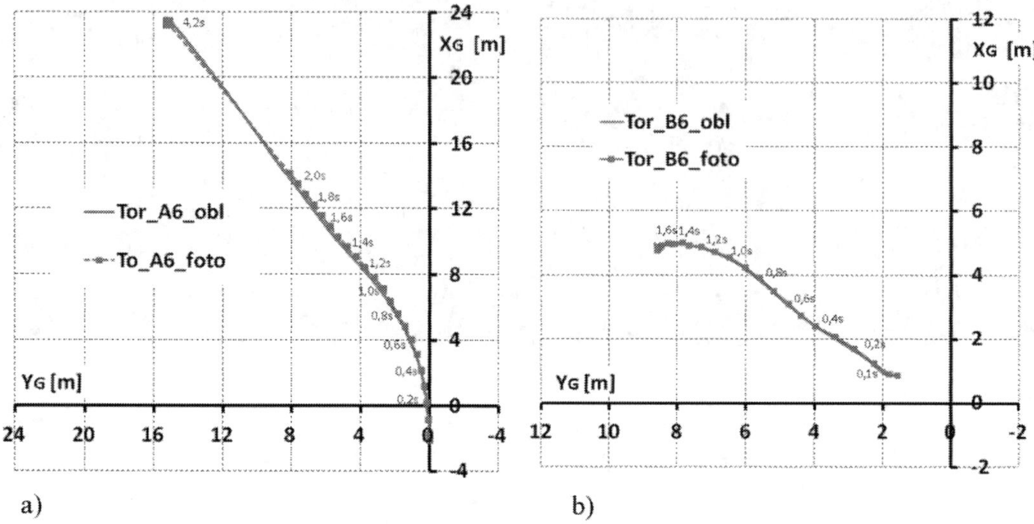

Figure 5. Trajectory of the collision vehicle: a) vehicle A; b) vehicle B

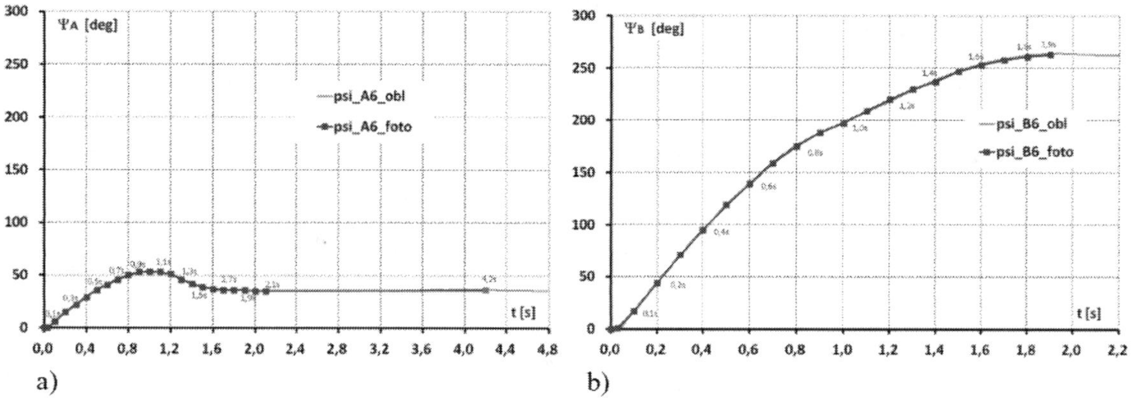

Figure 6. Time histories of the Ψ_i angle: a) vehicle A; b) vehicle B

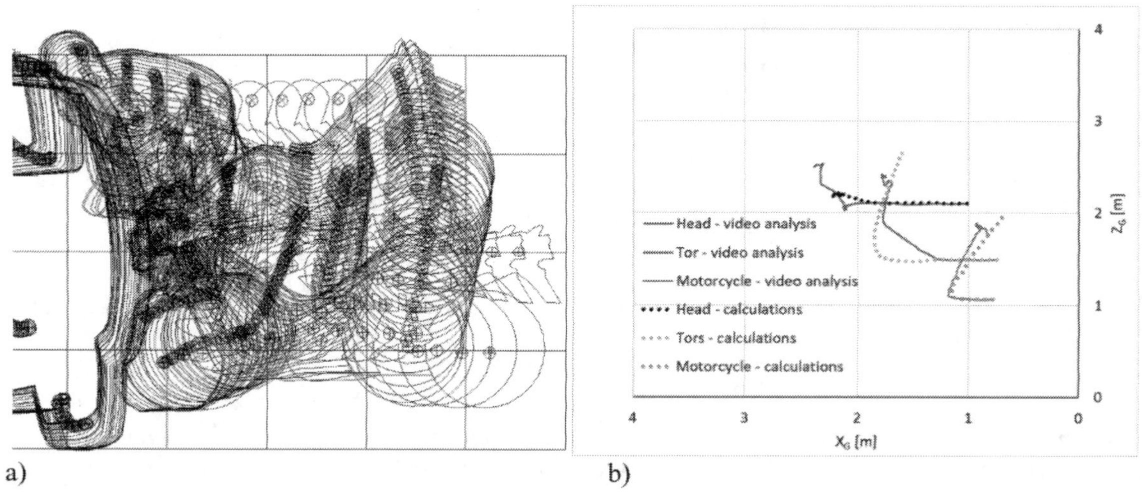

Figure 7. An example of the analysis results, a) results of video analysis, b) results of calculations

4. Summary

The method of reconstruction of changes in the positions of cars on the road during a collision, based on the integration of data from acceleration and angular velocity sensors, has given a satisfactory agreement with the results of the analysis "frame by frame" of movies recorded by cameras.
A small permanent error in the zero calibration of individual sensors can lead to large calculation errors. Therefore, the work confronted the results of the "frame by frame" analysis of films recorded by a camera with the results of calculations.
In all crash tests, the positions of the silhouettes resulting from the calculations were consistent with the positions recorded by the camera in the subsequent phases of the collision. This fact allowed the researcher to consider the results of measurements (acceleration and yaw velocity) as correct, and the applied procedures for determining changes in the location objects as correct and reliable, and therefore possible to use in other analyses.

References

[1] Buraczewski P, Jackowski J, Porównanie metod określania parametrów ruchu pojazdu podczas zderzenia z przeszkodą, XXXI Seminarium Kół Naukowych Wydziału Mechanicznego, Warszawa 2012
[2] Dąbrowski F, Gidlewski M, Front and Side Crash of Cars in Motion – comparative analysis of experimental and modelling test results, *Zeszyty Naukowe Instytutu Pojazdów Politechniki Warszawskiej*, 4/2015.
[3] Prochowski L, Dębowski A, Zuchowski A, Zielonka K, Evaluation of the influence of velocity on dynamic passenger loads during a frontal minibus impact against an obstacle. *IOP Conference Series: Materials Science and Engineering*, 148(1) 2016.
[4] Prochowski L, Ziubiński M, Szwajkowski P, Gidlewski M, Pusty T, Stańczyk T L, Impact of Control System Model Parameters on the Obstacle Avoidance by an Autonomous Car-Trailer Unit: Research Results, *Energies*, Vol 14(10), 2021.
[5] Żardecki D, Dębowski A, Non-smooth models and simulation studies of the suspension system dynamics basing on piecewise linear luz(…) and tar(…) projections. *Applied Mathematical Modelling*, 2021
[6] Dziewiecki M, Gidlewski M, Limitations of Use of an Inertial Positioning System in a Truck During a Maneuver of Avoiding a Suddenly Appearing Obstacle. ESV Conference Paper Number 15-0341. 24th ESV International Technical Conference on the Enhanced Safety of Vehicles. Goteborg, Sweden 2015.
[7] Jackowski J, Jakubowski J, Recognition of Moving Tracked and Wheeled Vehicles Based on Sound Analysis and Machine Learning Algorithms. *International Journal of Automotive and Mechanical Engineering*, 18(1) 2021.
[8] Guzek M, Lozia Z: Possible Errors occurring during Accident Reconstruction based on Car "Black Box" Records. 2002 World SAE Congress and Exposition. Detroit, Michigan USA. SAE TP 2002-01-0549, 2002.
[9] Guzek M, Lozia Z, Analiza dokładności rekonstrukcji wypadku drogowego wykorzystujących zapisy samochodowej „czarnej skrzynki". II Konferencja Naukowo-techniczna "Rozwój techniki samochodowej a ubezpieczenia komunikacyjne". Radom, Poland, 2004.
[10] Jurecki R, Jaśkiewicz M, Analysis of Road Accidents in Poland Over the Last Ten Years; *Scientific Journals Maritime University of Szczecin: Szczecin*, 32, 2012.
[11] Poliak M, Tomicova J, Jaskiewicz M, Rudawska A, Lakhmetkina N, Impact of neutralization of transport documents on the safety of the road carrier 12th International Science-Technical Conference AUTOMOTIVE SAFETY, Kielce, Poland, 2020.

IOP Conf. Series: Materials Science and Engineering 1247 (2022) 012028 doi:10.1088/1757-899X/1247/1/012028

The importance and directions of electromagnetic compatibility test methodologies development in the aspect of vehicle safety

W Zaklika[1], **D Sterniczuk**[1*]

[1] Electronics and Acoustics Laboratory, Łukasiewicz Research Network – Automotive Industry Institute, 55 Jagiellońska Str., 03-301 Warsaw, Poland

* daniel.sterniczuk@pimot.lukasiewicz.gov.pl

Abstract. In this paper the definition of electromagnetic compatibility in the context of testing vehicle components and complete vehicles was presented. Potential threats resulting from particular EMC phenomena in vehicles were presented. Analysis of the areas identified in obligatory and non-obligatory normative documents was made. The areas which, in the opinion of the authors, should be incorporated into the obligatory normative documents were indicated. The substantive justification for the need to conduct EMC tests and analysis at the design, prototyping stage, before implementation, and periodically during operation (for justified cases) was presented. Good engineering practices in the field of eliminating the negative effects of EMC phenomena were indicated. Authors described how the change in the approach to EMC issues increases the vehicles safety.

Keywords: safety, vehicle, automotive safety

1. Introduction

The fundamental problems related to electromagnetic compatibility (EMC) of automotive devices and vehicles have remained valid for many years. The general technological development and the encroachment of electronics into new and new application areas often require that these problems be dealt with in a new and modified way. This applies to both sides of the EMC. Namely the emission of electromagnetic disturbances by the tested devices and their immunity to such disturbances. Electromagnetic compatibility is the sum of elementary electromagnetic phenomena that defines the ability to operate undisturbed and not to interfere with the operation of other devices by a given electrical/electronic device.

Electromagnetic compatibility is of great importance in the context of vehicles safe road use. Imagine a situation in which a vehicle passing near a high power radio transmitter no longer functions properly. Malfunction of key components, such as driver assistance systems, can confuse the driver and make it difficult to steer, potentially putting other road users at risk. The source of the interference that may adversely affect the vehicle does not have to be the aforementioned radio transmitter by the road, but it can be any smallest device installed inside the vehicle. The effects of installing non-EMC components can be very similar as described above.

The assessment of such effects is performed mainly by accredited laboratories in semi-anechoic chambers, an example of an object in such chamber is presented in Fig. 1. The test object was an emergency vehicle extended with a large number of electronic components. The tests were carried out in the Electronics and Acoustics Laboratory, which is part of the Łukasiewicz Research Network - Automotive Industry Institute.

Figure 1. Type-approval radiated emission test of vehicle (test object blurred for confidentiality reasons)

2. Analysis of the problem status

Electromagnetic compatibility can be divide into four main areas [1]:
- radiated emissions,
- radiated immunity,
- conducted emissions,
- conducted immunity.

Depending on the case, the phenomena may be continuous or transient. Many test methodologies have been developed to precisely define them due to the wide spectrum of their occurrence. The currently used equipment allows to assess the phenomena, which impacts the vehicle safety. List of these phenomena and possible influence on safety is given in tab. 1.

Table 1. Potential influence of EMC phenomena on vehicle safety.

Phenomenon	Potential influence on safety
Emission of radiated electromagnetic disturbances	Interference with the functioning of other vehicles in close proximity
Immunity to radiated electromagnetic field	Malfunction of the vehicle sub-assemblies in the operating electromagnetic environment

Emission of conducted electromagnetic disturbances	Interference with the operation of other devices physically connected by electricity or using the same power source
Emission of conducted transients	Damage to other devices in the installation due to overvoltage
Emission of harmonics generated in AC power lines	Interference with the operation of other devices drawing power from the same electric grid
Emission of changes, fluctuations and flicker	Interference with the operation of other devices drawing power from the same electric grid, especially danger for lighting devices
Immunity to transients	Incorrect operation as a result of impulses appearing in the vehicle's electrical system (e.g. engine starting, switching on relays, turning on/off devices)
Immunity to BURST	Incorrect functioning or damage to the device as a result of e.g. arcing in the electromechanical connections in the installation
Immunity to SURGE	Incorrect functioning or damage to the device as a result of e.g. lightning
Immunity to voltage dips, short interruptions and voltage variations	Incorrect functioning of the device as a result of decays, dips and voltage changes in the power grid
Immunity to voltage fluctuations	Incorrect functioning of the device due to voltage fluctuations in the power grid
Immunity to electrostatic discharge	Incorrect functioning or damage to the device as a result of electrostatic discharge during transport, handling and service works

Many normative documents have been developed that differ mainly in the scopes of assessment of these phenomena, in order to describe the test methodologies allowing to determinate the above parameters. These documents are used in all industries (e.g. military, medical, railway, home and slightly industrialized environment, etc.)
The main obligatory area for every manufacturer of electrical/electronic components in the automotive industry and complete vehicles is the so-called type-approval. As part of the Geneva Agreement of 1958, the UNECE prepared a series of documents called UNECE Regulations, which included a list of test methodologies and criteria for assessing the parameters of all components and complete vehicles. The requirements for electromagnetic compatibility were laid down by the UNECE in Regulation No. 10 (R10) [2]. The test ranges described in R10 constitute the necessary minimum that devices and vehicles must meet in order to be allowed for sale on the European market and for use in public road traffic. Most of the test methodologies from R10 are derived from other normative documents, which covers much wider scopes, however, their requirements are not obligatory. They are used for testing electric devices dedicated to other areas of industry. There are also specific requirements of large automotive companies, which extend beyond the scope of the type-approval.

2.1. Type-approval field (required)
Type-approval area includes assessment of conducted and radiated disturbances. Tests for the emission of radiated disturbances (broadband/narrowband) are performed in accordance with the standards of the CISPR [3] series in frequency range of 30 MHz – 1 GHz. The most common reasons for their occurrence are: noise induced in the cables, ineffective decoupling of PCB, faulty cable termination, peripheral devices that do not meet the requirements, the wrong class of power supply, arrangement of components (including decoupling capacitors and their paths), large current loops.

Immunity to radiated electromagnetic field is tested according to the series of standards ISO 11451 (vehicle resistance) [4] and ISO 11452 (component resistance) [5] in frequency range of 20 MHz – 2 GHz. The tested object is subjected to an electromagnetic field with a specified modulation and intensity standards, which is to simulate the operation of the device in a characteristic electromagnetic environment. This serves to verify the correctness of its operation under the influence of disturbances. As in the case of emission, it is important here that the cables and the PCB are properly shielded, as well as the appropriate filtering of signals is used.
Conducted disturbance tests are mainly performed in accordance with the ISO 7637 series of standards as well as the EN 61000 series. These tests include:
- measurements of transient emissions during devices switching on and off - some devices generate voltage spikes during switching on/off, which, if they exceed the permissible limits, can damage the elements of the installation or other devices [5];
- testing the immunity to transients - this test simulates the electrical impulses appearing in the vehicle installation during, e.g. starting the vehicle, switching on relays [5];
- emissions of radio disturbances conducted along the power line - radio disturbances with relatively low frequencies are conducted to the power grid and may cause its malfunction and disrupt the operation of other devices connected to the installation at a short distance from the device generating disturbances [3];
- emissions of harmonics generated from the device to the power grid - too high values of harmonic currents may cause overheating and decrease in the efficiency of the power source and have an impact on the quality of energy supplied to other devices in the power grid [6];
- emissions of voltage changes, voltage fluctuations and flickers in AC power lines - the impact on light sources connected to the same power grid is determined [7];
- immunity to fast electrical transients conducted along the AC power line (so-called BURST) - low-energy, but high-voltage pulses generated, e.g. by: switching electrical switches nearby, capacitive couplings of interference in the wiring from switched loads, electric motors and electromechanical relays, fluorescent lamps ballasts [8];
- immunity to surges conducted along the AC power line - high-current, relatively long-lasting pulses generated, among others, by: switching power supply, insulation damage in the power network, switching passive loads (e.g. motors) nearby, blown fuses, nearby (indirect) lightning.
Testing in the field of type-approval do not exhausts the subject of electromagnetic compatibility by any means.

2.2. Other fields (not obligatory, e.g. OEM)

Large automotive companies, military industry or industry in general use the same test methodologies. However, they use requirements that go far beyond the scope of type-approval. For example, the emission of radiated disturbances according to CISPR or MIL-STD 461 standards should be measured also in frequency ranges from 9 kHz to 30 MHz and from 1 to 40 GHz, depending on the type of test object. The immunity to electromagnetic field is testes even up to 40 GHz [3, 10].
The tests of immunity to ESD and various types of impulse anomalies along the power and signal lines also do not coincide with the requirements of the tests for type-approval. There are normative procedures and documents, e.g. LV 124, LV 148, which describe the characteristic exposure pulses, which are not provided by the UNECE Regulation No. 10 [11, 12].
Nowadays in the automotive industry it is not reasonable to carry out measurements in the entire above-mentioned range, but at the same time it should be emphasized that the existing obligation seems to be insufficient for the correct assessment of the vehicle safety.

3. Fields not mentioned in obligatory normatives for type-approval procedures

Taking into account the design of electronic components used in the automotive industry, the communication interfaces used and radio-wave transmitting and receiving devices, it is reasonable to indicate areas of radiated disturbances tests that are not covered by the testing procedure for type-approval, and it seems that these phenomena may have significant influence on the proper functioning and safety of the vehicle. Examples of technologies widely used in automotive components are GPS

(1580 - 1650 MHz), Wi-Fi (2400 - 2500 MHz) or DSRC (5900 - 6000 MHz) [13]. The scope in which these technologies operate exceeds the test ranges of radiated emission by R10 (30 - 1000 MHz) and electromagnetic field immunity according to R10 (20 - 2000 MHz). That is why it should be considered to expand frequency range of R10 radiated tests to 20 – 6000 MHz.

The literature indicates individual works on the measurements of electromagnetic disturbance emissions inside electric vehicles [14]. Although this study was carried out on a passenger car with a low power drive, a very important problem was highlighted, which should be addressed especially to large electric vehicles with a high power drive, e.g. city buses. There are no normative documents that would describe the test methodology and the requirements for assessment of the electromagnetic field value inside such vehicles, especially in the passengers compartment. The power supply elements for the propulsion of such a vehicle are usually placed under and above the passengers compartment, surrounded by cables and high current components. Thus, it is suspected that the electromagnetic field due to the flowing current may reach intensity values that are harmful to the health of travelers. A research project on such measurements is carried out at the Łukasiewicz Research Network - Automotive Industry Institute, and its result is to be an assessment of the problem scale, possible harmfulness and guidelines for vehicle manufacturers for design solutions limiting the intensity of the field inside.

Another area that is becoming more and more important, especially in the context of autonomous vehicles, is the immunity of sensitive electronic components in a vehicle to electromagnetic attack (in the sense of deliberately interfering with an electromagnetic field). The assessment of this parameter and the appropriate refinement of devices, e.g. a decision computer, ambient sensors in the context of their immunity to electromagnetic field, is of key importance for the safe use of these vehicles. [15]

Looking towards electric or autonomous vehicles that use and will use more and more electronic devices for communication and the decision-making system, it is important to emphasize the importance of the safety problem of these vehicles related to the lack of information on the scale of the risk in the above-mentioned ranges.

4. Good practices in electronic components designing in the automotive industry

Currently, the topic of EMC tests is mainly touched upon at the stage of the finished product. Many manufacturers consider it to be just a problem of type-approval administrative procedure not considering the physical influence on vehicle safety. However, as described above, the topic is much more complicated and it should be taken much more seriously.

Due to the high impact on vehicle safety, the EMC assessment process should start much earlier. Literature [15] indicates list of activities which are related directly to functional safety of the electronic/electric systems. This list explains how to properly manage the levels of risks caused by electromagnetic disturbances at all stages of the product's life. It is not addressed to automotive directly, but it should be considered because of that vehicle is the perfect example of characteristic electronic/electric system.

To ensure the greatest functional safety of the vehicle, it is necessary to divide the EMC assessment into individual stages:

- development of the concept and initial specification:
  - ✓ making the list of EMC standards related to object under development,
  - ✓ specifying verification testing plan,
  - ✓ gathering the characteristic parameters, interfaces and functions of the device which results from its intended use,
  - ✓ defining which parameters impacts vehicle safety,
  - ✓ considering unusual risks of degrading device functions in scope of EMC (e.g. high impact, malicious EMI),
  - ✓ analyzing operational electromagnetic environment for the device,
- designing of the prototype:
  - ✓ using appropriate filtering and shielding for used interfaces,
  - ✓ using appropriate materials for housing,
  - ✓ specifying critical components and its paths arrangement on PCB,
  - ✓ specifying critical sub-assemblies and its wiring arrangement in vehicle,

- ✓ analyzing what coupling phenomena can occur in the device,
- ✓ defining intentional electromagnetic emission in terms of frequency range and transmitting power level,
- ✓ conducting EMC simulation in dedicated software,

- integration of the system, installation method:
  - ✓ specifying bill of materials with minimum quality level defined,
  - ✓ verification test of peripherals and sub-assemblies of device/vehicle,
  - ✓ integration of the most comprehensive variant of designed prototype (including all of peripherals and interfaces intended to use),
  - ✓ analyzing the consequences of reducing amount of peripherals and interfaces,
  - ✓ specifying detailed installation instruction in terms of EMC (including filters, components arrangements, materials, etc.),
- verification of integrated prototype:
  - ✓ verification according to the plan prepared at the initial stage,
  - ✓ conducting not standardized, simplified tests,
  - ✓ verifying correct assembly, wiring termination, length and arrangement,
  - ✓ conducting EMC tests before and after deliberate aging of key components,
  - ✓ verification should be finished with conclusion that the device is not prototype anymore and the producer gaining confidence of passing obligatory testing,
- testing of final device:
  - ✓ conducting tests according to obligatory normative,
  - ✓ conducting tests according to EMC standards list made at the initial stage,
  - ✓ specifying final technical documentation,
  - ✓ conducting control of production (COP) - ensuring the repeatability of the finished product quality,

- post implementation activities
  - ✓ conducting tests after: any modifications of PCB, any modifications in sub-assemblies arrangement in vehicle, any change in bill of materials (especially connected with degradation of components quality),
  - ✓ periodic tests of random sample from production line,

All above-mentioned points are really important to achieve complex compliance with EMC requirements. Example of the tests results corrected by tests at various stages of production is shown in Fig. 2.

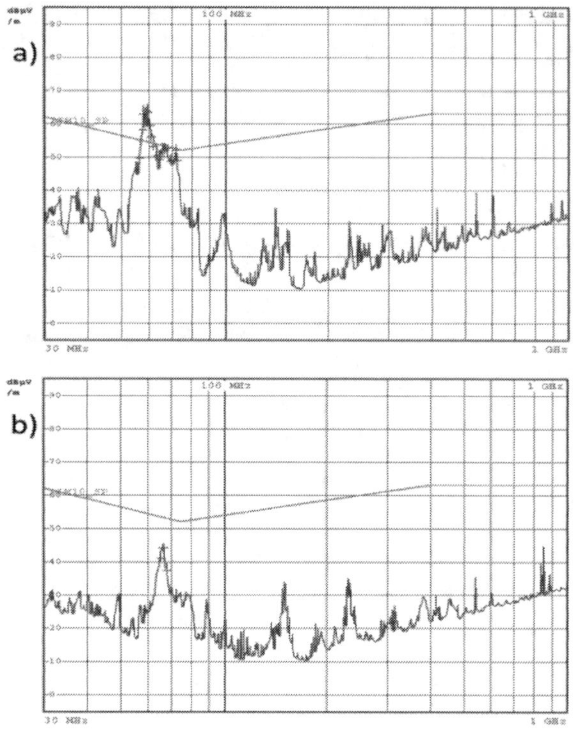

Figure 2. Tests results of electronic subassembly for vehicles: a) stage: integration, before adding filtering, b) stage: verification, after adding appropriate filtering.

5. Conclusion

Electromagnetic compatibility is an area that is directly related to vehicle safety. This safety may be influenced by a direct impact on the basic functions of the vehicle (steering, drive operation, etc.) but also by indirectly influencing the functioning of the vehicle components, causing e.g. confusion for the driver or other road users (e.g. external lighting disfunction). In modern vehicles, most of the functions are controlled by electronic/electrical devices that must meet the relevant requirements to ensure this safety. Unfortunately, it is impossible to ensure full safety by meeting the obligatory/type-approval requirements, because, as it has been shown, they do not cover the full range of EMC phenomena. Further development of vehicles (electric, autonomous) should confirm the belief that this scope should be extended in the near future.

The experience of EMC laboratories notified for testing in the automotive industry shows difficulties even in meeting the obligatory requirements, however, the activities indicated in the work and the procedures described for each stage of the device/vehicle life are to help meet the requirements even in the proposed-extended scope.

The current state of technology and easier access to advanced research equipment (e.g. in the Łukasiewicz Research Network - Automotive Industry Institute) allow manufacturers of equipment and vehicles to ensure vehicle safety in the comprehensive scope of electromagnetic compatibility.

Acknowledgments

The authors would like to thank the Łukasiewicz Research Network - Automotive Industry Institute for providing photos of the test object on the test stand and the opportunity to conduct tests, the results of which serve as an example in this paper.

References

[1] Charoy A 1999 *Zakłócenia w urządzeniach elektronicznych* (Polska: WNT)

[2] UN Regulation No.10.06 2019 *Uniform provisions concerning the approval of vehicles with regard to electromagnetic compatibility* (Geneva: UNECE)

[3] CISPR 25 2004 *Limits and methods of measurement of radio disturbance characteristics for the protection of receivers used on board vehicles* (IEC)

IOP Conf. Series: Materials Science and Engineering 1247 (2022) 012028 doi:10.1088/1757-899X/1247/1/012028

[4] ISO 11451 *Road vehicles - Electrical disturbances by narrowband radiated electromagnetic energy - Vehicle test methods* (ISO)

[5] ISO 7637-2 *Road vehicles - Electrical disturbance from conduction and coupling - Part 2: Electrical transient conduction along supply lines only on vehicles with nominal 12 V or 24 V supply voltage* (ISO)

[6] IEC 61000-3-2 *Electromagnetic Compatibility (EMC) - Part 3-2 - Limits for harmonic current emissions (equipment input current ≤ 16 A per phase)*

[7] IEC 61000-3-3 *Electromagnetic Compatibility (EMC) - Part 3-3 - Limits - Limitation of voltage changes, voltage fluctuations and flicker in public lowvoltage systems for equipment with rated current ≤ 16 A per phase and not subjected to conditional connection*

[8] IEC 61000-4-4 *Electromagnetic Compatibility (EMC) - Part 4-4 - Testing and measurement techniques - Electrical fast transients/burst immunity test*

[9] IEC 61000-4-5 *Electromagnetic Compatibility (EMC) - Part 4-5 - Testing and measurement techniques - Surge immunity test*

[10] MIL-STD-461 *Requirements of the control of electromagnetic interference characteristics of subsystem and equipment*

[11] ISO 10605 *Road vehicles — Test methods for electrical disturbances from electrostatic discharge*

[12] LV 124 / LV 148 *Electric And Electronic Components In Motor Vehicles Up To 3.5t*

[13] Mortazavi S, Schleicher D, Eremyan D, Gheonijan A, Jobava R, Sinai A, Gerfers F 2019 *Investigation of Possible EMC Interferences between Multi-Gig Communication Link and RF Applications in Vehicle* (Barcelona: EMC Europe 2019)

[14] Gumiela J, Sitnik L, Sztafrowski D 2019 *Pomiarowa identyfikacja emisji pola elektromagnetycznego przez samochód elektryczny* (Warszawa: Przegląd Elektrotechniczny 12/2019)

[15] IEEE Electromagnetic Society 2020 *IEEE Standard for Techniques and Measurement to Manage Functional Safety and Other Risk with Regards to Electromagnetic Disturbances* (IEEE Std 1848-2020)

Selection of child seats in terms of vibration comfort

D Frej[1]*, A Zuska[1], P Kubiak [2]

1 Department of Automotive Engineering and Transport, Kielce University of Technology, Avenue Tysiaclecia Państwa Polskiego 7, 25-314 Kielce, Poland;
2 Department of Transport, Warsaw University of Technology, Plac Politechniki 1, 00-661 Warsaw, Poland

* dfrej@tu.kielce.pl

Abstract. The article presents the results of laboratory and road tests of the impact of the method of fastening child seats on the vibration comfort of children transported in them. The tested child seats were mounted facing the direction of travel on the rear seat of a passenger vehicle. Two child seats were used in the work, the first (red) was attached to the rear seat behind the driver's seat using the ISOFIX base. The second child seat (grey) is attached to the rear seat behind the passenger seat with standard seat belts. During the measurements, the child seats were loaded with the mass of 5 kg, 10 kg and 15 kg, simulating the mass of a child fastened in a child seat. The child seats used can be mounted both forward and rearward facing. In the conducted research, they were installed in a passenger vehicle facing the direction of travel. Laboratory and road tests based on registered vertical accelerations showed a negative impact of using the ISOFIX base in terms of vibration comfort.

Keywords: vibration comfort, child seats, vehicle safety

1. Introduction

Choosing the right child seat is a difficult task. Before purchasing a child seat, one should measure it according to the child and the vehicle [1-5]. Otherwise, it may turn out that the selected seat is not compatible with one's vehicle. The problem may be too high an ISOfix system hitch, the design of the rear seat of the vehicle and the inability to remove the headrest in the rear seat [2,3]. A very important element in choosing a child seat is matching it to the vehicle seat. The height of the rear seat of a passenger vehicle has a great influence on the placement of the child's feet, which in turn translates into the overall comfort in the vehicle. Furthermore, the seat cushion of the child seat may be longer or shorter than the seat cushion of the rear vehicle seat, and consequently a situation may arise in which the child has no limited foot space. The second important issue in choosing a child seat is its backrest [2,3,5]. The upright backrest may cause the child to tilt over when traveling forward. In addition, in some vehicles, it is impossible to remove the headrest or raise it to the appropriate height, so the backrest of the child seat will stick out from the backrest of the passenger vehicles [2,3].

In terms of safety, there are other aspects to check. The length of the seat cushion must match the length of the child's legs. If the seat of the child seat is too long, the child will maybe slide out, which is very dangerous. In a forward-facing configuration, the child's legs must have a space of 15 to 20 cm between the extreme point of the child seat and the front vehicle seat [6,7]. The most common difficulties when fitting a child seat to the vehicle include [3,6]:

- short seat belts in the rear seat of the vehicle,

- the shape of the child's seat does not match the shape of the vehicle seat,
- too short support stabilizing the ISOFIX base of the child seat,
- the ISOFIX brackets in the vehicle are positioned at a different height than the brackets in the child seat.

The above problems in the installation of a child seat in a passenger vehicle mean that the child is not safe in the seat while driving and during a possible collision. In a road collision, the child seat may come out of the vehicle seat. Improper installation of the child seat will affect the child's posture during the journey, making it uncomfortable and unsafe [7-10].

The approval tests of child seats are mainly focused on the safety of the transported child during a vehicle accident and protection against the harmful effects of such events [7,9]. During the safety tests of ADAC (Allgemeiner Deutscher Automobil-Club) vehicle seats are carried out in laboratory conditions and simulating of a side impact at 50 km/h and a frontal collision at a speed of 72 km/h [10,11]. The child seats are only tested for overall comfort. The position of the child's seat is checked, the softness of the upholstery as well as whether the seat belt and other elements of the child seat do not put pressure on the child during the journey. Child seats are also checked and evaluated for use and cleaning. Unfortunately, approval tests and ADAC child seat tests do not include the impact of vibrations (vibration comfort) on the child's body [3,10,11].

The greatest threats to humans, caused by the impact of vibrations related to the movement of vehicles, occur in road transport [11,12]. The driver and passengers of the passenger vehicle assess the discomfort from driving based on the acceleration influencing them. The harmfulness of these vibrations depends mainly on the amplitude, frequency of vibrations and the duration of this phenomenon. The permissible values of vibrations acting on the adult human body are the subject of many studies, both in terms of comfort and ergonomics [7,9,12]. Vibration measurement methods and their values have been regulated in ISO 2631-1 [13] and British Standards BS 6841 [14]. Unfortunately, there are no relevant standards developed for measurements and vibration values affecting a child's body.

In the literature, one can find several normative acts and scientific publications describing the impact of vibrations on the human (adult) body [3,15]. Unfortunately, such studies are not carried out in the case of children. And the development of child seats is aimed solely at improving safety and ergonomics [3,4,9].

2. Research object and research methodology

The tests were carried out in laboratory conditions and in real road conditions. During laboratory tests, the wheels of the rear axle of the tested vehicle were made to vibrate using the measuring plate of the EUSAMA SA.640 device. This plate is used to test shock absorbers using the EUSAMA method and is part of the Bosch Beissbarth diagnostic line. The EUSAMA SA.64 device stimulates the road wheels to vertical vibrations with a constant amplitude of 6 mm and a frequency varying from the initial value of about 25 Hz to zero. The Bosch Beissbarth diagnostic line played the role of a vibration generator in the experimental tests.

Road tests were carried out on three sections of asphalt, gravel, and cobblestone roads. The speed of the passenger vehicle during the tests was kept constant at 50 km/h. The research used an Audi A6 passenger vehicle with air suspension, which could assume four positions: raised, comfortable, automatic, and dynamic. Two child seats were used in the experiment: Avionaut AeroFIX (Red) and Avionaut Pixel (Gray). The view of the tested child seats is shown in Figure 1. During road and laboratory tests, the child's seats were loaded with a mass imitating the mass of a child, which was 5 kg, 10 kg, and 15 kg, respectively.

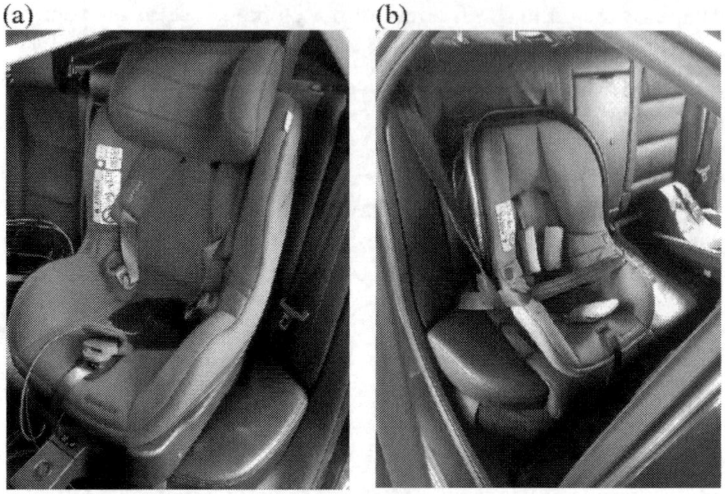

Figure 1. View of the tested child seats:
a) Avionaut AeroFIX seat (red) for transporting children up to 17.5 kg
b) Avionaut Pixel child seat (gray) for transporting children up to 13 kg

The aim of the experiment was to measure and analyse the distribution of vertical vibrations in two subsystems. The first consisted of a rear vehicle seat and a child seat with a classic fastening system. The second subsystem consisted of a rear passenger vehicle seat and child seat attached with the ISOFIX base. The view of the ISOFIX base is shown in Figure 2.

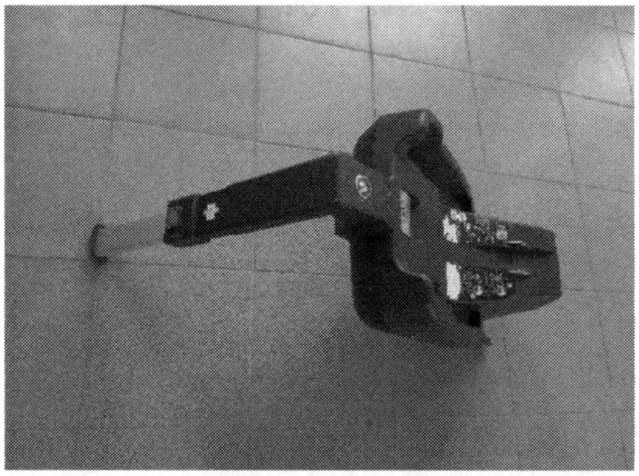

Figure 2. View of the ISOfix base with the stabilizing handle folded out

3. Analysis of the results

During the research, acceleration was recorded with a frequency of 1024 Hz using three-way acceleration sensors. The sensors were mounted on the seat of the rear seat of the vehicle, on the basis of ISOFIX, on the seats of child seats by means of semi-flexible measuring discs. The results of the measurements were used to determine the time courses of vertical accelerations determining the vibration comfort. Figures 3 to 8 show the vertical accelerations recorded on child seats during laboratory tests. The recorded waveforms of acceleration for the seat of the Avionaut Pixel seat and the Avionaut AeroFIX seat are similar in terms of the maximum acceleration values. It should be noted, however, that higher values of vertical acceleration on the child seat were recorded on the child seat fixed with the ISOFIX base.

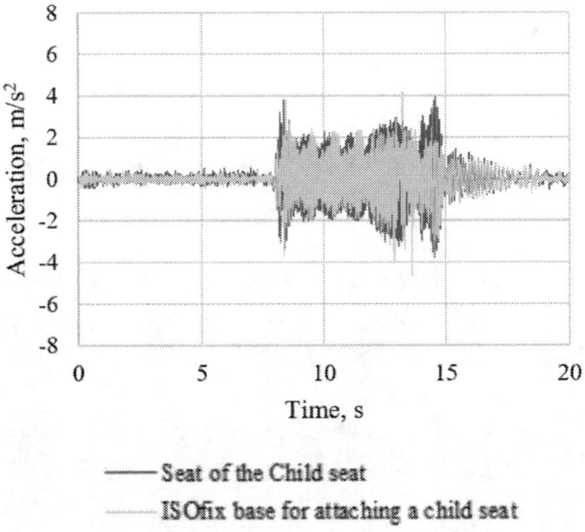

Figure 3. Vertical accelerations recorded on the basis of ISOfix and the seat of the Avionaut AeroFIX child seat, mass loading the child seat 5 kg

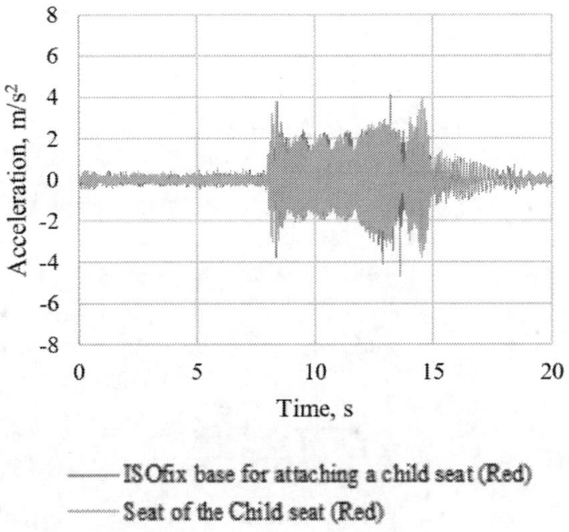

Figure 4. Vertical accelerations recorded on the rear seat of the passenger vehicle and the seat of the Avionaut Pixel child seat, mass loading the child seat 5 kg

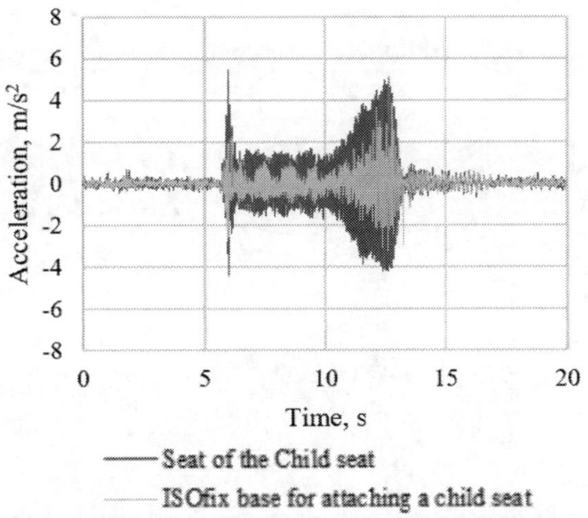

Figure 5. Vertical accelerations recorded on the basis of ISOfix and the seat of the Avionaut AeroFIX child seat, mass loading the child seat 10 kg

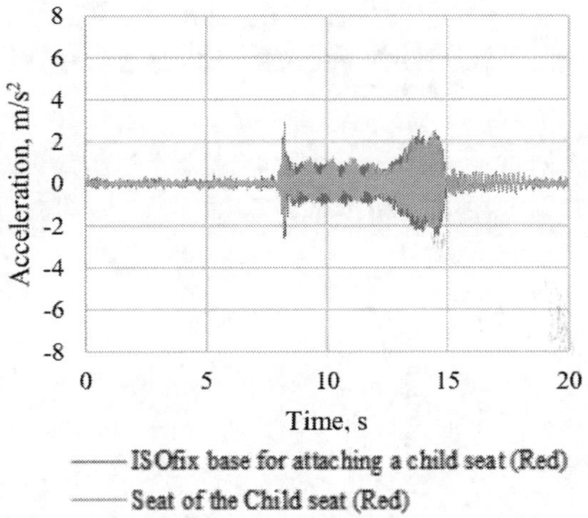

Figure 6. Vertical accelerations recorded on the rear seat of the passenger vehicle and the seat of the Avionaut Pixel child seat, mass loading the child seat 10 kg

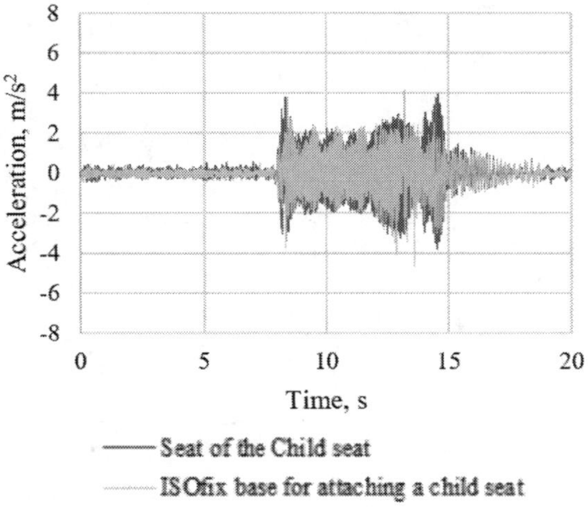

Figure 7. Vertical accelerations recorded on the basis of ISOfix and the seat of the Avionaut AeroFIX child seat, mass loading the child seat 15 kg

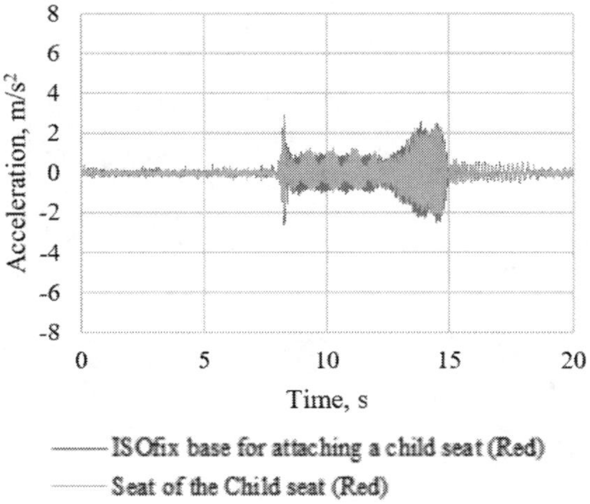

Figure 8. Vertical accelerations recorded on the rear seat of the passenger vehicle and the seat of the Avionaut Pixel child seat, mass loading the child seat 15 kg

Figures 9 to 14 show the vertical accelerations recorded on the tested child seats during road tests. During the tests on gravel and cobblestone surfaces, the highest acceleration values were recorded on the seat of the child seat fixed with the ISOfix base, they were significantly higher than the accelerations recorded on the seat of the child seat fixed with the classic method (with the use of seat belts).

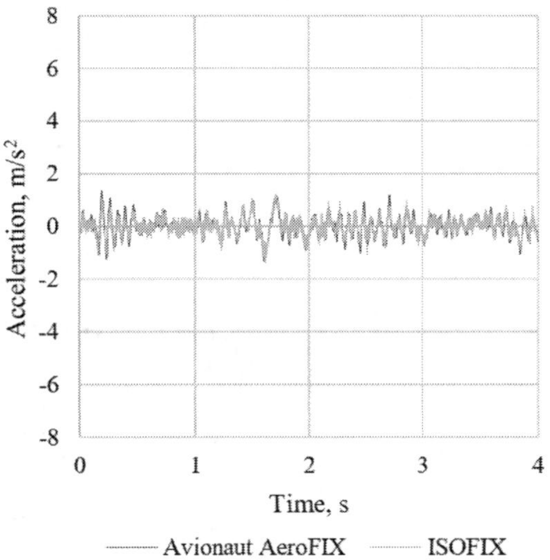

Figure 9. Vertical accelerations recorded on the basis of ISOfix and the seat of the child seat Avionaut AeroFIX (red), loaded with a mass simulating the mass of a child of 10 kg, vehicle traveling on an asphalt road

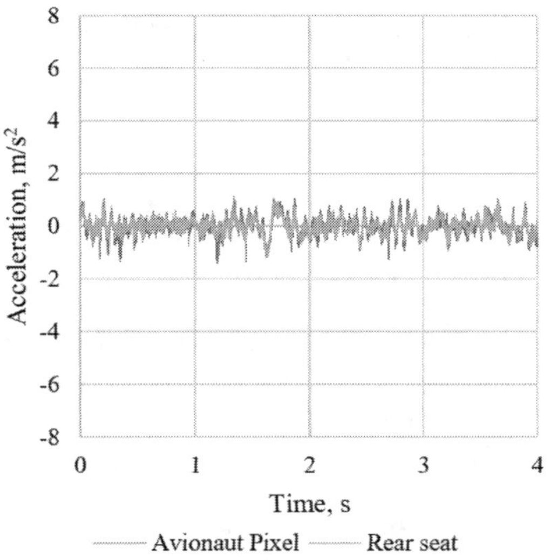

Figure 10. Vertical accelerations recorded on the rear seat of a passenger vehicle and the seat of the child seat Avionaut Pixe (grey), loaded with a mass simulating the mass of a child of 10 kg, vehicle traveling on an asphalt road

IOP Conf. Series: Materials Science and Engineering 1247 (2022) 012029 doi:10.1088/1757-899X/1247/1/012029

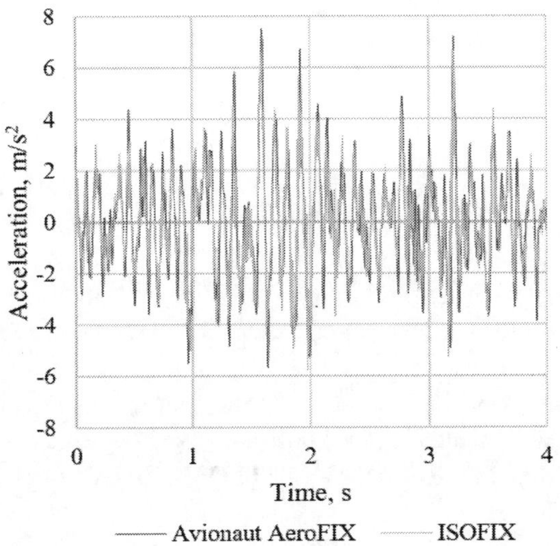

Figure 11. Vertical accelerations recorded on the basis of ISOfix and the seat of the child seat Avionaut AeroFIX (red), loaded with a mass simulating the mass of a child of 10 kg, vehicle traveling on a gravel road

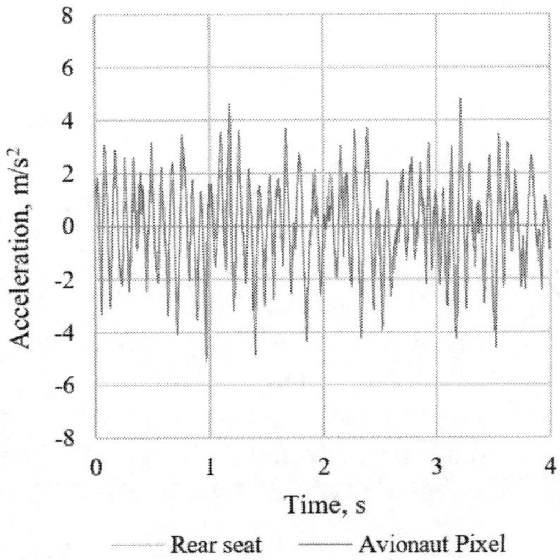

Figure 12. Vertical accelerations recorded on the rear seat of a passenger vehicle and the seat of the child seat Avionaut Pixe (grey), loaded with a mass simulating the mass of a child of 10 kg, vehicle traveling on a gravel road

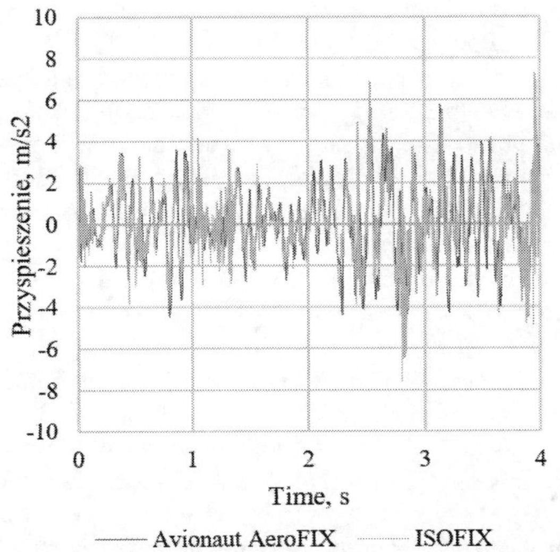

Figure 13. Vertical accelerations recorded on the basis of ISOfix and the seat of the child seat Avionaut AeroFIX (red), loaded with a mass simulating the mass of a child of 10 kg, vehicle traveling on a cobblestone road

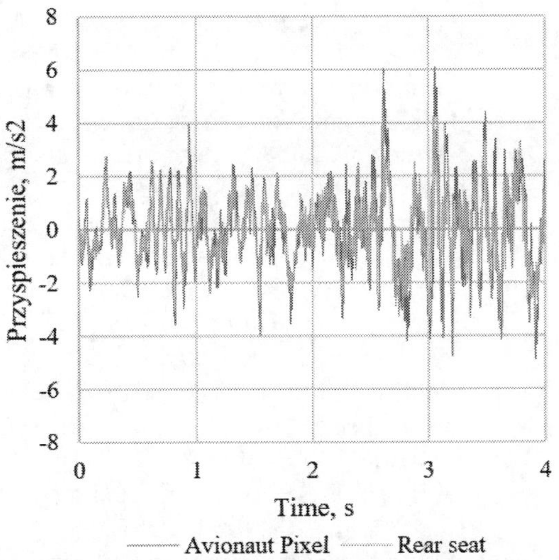

Figure 14. Vertical accelerations recorded on the rear seat of a passenger vehicle and the seat of the child seat Avionaut Pixe (grey), loaded with a mass simulating the mass of a child of 10 kg, vehicle traveling on a cobblestone road

4. Conclusions

The laboratory and road tests carried out have shown that the highest acceleration values occur on the base ISOFIX used to attach the child seats and they differ from the acceleration curves recorded in other areas. The use of the ISOFIX base for mounting a child seat is a great help for the parents of the child in terms of ergonomics. However, it should be noted that when traveling, the greatest attention to the manufacturers of child seats should be on the comfort of the child. A child placed in the child seat should feel safe and comfortable. The use of the ISOFIX base causes the child seat to be separated from the rear passenger vehicle seat. The vibrations transmitted from the road surface through the body to the seat of the child seat are not absorbed sufficiently in the case of the ISOFIX base, therefore higher vertical accelerations in the seat of the child seat are recorded than in the case of a seat secured with standard seat belts.

Literature:

[1] Lee G., Pope C.N., Nwosu A., McKenzie L.B., Zhu M., **2019**, Child passenger fatality: Child restraint system usage and contributing factors among the youngest passengers from 2011 to 2015, *Journal of Safety Research*, 70, 33-38.

[2] Frej D., Zuska A., Cadge K., **2019**, Analysis of vertical vibrations affecting a child transported in a child seat during a car passing over the release speed bump. *Archiwum Motoryzacji*. 86, 4, 111–125.

[3] Frej D., Zuska A., Szumska E., **2020**, Analysis of the impact of a child seat mounting method on vertical vibrations affecting a child in a child seat. *2020 XII International Science-Technical Conference AUTOMOTIVE SAFETY*. 1-8. doi: 10.1109/AUTOMOTIVESAFETY47494.2020.9293503.

[4] Cicchino J.B., Jermakian J.S., **2015**, Vehicle characteristics associated with LATCH use and correct use in real-world child restraint installations, *Journal of Safety Research*, 53, 77-85.

[5] Schwebel D. C., Tillman M.A., Crew M., Muller M., Johnston A., **2017**, Using interactive virtual presence to support accurate installation of child restraints: Efficacy and parental perceptions, *Journal of Safety Research*, 62, 235-243.

[6] Więckowski D., **2012**, An attempt to estimate natural frequencies of parts of the child's body (Próba oszacowania częstotliwości drgań własnych części ciała dziecka), *The Archives of Automotive Engineering - Archiwum Motoryzacji*, 55 (1), 61-74.

[7] Więckowski D., **2012**, Analysis domain of the time vertical vibration on account comfort child during ride in the car (Analiza w dziedzinie czasu drgań pionowych ze względu na komfort podróżowania dzieckaw samochodzie), *Czasopismo Techniczne. Mechanika*, 109, 10, 73-91.

[8] Krajnak K., **2018**, Health effects associated with occupational exposure to hand-arm or whole body vibration, *Journal of Toxicology and Environmental Health*, Part B, 21:5, 320-334.

[9] Qassem W., Othman M.O., Abdul-Majeed S., **1994**, The effects of vertical and horizontal vibrations on the human body, *Medical Engineering & Physics*, 16, 2, 151-161.

[10] Wicher J., Więckowski D., **2010**, Influence of vibrations of the child seat on the comfort of child's ride in a car (Wpływ drgań fotelika samochodowego na komfort podróżowania dziecka w samochodzie), *Eksploatacja i Niezawodność*, 4,102-110.

[11] Frej D., Grabski P., **2019**, The impact of the unbalanced rear wheel on the vibrating comfort of the child seat, *Transportation Research Procedia*, 40, 678-685.

[12] Troxel W.M., Helmus T.C., Tsang F., Price C.C., **2015**, Evaluating the impact of whole-body vibration (WBV) on fatigue and the implications for driver safety, *Santa Monica, Calif.: RAND Corporation,* RR-1057-BOSE.

[13] ISO 2631-1: **1997** *Mechanical vibration and shock. Evaluation of human exposure to whole-body vibration*. Part 1: General requirements.

[14] BS (British Standard) 6841: **1987** *Guide to measurement and evaluation of human exposure to whole-body mechanical vibration and repeated shock.*

[15] Jaśkiewicz M., Więckowski D., **2018**, Constructional solutions for active suspensions used in vehicles (Rozwiązania konstrukcyjne aktywnych zawieszeń stosowanych w pojazdach), *Autobusy – Technika, Eksploatacja, Systemy Transportowe*, 223, 9, 225- 229.

Simulation tests of the integrated lane change control system

M Gidlewski [1,3], **L Jemioł** [2] **and D Żardecki** [1]

[1] Military University of Technology (WAT), Poland
[2] University of Technology and Humanities in Radom, Poland
[3] ŁUKASIEWICZ Research Network – Automotive Industry Institute (ŁUKASIEWICZ-PIMOT), Poland

Abstract. This paper presents the concept of an integrated control system for lane change by a two-axle truck. In earlier works the authors used only a controller coupled to the steering system for sudden lane changes. Nowadays, the steering system controller is integrated with ESP (Electronic Stability Program) which causes momentary braking of chosen vehicle wheels when traffic destabilization threatens. A simulation study of the integrated control system operation during sudden lane change was performed for an unladen and fully laden car on a wet road in near limiting conditions. The test results were compared with those obtained when this maneuver was performed using only the steering controller.

Keywords – safety, vehicle, vehicle detection

1. Introduction

In the last twenty years, the number of road accidents and their victims in the EU (including Poland) has significantly decreased. This was despite the fact that the number of cars in operation has been increasing from year to year [1,2,3]. The decrease of the basic indicators of accident risk was influenced by many factors, including: among others, decisive improvement of passive and active car safety.

The driver of a modern car is assisted passively (information systems that warn of hazards or recommend necessary actions) and actively (systems that facilitate and correct certain driving activities, especially in boundary conditions). With the increasing use of various types of driver assistance systems in cars of many brands, their alternative name „assistant systems" has become popular [4].

Passive safety systems do not directly intervene in the car's movement control process, but only provide the driver with additional information. In dangerous situations, they warn of existing danger and suggest the initiation of designated defensive maneuvers. Active safety systems participate in controlling the vehicle together with the driver. They have the capability to assess the hazards caused by the driver's actions in order to determine when and how strongly to intervene. They generally accept the driver's actions, opposing them only if they lead to a collision or loss of stability of the car. Such systems therefore intervene automatically only when necessary to correct driver actions deemed unsafe. The interventions of safety systems should maintain the necessary balance between the level and frequency of interventions, i.e. do not change the interactions with the car controls too strongly, too early or too often [5].

Currently, there is a noticeable trend towards combining passive and active safety systems (Integrated Safety Technology).

Driving a road vehicle is a complex dynamic process in which the driver carries out successive emerging tasks. These tasks result from the current driving conditions and the road situation and they are characterized by varying degrees of difficulty and different ways of implementation. The relatively easy ones include stabilization of the vehicle movement in the presence of disturbances (bumps in the road, gusts of wind, etc.) or following the trajectory set by the lines limiting the roadway. More difficult and requiring the use of more advanced steering algorithms are maneuvers (e.g. parking). The most difficult are sudden maneuvers performed at high speeds. In such extreme dynamic conditions, the vehicle is at risk of losing directional stability, and in the case of a truck, also of tipping over. Practice shows that in such critical driving situations, an experienced driver is forced to act intuitively (almost „blindly", but according to proven and previously learned patterns). Many drivers (especially those with low experience) in critical situations are unable to correctly select and then safely and effectively execute the appropriate defensive maneuver, which in consequence leads to a collision or road accident [6]. It is therefore widely believed that the role of the driver in driving should be increasingly reduced aiming consistently at a fully autonomous vehicle in which the driver becomes unnecessary.

Today, many academic and industrial centers are conducting intensive research on autonomous vehicles. However, despite of the considerable investment in research in this field, the introduction of fully automated vehicles into general use is still an open question. Many problems remain to be solved, not only technical, but also social and legal.

The prolonged period of introduction of fully autonomous cars into service as well as the dynamic development of various types of controllers, actuators, sensors, monitoring devices, microprocessors, networks and telecommunication technologies are conducive to the emergence of more and more effective systems supporting the driver's actions related to driving the vehicle. In recent years, a significant effort has been made to develop assistant systems that are activated automatically when a critical driving situation occurs and that replace the driver in the selection and execution of an effective or at least the most advantageous defensive maneuver in the given situation. Advanced research work on such systems is carried out in many research centers, e.g. [7,8], and in the near future they will be used in serially produced cars. In December 2019, the European Parliament passed legislation mandating (as of May 2022) the mandatory equipping of new models of cars and vans with advanced safety features such as automatically activated emergency braking systems and lane-keeping systems in emergency situations.

As part of research project N N509 568439, extensive analytical work was undertaken on the application of Active Front Steering (AFS) in the automatic steering of a two-axle medium-duty truck in accident-prone traffic situations due to a suddenly appearing obstacle [9-12]. Extensive modeling studies were conducted based on avoiding a sudden obstacle in the shortest possible time and on the shortest possible path resulting from the prevailing road conditions and the traffic condition of the vehicle. This meant that the lane changes necessary to avoid the obstacle were made under extreme traffic conditions with low stability. Simulation studies were conducted for a very wide range of varying road and operating conditions, yet the obstacle avoidance maneuver was successfully completed in the vast majority of trials without the need to change previously made assumptions and established values of controller tuning parameters [13]. Only the unladen car proved to be very sensitive to changes in traffic conditions and parameters, which manifested itself as a loss of directional stability during the manoeuvre. In order to avoid this, the reserve of stability was increased by decreasing by 20% (in relation to the other considered states of car load) the value of the reference angle of steering wheel rotation given to regulators [14]. Unfortunately, this resulted in an increase in the distance required to avoid the obstacle. In the literature, e.g. in [15, 16], a thesis is presented that the steering system provides good handling properties only when the lateral acceleration is low, the vehicle drift angle is small, and the lateral tire drift characteristics are linear. However, in emergency situations accompanied by high lateral acceleration (such as occurs during sudden lane changes), the characteristics of the tires become nonlinear and the desired effect of controlling the motion of the car cannot be achieved only by steering wheel control. In this case, the additional integration of the braking system as well as the drive train (ESP or DYC - Direct Yaw Control) into the vehicle's motion control can significantly improve the driving characteristics.

IOP Conf. Series: Materials Science and Engineering 1247 (2022) 012030 doi:10.1088/1757-899X/1247/1/012030

It was decided to modify the developed active steering system ASF by supplementing it with an electronic trajectory control system ESP. This paper presents the concept of the integrated steering system. Simulation tests of the integrated system functioning were carried out during the execution of sudden lane change for an unladen and laden car on a wet road, in conditions close to the limit. The results of the tests were compared with the results obtained when performing this maneuver only by the ASF system.

2. Concept and mathematical model of an integrated automatic lane change assistant system

The integrated assistance system consists of the ASF active steering system inducing preset front wheel steering angles and the ESP electronic control system causing momentary braking of selected vehicle wheels. Integration of these systems allows simultaneous control of these subsystems and mutual exchange of information from individual sensors and actuators. Coordination, on the other hand, ensures harmonious functioning of the systems and synchronization of the partial actions of the individual systems, which enable proper realization of the main goal, i.e. sudden lane change. The block diagram of the integrated assistant system is shown in Figure 1.

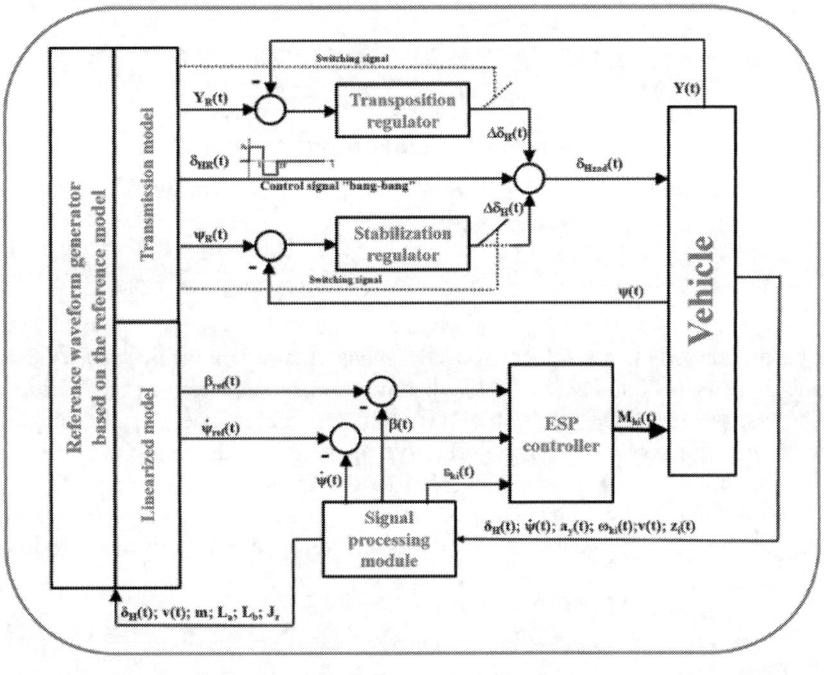

Figure 1. Block diagram of an integrated assistant system in the lane change process.

The motion of a two-axle truck traveling at a fixed speed on a straight section of a one-way roadway having at least two lanes in one direction is analyzed. At some point, an obstacle appears in the vehicle's lane at a distance less than the distance needed to stop the vehicle but sufficient to avoid the obstacle. The assistant system initiates an abrupt lane-change maneuver, initially by turning the front wheels only, so that the car is moved into the adjacent lane in the shortest possible time and over the shortest possible distance. In practice, this maneuver requires the steering wheel to be turned very quickly in either direction (the ASF is solely responsible for this), while of course maintaining traffic stability conditions. ESP is a system that overrides the ASF, and when the vehicle approaches the limit of adhesion, brakes the appropriate wheels causing an increase in the radius of the trajectory and a reduction in speed, which further protects the vehicle. and reduce the speed of the car which additionally secures stable movement of the vehicle.

The two systems forming the assistant system use as a reference model the flat, single-track „bicycle model" of a car known from numerous publications [17, 18] (Figure 2) describing the dynamics of motion of a vehicle traveling at a constant speed under not too strong perturbations of the steady-state motion. However, the classical „bicycle model" was modified by performing simple transformations to express the vehicle motion in the global (OXY) system and by linearizing the equations of motion. It has been shown in [10] that for small and short-lived disturbances of motion (such forcing we deal with during obstacle avoidance), linearization of the equations of motion of the vehicle is acceptable.

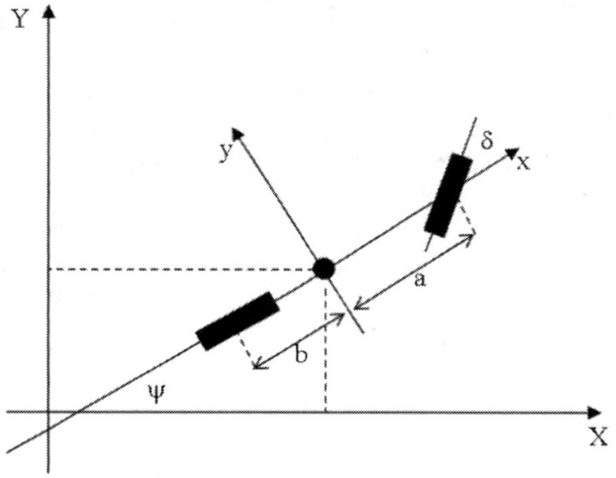

Figure 2. The idea of „bicycle model".

The equations of motion of the reference car model used are shown below:

$$m\ddot{Y}(t) + \frac{k_A+k_B}{V}\dot{Y}(t) + \frac{k_Aa-k_Bb}{V}\dot{\Psi}(t) - (k_A + k_B)\Psi(t) = k_A\delta(t)$$
$$J\ddot{\Psi}(t) + \frac{k_Aa^2+k_Bb^2}{V}\dot{\Psi}(t) - (k_Aa - k_Bb)\Psi(t) + \frac{k_Aa-k_Bb}{V}\dot{Y}(t) = k_Aa\delta(t) \quad (1)$$

There are seven parameters in the model: car speed V, mass m, moment of inertia J, wheel drift resistance coefficients of the front axle k_A and rear axle k_B, distance from the center of mass of the front axle a and rear axle b. The reference model of vehicle motion dynamics specifically has a rather simple mathematical form, which is essential for the effective operation of the real-time assistance system.
Equations (1) are linear, so they can be subjected to the Laplace transform and the corresponding transmittances binding the waveform transforms can be determined. With zero initial conditions on the variables, one then obtains an equivalent notation of the reference model in the transfer functions form (2).

$$Y(s) = G_{Y\delta}(s)\delta(s); \quad \psi(s) = G_{\psi\delta}(s)\delta(s) = \frac{G_{\dot{\psi}\delta}(s)}{s}\delta(s) \quad (2)$$

The active steering ASF uses a reference model in the transmittance form (2) and operates in the structure of an optimal tracking system [19]. The operation strategy of the system is based on the time decomposition of the lane change process. The process of steering angle control is divided into two consecutive phases which carry out the task of rearranging the lateral center of mass and then the task of stabilizing the car path. The signal generator controlling the steering wheel angle sends to its actuator the set reference waveform δ_{HR}, which is corrected by a system of two successively switched Kalman regulators. An important distinction of the developed control system (in comparison with other similar studies) is that the reference signals are determined on the basis of the vehicle reference model in the transmittance form when the steering angle is forced by the „bang-bang" type (nomenclature used in automation).
The base steering angle reference signal $\delta_{HR0}(t)$ results from rescaling the steering angle $\delta_R(t)$ ($\delta_{HR0}=p\delta_R$, where p is the kinematic steering ratio) and is described by the formula:

$$\delta_{HR0}(t) = p\delta_R(t) = p(\delta_0(1(t) - 2\cdot 1(t-T) + 1(t-2\mathrm{T}))) \quad (3)$$

where: 1(t) – Heaviside function.

The parameter values of steering angle amplitude δ_0 and maximum steering angle holding time T are described by the expressions:

$$\delta_0 = \frac{\dot{\Psi}_{gr}}{G_{\dot{\Psi}\delta 0}} = \frac{a_{ygr}}{V G_{\dot{\Psi}0}}; \quad T = \sqrt{\frac{Y_0}{V \delta_0 G_{\dot{\Psi}\delta 0}}} = \sqrt{\frac{Y_0}{V \dot{\Psi}_{gr}}} \tag{4}$$

where:

$$G_{\dot{\Psi}\delta 0} = \frac{k_A k_B (a+b) V}{k_A k_B (a+b)^2 - m V^2 (k_A a + k_B b)} \tag{5}$$

is transfer function parameter

Expressions (4) and (5) were derived using the reference car model in the transmittance form [9]. They guarantee stable motion of the reference model for the given road and operating conditions. This means that for a given course of the steering angle according to (3), the transient courses of the lateral acceleration $\ddot{Y}(t)$ and the yaw rate $\dot{\Psi}$ will not exceed the preset limit values a_{ygr} and $\dot{\Psi}_{gr}$. It has been shown in [20] that the limit values of a_{ygr} and $\dot{\Psi}_{gr}$ can be determined using the distance from the obstacle, and in a critical situation [21] using the maximum value of the coefficient of adhesion of tires to the road surface μ in specific road conditions $\left(a_{ygr} = \mu g\right)$ or the maximum values of the considered quantities obtained during the simulation of driving a real vehicle on a curve in steady or transient conditions. Increasing the values of a_{ygr} and $\dot{\Psi}_{gr}$ results in an increase in the steering amplitude δ_0 and a decrease in the time of holding the maximum steering angle T, which consequently leads to a decrease in the time and distance needed to avoid an obstacle, but also to a decrease in the stability reserve of the car during lane change.

ESP uses a vehicle reference model in the form of linearized equations of motion (1). In critical situations, ESP reduces the risk of the vehicle skidding or rolling over. ESP performs its task by braking the relevant vehicle wheels, thus generating an optimally adjusted additional torque to bring the vehicle back on track or to reduce the value of the lateral acceleration that could cause the vehicle to roll over. The course of the real steering angle $\boldsymbol{\delta_{Hzad}(t)}$ after scaling to the steering angle $\left(\boldsymbol{\delta(t) = \delta_{Hzad}(t)/p}\right)$ is the forcing acting on the equations of motion (1). Solving these equations allows us to calculate the reference signals for the ESP: yaw rate $\boldsymbol{\dot{\Psi}_{ref}(t)}$ and center-of-mass drift angle $\boldsymbol{\beta_{ref}(t)}$.

3. Methodology for simulation studies

The adopted concept of car control with the use of integrated control system and the correctness of operation of controllers were verified by conducting model tests. Such tests are currently the most efficient and effective method for testing various control systems. An extensive model of a truck (24 degrees of freedom), thoroughly verified experimentally in bench and road tests, was used as a virtual control object. Simulation studies consisted in avoiding a suddenly appearing obstacle (single lane change) on the shortest possible route. This maneuver is one of the basic maneuvers used in car driving. It is particularly dangerous when performed suddenly at high speed. The automation of this maneuver is particularly important for automating car steering. Simulation test conditions:

- the car is moving in a steady, rectilinear motion on a one-way roadway having two lanes in one direction, in the middle of the right lane;
- the width of one lane is 3 m and the width of the car is d=2,4 m;
- in each test the vehicle speed and the type of road surface on which the vehicle is travelling (the value of the coefficient of adhesion of the wheel to the road surface) are known;
- an obstacle suddenly appears in the car lane blocking the whole width of the right lane, the left lane is free;
- the distance of the emerging obstacle S_p from the centre of mass of the car is not sufficient to stop the car in front of the obstacle but is greater than or equal to the minimum distance S_{ou} sufficient to avoid the obstacle under the current operational and driving conditions;
- the controller initiates intensive braking of the vehicle. When the vehicle reaches a distance from the obstacle equal to S_{ou} the vehicle stops braking and an automatic obstacle avoidance

manoeuvre is initiated for the existing road conditions and operating conditions. The reference waveforms of the lateral displacement of the centre of mass, the car yaw angle and the steering wheel rotation angle are generated using the reference model;

- passing an obstacle is considered to be correct if, after driving a distance S_{ou} the lateral displacement of the car's centre of mass slightly exceeds 3m and the car passing the obstacle has an angle of deflection close to 0° (measured in relation to the road axis). In such a case, the vehicle passing the obstacle will travel in the centre of the left lane parallel to the axis of the carriageway, at a distance of Δd=0.3m measured across the carriageway both from the obstacle and from the left edge of the carriageway.

In order to assess and compare the quality of the waveforms of selected quantities (obtained during model tests, significant for obstacle avoidance), special integral indices [22] were introduced, expressing the percentage discrepancy between the reference waveforms and the actual waveforms obtained as a result of simulation, which is shown in Figure 3.

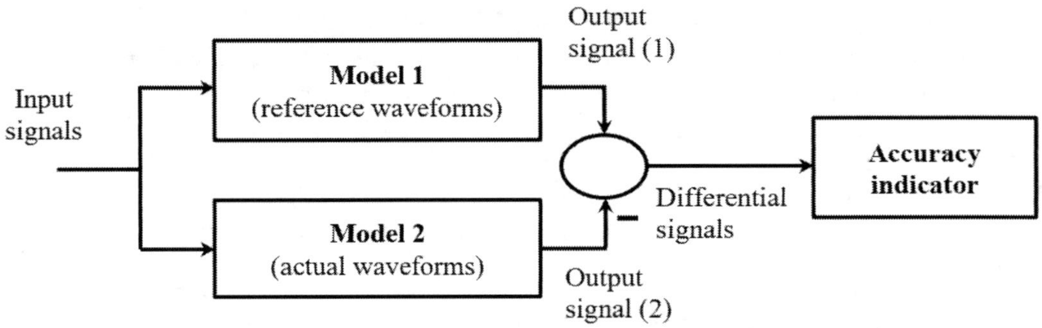

Figure 3. Method for determining the integral index of the accuracy of the realisation of reference runs during obstacle avoidance.

The integral accuracy index for signals $S_1(t)$ and $S_2(t)$ is described by the formula:

$$W_S = 100\% \frac{\int_0^T (S_1(t)-S_2(t))^2 dt}{\int_0^T (S_1(t))^2 dt} \tag{6}$$

In the research on the accuracy of the realisation of the reference signals, the set reference waveform and the real waveform obtained as a result of the simulation were compared for several quantities. Using expression (6) the values of accuracy index W_S were calculated for:

- steering wheel rotation angle δ_H – accuracy index $W_{\delta H}$;
- the lateral displacement of the centre of mass of the vehicle Y – accuracy index W_Y;
- vehicle yaw angle ψ – accuracy index W_ψ.

In addition, the evaluation was subject to:

- distance needed to avoid the obstacle (lateral displacement of the vehicle's centre of mass $Y \geq 3$m, longitudinal axis deviation angle $\psi \approx 0$);
- maximum excess of lateral displacement of the vehicle centre of mass ΔY of the set value Y=3m.

An obstacle avoidance manoeuvre was considered successful in a given test only if the following conditions were met:

- the distance needed to avoid an obstacle S_o should be less than or equal to the set distance, obviously the smallest possible values of S_o were preferred;
- the maximum lateral displacement of the vehicle centre of mass ΔY shall not exceed by more than 0.3m the set reference value ($\Delta Y \leq 0.3$m);
- the values of the W_s indicators should be as low as possible.

4. Results of simulation studies

Extensive simulation studies of AFS and integrated system performance were carried out under identical operating and road conditions. The behaviour of the unladen and laden car on wet road surfaces was studied. In addition, three different driving speeds of the car during obstacle avoidance were considered V=60, 70 and 80 km/h.

Figure 4 shows an example of the waveforms of quantities describing the motion of an unloaded and loaded car while avoiding a suddenly appearing obstacle on a wet pavement (μ=0.3) at a speed of V=70km/h, for the amplitude of the reference steering wheel rotation angle δ_{HR} equal to the base value (3). Only AFS was used to steer the cars. In most of the graphs shown in the figure, there are two waveforms that differ in line colours. The continuous blue line describes the responses of the reference vehicle model to „bang-bang" steering wheel rotation. The black continuous line describes the response of the virtual vehicle model to steering wheel rotation generated by the steering system controls. From the graphs shown in the left column of Figure 4, it can be seen that the AFS failed to avoid the obstacle. The unladen car lost directional stability while performing this manoeuvre. Although the trajectory of the vehicle's centre of mass crossed the reference trajectory quite quickly at a considerable distance from the obstacle, the car lost directional stability a moment later. The AFS system, in an attempt to establish a trajectory of the vehicle on the left lane (parallel to the road axis), generated excessive values of the steering wheel rotation angle to the left, which led to exhaustion of the adhesion reserve on the wheels of the rear axle of the vehicle. As a consequence, the wheels of the rear axle of the car skidded completely. The vehicle started to rotate to the right and its centre of mass moved to the left. The loss of directional stability of the unladen car controlled by the AFS was found for all analysed driving speeds (V=60÷80km/h).

From the graphs shown in the right column of Figure 4, it can be seen that the loaded vehicle controlled by the AFS avoided the obstacle on the S_{ou} road without any problems, even showing a certain reserve of lateral adhesion (course of lateral acceleration $a_y(t)$). The tests showed that the fully loaded vehicle was a stable system in all considered and it was very difficult to make it lose directional stability even when the amplitude of the reference steering angle was increased by more than 50% compared to the base value.

Unladen car – ASF system Laden car – ASF system

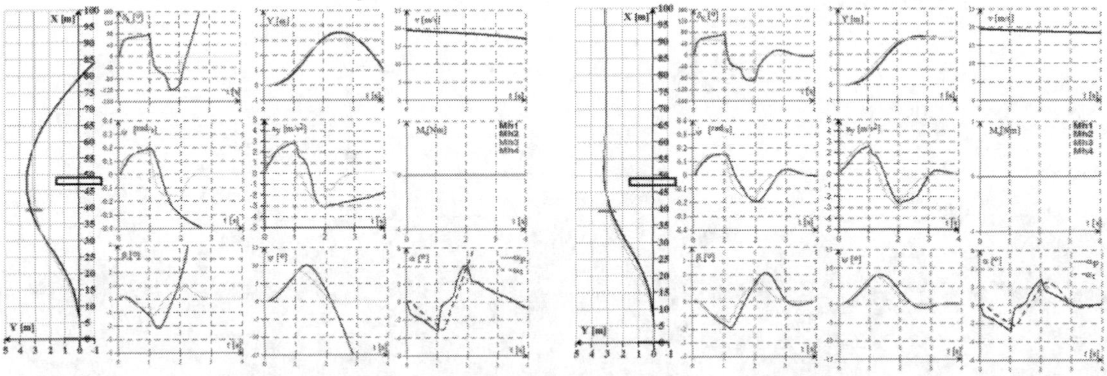

Figure 4. AFS active steering system. Curves of changes of selected values describing vehicle motion during obstacle avoidance (μ=0.3, V=70km/h).

Figure 5 compares the waveforms describing the movement of the unladen car steered by the AFS (left column diagrams) and by the integrated system (right column diagrams).

From the waveforms shown in the right column, it can be seen that the car steered by the integrated control system correctly avoided the obstacle without using all of its reserves of grip. Braking of the vehicle wheels indicated by the control system at appropriate moments limited the maximum values of yaw rate (graphs $\dot{\psi}(t)$), lateral acceleration (graphs ay(t)), yaw angles (graphs ψ(t)) and centre of mass drift angles (graphs β(t)) in relation to the values of these quantities when the vehicle was steered by the ASF system. Therefore, the car did not lose directional stability.

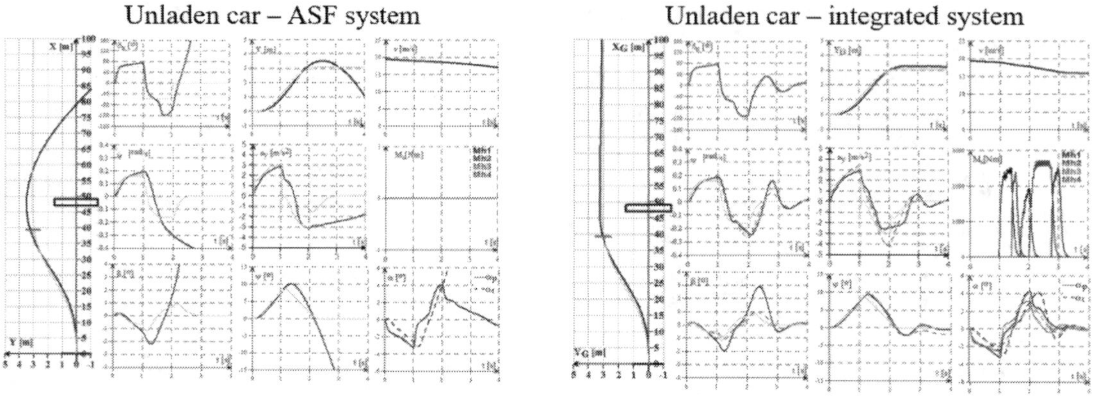

Figure 5. Comparison of the waveforms of selected AFS-controlled and integrated control system-controlled unladen vehicle motion during obstacle avoidance on a wet road (μ=0.3, V=70km/h).

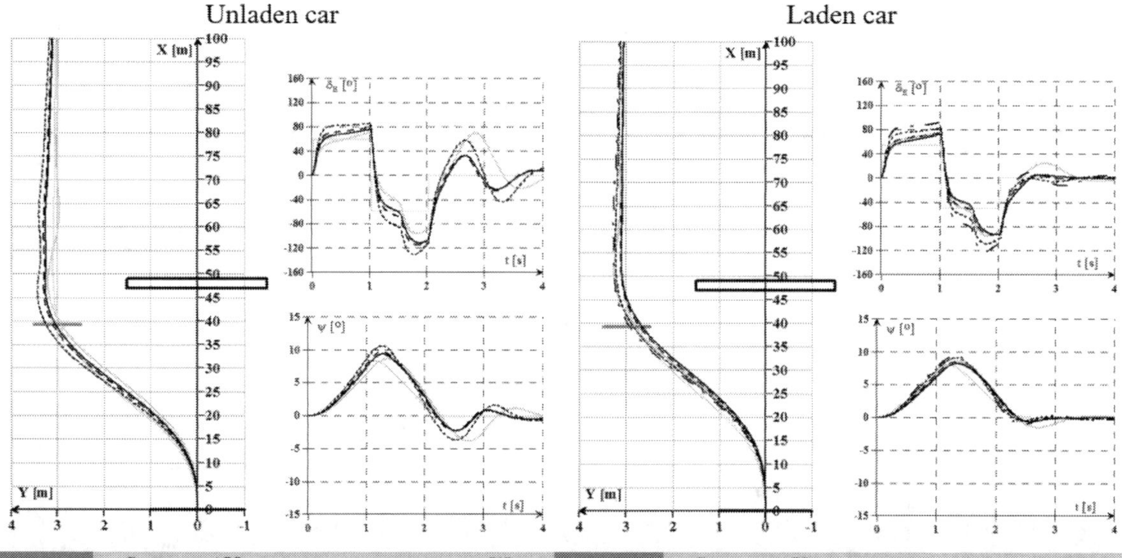

δ_{HR} [%]δ_{HR0}	S_o [m]	ΔY [m]	$W_{\delta H}$ [%]	W_Y [%]	W_ψ [%]	δ_{HR} [%]δ_{HR0}	S_o [m]	ΔY [m]	$W_{\delta H}$ [%]	W_Y [%]	W_ψ [%]
	ASF system						ASF system				
80	42.9	0.20	35.7	0.73	17.5	110	44.3	0.25	10.3	0.40	6.3
	Integrated control system						Integrated control system				
100	41.1	0.26	16.2	0.45	7.5	100	45.9	0.10	12.5	0.74	5.4
110	40.5	0.30	13.1	0.44	7.0	110	45.0	0.11	13.8	0.52	5.3
130	39.1	0.43	16.6	0.73	12.1	130	42.5	0.22	14.9	0.31	3.6
						150	41.1	0.27	19.7	0.30	4.3

Figure 6. Comparison of changes in selected quantities and values of indicators used for their evaluation for a car steered by the AFS and by the integrated steering system during avoiding a suddenly appearing obstacle on a wet road surface for different values of steering wheel rotational amplitude δ_{HR} (μ=0.3, V=70km/h).

Figure 6 shows selected waveforms describing the motion of the car steered by the integrated control system obtained for different values of the reference amplitude of the steering wheel rotation angle ($\delta_{HR}=\delta_{HR0}$ – solid black line, δ_{HR}=110%δ_{HR0} – dashed black line, δ_{HR}=130%δ_{HR0} – dotted black line, δ_{HR}=150%δ_{HR0} – puncted black line). The waveforms are presented against the reference waveforms (continuous blue line) and the waveforms obtained during vehicle control by the AFS system (δ_{HR}=80%δ_{HR0} – continuous green line). From the graphs presented in Figure 6 and the table in the figure below, it can be seen that the integrated vehicle control system controlled the vehicle much better and safer than the unladen vehicle during its avoidance of a suddenly appearing obstacle. Increasing the δ_{HR} even by 30% compared to the baseline value did not cause the vehicle to lose directional stability while reducing the distance needed to avoid the obstacle. For the laden car, increasing the δ_{HR} value resulted in a reduction of the distance needed to avoid the obstacle, but, at the same time, it increased the values of the other indices used to assess the quality of the considered runs.

5. Conclusions

The paper presents an assistant integrated control system for automatic lane changing in boundary conditions. The construction and functioning of the system is described.

The system proved to be particularly effective when the unladen vehicle was avoiding a suddenly appearing obstacle. It guaranteed maintenance of the vehicle's directional stability in all considered road and operating conditions. In addition. it shortened the distance needed to bypass an obstacle by 4-7% in comparison to the AFS system. at the same time lowering the values of most W_s indices describing the accuracy of reference course reproduction.

For a laden car. the integrated system was not as useful. as it did not have to maintain directional stability and. in addition. slightly increased the distance required to avoid an obstacle. The 7% reduction in obstacle avoidance distance was only achieved when the amplitude of the reference steering angle was increased by 50% compared to the base value. However. this resulted in an increase in all of the W_s indices describing the accuracy of reference course reproduction.

Works improving the functioning of the integrated control system will continue.

References

[1] http://www.obserwatoriumbrd.pl/pl/ (access: 15.02.2022r)

[2] https://ec.europa.eu/transport/road_safety/ (access: data 15.02.2022r.)

[3] Jurecki R; Jaśkiewicz M 2013 Analysis of Road Accidents in Poland Over the Last Ten Years; Scientific Journals Maritime University of Szczecin: Szczecin. Poland. 2012, **32**

[4] Stańczyk T L 2013 Działania kierowcy w sytuacjach krytycznych. Badania eksperymentalne i modelowe. Wyd. Politechniki Świętokrzyskiej. Kielce

[5] Anderson S J Peters S C Pilutti T E and Iagnemma K 2010 An optimal-control-based framework for trajectory planning. threat assessment. and semi-autonomous control of passenger vehicles in hazard avoidance scenarios". Int. J. Vehicle Autonom. Systems. **8**

[6] Markkula G Benderius O Wolff K et al. 2013 Effects of experience and electronic stability control on low friction collision avoidance in a truck driving simulator. Accident Analysis and Prevention. **50**

[7] Hayashi R et.al. 2012 Autonomous collision avoidance system by combined control of steering and braking using geometrically optimized vehicular trajectory. Vehicle System Dynamics. **50**

[8] Jiménez F 2015 Autonomous collision avoidance system based on accurate knowledge of the vehicle surroundings. IET Intelligent Transport Systems. **9(1)**

[9] Gidlewski M. Żardecki D 2015 Automatic Control of Steering System During Lane Change. ESV Paper Number 15-0106. Proceedings of 24th ESV' in Geteborg Szwecja. available on the Internet www.nhtsa.gov/ESV.

[10] Gidlewski M. Żardecki D 2017 Linearization of the lateral dynamics reference model for the motion control of vehicles. Mechanics Research Communications. **82**

[11] Gidlewski M. Jankowski K. Muszyński A. Żardecki D 2017 Vehicle lane change automation with active steering – theoretical studies and numerical investigations. SAE Paper 2017-01-1555.

IOP Conf. Series: Materials Science and Engineering 1247 (2022) 012030 doi:10.1088/1757-899X/1247/1/012030

[12] Gidlewski M. Jemioł L. Żardecki D 2019 Sensitivity investigations of the automated lane-change manoeuvre – selected issues. Journal-article. Proceedings of the Institution of Mechanical Engineers. Part I: Journal of Systems and Control Engineering. **233 (4)**

[13] Gidlewski M. Jemioł L. Żardecki D 2016 Simulation Investigation of the Dynamics of the Process of Sudden Obstacle Avoiding by a Motor Vehicle. The Archives of Automotive Engineering. **73(3)**

[14] Gidlewski M. Jemioł L 2016 Badanie własności jezdnych nieobciążonego samochodu ciężarowego średniej ładowności. Autobusy **12**

[15] He J 2005 Integrated vehicle dynamics control using active steering driveline and braking [Ph.D. thesis]. University of Leeds.

[16] Doumiati M. Sename O. Dugard L. Martinez-Molina J-J. Gaspar vSzabo Z 2013 Integrated vehicle dynamics control via coordination of active front steering and rear braking." European Journal of Control. **19(2)**

[17] Mitschke M 1990 Dynamik der Kraftfahrzeuge. Band C: Fahrverhalten. Springer-Verlag Berlin Heideberg New York London Paris. Tokyo Hong Kong.

[18] Andrzejewski R. Awrejcewicz J 2003Nonlinear Dynamics of Wheeled Vehicle. Springer New York.

[19] Athans M.. Falb P L 1966 Optimal Control. An Introduction to the Theory and Its Applications. McGraw-Hill.

[20] Gidlewski M. Jemioł L. Żardecki D 2018 Selected problems of automatic obstacle avoiding. Conference: International Automotive Conference Konmot IOP Conf. Series: Materials Science and Engineering 421 No. 032009.

[21] Gidlewski M. Jemioł L. Żardecki D 2018 Concept of the Automatization of Dangerous Driving Manoeuvres. Journal of Kones Powertrain and Transport. **25(1)**

[22] Żardecki D 1980 The λ-sensitivity analysis and its application in simulation studies of dynamical systems. Archiwum Automatyki i Telemechaniki. **15(3)**

Military special vehicle multifunctional seat as an element of vehicle passive safety

T Wróbel[1*], M Szudrowicz[1], W Luty[1], J Seńko[2]

[1] Sieć Badawcza Łukasiewicz - Przemysłowy Instytut Motoryzacji, ul. Jagiellońska 55, 03-301 Warszawa, Polska
[2] Politechnika Warszawska, Wydział Samochodów i Maszyn Roboczych, ul. Ludwika Narbutta 84, 02-524 Warszawa, Polska

* tomasz.wrobel@pimot.lukasiewicz.gov.pl

Abstract. Road safety depends on the interaction in the human - vehicle - road system. The proper functioning or condition of each of these three elements is a prerequisite for ensuring the maximum possible level of safety. The most important requirement that is placed on special vehicles is to ensure crew protection during the implementation of operational tasks. Unfortunately, often when putting operational tasks overriding during designing vehicles, aspects related to driving comfort are ignored. The article presents a proposal for the description of light construction of seats for the crews of special vehicles, which, apart from the functions related to the ergonomic use of seats by the crew, in terms of safety ensure a high probability of protection of health and life of the crews during a collision or near detonation of an explosive charge. Proposed modular structure of the seats, in addition to the repeatability of elements, which reduces production costs and easy configuration of seats with various functions, will also contribute to a faster process of restoring their functionality in the event of damage. Modules should be selected and developed in such a way that they can be used in various configurations of seats, the functions of which can be determined due to: location in the vehicle, i.e. driver's, troops, system operators' seats, installation method, i.e. to the wall, floor, ceiling, direction assembly, i.e. side mount, rear mount, bottom mount.
Keywords: Explosion-proof seat, safety in transport, equipment of means of transport, explosion energy absorption, modularity, vehicle

1. Introduction

Human safety in transport is one of the elements of safety that is increasingly present in his daily activities. It results, inter alia, from the development of road communication networks, urban development of cities and the logistics supply chain. Generally, it can be argued that this safety depends on mutual interactions in the human (driver, passenger) - vehicle - road conditions and the correct condition of each of these elements. In general, it can also be argued that with respect to the component vehicles, safety is ensured by the proper operation of their functional systems [1].

The safety of vehicle use has not been directly defined, however, active and passive safety of vehicles are defined. Active safety is a set of factors that reduce the likelihood of a collision or a road accident and are provided by those elements that enable the driver to reduce the risk of an occurrence or avoid a hazard [2].

These are devices that allow the driver to take action before the occurrence of a road accident. These include factors such as ensuring good visibility from the vehicle (lighting, windows, mirrors, wipers), ergonomics (thanks to which the driver does not have to take his eyes off the road and his hands off the steering wheel), stability of the vehicle, installation of the brake system with supporting systems, braking force correctors as well as vehicle traction control systems, steering system, suspension and tires ensuring steerability, proper interaction between the vehicle wheels and the road and adhesion to the road. Safety factors also include a wide range of engine power, which enables the maneuver to be performed, e.g. to accelerate in the event of an emergency. Active safety is also influenced by the driver - his skills, mental and physical condition.

Passive safety is the vehicle's features aimed at reducing the effects of a collision or road accident from the point of view of all its participants, i.e. events occurring at a time when the driver cannot influence the development of events and the nature of the vehicle's movement. The elements influencing passive safety in the structure are side reinforcements, crush zones, a safety cage, the appropriate shape of the body, a broken steering column, seats with a special design, headrests, seat belts, seat belt tensioners, airbags, elements made of non-flammable materials [2].

Compared to vehicles of the "civilian" sphere, there are several fundamental differences in the construction of military vehicles and in the conditions of their use that have an impact on human safety. This applies especially to military special vehicles, which are a special subset of the group of special vehicles. Some elements specific to active and passive safety may not be present, and at the same time, vehicle safety should be viewed much more wider [3].

A special vehicle is defined as a vehicle or trailer intended for special function which causes the body adapting or having special equipment [1]. A vehicle used for special purposes is a vehicle specially adapted for the transport of people or goods used by the Armed Forces. Therefore, a special military vehicle is a vehicle designed to perform special functions and having features that distinguish them from the group of special vehicles or those used for special purposes. Features that distinguish them from the group of special vehicles are: high center of mass, curb weight close to the permissible total weight, high moments of inertia of the body structure, tires with off-road tread, often lack of typical passive and active safety support systems, additional equipment mounted outside body, smaller window area [2,3].

From the point of view of functionality, military special vehicles also have special features indicating their separateness in the group of special vehicles, such as universality - the possibility of using the platform in both local and global conflicts, integrity - the need for connection with other types of equipment, primarily in terms of logistics or communication, modularity - assumes the existence of a base platform and the possibility of changing the purpose of the vehicle, transportability - the possibility of transport by rail, sea and air, mobility - the ability of the vehicle to negotiate the terrain, survivability - the ability to survive, continue tasks despite damage to the vehicle, lethality - ability to destroy the enemy (weapons, firepower) [2].

2. Seats as an element of passive safety

When talking about the safety of military special vehicles, this article will discuss the safety element, which is a seat, the functions of which, and thus also the structure, result from the conditions of "normal" vehicle operation, which include, among others, ergonomics, and on the other hand, the need to counteract threats from the enemy, i.e. such forms of his activity that are aimed at reducing the efficiency of the crew or damaging the vehicle.

The most important threats in this area include [2]:

- low level of anti-mine ballistic protection,
- injuries caused by dynamic loads while driving,
- injuries resulting from the organization of the interior (e.g. sharp edges, no escape routes),
- loss of health as a result of long-term work in a harmful environment (noise, vibrations),
- fatigue, poor mental and physical condition of the crew,
- loss of lateral stability (rollover).

It should be emphasized that seats, regardless of the type of vehicle, whether they are used in the civil or military sphere, are classified as passive safety elements. The appropriate shape and hardness of the seat and backrest with a headrest play a very important role in ensuring safety while driving, and above all in the event of an accident. The seat safety system is complemented by seat belts with tensioners and force limiters, which ensure the correct position in times of high impulse overloads.

In the event of a road collision, approximately 150 milliseconds from the moment of the collision, the acceleration value should automatically decrease. It takes approximately 120-150 milliseconds to stop a vehicle traveling at 50 km / h in a frontal collision. In such a short time, the driver has no chance to

IOP Conf. Series: Materials Science and Engineering 1247 (2022) 012031 doi:10.1088/1757-899X/1247/1/012031

react. Thanks to the seat belt and its belt tensioner, the body does not move forward, and the torso is close to the backrest [1].

In addition, the above-mentioned assumptions overlap with the requirements of the ergonomics of the seat. The tasks of ergonomics are adapting technical objects to the dimensions and shapes of a person, securing the safety and comfort of using a technical object, eliminating the negative impact of the product on the conditions of the human environment, ensuring the functionality of a technical object (e.g. efficiency, reliability, susceptibility to adjustment and repair.

The seat should be able to be adapted to the individual dynamic and anthropometric characteristics of the crew. The term "anthropometric features" includes height, weight and measurable features of the anatomy of the human body, which distinguish in particular the spine, etc. Correct sitting position minimizes the strain on the muscles of the spine, abdomen, pelvis and thighs. The pressure on the backrest is much lower than on the seat surface and from an orthopedic point of view, it is very important to properly support the spine, especially the lumbar vertebrae. The backrest of the seat should be properly adjusted to the natural curvature of the spine.

To prevent lateral displacement of the body when driving on uneven surfaces, it is important to properly adjust the shape of the entire seat. However, in the lower part of the backrest, around the center of gravity of the body, as well as in the rear part of the seat, there should be sufficient clearance. The upper part of the backrest should be constructed in such a way as to ensure freedom of arm movements while e.g. driving a vehicle [4].

Another issue to consider is the universality of the workspace (anthropometry). The percentile of the p order (Cp) is the value for which p% of the population has the feature value lower, and the remainder of the population has the feature value greater than Cp. The ergonomics of the seat should be kept from 5 to 95 Cp. In general, it can be stated that the spatial structure of the seat should ensure safe and comfortable work for 90% of the user population, be adapted to their extreme dimensional characteristics, allow for the adjustment of some spatial parameters of the seat to the individual needs of users, introducing the possibility of adjustment, prevent the emergence of accident and health hazards.

The general requirements of the seat structure are specified, which should ensure the stabilization of the user's torso, limbs and head, the stability and durability of the seat itself, the possibility of its adjustment and easy operation.

Comfort angles have been developed (in the subject standards), i.e. optimal angles determining the sitting position ensuring maximum comfort and minimum fatigue. These angles characterize the bends and tilts of individual body segments, which ensure freedom and relaxation of the muscles and require a minimum of physical effort - Figure 1.

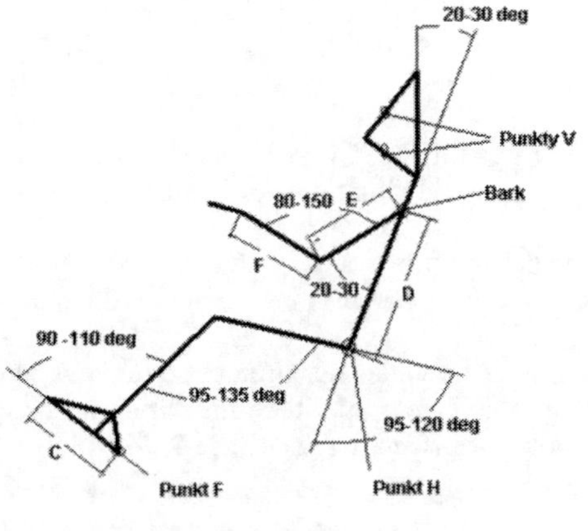

Figure 1. The angles required in a sitting position for optimal comfort

This can be achieved by using profiling and tilting the seat plate (for the working condition, the tilt angle should be between 3 ÷ 15 degrees), shaping the side parts and backrests (support radius > 101,5 cm), supports for the feet, elbows, head (depending on the function).

The standards also provide the so-called curves of the sensitivity of the human body to vertical accelerations, which is related to the pressure distribution of a sitting person. These curves limit the exposure time that is comfortable. They are also a guide to choosing the shape of the seat, proper covering of the seat surface due to their hardness (deflection during dynamic movements), heat dissipation rate, friction coefficient, electrifying properties. A soft seat strains the spine, which causes pain. Lack of support for the thighs also causes fatigue.

Correct sitting position in the car affects the functioning of the musculoskeletal and circulatory systems, but also significantly increases the level of safety of the driver and people who are currently in the vehicle.

It is generally accepted that, depending on the type of car, various elements are movable and allow them to be adjusted depending on the height and dimensions of the driver. The more such elements, the better. Practically in every car, we can set the height of the seat, the angle of the backrest, and the height of the headrest. It is important to sit on the seat in such a way that the back rests completely against the back of the seat, and the body are inserted as far as possible to the entire depth of the seat, at the same time the edge of the seat should not press against the knee pits, the headrest should support the head, it is important that it is neither too low - on the neck, nor too high - above the head.

The acceptable criteria for periodic vibrations for humans are the time of operation, frequency of vibrations and the direction of vibrations - vertical ones cause the greatest injuries in the skeletal system - Figure 2.

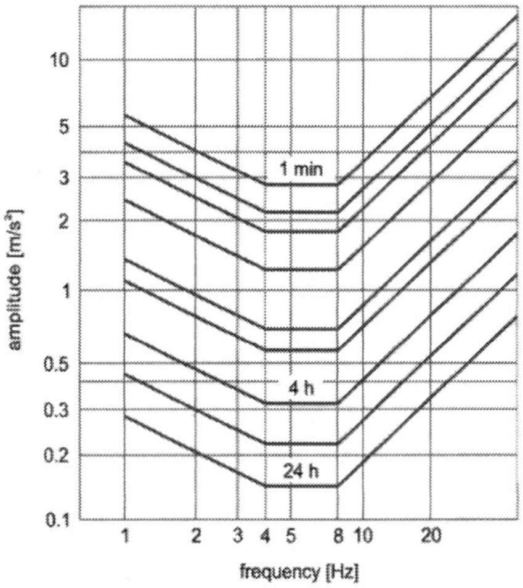

Figure 2. The acceptable criteria for periodic vibrations for humans [4]

However, in the case of special military vehicles, the seat's functions, apart from those listed above, which are typical for driving on the road, are much wider and result from the probability of negative enemy influences.

This challenge comes down to the simultaneous protection of crew members, in addition to ergonomic factors, against force impulses of different amplitude and duration, acting on the vehicle in the horizontal and vertical directions, resulting from a collision and the action of an explosive charge under and on the side of the vehicle (table 1).

Table 1. Comparison of the value of the maximum acceleration and the duration of a frontal car collision at a speed of 30 mph and an explosion under the vehicle [6].

	Acceleration (g)	Impulse duration (ms)
Frontal car collision	25 g to 50 g	70 ms to 120 ms
Explosion under the vehicle	100 g to 400 g	3 ms to 30 ms

In modeling, due to the multifunctionality of the seat for the driver, commander and the troops, in connection with the above-described requirements, conditions and functions, the assumption was made of the modular structure of the seat.

The seat modules are its components, where apart from its main elements (seat, backrest, headrest, footrest), there are also fixing modules, or modules of movable joints, adjustment and shock absorption elements. The modules can be used in other construction configurations of seats, the functions of which can be determined due to: location in the vehicle, i.e. the seats of the driver, troops, system operators, mounting method, i.e. to the wall, floor, ceiling, mounting direction, i.e. side mounting, mounting from the back, from the bottom.

The particular location of the seats imposes additional, individual functionalities on them, e.g. a reclining backrest (in the driver's seat due to the possibility of evacuation), seat rotation (for system operators), folding seat and backrest (for a troops), a platform for standing in the hatch, integrated fragments protection (in the backrest, seat and additional side covers).

Modularity, in addition to the repeatability of elements, which reduces production costs and facilitates the configuration of seats with different functions, also contributes, in the event of damage to the seats, to a faster process of restoring its functionality during operation (module replacement).

The challenge from the point of view of technology is the selection of materials for the production of modules, where the common feature is the strength of the materials with their low weight, selection of the strength, dimensions and mass characteristics of the main components of the seats, and the characteristics of the damping elements and their materials, i.e. such shaping of the characteristics of the damping element with influence the overall movement of the seat to reduce the forces affecting the human body, including those caused by an explosion, to a tolerable one. In addition, on the one hand, it should be avoided that the deformed floor of the vehicle will hit the legs of the crew member and, on the other hand, that it will hit the roof with its head, while reducing the acceleration of the seat. The seat materials should meet other requirements specific to the materials used in means of transport (non-flammability, corrosion resistance).

3. Modeling of seat properties in the case of a mine blast under the vehicle

It was assumed that the resistance to mechanical vibrations and impacts of the entire structure of the seat is to meet the assumptions of the document Defense Standard NO-06-A103 "Weapons and military equipment", the requirements of UNECE Regulation No. 14, in relation to seat belt anchorages, UNECE Regulation No. 17, checking the strength of the backrest and headrest.

Particular attention was paid to his energy consumption, i.e. the ability to dissipate the energy of an impact, the requirements of the NATO STANAG 4569 standardization agreement "Levels of protection of persons in logistic and light armored vehicles" and the accompanying AEP-55 Volume 2 procedure. The seat model was prepared for numerical calculations - Figure 3.

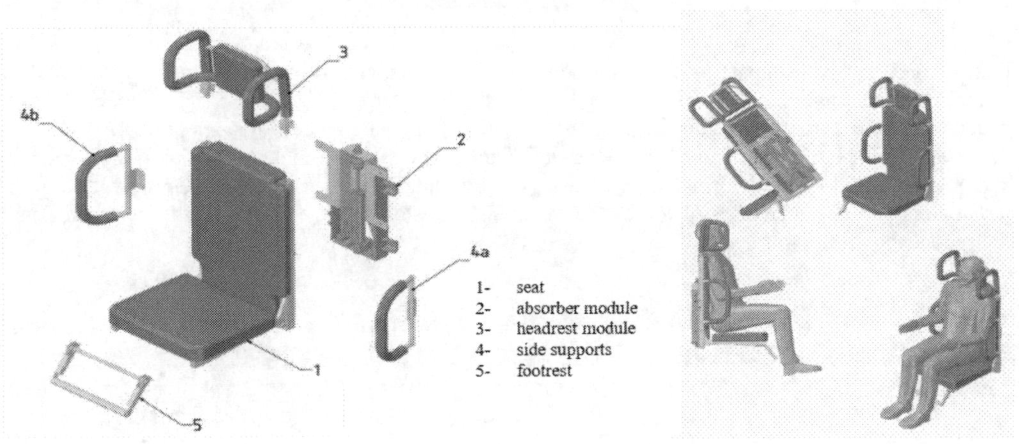

Figure 3. Model of the modular seat

IOP Conf. Series: Materials Science and Engineering 1247 (2022) 012031 doi:10.1088/1757-899X/1247/1/012031

Vertical impacts on the vehicle seat from explosions are short-term forces. The models described in [7] are used to determine the Dynamic Response Index (DRIz) introduced by Stech and Payne [8]. In this model, the assessment of the organism response of a vehicle crew member is determined using the simple single-mass model shown in Figure 4.

The upper body is lumped together and is represented by the analogy of a single mass m resting on the top of the lumbar vertebrae. The stiffness of the vertebral column is represented by the massless spring with spring constant k. The damping behavior of the body is captured by the damping coefficient c.

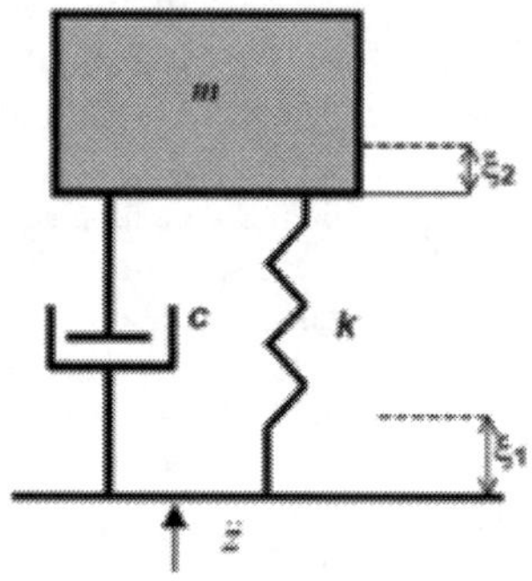

Figure 4. Model diagram for determining the DRIz coefficient [9]

The equation of motion for this model is:

$$\ddot{z}(t) = \ddot{\delta} + 2 \cdot \zeta \cdot \omega_n \cdot \dot{\delta} + \omega_n^2 \delta \qquad (1)$$

where
$\ddot{z}(t)$ – is the acceleration in the vertical direction measured at the position of initiation;
δ – is the relative displacement of the system with, $\delta = \xi_1 - \xi_2$, and $\delta > 0$ => compression;
ζ – is the damping coefficient with, $\zeta = \frac{c}{2 \cdot m \cdot \omega_n}$
ω_n – is the natural frequency with, $\omega_n = \sqrt{\frac{k}{m}}$
The DRIz is calculated by the maximum relative displacement δ_{max}, ω_n and the gravity acceleration g:

$$DRIz = \frac{\omega_n \cdot \delta_{max}}{g} \qquad (2)$$

Designing a seat with impact protection systems is usually supported by simulations in the LSDYNA program. Simulation models of explosions developed in programs such as MES Explicite, enabling the observation of deformation of the seat structure. There are special mechanisms and protective structures in the explosion-proof seat structures. Designing mechanical systems requires mapping their geometry and parametrization of deformable structures during the simulation process. Optimization of the parameters and characteristics of the elements in an explosion-proof chair should be optimized using the model shown in Figure 5.

IOP Conf. Series: Materials Science and Engineering 1247 (2022) 012031 doi:10.1088/1757-899X/1247/1/012031

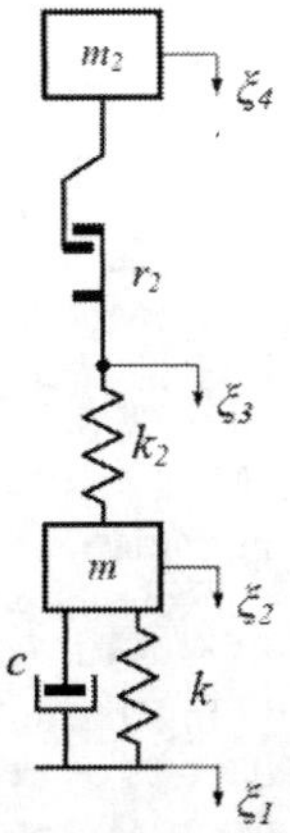

Figure 5. Model of the seat including seat belts $\xi_{1,2,3,4}$ - displacements of the system's elements, m - seat mass, m_2 - body mass, c - damping coefficient (seat), k - spring constant (seat), r_2 - movement limiter (belts)

Separating the mass of the passenger from the mass of the seat enables a simulation examination of the influence of the stiffness of the seat and seat belts on the DRIz coefficient. Using Multibody System (MBS) software, it is possible to select the design parameters of the springs and dampers used in the seat in order to minimize the acceleration values acting on the passengers. The dynamic model of the seat with seat belts and the mass representing the passenger was developed in the MSC.ADAMS program - Figure 6.

Figure 6. Parametric simulation model of the seat in MSC.ADAMS

The model developed in the MSC.ADAMS program has play and vulnerability in the passenger seat system. This model is used to optimize the parameters of the dampers of the explosion protection mechanism. A particularly important non-linear element controlled and verified during the simulation is the free movement of the passenger in relation to the seat. Passenger displacements significantly change DRIz values in relation to the model favored by NATO [7].

4. Summary

None of the manufacturers currently offering mine-blast proofing seats offers such a solution that allows the optimal adjustment of the seat to the requirements of the end user, who must compromise between the level of protection and ergonomics.
Due to the current market demand and the modernization plan of the armed forces, there is a need to develop a multifunctional seats to protect the crew members of a special military vehicles against the effects of short-term dynamic loads of significant values. A product should be created that currently does not have a serial counterpart on the domestic market.

The development of the seat structure, in particular the selection of the parameters of the damping elements, was carried out with the use of analytical models with one degree of freedom (Figure 4). The implementation of subsequent iterations of the seat project required the detailed specification of parameters related to energy dissipation during the explosion. The authors proposed the use of simulation models with a greater number of degrees of freedom (Figure 5), and developed a simulation model in the Multibody System ADAMS software (Figure 6).

In the coming years, in the construction of military combat vehicles, activities aimed at ensuring the safety of the crew will be further developed, e.g. by modernizing or adding new systems.

5. References

[1] Jasztal M, Nadra A and Milczarek M 2014 Modelowanie i symulacja funkcjonowania wybranych urządzeń bezpieczeństwa biernego pojazdów samochodowych *MECHANIK* **7/2014** (XVIII Międzynarodowa Szkoła Komputerowego Wspomagania Projektowania, Wytwarzania i Eksploatacji)
[2] Simiński P 2011 Bezpieczeństwo pojazdów wojskowych *Zeszyty Naukowe WSOWL* **1** (159) 2011 pp 224-238
[3] Simiński P 2012 Bezpieczeństwo pojazdów w badaniach WITPiS *Nowa Technika Wojskowa* **9/2012**
[4] International Organization for Standardization: ISO 2631-1: Mechanical vibration and shock - Evaluation of human exposure to whole-body vibration - Part 1: General requirements, 1997
[5] Zuska A, Jurecki R and Jaśkiewicz M 2016 Laboratory study about pressure distribution on car seats with people of different anthropometric features participation *Archiwum Motoryzacji* Vol 74 **nr 4** pp 115-127
[6] Nordstrom P 2019 Simulations of Safety Aspects for Occupants in Mine Blast Events *Master's Thesis in Engineering Physics* (Department of Physics Umea University)
[7] Research and Technology Organisation of NATO 2007 *Technical Report TR-HFM-090 Test Methodology for Protection of Vehicle Occupants against Anti-Vehicular Landmine Effects*
[8] Payne P R, Stech E L 1969 Dynamic Models of the Human Body *AMRL-TR-66-157* (Wright-Patterson Air Force Base)
[9] NATO Standardization Agency, 2014, Procedures for Evaluating the Protection Level of Armoured Vehicles: Volume 2 for Mine Threat, AEP-55 Volume 2

Hazards arising during road accidents in cases of vehicle adaptation for the person with special needs

L Prochowski[1,2], M Ziubiński[1*] and T Mantur[1]

[1] Military University of Technology (WAT), Faculty of Mechanical Engineering, Institute of Vehicles and Transportation, ul. gen. Sylwestra Kaliskiego 2, 00-908 Warsaw, Poland
[2] Łukasiewicz Research Network – Automotive Industry Institute (Łukasiewicz-PIMOT), ul. Jagiellońska 55, 03-301 Warsaw, Poland

* mateusz.ziubinski@wat.edu.pl

Abstract. Independent transport of people with special needs often requires the adaptation of a motorcar to increase the comfort of use and enable proper operating of a vehicle. This adaptation involves, for example, mounting special devices on the steering wheel. This violates the so-called survival zone, defined by (UN/ECE) regulation No. 21. and can deteriorate the occupant's safety. This study shows the risk for the driver caused by an additional knob mounted on a steering wheel. Model tests were prepared and carried out with the use of the PC-Crash program environment. The simulations of the frontal collisions into the obstacle were computed and the dynamic loads affecting the driver with and without additional adaptation elements on the steering wheel were assessed. Obtained results portray that the impact of the driver's head into the adaptation element during the collision is possible. This causes a visible increase in the HIC15 head injury indicator by up to 80% for a collision speed of 65 km/h compared to a car without an adaptation device. It was found that the risk of head injury is on average 15% greater compared to the variant without an adaptation device.

Keywords: automotive safety, passive safety, individual transport of the people with special needs, adaptation of vehicles, frontal impact collision, people with special needs, people with disabilities

1. Introduction

Safe and comfortable operating of a motorcar by a person with special needs often requires special preparation of a vehicle with the use of adaptation devices. Among such devices, we can stand out:

- knobs supporting turning of the steering wheel;
- devices facilitating the braking and accelerating process;
- devices that improve the use of vehicle equipment controls.

Solutions that concentrate on assisting the steering process are the ones that draw the most attention to themselves. There are several diverse types of these devices and their ergonomics and functionality are tailored to the drivers with special needs. Examples of such steering knobs has been shown in Figure 1.

IOP Conf. Series: Materials Science and Engineering 1247 (2022) 012032 doi:10.1088/1757-899X/1247/1/012032

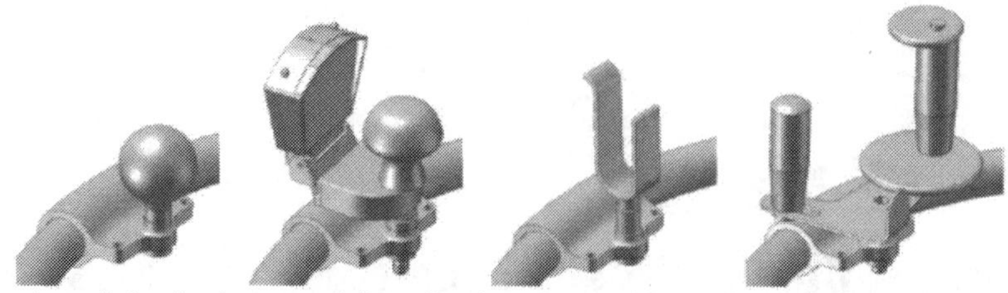

Figure 1. Examples of the devices used for adapting the steering wheels [1]

Installing additional devices with rigid and angular structures can cause interference in the vehicle survival space. This zone is the area where the driver's body moves during the collision [2 - 4].

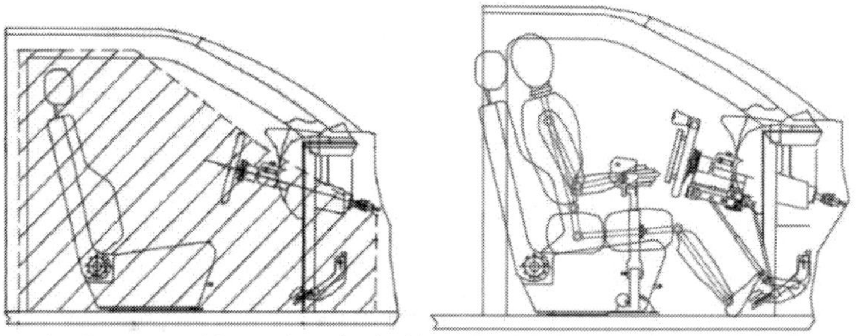

Figure 2. Vehicle survival space without (on the left) and with mounted special adaptation devices (on the right) [2]

Figure 2 presents an example of the interference in the survival space (marked with the hatch pattern on the drawing) arising from the mounting of the special adaptation devices. During the collision, the steering assistance equipment can be in the impact zone of the driver's head, which can greatly increase the risk of injury [5 - 7].

Legal regulations normalizing the shape and equipment of the vehicle cabin can be found in Regulation No. 21 of the Economic Commission for Europe of the United Nations (UN/ECE) — *Uniform provisions concerning the approval of vehicles with regard to their interior fittings*. Point 5.1.1 of the regulation states that the head impact zone must not contain any dangerous roughness or sharp edges likely to increase the risk of severe injury to the occupants [8]. Some of the adapting devices, and in particular additional knobs mounted on the steering wheel, may be a source of hazard to occupants constituting devices defined as dangerous in the previously mentioned regulations.

Research results which can be found in the literature show what risks can be caused by mounting adaptation devices on the steering wheel. The adaptation knobs shown in Figure 1 are the subject of research conducted in [1]. It was stated that the use of the forked knob and double knob is dangerous because they can severely interfere with the survival zone and can disturb the process of airbag deployment. In [9] it was established that the tri-pin adaptation knob interferes with the process of airbag deployment as well and is also causing the risk of a driver hand injury because of it being wedged in the knob. We can find similar conclusions in [10, 11].

The issue analysis confirms that adaptation devices, whose task is to improve the ergonomics of vehicles operated by people with special needs, can significantly affect safety during road collisions. The result of the airbag malfunctioning is reduced protection of the driver, which can lead to head impact into the adaptation device and steering wheel. It is therefore an additional risk during the collision. There are no research results to assess this risk.

This study shows the risk that arises from mounting the adaptation knob on the steering wheel. The risk of injury as a result of dangers arising during the road accident was analyzed. A similar issue was analyzed by the authors in [12], where the risk of a passenger's head impact into the mobile computer during a road accident of a touristic coach was analyzed.

2. Head-on collision model and its validation

The PC-Crash 13.0 program environment was used for the tests [13]. A model of biomechanics of the driver's body during the frontal collision into a rigid obstacle (eg. side of a heavy truck) was prepared. The model has been shown in Figure 3. Two states of model motion are shown – the beginning of the motorcar impact into the obstacle ($t = 0$) and during the collision ($t = t_1$). The movement in the flat consideration in the global plane $O_E X_E Z_E$ related to the ground is analyzed. In Figure 3, the local coordinate system $\{O_V\}$ with its beginning in the motorcar mass center is marked. The model was completed with the multi-body layout of a dummy on a driver's seat and a dashboard with elements of the steering wheel. The model includes the interaction of the dummy with an airbag mounted on the steering wheel. The bodies imitating the properties of the driver's seat, the elements of the dashboard and the steering wheel move with the motorcar (their position in the $\{O_V\}$ coordinate system does not change). Multi-body system of the driver dummy is attached to the vehicle by elastic-damper connections, which reproduce the characteristics of the three-point safety belts (green lines in Figure 3). Additionally, the dummy interacts with the seat and other bodies. In Figure 3 the example of the additional adaptation knob mounted on the steering wheel is shown (shaded with the red). The airbag is marked with gray shading.

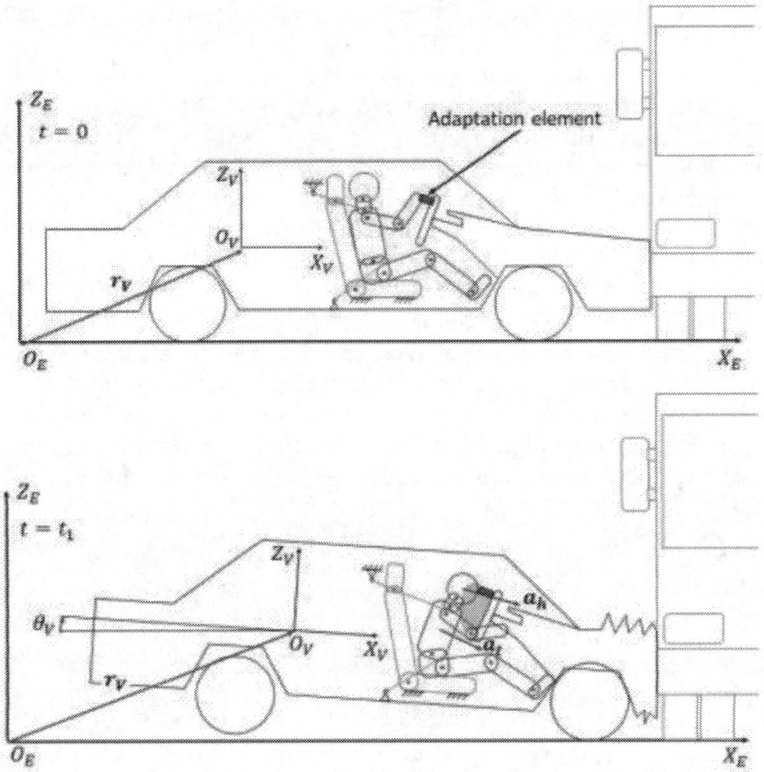

Figure 3. Physical model of a head-on motorcar impact into a rigid obstacle with a multi-body system of the driver dummy on a seat, beginning of the impact ($t = 0$) and during the impact ($t = t_1$)

In Figure 3, the following indications are used:

- $\boldsymbol{r_V}$ – vector from the beginning of the global coordinate system $\{O_E\}$ to the mass center O_V; $\boldsymbol{r_V} = [x_V \quad z_V]^T$;
- θ_V – motorcar body rotation angle;
- $\boldsymbol{a_h}$, $\boldsymbol{a_t}$ – the resultant acceleration vectors of the solids reproducing parameters of a dummy's head and torso.

The motion of a motorcar is described by the following equation [14]:

$$m_V(\ddot{\boldsymbol{r}}_V + \boldsymbol{\Omega}_V \times \dot{\boldsymbol{r}}_V) = \sum_i^n \boldsymbol{F}_i \quad (1)$$

$$T_V\dot{\boldsymbol{\Omega}}_V + \boldsymbol{\Omega}_V \times T_V\boldsymbol{\Omega}_V = \sum_j^k \boldsymbol{M}_j \quad (2)$$

where:
m_V - motorcar mass;
$\boldsymbol{F}_i$, $\boldsymbol{M}_j$ - generalized external forces and moments affecting the vehicle;
T_V - tensor of the vehicle inertia about the center of mass in the local coordinate system $\{O_V\}$;
$\boldsymbol{\Omega}_V$ - vector of the angular velocity of the vehicle body s in the local coordinate system $\{C_s\}$; $\boldsymbol{\Omega}_V = \dot{\theta}_V$.
The following external forces are taken into account:

- force of impact on the area of contact of a motorcar with an obstacle, calculated according to the force-based impact model [15];
- tangential reactions at the contact of road wheels with the surface, calculated in the non-linear TM-Easy model [16-18];
- normal reaction from the road surface.

The excitation in the model is the initial speed of a motorcar $\dot{x}_V(t = 0) = v_0$. The model for calculating the contact forces between solids is based on [19, 20].
During the validation of the model, the results of the NHTSA experimental research were used [21]. Crash test number 5062 was chosen [21], in which vehicle Honda Accord (year of production 2004) head-on crashed into a rigid obstacle at the speed of 56,6 km/h. In the driver's seat, the Hybrid III 50th Male dummy was seated. The vehicle was equipped with an airbag mounted on the steering wheel.
The initial positions of the dummy, dashboard and steering wheel were determined in accordance with the results of the experimental research. The validation was carried out in two stages. First, the process of impact of the motorcar into the obstacle was analyzed and according to this, the stiffness and damping parameters describing the frontal part of a car body were determined. The result of this validation is shown in Figure 4. The horizontal part of the vector $\boldsymbol{r_V}$ and its derivatives (acceleration $\ddot{x}_V$, speed $\dot{x}_V$ and displacement x_V of the motorcar mass center (compare: Figure 3)) was shown. Values from the experimental research are marked with a dashed line and the ones from the model are marked with a solid line.

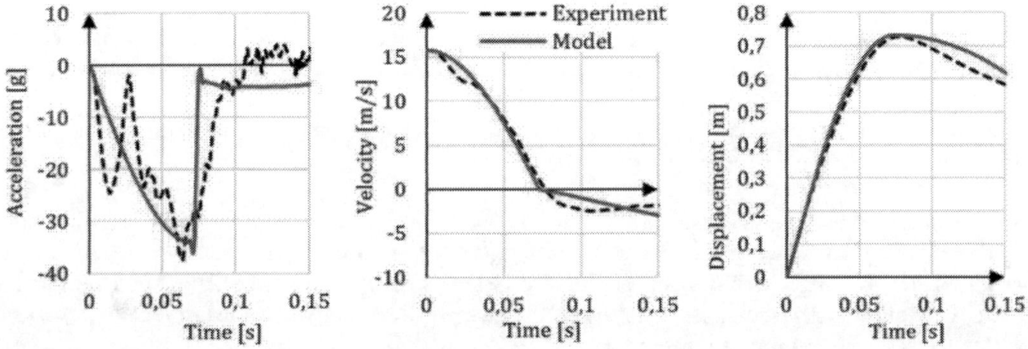

Figure. 4. Acceleration $\ddot{x}_V$, speed $\dot{x}_V$ and displacement x_V of the motorcar mass center during the impact into the rigid obstacle

In the second stage of the validation, the parameters describing the biomechanical human body model and seatbelt parameters were determined. The selection process was based on the use of head and torso acceleration charts. The result of this validation is shown in Figure 5.

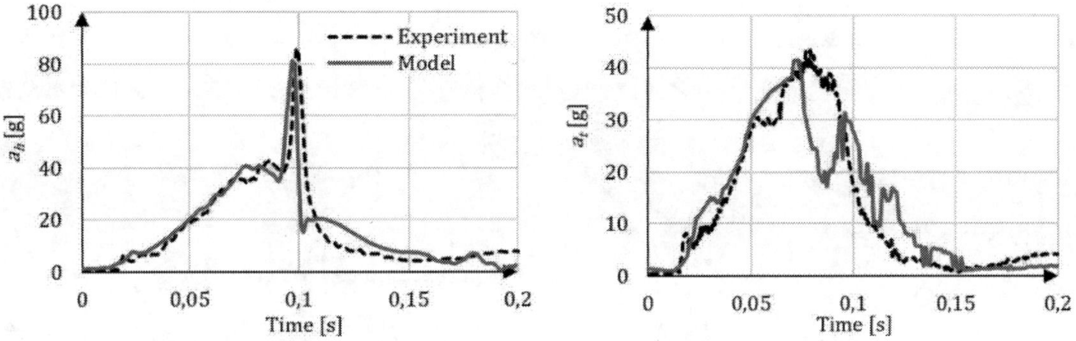

Figure. 5. Resultant head and torso acceleration

The second stage of the validation was extended with the Head Injury Criterion (HIC) parameter calculations:

$$HIC = max\left\{\left[\frac{1}{t_2 - t_1}\int_{t_1}^{t_2} a_h(t)dt\right]^{2,5} \cdot (t_2 - t_1)\right\} \tag{3}$$

where: $t_2 - t_1$ is the period in which highest values of acceleration $a_h(t)$ occur. This indicator is the basis to assess the consequences of the road accidents in the form of head injury risk. Calculations were carried out for periods 0.015 s (HIC15) and 0.036 s (HIC36). The results are shown in Table 1.

Table. 1. HIC indicator calculation results

Indicator	Experimental research	Model calculations
HIC15	312	311
HIC36	457	456

Results of the validation are the basis to conclude about good compatibility of the obtained results of experimental and model research. The waveform of the resultant head acceleration is important (Figure 5). For the period 0,09-0,1 s we can see the peak of this acceleration. It is caused by a head impact into the steering wheel. This process is important for this paper and assessing the influence of risks caused by adaptive devices mounted on the steering wheel.

3. Model research and their results

Model research concentrating on the analysis of the influence of mounting the additional element (compare: Figure 1) on the steering wheel on the process and effects of the frontal collision into the obstacle was prepared. The following variants of mounting of the adaptive element were considered:

- W1 – at the top of the steering wheel;
- W2 - 30 degrees from the vertical axis of the steering wheel;
- W3 – 60 degrees from the vertical axis of the steering wheel.

Variants are shown in Figure 6 in which the red dot represents the location of the adaptation device. Additionally, variant W0 that does not have an additional knob mounted on the steering wheel is considered. For every variant, the simulation of the collision into the obstacle with speed varying from 45 to 75 km/h was carried out.

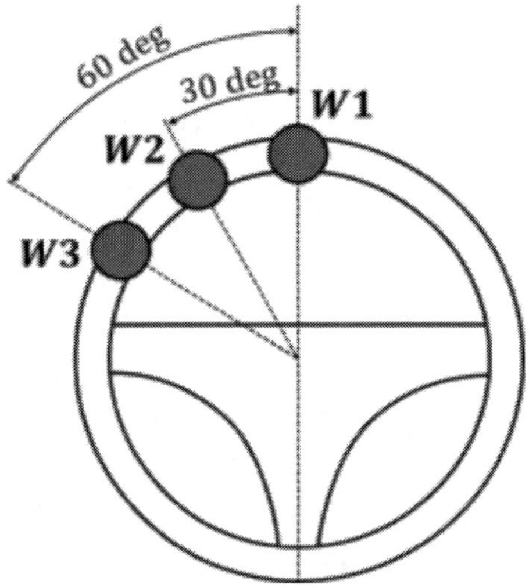

Figure. 6. Variants of adaptive element mounting on the steering wheel

Charts of the resultant dummy's head acceleration and HIC values (calculated according to (3)) are the basis for assessing the risk of injury in the Abbreviated Injury Scale [18] in Figures 7 - 10 and their values are shown in Table 2 and 3. In Figure 7, resultant head acceleration in variants W0 - W3 is shown. These charts portray the influence of mounting of the adaptation knob on the steering wheel on the dummy's head acceleration, especially in variants W1 and W2. Head impact into the adaptive element can be observed around t = 0,08 s, thus by 0,01 s earlier than in the variant W0 (without additional knob, because distance from head to steering wheel has decreased). For variant W1, we can observe the biggest maximum head acceleration value. Based on the charts from Figure 6, the calculations of HIC values were carried out and based on it the injury risk in AIS2 and AIS3 scales was assessed [22].

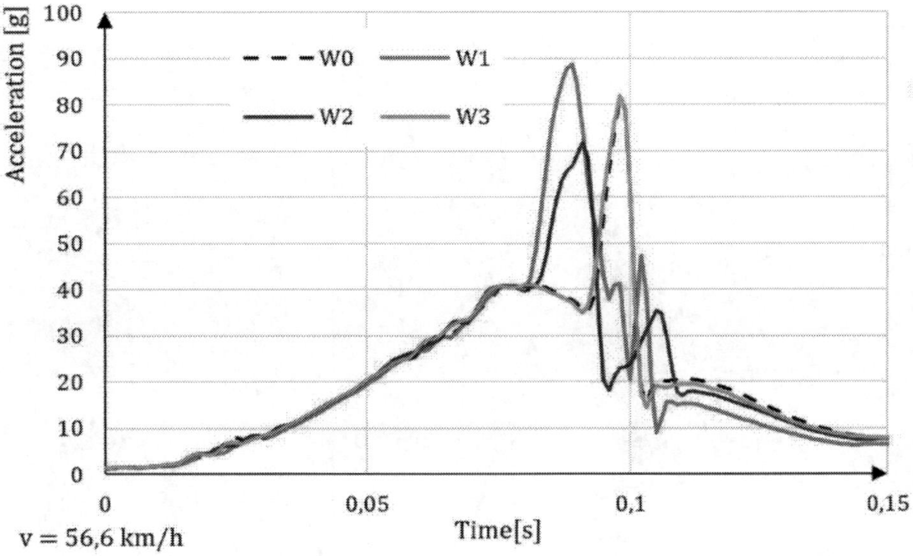

Figure. 7. Head acceleration during the collision for various adaption devices mount variants in time function

Table 2. Results of HIC and injury risk assessment for speed v_0 = 56,6 km/h

Indicator	W0	W1	W2	W3
HIC15	311	603	376	327
HIC36	456	644	447	479
AIS2	0,16	0,28	0,15	0,17

Mounting an additional knob according to variant W1 increases the value of HIC by 41% compared to variant W0. In variants W2 and W3 value of HIC36 are comparable to W0. Values of HIC15 which is estimated for a shorter period (0,015 s) are interesting. The additional device causes a significant increase in acceleration values. In variant W1 we can observe HIC15 almost doubling the value in comparison to its value in W0. Bigger values of HIC36 in variant W1 translate to higher injury risk which in AIS2 scale is 12% higher than in variant W0. In turn the risk in AIS3 scale doubles (from 0,04 to 0,09).

In Figure 8 the waveforms of head acceleration for various collision speeds are compared. As the speed of collision increases, we can observe a logical increase in the dynamics loads affecting the head. There is no head impact into the steering wheel for a collision speed of 45 km/h in variant W0. On the other hand, the contact between head and adaption device can be observed in variant W1. Maximal accelerations achieved in variants W0 and W1 are comparable for various collision speeds. However, for the W1 variant the maximum acceleration time duration is greater because head before impacting the steering wheel impacts the adaption device as well, which is the source of greater injuries. This leads to higher values of HIC15 and HIC36 indicators, which are presented in Table 3.

IOP Conf. Series: Materials Science and Engineering 1247 (2022) 012032 doi:10.1088/1757-899X/1247/1/012032

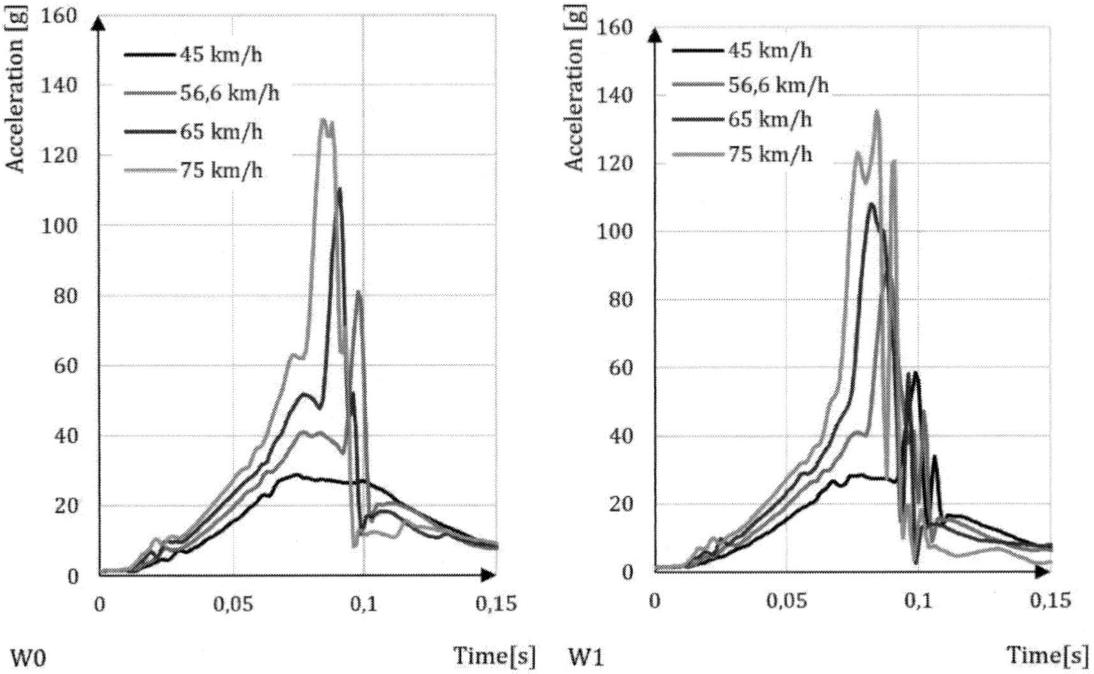

Figure. 8. Head accelerations during the collision for various speeds for W0 (on the left) and W1 variant (on the right)

Table. 3. Values of HIC indicators and injury risk for W0 and W1 variants

Indicator	45 km/h	56,6 km/h	65 km/h	75 km/h
			W0	
HIC15	72	311	735	1772
HIC36	149	456	848	1687
AIS2	0,01	0,16	0,40	0,71
AIS3	0,00	0,04	0,17	0,49
			W1	
HIC15	159	603	1320	2209
HIC36	220	644	1114	1966
AIS2	0,03	0,28	0,52	0,77
AIS3	0,00	0,09	0,28	0,57

With increasing collision speed values, we can observe an increase of the HIC indicator for W0 (for 75 km/h speed HIC15 value is more than twenty-four times greater than for 45 km/h speed) and W1 variants (for 75 km/h speed HIC15 value is more than fourteen times greater than for 45 km/h speed). A comparison of various variants indicators shows the important influence of the mounted adaption device on the passive safety of the vehicle. Values of HIC15 and HIC36 indicators for the W1 variant are higher compared to variant W0. This translates into a greater risk of head injury in the AIS2 and AIS3 scales.

In Figure 9, the waveforms of the HIC15 and HIC36 values are portrayed for various variants W0 – W3 according to motorcar collision speeds. These results confirm the adverse effect of mounting an adaptation device on top of the steering wheel (W1) at a wide range of collision speeds. This confirms the validity of regulations included in Regulation No 21 of the Economic Commission for Europe of the United Nations [4], which characterize the area of occupant head impact during the road collision. Model research which was carried out shows an important increase in head injury risk in AIS2-4 scale which is portrayed in Figure 10. This applies especially to the W1 variant, where the risk is on average 15% greater, regardless of the motorcar speed hitting the obstacle.

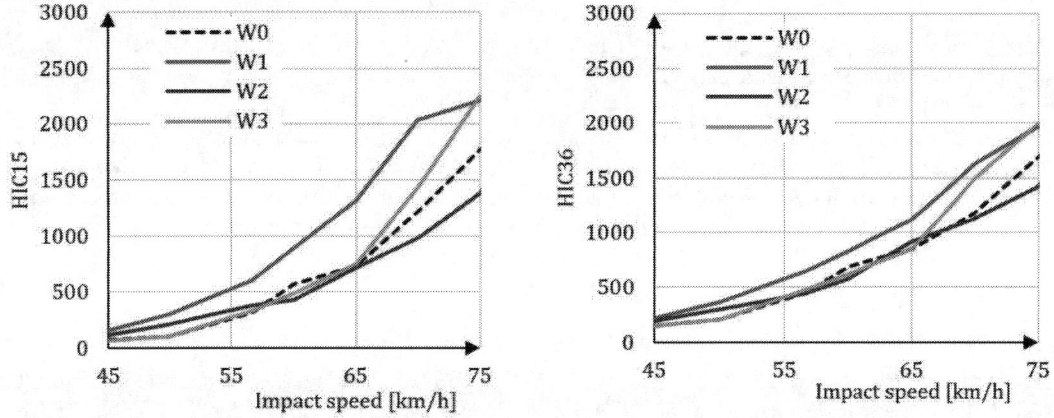

Figure. 9. Values of the HIC15 (on the left) and HIC 36 indicators (on the right) for various adaptation device mounting variants in the function of the collision speed

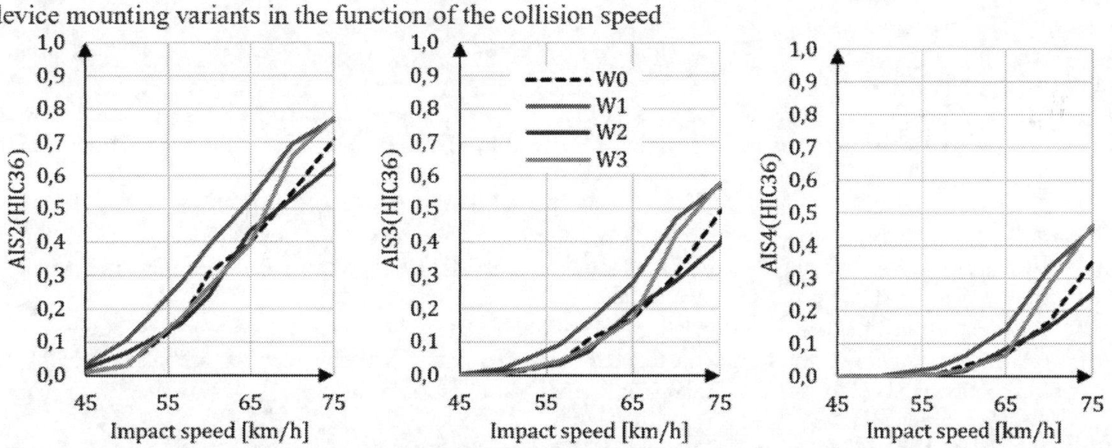

Figure. 10. Values of AIS2 (on the left), AIS3 (in the middle) and AIS 4 (on the right) indicators for various adaptation device mounting variants in the function of the collision speed

4. Summary and conclusions

Meeting the people with special needs transport needs often requires adaptation of a motorcar increasing the comfort of use and enabling proper driving. Usually, it is achieved by mounting additional devices which interfere in the so-called survival zone inside a motorcar. As a part of this paper, the model research was prepared, which enabled the assessment of the influence of installing the exemplary adaptation device (in form of a knob) on the steering wheel on the driver's injury risk during collisions. The model of the driver's body biomechanics during the frontal impact into the rigid obstacle was prepared and the calculations were carried out. The results of this research allowed to draw the following conclusions:

- head impact into the adaptation device during the frontal collision increases values of HIC head injury indicator and increases the risk of injury (Table 2);
- installing the adaptation device on the steering wheel caused the HIC15 indicator to increase by 80% and HIC36 by 31% for 65 km/h collision speed (Figure 9);
- there is a visible influence of the location of the additional device on the steering wheel on the injury risk values at the beginning of the collision (Figure 7, 9);
- the greatest risk arises when the knob is in the top position on the steering wheel (variant W1, compare: Figure 6), regardless of the speed of the motorcar collision.

Research has shown that the installation of an adaptation device that facilitates operating a motorcar by a person with special needs, can be a source of significant risk in a road collision. Therefore, in such situations, it is a worthy idea to combine the installation of adaptation devices in a motorcar with the need to limit its maximum speed or a special airbag on the steering wheel. Separate research in this area may result in the development of design guidelines, the implementation of which could improve the passive safety of motorcar adapted to the needs of people with special needs.

Founding
The article was prepared as part of the project no. POWR.03.05.00-00-CW07/20-00 entitled: "Knowledge Centre on Accessibility to Transport and Mobility of People with Special Needs" co-financed by the European Social Fund under the Operational Program Knowledge Education Development 2014-2020.

Literature:

[1] Masia J, Eixeres B, Dols J F and Esquerdo T V 2009 Finite Element Analysis of Folded Airbag in Frontal Impact of Adapted Vehicles for Disabled Drivers Third Manufacturing Engineering Society International Conference
[2] Dols J, Garcia O, Font J and Sánchez S 1999 Evaluation of Injury Risks in Vehicles Adapted to be Driven by Handicapped People Automobile Laboratory – Mechanical and Materials Engineering Department Polytechnic University of Valencia (Spain) Proceedings of EAEC 99 Barcelona
[3] Posuniak P, Jaśkiewicz M, Kowalski , Dabrowski F 2018 Child restraint systems: Problems related to the safety of children transported in booster seats (without integral safety belts) 11th International Science and Technical Conference Automotive Safety, AUTOMOTIVE SAFETY 2018, pp. 1–7.
[4] Frej D, Jaśkiewicz M 2021Test Stand for Experimental Crash Tests at Low Speeds. Transportation Research Procedia, 2021, 55, pp. 1627–1634
[5] Turner-Stokes L. Etchell L. Glonys P. and Rattenbury S. 1996 Secondary Safety of Car Adaptations for Disabled Motorists Disability and Rehabilitation Vol. 18
[6] Jackowski J, Zmuda M 2016 Using different study objects during researches at low speed Crashes. Transport Means - Proceedings of the International Conference, 2016, 2016-October, pp. 402–407
[7] Jaśkiewicz M, Frej D, Matej J, Chaba R 2021 Analysis of the head of a simulation crash test dummy with speed motion. Energies, 2021, 14(5), 1476.

IOP Conf. Series: Materials Science and Engineering 1247 (2022) 012032 doi:10.1088/1757-899X/1247/1/012032

[8] Regulation No 21 of the Economic Commission for Europe of the United Nations (UN/ECE) — Uniform provisions concerning the approval of vehicles with regard to their interior fittings 2008
[9] Masiá Vañó J. Eixerés Tomás B. and Dols Ruiz JF. 2011 Evaluation of the passive safety in cars adapted with steering control devices for disabled drivers International Journal of Crashworthiness
[10] Eixeres B. Masia J. Dols J.F. and Colomina FJ 2009 Safety Simulation Model for Drivers of Adapted Cars Third Manufacturing Engineering Society International Conference Spain
[11] Eixeres B. Masia J. Dols J.F. and Esquerdo T.V. 2009 Evaluation of Steering Control Devices in Adapted Cars Using Sled Deceleration Tests Third Manufacturing Engineering Society International Conference Spain
[12] Dziewiecki K. Prochowski L. Ziubiński M. and Zielonka K. 2018 Assessment of the danger arising from an impact against a mobile computer present in the space between bus seats during a road accident Archiwum Motoryzacji 79(1)
[13] Steffan H. and Moser A. 1996 The collision and trajectory models of PC-CRASH (No. 960886). SAE Technical Paper
[14] Datentechnik, S. (2008). PC-CRASH-A Simulation Program for Vehicle Accidents. Linz, Austria.
[15] Steffan, H. (2009). Accident reconstruction methods. Vehicle system dynamics, 47(8), 1049-1073.
[16] Rill G., Simulation von Kraftfahrzeugen, Vieweg & Sohn Verlag GmbH, Braunschweig/Wiesbaden 1994.
[17] W. Hirschberg, G. Rill & H. Weinfurter (2007) Tire model TMeasy, Vehicle System Dynamics, 45:sup1, pp. 101-119, DOI: 10.1080/00423110701776284
[18] G Rill, TMeasy–A Handling Tire Model based on a three-dimensional slip approach, Proceedings of the XXIII International Symposium on Vehicle Dynamics…, 2013 - researchgate.net
[19] Steffan H. Moser A. Geigl B. C. and Motomiya, Y. 2000 Validation of the coupled PC-CRASH-MADYMO occupant simulation model (No. 2000-01-0471) SAE Technical Paper
[20] Steffan H. Geigl B. C. and Moser A. 1999 A New Approach to Occupant Simulation Through the Coupling of PC-Crash and MADYMO SAE transactions 785-793.
[21] https://www.nhtsa.gov/research-data/research-testing-databases#/vehicle/5062
[22] MacKenzie, E. J., Shapiro, S., & Eastham, J. N. (1985). The Abbreviated Injury Scale and Injury Severity Score: levels of inter-and intrarater reliability. Medical care, 823-835.

Comparison of freight forwarder liability in selected countries of the European Union and selected countries of the Commonwealth of Independent States (CIS)

M Poliak[1], E Salamakhina[1], N Lakhmetkina[2], N Zhuravleva[3]

[1] Department of Road and Urban Transport, University of Zilina, Univerzitná 1, 010 26 Žilina, Slovakia
[2] Transportation Economics Department, Petersburg State Transport University, Moskovskij prospect, 9, St. Petersburg 190031, Russian Federation
[3] Emperor Alexander I St. Petersburg State Transport University, 9 Moskovsky pr., Saint Petersburg, 190031, Russian Federation

* salamakhinaekaterina@gmail.com

Abstract. The continuous flow of movement of goods on the world market forces manufacturers to turn to intermediaries who are able to organize cargo transportation services. At the moment, the most common intermediary is a freight forwarder. This article defines the understanding of the term freight forwarder, its role in the supply chain, as well as the risks associated with transportation. This article analyses the part of forwarding contract which deals with a limitation of the forwarder's liability. The limitation of the forwarder's liability allows predicting the amount of compensation to the cargo owner in case of damage or loss of goods. In addition to the previously conducted analysis of the freight forwarder's liability in the selected countries of the European Union, an analysis of the responsibility of freight forwarders in the selected CIS countries was carried out. Among other things, it was determined which national regulations regulate freight forwarding activities in these countries. The limitations of the liability of freight forwarders proposed by the conditions of the expedition of the national Associations of logistics and forwarding, compiled on the basis of the standard rules of the FIATA organization, are also considered.

Keywords: road safety, safety, road accident

1. Introduction

Since ancient times, mankind has embarked on the path of international trade. Naturally, with the geographical expansion of the market came the development of transport. In the conditions of modern globalization and large flows of goods, the demand for forwarding companies is huge.

The freight forwarder is an intermediate link between the client and the cargo carrier. The tasks of the forwarder may include such tasks as the organization of loading and unloading of goods, storage, consolidation, packaging, labelling, selection of performers, documentation, carrying out operations during customs control.

In addition to attracting third-party performers, freight forwarders can carry out cargo transportation by their own forces and vehicles. In the process of transportation, road, rail, water, and air transport can be used separately or in any combination thereof [1].

In addition to the service role, it is necessary to study the legal role of the forwarder. The main factor determining the legal role of the forwarder is the entire course of business relations between the forwarder and the shipper, including the contractual terms of the contract. This means that in any specific claim against the freight forwarder after damage or loss of cargo, the court is likely to consider a general or comprehensive agreement(s) between the shipper and the freight forwarder and a specific contract(s) of carriage related to the carriage of the relevant cargo. The freight forwarder may act contractually as a principal, agent, carrier, or some other legal mixture or hybrid [2].

Sometimes it may be unclear how the forwarder differs from the intermediary. The answer may be as follows: if a logistics service provider acts as an intermediary for any type of cargo transportation services, it can be classified as a freight forwarder recognized in the market and can join the relevant national or international associations. But if a logistics service provider offers only basic transportation, warehousing, packaging, or other logistics services, it cannot be recognized as a freight forwarder [3].

The freight forwarder undertakes, for remuneration and at the expense of the client, to perform, or arrange for the performance of services related to cargo transportation specified in the expedition contract. In case of realization of the service by the forwarder, a forwarding contract is concluded between him and the client. In turn, a freight forwarder and a cargo carrier conclude a contract on cargo transportation.

In the case of the provision of services by an intermediary, a mediation agreement, a mandate agreement, or a representation agreement is concluded between him and the client, which obliges the intermediary to provide the service of finding a suitable carrier or other third party whose services are needed by the client. Within the framework of these relationships, the contract of carriage will be concluded directly between the client and the cargo carrier [1].

Recently, multimodal cargo transportation has been developing a lot, connecting together almost all transport industries and transport technologies in international transport corridors. As part of this type of transportation, freight forwarders act as multimodal transportation operators. The role of the forwarder in this case changes. The responsibility of the multimodal transport operator is complicated due to the fact that it acts on behalf of its customer. His legal position in the multimodal transportation contract makes him even more responsible due to the fact that the multimodal transportation operator is also responsible for the work of all other persons who were hired by him [4].

The network system of responsibility links the multimodal transport operator in accordance with the laws and institutions for the part of transit where the incident occurred. This means that the multimodal operator is bound by the relevant laws, depending on which branch of transport the accident occurred in: road laws cover road parts of transport, marine laws - marine parts, etc. Also, the situation of multimodal transport operators is complicated by the fact that there is no single liability insurance for MTO [5].

Regulation of the forwarder's access to the market is carried out by each state separately. In some countries, entrepreneurs wishing to carry out forwarding activities must obtain an appropriate license, but in some countries, it is sufficient to meet the general requirements that allow conducting unlicensed business activities [6].

In Slovakia, access to the implementation of forwarding activities can be granted only if the entrepreneur meets not only the general requirements of entrepreneurship, but also meets the requirements necessary to obtain the appropriate license:

- higher specialized education

or

- secondary education in this field and one year of work experience in this field

or

- certificate of completion of an accredited training program in this field and two years of work experience in this field [7].

According to the results of the analysis of some countries of the European Union, it was revealed that a company or a person who does not have professional training or professional representation in seven of the analysed countries, such as Bulgaria, Croatia, Czech Republic, Germany, the Netherlands, Sweden, and the United Kingdom, can be a freight forwarder. In these countries, freight forwarding is an unregulated trade, which means that anyone who meets the general conditions of doing business can do business in the field of freight forwarding.

The mandatory conditions for obtaining a business license in other countries vary, but the requirements are reduced to:

- Availability of education - secondary or higher. In some countries, education should be specialized. Sometimes the alternative may be successful completion of certain professional courses.
- Successful completion of professional tests and exams.
- Experience in the field of forwarding in a management position.
- Financial guarantees of the entrepreneur.
- Other responsibilities include, for example, a certificate of professional experience, which is issued by a person who provides daily management of the company
- Registration in the freight forwarder's register.
- Another important condition is proof of honesty. This means that the applicant has no managers, members of the board of directors and technical directors with a criminal record and so on [8].

2. Risks and responsibilities of freight forwarders.

Cargo transportation is an activity involving a large number of associated risks. The main risk factors are:

- Technical risk that may be associated with the condition of the vehicle and the condition of the loading equipment, which are in unsatisfactory condition. This risk can lead to a traffic accident and an accident at work, which will affect not only the condition of the cargo, but also the life of the driver and loader.
- Infrastructure risk is associated with the condition of the road. This factor can affect both the safety of the transported cargo and the planning of the transportation route.
- The risk of natural disasters may lead to deterioration of visual characteristics during transportation and may lead to a traffic accident, which will undoubtedly affect the condition of the cargo.
- The risk of an unpredictable event may arise due to the risks of theft, which directly affect the loss of cargo and damage to cargo as a result of the actions of robbers [9].

Each of the above risks can lead to consequences such as delayed delivery of goods, partial or complete damage to the cargo or its loss. It is important to understand what responsibility the freight forwarder bears for these consequences.

An urgent problem is the absence of any international treaties or conventions that would regulate forwarding activities.

In each country, the responsibility of the forwarder is regulated in accordance with national legislation, which creates certain difficulties in the implementation of international forwarding activities. Clients are forced to study the laws of different countries regulating the liability of freight forwarders if they work with freight forwarders from other states. A similar problem arises for freight forwarders, if they contract freight forwarders from other countries to carry out freight forwarding activities.

The freight forwarder has the right to limited liability, which ensures economic stability and allows predicting possible monetary compensation to the cargo owner. These restrictions are usually not established by law. The maximum amount of the freight forwarder's responsibility is established when signing the forwarding contract. One of the exceptions is the legislation of the Russian Federation, which in the Federal Law of 30.06.2003 N 87-FZ (ed. of 18.03.2020) "On freight forwarding activities" directly specifies the limitation of the forwarder's liability. However, article 11 of this law states that the freight forwarding contract may provide for a higher amount of the forwarder's liability in comparison with the amount of liability established by this Federal Law or an international agreement of the Russian Federation [10].

More and more forwarding companies are joining freight forwarders' associations, the purpose of which is to represent and protect the interests of freight forwarders who are members of these associations. Among other things, these associations provide an opportunity for their members to use standard expedition contracts, forwarding conditions, and so on, drawn up to supplement the existing national legal norms and not contradict them.

In Slovakia, this association is Association of Logistics and Freight Forwarding of the Slovak Republic, which provides General Freight Forwarder's conditions of Association of Logistics and Freight Forwarding of the Slovak Republic. Their purpose is to further amend and supplement the provisions of sections 601 - 609 of the Commercial Code on the forwarding contract.

As a rule, the forwarding conditions proposed by the associations of freight forwarders are based on the FIATA model rules governing forwarding activities. FIATA is an international federation of freight forwarders' associations. The documents of this organization contain important terms and conditions in terms of obligations that are accepted by the transport agent issuing this FIATA document [11].

These rules apply only when they are included in the forwarding contract in writing or in any other form as a reference to the standard FIAT conditions governing forwarding activities [12].

The limitation of liability can be specified both in the currency of the country and in units of Special Drawing Right (SDR/XDR). Special drawing rights were created as an additional international reserve asset in the context of the Bretton Woods fixed exchange rate system. The SDR is neither a currency nor a claim on the IMF. Rather, it is a potential claim on the freely usable currencies of IMF members. SDRs can be exchanged for these currencies [13].

3. Responsibility of freight forwarders in the EU countries.

Previously, the size of the maximum liability in selected EU countries was investigated. In Slovakia, according to the General Freight Forwarder's conditions of Association of Logistics and Freight Forwarding of the Slovak Republic, the freight forwarder's responsibility is:

• in case of damage to the cargo during transportation by a vehicle or during loading and unloading operations related to transportation (during loading, transportation, loading, unloading), the amount of damage is limited to 8.33 XDR per 1 kg of gross weight of damaged, destroyed or lost goods, the maximum amount is 20 000 XDR per case of damage.,

• if the damage is caused by late delivery of the shipment, the freight forwarder's obligation to compensate the damage is limited to the agreed amount of remuneration to the carrier.

• in the case of warehouse services, the freight forwarder's liability for lost, damaged or destroyed goods is limited to the amount:

• 3.925 XDR per 1 kg of gross weight of damaged, lost or destroyed goods; however, the maximum amount is 3 925 XDR one damage case (one party).,

• 19 625 XDR, if the damage caused to the investor (client), consists of the difference between the desired and actual state of the stored goods.,

• in the cases of other damages liability of the freight forwarder for damages will be limited to 20 000 XDR per one case of damaged.

General Freight Forwarder's conditions of Association of Logistics and Freight Forwarding of the Slovak Republic applies to commercial relations between principals and freight forwarders in the field of national and international freight forwarding, provided that they become part of the freight forwarding agreement [14].

Analysis of other selected EU countries showed that the freight forwarder's liability is limited in accordance with the forwarding conditions and can range from 1.09 EUR/kg (Austria) to unlimited freight forwarder's liability (Croatia). In most of the analysed countries, the freight forwarder's liability is limited to 8.33 XDR per kilogram of damaged or destroyed goods, as in the Slovak Republic (Slovakia, Italy, Estonia, Czech Republic, Finland, Norway, Denmark, and Sweden). The freight forwarder's liability is maximally limited from 1.09 EUR/kg for damage in Austria to unlimited liability (in Croatia, Italy). In most of the analysed countries, the maximum liability of the forwarder is 50 000 XDR for one case of damage, which corresponds to the cost of 56 200 EUR [15].

4. Responsibility of freight forwarders in selected CIS countries

The purpose of this article is to analyse the responsibility of freight forwarders in the CIS countries. Without regard to the fact that the CIS is a separate economic zone, trade relations between these countries and the EU countries are quite intense (Figure 1). This fact implies frequent interaction of market participants, which leads to the need to consider the possibility of combining the conditions of international forwarding.

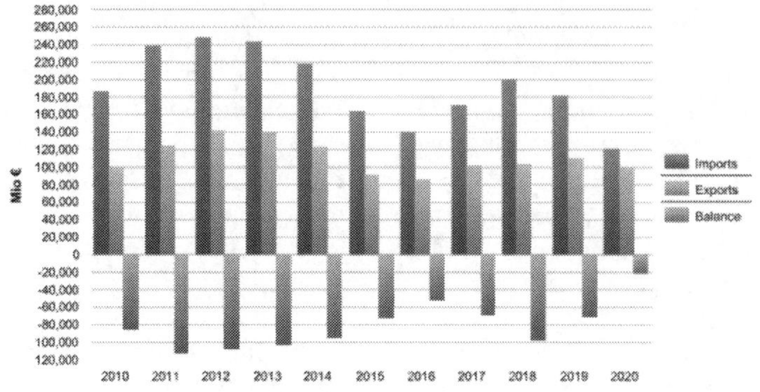

Figure 1. EU-CIS trade. Annual data 2010-2020 [16].

4.1. Responsibility of freight forwarders in the Russian Federation

In the Russian Federation, the activities of freight forwarders are regulated by the Civil Code and the Federal Law of 30.06.2003 N 87-FZ (ed. of 18.03.2020) "On freight forwarding activities". Chapter 41 of the Civil Code of the Russian defines the freight forwarding contract, responsibility, documents provided to the forwarder, etc. [17]. The amount of the freight forwarder's responsibility in the Russian Federation is determined by the Federal Law "On Freight Forwarding Activities" dated 30.06.2003 No. 87-FZ (latest edition). The freight forwarder's liability is limited to 2 SDR per kilogram of gross weight of lost or damaged cargo or 666.67 XDR per place or other unit of shipment [18]. The Russian Association of Forwarding and Logistics Organizations announces the FIATA model rules, and also duplicates information on the limitation of liability of freight forwarders, defined by the law on freight forwarding activities [19].

4.2. Responsibility of freight forwarders in Ukraine

In Ukraine, forwarding activity is regulated by Chapter 65 of the Civil Code of Ukraine [20], as well as the Law of Ukraine on Freight Forwarding Activities of July 5, 2012, No. 5060-VI, which describes in more detail the conditions of freight forwarding, but does not establish a certain amount of responsibility of the forwarder [21]. The amount of liability can be established in accordance with the freight forwarding agreement concluded between the freight forwarder and the client. The International Association of Freight Forwarders of Ukraine offers conditions for freight forwarding activities. In accordance with these conditions, the liability for loss/damage of the goods cannot exceed the equivalent of 2 SDR per kilogram of gross weight. The liability of the freight forwarding company for losses of any kind should not exceed the total amount of 10 000 SDR for each case. In case of a delay in the delivery of goods, the forwarder returns to the client the remuneration received earlier for the implementation of freight forwarding activities [22].

4.3. Responsibility of freight forwarders in the Republic of Belarus

Regulation of freight forwarding activities in the Republic of Belarus is carried out by the Civil Code of the Republic of Belarus [23] and the Law of the Republic of Belarus on June 13, 2006, No. 124-ZO of freight forwarding activities [24].

In the case of using a contract concluded by members of the Belarusian Association of Freight Forwarders, the following limitation of the forwarder's liability is established:

- the freight forwarder's liability is limited to an amount not exceeding 666.67 XDR per piece or other unit of shipment, or 2 XDR per one kilogram of gross weight of lost or damaged cargo, whichever is higher.
- if the cargo delivery does not include transportation by sea or by inland waterways, the freight forwarder's liability is limited to an amount not exceeding 8.33 XDR per kilogram of gross weight of the lost or damaged cargo.
- the freight forwarder's liability for damage resulting from a delay in the delivery of the cargo is limited to an amount not exceeding the freight charges due for the cargo delayed by delivery.
- the freight forwarder loses the right to limitation of liability if it is proved that the losses caused by loss, damage or delay in delivery were the result of actions or omissions of the freight forwarder committed with the intent to cause such losses [25].

4.4. Responsibility of freight forwarders in the Republic of Kazakhstan

The activity of freight forwarders in Kazakhstan is determined by the Civil Code of the Republic of Kazakhstan [26].

When concluding a forwarding contract with members of the Association of National Freight Forwarders of the Republic of Kazakhstan, the freight forwarder cannot be held responsible for the loss of goods and damage caused to the goods in an amount exceeding the equivalent of 2 XDR per one kilogram of gross weight of the lost or damaged cargo [27].

4.5. Comparison of freight forwarder's liability limitations in the CIS countries

Table 4 shows the legislation regulating freight forwarding activities, associations of freight forwarders, as well as the limitation of liability of freight forwarders in selected countries.

Table 4. Comparison of freight forwarders' responsibilities in selected countries.

A country	Legislation regulating freight forwarding activities.	The amount of the freight forwarder's liability in accordance with the legislation.	Association of Freight Forwarders operating in the country.	The amount of liability determined by the contract of the association of freight forwarders.
Russian Federation	Civil Code of the Russian Federation Federal Law No. 87-FZ of 30.06.2003 "On Freight Forwarding Activities"	2 XDR per kilogram of gross weight 666.67 XDR per piece or other unit of shipment	Russian Association of Forwarding and Logistics Organizations	In accordance with the current legislation.
Ukraine	The Civil Code of Ukraine The Law of Ukraine on Freight Forwarding activities	In accordance with the transport forwarding agreement	Association of International Freight Forwarders of Ukraine	2 XDR per kilogram of gross weight 10 000 XDR for each case.
Republic of Belarus	Civil Code of the Republic of Belarus LAW OF THE REPUBLIC OF BELARUS No. 124-ZO of Freight Forwarding activities, on June 13, 2006	Amount by which the value of the cargo has decreased, and if it is impossible to restore the damaged cargo – in the amount of its value	Association of International Freight Forwarders and Logistics	2 XDR per kilogram of gross weight 8.33 SDR per kilogram of gross weight (if cargo delivery does not include transportation by sea or by inland waterways) 666.67 XDR per piece or other unit of shipment

IOP Conf. Series: Materials Science and Engineering 1247 (2022) 012033 doi:10.1088/1757-899X/1247/1/012033

Republic of Kazakhstan	Civil Code of the Republic of Kazakhstan	Amount by which the value of the cargo has decreased, and if it is impossible to restore the damaged cargo – in the amount of its value	Association of National Freight Forwarders of the Republic of Kazakhstan	2 XDR per kilogram of gross weight

The general indicator of limitation of liability is 2 XDR per one kilogram of gross weight.

In the study Poliak M. and Šimurková, P. [15] a solution was proposed to combine the responsibility of freight forwarders in selected EU countries by averaging the data obtained:

- 5.10 €/kg gross weight of damaged, lost or destroyed goods
- 23 000 EUR for one case of damage.

This proposal has not yet been put into practice. Given that freight forwarders not only provide their activities within EU member states but also provide their services within non-EU countries, it is desirable that an international legal standard be adopted that unifies the responsibility of freight forwarders. Difference in the restriction of shippers is not unified even in the CIS member countries. However, it can be concluded that the limitation of liability in these countries is at a higher value when compared with the proposed limit of liability in Poliak M. and Šimurková, P.

5. Conclusion

The role of the freight forwarder is to represent the interests of his client in the activities related to the transportation of the shipment. The freight forwarder acts as an expert and it is assumed that the freight forwarder will be ordered by a client who is not an expert in the field of transport services. Although it is assumed that the client is not an expert in the field of transport services, the current system of different liability of freight forwarders in different countries creates a requirement for the client to monitor the difference in liability and take this into account in the selection of a suitable freight forwarder. It is possible that several clients are not aware of these differences. However, such ignorance creates the preconditions that the client will not decide when choosing a freight forwarder for the freight forwarder with the best conditions. For this reason, legislation is needed that would unify the responsibility of freight forwarders not only in the EU countries but in all FIATA countries.

It is obvious that there is a need to create a single international treaty or convention that would regulate freight forwarding activities, as well as determine the limitation of liability of freight forwarders. When creating this contract, it would be reasonable to refer to the experience of creating existing contracts, such as the CMR Convention, which limits the liability of road carriers in the amount of 8.33 XDR per gross kilogram of lost or damaged cargo, and also obliges cargo carriers to insure their liability. It would be reasonable to propose the task of creating this agreement to such a large-scale organization as FIATA, which currently has rich experience in regulating forwarding conditions at the non-state level.

Further work in this area should also identify the relationship of responsibility of the freight forwarder and the carrier. The freight forwarder procures transportations not only by road transport, but also by rail, water or air transport. However, each mode of transport has a different limitation of liability.

6. References

[1] Gnap J, Poliak M, Sosedová J and Jagelčák J., 2021 *Zasielatelstvo* (Žilina: EDIS-vydavateľské centrum ŽU) p 262

[2] Cain P., 2014 Complexity, confusion and the multifaceted legal roles of the international freight forwarder *Macquarie law journal,* vol 13-14

[3] Vidovic M, Kilibarda M, Zecevic S, Radivojevic G and Miljus M., 2019 Freight forwarding industry – the contemporary role and development trends in Serbia *Proc. of the 4th logistics Int. Conf.(Belgrad), ed Vidović M et al.* p 335

[4] Jashari N., 2017 Liability of the multimodal transport operator – the case of Kosovo *Pomorstvo-scientific journal of maritime research*, vol 31(1) pp. 77-81
[5] Zelenika R, Lattice T and Buzan E., 2011 Multimodal transport operator liability insurance model *Promet-traffic & transportation,* volume 23(1) pp.25-38
[6] Kholopov K V and Golubchik A M., 2019 O mezhdunarodnom i rossiyskom opyte litsenzirovaniya deyatel'nosti ekspeditorov i perevozchikov [About the international and Russian experience of licensing the activities of freight forwarders and carriers] *Russian Foreign Economic Bulletin* vol 11 pp. 45-55
[7] Act of 2. October 1991 on trade licensing (trade Licensing Act). Available online: https://www.slov-lex.sk/pravne-predpisy/SK/ZZ/1991/455/20220201 (25.02.2022)
[8] Poliak M, Šimurková P, Mrníková, M and Komačková L., 2016 Market access in freight forwarding and responsibility of freight forwarder *Transport And Communications vol 2*
[9] Panjee1 P, Amornsawadwatana S and Vongmanee V., 2021 Risk factors affecting cargo damage in road transportation *Turkish Journal of Physiotherapy and Rehabilitation* 32(3)
[10] Federal Law "On Freight Forwarding Activities" dated 30.06.2003 N 87-FZ (latest edition). Available online: http://www.consultant.ru/document/cons_doc_LAW_43006/dba7b54ef4a85aa5a04936c339ea6c33fc403657/ (25.02.2022)
[11] Parhammehr H R and Zeajeldi I 2016 The role of FIATA multimodal transport bill of lading in business exchange in international transport *Social Sciences* 8(2) pp. 661-665
[12] Explanatory Note to the Revision of and Guidelines for the Usage of FIATA Model Rules for Freight Forwarding Services. Available online: https://fiata.org/fileadmin/user_upload/documents/Explanatory_Note_to_the_Revision_of_FIATA_Model_Rules_for_Freight_Forwarding_Services.pdf (25.02.2022)
[13] Special Drawing Rights (SDR). Available online: https://www.imf.org/en/About/Factsheets/Sheets/2016/08/01/14/51/Special-Drawing-Right-SDR (25.02.2022)
[14] General Freight Forwarder's conditions of Association of Logistics and Freight Forwarding of the Slovak Republic. Available online: https://zlz.sk/informacie-o-zvaze/vseobecne-zasielatelske-podmienky/ (25.02.2022)
[15] Poliak M and Šimurková P 2017 Harmonization of market conditions in provision of freight forwarding *MATEC Web of Conf.vol 4*
[16] European Union, Trade in goods with CIS Commonwealth Of Independent States. Available online: https://webgate.ec.europa.eu/isdb_results/factsheets/region/details_cis-commonwealth-of-independent-states_en.pdf (25.02.2022)
[17] "The Civil Code of the Russian Federation (Part Two)" dated 26.01.1996 N 14-FZ (ed. from 01.07.2021, with amendments. from 08.07.2021) (with amendments and additions, intro. effective from 01.01.2022). Available online:http://www.consultant.ru/document/cons_doc_LAW_9027/dfe46eb17d8b852987049332ca429b48b665b8a9/ (25.02.2022)
[18] Federal Law "On Freight Forwarding Activities" dated 30.06.2003 N 87-FZ (latest edition). Chapter 3. Responsibility of the forwarder and the client. Available online: http://www.consultant.ru/document/cons_doc_LAW_43006/1c7c5ac903e06cd5dbd340c9ecdde4d8d7201bc1/ (25.02.2022)
[19] The Russian Association of Freight Forwarding and Logistics Organizations. Available online: http://www.far-aerf.ru/ENGLISH/index.htm (25.02.2022)
[20] Article 934 of the Civil Code of Ukraine. Freight forwarder's responsibility under the freight forwarding agreement. Available online: https://urst.com.ua/ru/gku/st-934 (25.02.2022)
[21] The Law of Ukraine on Freight Forwarding activitiesю Available online: https://www.mcs.od.ua/zakon-ukrainy-o-transportno-ekspeditorskoj-deyatelnosti/ (25.02.2022)
[22] The Association of International Freight Forwarders of Ukraineю Available online: https://ameu.org.ua/ru/2-uncategorised/38-generalnye-usloviya-ted (25.02.2022)

[23] Civil Code of the Republic of Belarusю Article 756. Freight forwarder's responsibility under the freight forwarding contractю Available online: https://kodeksy-by.com/grazhdanskij_kodeks_rb/756.htm (25.02.2022)
[24] Law of the Republic of Belarus No. 124-Z of June 13, 2006 on Freight Forwarding Activities. Available online: https://pravo.by/document/?guid=3871&p0=H10600124 (25.02.2022)
[25] Association of International Freight Forwarders and Logistics. General conditions of activity of Belarusian freight forwarders. Available online: http://www.baifby.com/page/27 (25.02.2022)
[26] Civil Code of the Republic of Kazakhstan (Special part) Article 713. Freight forwarder's responsibility under the freight forwarding contract. Available online: https://kodeksy-kz.com/ka/grazhdanskij_kodeks_osobennaya_chast/713.htm (25.02.2022)
[27] FIATA Model Rules Governing the Provision of Freight Forwarding Services (October 2019). Available online: http://kffanek.kz/f/tipovye_pravila_fiata_2019_otformatir.pdf. (25.02.2022)

Acknowledgments
This paper has been developed under support of project: VEGA No. 1/0245/20 Poliak: Identification of the impact of a change in transport related legislation on the competitiveness of carriers and carriage safety

IOP Conf. Series: Materials Science and Engineering 1247 (2022) 012034 doi:10.1088/1757-899X/1247/1/012034

Driving Simulator as a Tool for Reaction Time Measurement

K. Čulík [1*], V Štefancová [2], A. Kalašová [1]

[1] Department of Road and Urban Transport, Faculty of Operation and Economics of Transport and Communications, University of Žilina, Univerzitná 1, Žilina, Slovakia
[2] Department of Railway Transport, Faculty of Operation and Economics of Transport and Communications, University of Žilina, Univerzitná 1, Žilina, Slovakia

* kristian.culik@fpedas.uniza.sk

Abstract. The reaction times of a driver were measured in a driving simulator environment in this article. Young drivers under the age of 26 were the focus of the study. They cause many accidents. Each participating driver provided basic information later used for mathematical-statistical analysis. The main advantage of driving simulators is limitless usage. It is possible to simulate situations that would be unacceptable in real road traffic. Therefore, this study could focus also on drivers influenced by alcohol. The main goal of the article is to evaluate the reaction times. Then it was possible to evaluate if gender, practice, and alcohol, statistically significantly affect the reaction time. We also focused on drinking before driving for a smaller number of drivers. For these mathematical-statistical purposes, we used a one-sample t-test, paired-samples t-test, and independent-sample t-test.

Key words: simulator, driver behavior, automotive safety

1. Introduction

In general, human error is a source of more than 90% of road accidents [1; 2; 3]. It is possible for a pedestrian, cyclist, or another road user to make a mistake. The driver is the one who causes the most accidents [4]. Road accidents are also caused by fatigue and stress [5]. Many researchers have focused on professional drivers' accidents [6; 7]. Road accidents caused by alcohol are also dangerous [8; 9]. They frequently include a young driver who is inattentive and prone to drink-driving. Scientific research has diverse interpretations of the term "young drivers." Most of these drivers are between the ages of 18 and 26. However, in some research, the highest age limit may be different. For example, in [10], the authors concentrated on a group of 18 to 20-year-olds, in [11], the Romanian study focused on 18 to 24-year-old drivers, and in [12], the Belgian study focused on 17 to 24-year-old drivers. On the other hand, [13] concentrates on 16 to 24-year-old drivers, among other things. The mix of young drivers and alcohol has been the subject of several investigations. For example, in Greece, the legal age for driving while intoxicated is 18. A study of 241 young Greek drivers (aged 18-24) indicated that young drivers who often consumed alcohol in their life were more likely involved in an accident [14]. In addition to alcohol, poor driving experience also contributes to higher accident rates for young drivers [15].

Men are more prone than women to drink alcohol, according to another study [16]. Other researchers [17; 18; 19] claim that after drinking, people are more likely to engage in dangerous behaviour and aggressiveness. Other research, such as [20], has found that men and women have different characteristics. Male risk perception is lower, implying that they drive more carelessly [21; 22]. The effects of alcohol on drivers were investigated in a study [23]. The main goal of this study was to investigate the impact of alcohol consumption on drivers' behaviour at STOP signs and at traffic light intersections. It was a lab experiment that included the driving simulator. The results revealed a considerable difference between sober and drunk drivers. A driving simulator will be used also this study. A driving simulator is a device that models the operation of a road vehicle in a virtual environment that is as realistic as possible [24; 25; 26].

The driving simulator is a complicated device that consists of many parts. Its software is a vital component. It influences the simulation's quality and the simulator's ability to be used for research [27]. The University Science Park at University of Žilina has a training driving simulator available for research purposes. The essential difference between the research driving simulator and the training driving simulator is in the purpose of use:

a) **Training driving simulators** are used to train new drivers and are commonly found in driving schools. In Methodical Instruction number. 22/2005 on technical requirements for training driving simulators [28] dated September 26, 2005, the name 'training driving simulator' appears. The basic requirements for teaching driving simulators are outlined in this guideline. For example, there is a requirement for a projection area of at least 180 x 130 mm, as well as various sound, vehicle dynamics, and virtual environment requirements. Researchers use training driving simulators very rarely, such as in the study [29], or to assess the efficiency of training at driving schools [30].

b) **Research driving simulators** are used in research institutes, universities, and the automotive industry. Their key advantage is their ability to adjust to the experiment's present requirements. It means they need an open system in which the virtual environment, vehicle, and its features may all be changed. They were used in studies on fatigue research [31; 32], crossing intersections [33], lane change behaviour [34; 35; 36], driver error rate [37], driver glare [38] or human - vehicle interaction research in general [39]. Simulators also have great potential in autonomous vehicle research [40; 41; 42].

2. Used equipment and methodology of the research

The goal of the study was to measure and evaluate young drivers' reaction times. We aimed to find if there were differences in reaction times between men and women. The difference in reaction time before and after drinking alcohol was also measured. The experimental driving simulation was ensured by the SNA - 211 REN training driving simulator [43; 44]. The people who took part in this study had never used a driving simulator before. As a result, they had to be retrained before their timed ride. The drivers got around ten minutes to become used to the setting and simulator controls. Each driver was allowed to practice starting, braking, turning, and shifting gears.

Other used equipment is shown in the Figure 1.

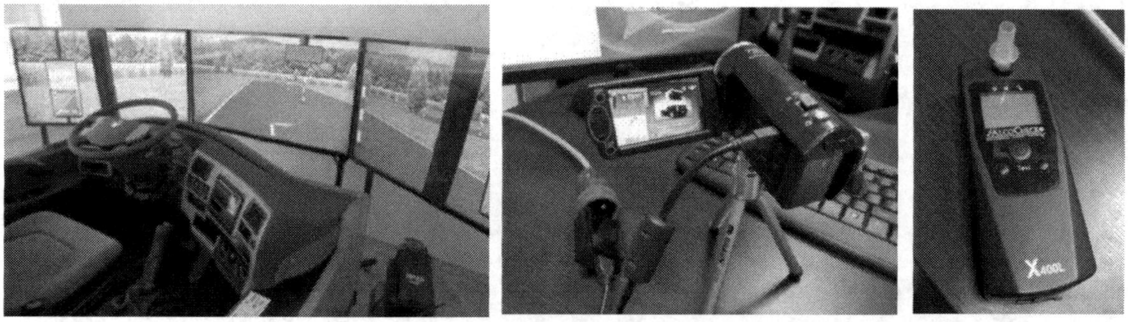

(a) (b) (c)

Figure 1. Used equipment: (a) Training driving simulator in SNA – 211 REN; (b) External video camera for recording; (c) AlcoCheck X400L. Source: Processed by authors.

During the individual measurements, three people were in the laboratory: the driving simulator supervisor, the recording person, and the tested driver. During the measurements, the other participants in the research task were in another room. Before their ride, it was vital to ensure that they had not seen the obstacles or virtual environment.

The drivers' reactions lasted from the moment they were confronted with an unexpected situation until they pressed the brake pedal. The scenery and braking were recorded using the computer program Corel VideoStudio. It was installed on the PC that was used for the driving simulator. An external camera was also used to record the footage. The AlcoCheck X400L was also used to test the drivers' breath for alcohol. The results obtained were not used in the statistical analysis.
The measurement process was divided into three sections described in the following lines:

1. Unexpected obstacle
During the first measurement, the drivers were focused but did not expect an obstacle. The obstacle in the virtual environment was the running animal from behind the tree across the road. The reaction time of the driver was recorded as the time interval between the obstacle animation trigger and the moment of activating the brake. We did not evaluate the success of the obstacle stop in this article.

2. Expected obstacle
The second measurement continued after the first. The scene for evaluating reaction times went on, but the drivers were already known what the purpose of their driving is. As a result, the reaction time should be even shorter than it was for the first unexpected obstacle. The drivers were more cautious during this test. They took a quick glance at the edge of the road.

3. Drunk-driving
Not all drivers completed additional tests due to technological issues. As a result, only ten drivers were picked for the last portion of the test. In the fourth and fifth parts of the research, all the drivers agreed to drink alcohol. Regardless of weight or other circumstances, each driver drank 200 ml of 35 percent alcohol. Each driver was given ten minutes to complete this task. His journey started in a simulated environment after a 10-minute break.

3. Evaluation of measured data

During two reaction time measurements, we were able to measure the reaction times of 30 drivers in total. In the following paragraph, all data collected throughout these experiments will be presented in the separate tests. In this article, we will use the t-test to test the hypotheses described in the following lines. First, we verify that the mean value of the reaction times in the first measurement is equal to 0.8 seconds, which is the middle of the table value for the concentrated driver who does not expect the obstacle. We assume that the mean value will be shorter than the table data, but we will verify this hypothesis at the significance level $\alpha = 5\%$.

3.1. One-Sample t-Test

We will divide the individual tests into points of the procedure so that, in addition to the results themselves, we will also point out the procedure of statistical testing. We per-formed the One-Sample t-Test from the data in Table 1.

Table 1. Data used as input of One-Sample t-Test

Driver	1	2	3	4	...	27	28	29	30
Reaction time	0.60	0.73	0.95	0.72	...	0.67	0.68	0.65	0.46

IOP Conf. Series: Materials Science and Engineering 1247 (2022) 012034 doi:10.1088/1757-899X/1247/1/012034

One-Sample t-Test has the following procedure:
First, it is necessary to determine the hypotheses: H_0: The mean value of the reaction times of the concentrated drivers is 0.80 seconds: μ = 0.80 s. H_1: The mean value of the reaction times of the concentrated drivers is less than 0.80 seconds: $\mu < 0.80$ s.
Calculation of test criterion (1), where x is the arithmetic mean of all measured reaction times (0.7317) and μ_0 is the mean value – we have chosen 0.80 s. The parameter is n - the number of all measurements (30), and S is standard deviation, which in this case is (0.1661).

$$t = \frac{\bar{x} - \mu_0}{S} \cdot \sqrt{n} \tag{1}$$

We computed that the test criterion has a value of -2.2522 using the formula. In this example, the critical field is provided by relation (2), where 0.05 is the level of significance. As a result, we check for the value $t_{0.95}(29)$, which is 1.699, in the quantile tables of the Student's distribution.

$$W_\alpha = \{t \leq -t_{1-\alpha}(n-1)\} \tag{2}$$

Subsequently, we can add to the formula itself as follows (3):

$$W_\alpha = \{-2.2522 \leq -1.699\} \tag{3}$$

As a result, the critical field is met. Therefore, we reject the original hypothesis H_0 in favour of the alternative hypothesis H_1. The answer in this example is: At a significance level of 5%, the mean value of the reaction times of the concentrated drivers is less than 0.80 seconds.

3.2. Independent-Sample t-Test

We used the also the Independent-Sample t-Test. This test for independent selections is a widely used approach for determining the difference between the two groups' averages. In Table 2, there is a dataset used for this type of test.

Table 2. Data used as input of Independent-Sample t-Test

Gender/driver	1	2	3	4	...	12	13	14	15
Male	0.48	0.63	0.82	0.48	...	0.75	0.71	0.59	0.43
Female	0.88	0.64	0.94	0.88	...	0.89	0.75	0.64	0.62

Independent-Sample t-Test has a similar calculation process as One-Sample t-Test:
We determinate two hypotheses: H_0: the mean reaction time of men and women is the same: $\mu_M = \mu_W$. H_1: the mean value of the reaction time of men and women is not the same (it is different): $\mu_M \neq \mu_W$.
In the test criterion (4) $\overline{x_1}$ is the arithmetic mean calculated from all measured reaction times of men and $\overline{x_2}$ of women. The number of measurements is n_1 (men) and n_2 (women). In the case of this test, the measurements may be different, as they are not paired S_1 and S_1 are standard deviations, which in this case are at the level $S_1 = 0.1472$ and $S_2 = 0.1356$. After substituting, we find that the test criterion has a value +1.910. After substituting, we find that the test criterion has a value +1.910.

$$u = \frac{\overline{x_1} - \overline{x_2}}{\sqrt{\frac{S_1^2}{n_1} + \frac{S_2^2}{n_2}}} \tag{4}$$

The critical field in this case is given by (5), α is the level of significance, in our case 0.05. It follows from the quantile tables of the normal distribution *N (0, 1)* we look for the value $u_{0.975}$, which is 1.960.

$$W_\alpha = \left\{|u| \geq u_{1-\frac{\alpha}{2}}\right\} \tag{5}$$

(5) Subsequently, we can add to the formula itself as follows (6):

$$W_\alpha = \{1{,}910 \not\geq 1.960\} \tag{6}$$

From this, we can conclude that the critical field is not met and therefore we accept the original hypothesis H_0. The answer in this case is: The mean reaction time of men and women is the same at a significance level of 5%. However, as can be seen, the test criterion is very close to the critical range.

3.3. Paired-Sample t-Test

The Paired-Sample t-Test was the third one we used. This test compares the values of a variable for the same respondent in two different experimental conditions. In our case, we will use this test to compare the reaction time before and after drinking alcohol. The structure of the input data is in the following table (Table 3).

Table 3. Data used as input of Independent-Sample t-Test

Driver	1	2	3	...	8	9	10
3rd attempt (alc.)	0.75	0.91	1.13	...	0.69	0.83	0.77
4th attempt (alc.)	0.85	1.28	0.95	...	0.56	1.19	0.75

The process of testing with the Paired-Sample t-test is as follows:
We determined two hypotheses: H_0: The mean value of the reaction times of the drivers in the sober state and under the influence of alcohol is the same: $\mu_S = \mu_D$. H_1: The mean value of the reaction times of drivers in a sober state and under the influence of alcohol is not the same (it is different): $\mu_S \neq \mu_D$. Calculation of test criterion (7), in which $\overline{D}$ is the arithmetic mean of all mutual deviations (differences) between two experiments. The number of measurements is denoted as *n*. It is also necessary to calculate the standard deviation S_D from all values of the mentioned differences for the calculation of the test criterion.

$$T = \frac{\sqrt{n} \cdot \overline{D}}{S_D} \tag{7}$$

After substituting, we find that the test criterion has a value -2.6181. In this case, the critical field is given by (8), α is the level of significance, in our case 0,05. It follows from the above that in the tables in the quantile tables of the Student's distribution we look for the value t_0.975 (9), which is 2.2620.

$$W_\alpha = \left\{|t| \geq t_{1-\frac{\alpha}{2}}(n-1)\right\} \tag{8}$$

Subsequently, we can check (9) the fulfilment or non-fulfilment of the critical field:

$$W_\alpha = \{2.6181 > 2.2620\} \tag{9}$$

From (9) we can conclude that the critical field is fulfilled and thus we reject the original hypothesis H0 and accept the alternative hypothesis H1. The answer in this case is: At a significance level of 5%, it was shown that H1: The mean value of the reaction times of drivers in the sober state and under the influence of alcohol is not the same (it is different).

4. Conclusion

From the point of view of the experiment itself, in this study, we were able to evaluate the reaction times of 30 drivers, which is not a sufficiently representative sample. In Slovakia, there are 244,663 registered drivers in the 17-24 age group. According to Sample Size Calculator, with a population of 244,663, Confidence Level 95%, Confidence Interval 5, we would need a sample size of 384 people. However, this was not possible in our case. For this reason, we verified the statistical significance by testing hypotheses.

Another goal of the paper is educational purpose. This article shows how it is possible statistically evaluate data. Therefore, students at the university can use described methods for their thesis. For this reason, all four tests are performed not by software (through the so-called p-values), but by traditional calculation.

This research brought several results. In the first test, the tabular value of the reaction time of the concentrated driver, which does not expect a stimulus, was not completely confirmed. Reaction times were shorter, probably due to the lack of distractors during the ride.

Statistical testing also confirmed that the reaction time of men and women is approximately the same. However, in this case, it is possible that similar reaction times occurred due to similar people in the selected sample. All the young drivers were students or graduates of a technical university. It means that this could have an impact on the results.

With the Paired-Samples t-Test, we tested the hypothesis of prolonging the reaction times of drivers under the influence of alcohol. The reaction times were indeed even shorter during some attempts. However, in general, at a significance level of 0.05, it can be stated that the times are different and, of course, shorter in the sober state.

From this article, the danger of drunk driving is evident. However, it is also clear from many other studies [45; 46; 47]. In terms of statistics, we have pointed out that the basic characteristics (median, mode, arithmetic mean, or standard deviation) are insufficient in similar research. It is necessary to assess statistical significance.

Funding

Funding: This paper was developed with support of a project: MŠVVŠ SR - VEGA No. 1/0178/22 Kalašová, A.: Basic research of the sharing economy as a tool for reducing negative externalities.

References

[1] De Felice, F., & Petrillo, A. (2011). Methodological approach for performing human reliability and error analysis in railway transportation system. International Journal of Engineering and Technology, 3(5), 341-353.

[2] Jurecki, R. S., Jaśkiewicz, M., & Zuska, A. (2013). The variety of the behaviour of drivers at risk accident situations. Zeszyty Naukowe/Akademia Morska w Szczecinie.

[3] Poliak, M., Križanová, A., Semanová, Š., & Štefániková, L. (2014). The influence of contract form choice of the transport services ensuring on performance contracting entity requirement. Problemy Transportu, 9(4), 153-161.

[4] Massie, D. L., Campbell, K. L., & Williams, A. F. (1995). Traffic accident involvement rates by driver age and gender. Accident Analysis and Prevention, 27(1), 73-87. https://doi.org/10.1016/0001-4575(94)00050-V.

[5] Useche, S., Cendales, B., & Gómez, V. (2017). Work stress, fatigue and risk behaviors at the wheel: Data to assess the association between psychosocial work factors and risky driving on bus rapid transit drivers. Data in Brief, 15, 335-339. https://doi.org/10.1016/j.dib.2017.09.032.

[6] CENDALES, B., USECHE, S., & GÓMEZ, V. (2014). Psychosocial work factors, blood pressure and psychological strain in male bus operators. Industrial Health, 52(4), 279-288. https://doi.org/10.2486/indhealth.2013-0156.

[7] Useche, S. A., Cendales, B., Alonso, F., Montoro, L., & Pastor, J. C. (2019). Trait driving anger and driving styles among colombian professional drivers. Heliyon, 5(8), e02259-e02259. https://doi.org/10.1016/j.heliyon.2019.e02259.

[8] Mann, R. E., Stoduto, G., Vingilis, E., Asbridge, M., Wickens, C. M., Ialomiteanu, A., Sharpley, J., & Smart, R. G. (2010). Alcohol and driving factors in collision risk. Accident Analysis and Prevention, 42(6), 1538-1544. https://doi.org/10.1016/j.aap.2010.03.010.

[9] Williams, A. F. (2006). Alcohol-impaired driving and its consequences in the United States: The past 25 years. Journal of Safety Research, 37(2), 123-138. https://doi.org/10.1016/j.jsr.2006.01.001.

[10] Rolison, J. J., & Moutari, S. (2020). Combinations of factors contribute to young driver crashes. Journal of Safety Research, 73, 171-177. https://doi.org/10.1016/j.jsr.2020.02.017.

[11] Gheorghiu, A., & Havârneanu, C. (2012). Driving behaviour of a sample of young romanian drivers. 33 697-701. https://doi.org/10.1016/j.sbspro.2012.01.211.

[12] Jongen, E. M. M., Brijs, K., Komlos, M., Brijs, T., & Wets, G. (2011). Inhibitory control and reward predict risky driving in young novice drivers - A simulator study. Procedia - Social and Behavioral Sciences, 20 604-612. https://doi.org/10.1016/j.sbspro.2011.08.067.

[13] Dénommée, J. A., Foglia, V., Roy-Charland, A., Turcotte, K., Lemieux, S., & Yantzi, N. (2020;2019). Cellphone use and young drivers. Canadian Psychology = Psychologie Canadienne, 61(1), 22-30. https://doi.org/10.1037/cap0000175.

[14] Chliaoutakis, J. E., Darviri, C., & Demakakos, P. T. (1999). The impact of young drivers' lifestyle on their road traffic accident risk in greater athens area. Accident Analysis and Prevention, 31(6), 771-780. https://doi.org/10.1016/S0001-4575(99)00040-8.

[15] Williams, A. F. (2003). Teenage drivers: Patterns of risk. Journal of Safety Research, 34(1), 5-15. https://doi.org/10.1016/S0022-4375(02)00075-0.

[16] Nicolai, J., Moshagen, M., & Demmel, R. (2012). Patterns of alcohol expectancies and alcohol use across age and gender. Drug and Alcohol Dependence, 126(3), 347-353. https://doi.org/10.1016/j.drugalcdep.2012.05.040.

[17] Graham, K., Bernards, S., Knibbe, R., Kairouz, S., Kuntsche, S., Wilsnack, S. C., Greenfield, T. K., Dietze, P., Obot, I., & Gmel, G. (2011). Alcohol-related negative consequences among drinkers around the world. Addiction (Abingdon, England), 106(8), 1391-1405. https://doi.org/10.1111/j.1360-0443.2011.03425.x.

[18] Jurecki, R. S., & Stańczyk, T. L. (2011). The test methods and the reaction time of drivers. Eksploatacja i Niezawodnosc, 51(3), 84-91.

[19] Byrnes, J. P., Miller, D. C., & Schafer, W. D. (1999). Gender differences in risk taking: A meta-analysis. Psychological Bulletin, 125(3), 367-383. https://doi.org/10.1037/0033-2909.125.3.367.

[20] Useche, S. A., Hezaveh, A. M., Llamazares, F. J., & Cherry, C. (2021). Not gendered… but different from each other? A structural equation model for explaining risky road behaviors of female and male pedestrians. Accident Analysis and Preven-tion, 150https://doi.org/10.1016/j.aap.2020.105942.

[21] Oppenheim, I., Oron-Gilad, T., Parmet, Y., & Shinar, D. (2016). Can traffic violations be traced to gender-role, sensation seeking, demographics and driving exposure? Transportation Research. Part F, Traffic Psychology and Behaviour, 43, 387-395. https://doi.org/10.1016/j.trf.2016.06.027.

[22] Ondruš, J., Vrábel, J., & Kolla, E. (2018). The influence of the vehicle weight on the selected vehicle braking characteristics. In Transport Means-Proceedings of the International Conference (pp. 384-390).

[23] Wan, J., Wu, C., Zhang, Y., Houston, R. J., Chen, C. W., & Chanawangsa, P. (2017). Drinking and driving behavior at stop signs and red lights. Accident Analysis and Prevention, 104, 10-17. https://doi.org/10.1016/j.aap.2017.04.008.

[24] Konstantopoulos, P., Chapman, P., & Crundall, D. (2010). Driver's visual attention as a function of driving experience and visibility. using a driving simulator to explore drivers' eye movements in day, night and rain driving. Accident Analysis and Prevention, 42(3), 827-834. https://doi.org/10.1016/j.aap.2009.09.022.

[25] Blana, E. (1996). A Survey of Driving Research Simulators Around the World.

[26] Carsten, O., & Jamson, A. H. (2011). Driving simulators as research tools in traffic psychology. Handbook of traffic psychology (pp. 87-96) https://doi.org/10.1016/B978-0-12-381984-0.10007-4.

[27] Auberlet, J., Pacaux, M., Anceaux, F., Plainchault, P., & Rosey, F. (2010). The impact of perceptual treatments on lateral control: A study using fixed-base and motion-base driving simulators. Accident Analysis and Prevention, 42(1), 166-173. https://doi.org/10.1016/j.aap.2009.07.017.

[28] Methodical instruction no. 22/2005 on technical requirements for simulators, Ministry of Transport, Posts and Telecommunications of the Slovak Republic, 26.9.2005, Bratislava.

[29] Martín-delosReyes, L. M., Jiménez-Mejías, E., Martínez-Ruiz, V., Moreno-Roldán, E., Molina-Soberanes, D., & Lardelli-Claret, P. (2019). Efficacy of training with driving simulators in improving safety in young novice or learner drivers: A systematic review. Transportation Research. Part F, Traffic Psychology and Behaviour, 62, 58-65. https://doi.org/10.1016/j.trf.2018.12.006.

[30] van Leeuwen, P. M., Happee, R., & de Winter, J. C. F. (2015). Changes of driving performance and gaze behavior of novice drivers during a 30-min simulator-based training. Procedia Manufacturing, 3, 3325-3332. https://doi.org/10.1016/j.promfg.2015.07.422.

[31] Rossia, R., Gastaldia, M., & Gecchelea, G. (2011). Analysis of driver task-related fatigue using driving simulator experiments. Procedia - Social and Behavioral Sciences, 20 666-675. https://doi.org/10.1016/j.sbspro.2011.08.074.

[32] Meng, F., Wong, S. C., Yan, W., Li, Y. C., & Yang, L. (2019). Temporal patterns of driving fatigue and driving performance among male taxi drivers in hong kong: A driving simulator approach. Accident Analysis and Prevention, 125, 7-13. https://doi.org/10.

[33] Matsumoto, Y., & Peng, G. (2015). Analysis of driving behavior with information for passing through signalized intersection by driving simulator. Transportation Research Procedia, 10, 103-112. https://doi.org/10.1016/j.trpro.2015.09.060.

[34] Hess, S., Choudhury, C. F., Bliemer, M. C. J., & Hibberd, D. (2020). Modelling lane changing behaviour in approaches to roadworks: Contrasting and combining driving simulator data with stated choice data. Transportation Research. Part C, Emerging Technologies, 112, 282-294. https://doi.org/10.1016/j.trc.2019.12.003.

[35] Calvi, A., D'Amico, F., Ferrante, C., & Bianchini Ciampoli, L. (2020). A driving simulator validation study for evaluating the driving performance on deceleration and acceleration lanes. Advances in Transportation Studies, 50, 67-80. https://doi.org/10.4399/97888255317325.

[36] Yuan, J., Abdel-Aty, M., Cai, Q., & Lee, J. (2019). Investigating drivers' mandatory lane change behavior on the weaving section of freeway with managed lanes: A driving simulator study. Transportation Research. Part F, Traffic Psychology and Behaviour, 62, 11-32. https://doi.org/10.1016/j.trf.2018.

[37] Poliak, M., Mrnikova, M., Simurkova, P., Medvid, P., Poliakova, A., & Hernandez, S. (2018). Social law in road transport like tool safety road transport. Paper presented at the 1-7. https://doi.org/10.1109/AUTOSAFE.2018.8373317.

[38] Haycock, B. C., Campos, J. L., Koenraad, N., Potter, M., & Advani, S. K. (2019). Creating headlight glare in a driving simulator. Transportation Research. Part F, Traffic Psychology and Behaviour, 61, 93-106. https://doi.org/10.1016/j.trf.2017.10.006.

[39] Weir, D. H. (2010). Application of a driving simulator to the development of in-vehicle human–machine-interfaces. IATSS Research, 34(1), 16-21. https://doi.org/10.1016/j.iatssr.2010.06.005.

[40] Jang, J., Lee, H., & Kim, J. (2022;2021). Carfree: Hassle-free object detection dataset generation using carla autonomous driving simulator. Applied Sciences, 12(1), 281. https://doi.org/10.3390/app12010281.
[41] Riegler, A., Riener, A., & Holzmann, C. (2019). AutoWSD: Virtual reality automated driving simulator for rapid HCI prototyping. Proceedings of Mensch und Computer 2019, 853-857. https://doi.org/10.1145/3340764.3345366.
[42] Gnap, J., Konečný, V., Slávik, R., Beňová, D., & University of Žilina Faculty of Operation and Economics of Transport and Communications Department of Road and Urban Transport Slovak Republic. (2018). Possible impacts of regulating the weekly rest of road freight drivers on logistics in EU countries. Naše More Znanstveni Časopis Za More i Pomorstvo, 65(4), 259-265. https://doi.org/10.17818/NM/2018/4SI.18
[43] Driving simulator SNA - 211 REN: Operating instructions. Olomouc: JKZ spol. s r.o., 2016. 25 p.
[44] Ondruš, J., & Hockicko, P. (2015). Braking deceleration measurement using the video analysis of motions by sw tracker. Transport and Telecommunication, 16(2), 127-137. https://doi.org/10.1515/ttj-2015-0012
[45] Vrábel, J., Sarkan, B., Vashisth, A. (2020). Change of driver's reaction time depending on the amount of alcohol consumed by the driver - the case study. The Archives of Automotive Engineering – Archiwum Motoryzacji, 87(1), 47-56. https://doi.org/10.14669/AM.VOL87.ART4.
[46] Yadav, A. K., & Velaga, N. R. (2019). Modelling the relationship between different blood alcohol concentrations and reaction time of young and mature drivers. Transportation Research. Part F, Traffic Psychology and Behaviour, 64, 227-245. https://doi.org/10.1016/j.trf.2019.05.011.
[47] Dirnbach, I., Kubjatko, T., Kolla, E., Ondruš, J., & Šarić, Ž. (2020). Methodology designed to evaluate accidents at intersection crossings with respect to forensic purposes and transport sustainability. Sustainability (Basel, Switzerland), 12(5), 1972. https://doi.org/10.3390/su12051972.

IOP Conf. Series: Materials Science and Engineering 1247 (2022) 012035 doi:10.1088/1757-899X/1247/1/012035

Comparison of driver awareness in real traffic and driving on a simulator

J. Beňuš[1], M. Poliak[1*]

[1] Department of Road and Urban Transport, University of Zilina, Univerzitná 1, 010 26 Žilina, Slovakia

* milos.poliak@fpedas.uniza.sk

Abstract. A driver's awareness is a critical aspect that must be closely monitored at all times. This article proposes a number of experiments that will be carried out in the near future, the findings of which may aid in improving the existing condition. The authors feel that the results of individual measurements can help to improve road freight transportation safety. Furthermore, the contribution focuses on driver drowsiness as a significant factor and mechanism of major and fatal traffic accidents, as well as extensive damage to goods or property in road freight transit. Among other things, detection of driver drowsiness was determined. The individual measurements are described and the result of a driver on the simulator are compared with a motorist driving in real traffic. With this information obtained from the individual measurements, we can evaluate number of driver mistakes and different signs of tired behaviour per one shift and then we could be able to estimate a maximum driving duration of one day for twenty-four hours day.

Keywords: road safety, safety, road accident

1. Introduction

A driver's awareness is an extremely vital component that must be constantly monitored. A drowsy driver can create a variety of mistakes and accidents on highways, resulting in financial loss, physical damage, and, most importantly, human life loss [1]. Accidents caused by fatigue are more prevalent than those caused by drunk driving. Lack of sleep, frequent driving, narcotics, and prescription can all cause drivers to feel sleepy. If a motorist falls asleep for four seconds while driving at 100 km/h, the automobile will have travelled 111 meters without a driver attention. When traveling at high speeds, there is a greater chance of death or serious injury in the event of a collision. Driver drowsiness has a role in around 20% of fatal traffic accidents. Fatigue is a leading cause of collisions in the Australian state of Victoria, with over fifty people killed and 300 seriously injured each year [2]. We must also address sleepiness, since we have convincingly proved that it is a major cause of accidents [3].

The intended research to determine driver sleepiness in road freight transport is detailed in this article. In the Slovak Republic and the European Union, the rules on driving times, breaks and rest periods for drivers engaged in road haulage and passenger transport are laid down in Regulation (EC) No 561/2006 of the European Parliament and of the Council. The aim of this Regulation is to improve working conditions and road safety, to promote better monitoring and enforcement by Member States and to improve working practices in the road transport sector [4].

We think that by organizing measures of driver sleepiness in real-world traffic conditions, as well as measurements on a simulator, we will be able to confirm or contribute to the improvement of present regulations, particularly in the sector of international road freight transport. The proposed study will be divided into two parts and will include a variety of measurements that will offer us with diverse views on the challenges at hand. Methodology of measurements are presented in the first section, with the findings being assessed and outputs being given in the second part name results. The initial set of measures will be carried out on a simulator, with the driver's face being observed by a camera. The second measurement will take place in real-time traffic, with the faces of the chosen driver being watched by a camera while driving in real-time traffic. The final section of the report focuses on assessment and future measuring recommendations. This research is being used as a pilot test for more extensive measures to be conducted in the future.

2. Detection of driver drowsiness

The raw measurement and the processing steps used to convert measurements into characteristics can be used to identify input sources for drowsiness detection algorithms. Measures explored in the literature include heart rate [5], brain activity [6, 7], eye closure and tracking [6, 8, 9], lane position [10], and steering-wheel angle [11, 12]. Although most previous algorithms focus on a one type of measure, several employ a combination of measures [8, 10, 13, 14]. The most commonly applied and theoretically rigorous measures are electro-encephalogram (EEG), percent eye-closure over a fixed time window (PERCLOS), and steering-wheel angle [15]. EEG has the benefit of having a well-established correlation between spectral patterns in the signal and the transition between awake and sleep [7]. The quantity of pre-processing necessary before categorization, the sensitivity to artifacts, and the feasibility of collecting EEG from drivers in real-world scenarios are all factors that restrict EEG. PERCLOS, developed by Wierwille et al. (1994) [9], is the gold standard measure for drowsiness detection. PERCLOS predicts drowsiness based on the percentage of time an individual's eyes are more than 80% closed over a 2-min period. Dinges et al. (1998) [6] demonstrated that the PERCLOS algorithm had over 90% accuracy in detecting degraded performance during a vigilance task, which was more reliable across drivers than EEG, blinks, and head position in the study. PERCLOS is a ground truth sleepiness measure that has been integrated into aftermarket equipment such as the Co-pilot [9]. PERCLOS, despite its widespread adoption, has a number of practical drawbacks. Because the existing camera technology necessary for its measurement is expensive, has not been fully verified, and may be inaccurate when the driver wears sunglasses or when weather conditions generate large quantities of glare, PERCLOS for real-time detection is limited [15]. Despite these drawbacks, the extensive evidence of PERCLOS' effectiveness implies that it might be beneficial for testing new algorithms [16]. Nor can we forget to use Subjective Methods, which can help improve the level of experiments performed, as well. The Stanford sleepiness scale [17] and Karolinska sleepiness scale [18] are the two most commonly used subjective methods. The Stanford scale is a seven-point scale that describes a person's present level of sleepiness. On the other hand, the Caroline Sleepiness Scale is a 9-point scale. This scale is said to be more comprehensive, as it can classify driver drowsiness into various degrees [19]. The authors are considering applying the Stanford sleepiness scale in future studies, in which the level of tiredness of the driver would be assessed using a questionnaire and expert observation.

Table 1. THE STANFORD SLEEPINESS SCALE. [17]

Value	Description
1	Feeling active, vital, alert, or wide awake
2	Functioning at high levels, but not at peak; able to concentrate
3	Awake, but relaxed; responsive but not fully alert
4	Little foggy; not at peak
5	Foggy; losing interest in remaining awake; slowed down
6	Sleepy; woozy; fighting sleep; prefer to lie down
7	No longer fighting sleep; sleep onset soon; cannot stay awake

3. Methodology

The first experiment will take place in a driving simulator, as its shown in Figure 1. The data acquired while driving on the simulator will form the basis of the first portion of the findings. Throughout the journey, data will be captured and saved utilizing the simulator's software and video camera. The simulator keeps track of the vehicle's condition and the driver's actions (e.g., lane position, maximum and average speed, distance travelled, driving time, shifting without clutch, car collision with external object, clutch pedal position ...). It will take eleven hours to complete the tour, which will feature a genuine driving experience. To avoid the potential learning effect, four alternative driving conditions were employed, one for each ride. The same simulator events appear in each scenario, but they are presented in a different order. Data from each drive will be analysed in a variety of environments. Video inputs such as yawning, head loss, eye movements, blinking, and closure of the eyelids are recorded on the memory card installed inside the camera. Before the measures, the first visit will be a screening, introduction visit, and the second visit will be the measurements, which is expected to be a one-day trip starting early in the morning. The measurement and testing portion of the equipment has already been completed. We will be able to compare the driver's concentration while driving in the simulator to the result of the driver's concentration derived from camera recordings of the driver's driving in actual traffic based on the findings of this measurement

Figure 1. Driving simulator [own processing]

The measurements will be completed in a real vehicle with a professional driver for the second half of results. The second measurement will be based on video footage collected from a motorist who is being observed while driving in actual traffic, as shown in Figure 2. The information will be gathered during the day and will comprise a variety of truck driver journeys from Slovakia to various European nations. The driver's level of driving safety may vary depending on the travel trip. Several indicators are used to reflect various characteristics of dangerous driving behaviour [20]. The driver will provide extensive data, which will be analysed to determine if the present ten-hour driving law is safe and effective.

Figure 2. Video record of driver in real traffic [own processing]

At the end of the presented methodology is necessary to report that for evaluation of drivers' awareness will be used subjective method because software with PERCLOS is not available, yet. Authors are aware of possibility that this evaluation could lead to different results than PERCLOS method, which was and is also planning to be used for future measurements. In the future evaluation of extensive data, Junaedi and Akbar drowsiness detection method is expected to be used [21]. To do this, the conventional confusion matrix will be used to generate the positive and negative prediction scores of each model and then deduct several performance measures from those scores [22]. On the other side, it should be noted that using a subjective method can be consider as accurate as PERCLOS due to the detailed and responsible evaluation of authors.

4. Results

The first study we conducted was to see how conscious a driver is when driving in a simulator. With a maximum driving duration of four and a half hours and a break of at least forty-five minutes, the driver complied with existing valid social norms. The identification of yawning, tears, not concentrating, and taking hands off the driving wheel was the result of this investigation. The motorist first drove for four and a half hours on the highway before taking a 45-minute rest. The second time of the ride took place in the city lasting two hours. After a fifteen-minute break, the participants spent driving two and a half hours in a mountain setting. The driver had to take the second portion of the thirty-minute rest before starting the last driving segment. The remaining two hours of driving were spent in hill and curve mode. The overall driving duration for the driver was eleven hours, as intended, and as permitted by a new update to Regulation (EC) No 561/2006 enacted on August 20, 2020. The driver's face and upper body were scanned by a video camera during the simulation, and the simulator program also documented driver errors. Picture of the driver driving on the simulator is shown in Figure 3.

Figure 3. Video record of driver driving on a simulator [own processing]

Not just yawning, but also cases of loss of head and support with one hand, as well as reduced response time, were shown to be more prevalent in the first portion of the assessment than in the second. On the other hand, we can detect multiple errors in the second portion of the assessment that may have resulted in traffic accidents, but the majority of the errors were due to distractions, such as phone calls or SMS messages, rather than indicators of exhaustion. The driver has lost attention to the traffic condition as a result of this major present problem when using a cell phone, and this might jeopardize traffic safety. The evaluation of driver errors and the symptoms of fatigue behavior are shown in Figure 4.

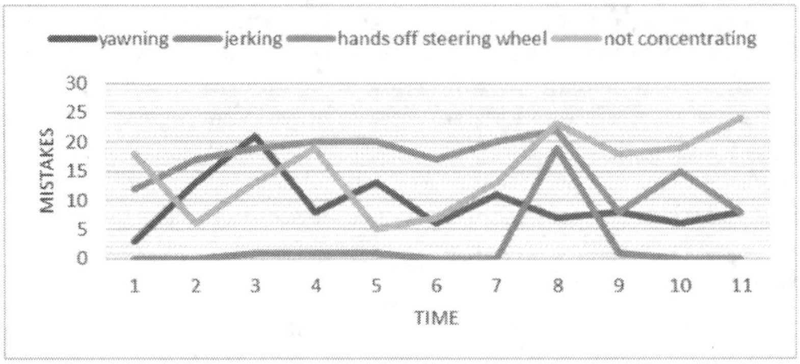

Figure 4. Evaluation of driver on the simulator [own processing]

We ran second research to determine drivers' attention while driving in regular traffic. With a maximum driving time of ten hours every classic day, the driver respected all applicable social requirements. The results of this measurement were the same as in the simulated measurement. We looked for indicators of exhaustion such as yawning, sleepy behavior, loss of concentration, and handshaking behind the wheel. The driver initially drove for half an hour on a standard road in Slovakia, then took a fifteen-minute rest before crossing the border. The driver then drove for over four hours before taking the second half of the thirty-minute rest period. After the break, the route was driven for four and a half hours. He had to take a second 45-minute rest before the driver could begin the final part of the journey. The last hour of the journey took place off the highway at the unloading spot when the driver also stopped for a minimum of nine-hour daily rest. The driver's total driving time and performance on that day was about ten hours, and it took place during sunny days and early evenings. During the whole journey, the driver was observed by a camera that recorded his face and a portion of his body from the waist up, but the camera was switched off during the driver's breaks. During the whole trip, we saw notable indicators of tired behavior. We can see that before the last hour of driving, indicators of exhaustion such as yawning and head dropping began to appear in the second portion of the assessment, but after a 45-minute rest, all symptoms disappeared. On the other side, we can notice a number of mistakes throughout the measurement that might threaten road safety, and the majority of the errors were due to phone distraction, either by calling or texting, as was seen in the simulator measurement, as well. Evaluations of driver errors and fatigue behavioral symptoms are shown below in Figure 5.

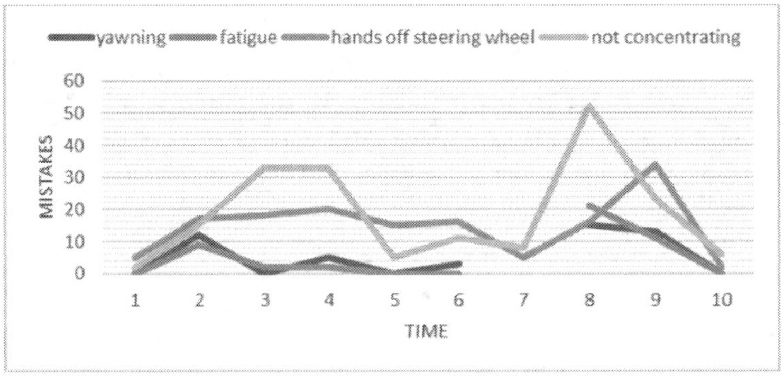

Figure 5. Evaluation of driver in real traffic [own processing]

5. Discussion

According to the World Health Organization's (WHO) road safety study, nearly 1.2 million people died each year due to driver drowsiness between 2001 and 2013 [23]. Because of the sensitivity of this problem, multiple kinds of study have been conducted to prevent tired drivers from operating a motor vehicle. One option is to educate drivers about the hazards of driving while sleeping. This method entails recognizing the consequences of lack of sleep, exhaustion induced by lengthy periods of repetitive driving, as well as other types of labour [24].

Increasing the safety of goods transport and hauliers themselves is a priority of various international associations [25]. There are around 6 million trucks in use in the EU. Since 2010, the number of trucks has increased by less than 2% per year [26]. Parking is possible for a smaller number of vehicles because of inappropriate parking and more and more drivers face a problem with full parking areas [27]. In addition, it is important to remember that drivers and hauliers must consider the type and size of the vehicle when planning and finding places to take breaks and rest periods, as well as the type of fuel needed [28]. On the other hand, the equipment of parking areas include safety, information and cleanliness are important for drivers' comfort [29].

One of the goals of this research is to develop educational materials for training centres that provide professional driver training on the effects of fatigue, distraction, and rest on drivers while driving by implementing and presenting individual measurements, as well as detailed documentation and evaluation. The driver, with his driving technique, can significantly influence the vehicle's consumption [30]. Different measures are planned to integrate subjective, physiological, and behavioural methodologies for a comprehensive approach to study, which should provide us with a better complete image of the situation. The most accurate approaches include physiological measures such as EEG, ECG, and electrooculogram (EOG). The transmission of electrical impulses from the driver within the body, which can detect immediate changes in the driver's awareness, ensures their great accuracy. These existing technologies are unworkable in their present implementations because they need invasive electrodes, expensive cameras, and complicated equipment that can only be used in laboratory conditions [31]. As a result of this issue, this device will only be utilized to take measurements on the simulator. Simulation tests are conducted worldwide by various institutes and organizations [32].

Our goal is to evaluate the driver's level of awareness while driving a truck for ten hours, which is the maximum daily driving time allowed according to Regulation 561/2006. In this experiment, we extended driving time up to eleven hours and from the result we can review that driver was able to drive safely. In future simulator trials, we would like to prolong daily driving time to maximum levels at which the driver can still drive safely and respond to external stimuli such as answering questions. With this information, we can estimate a maximum driving duration of one day for twenty-four hours day. This will be augmented with data collected in the simulator utilizing a behavioural technique such lane departure, eye opening and closing time, and so on. Behavioural techniques should assist in improving the picture of driver weariness that results. The driver's face will be taken during the two-week operating mode in the second future experiment, which will take place on real trucks, as contrasted to the one-day measures that were done this time as a pilot experiment, right now.

In contrast to the previous experiment, which applied a subjective way to quantify driver weariness, the next assessment will use the most well-known behavioural method and the PERCLOS methodology. When compared to EEG, PERCLOS produces almost identical findings, but it is easier to install and has a lower purchase price. We should be able to recognize and quantify yawns, head loss, attention loss, and concentration based on each driver's facial camera. We think that a variety of investigations and methodologies will generate data that we will be able to maximize in the future to improve driver training and modify present legislation regulating driver shifts in road freight transportation. With our initial simulator study, we hope to demonstrate that drivers are capable of driving safely for greater periods of time than currently defined by law.

We would like to confirm that the existing regulation is insufficient and that drivers may drive for longer than is permitted in a second experiment conducted in partnership with a professional driver. Compared to current social legislation in the United States, European Union drivers can drive four hours less per week and up to thirty hours in two weeks [33]. Based on the analysis of accident statistics from the EU and the USA, it was found that the situation in the EU is better than in terms of accident rates over a period of four years from 2009 to 2012. Even though the total number of fatal accidents in the EU and the USA is similar [34-36]. We consider that detailed study might provide a solid foundation for changing present regulations without increasing the number of traffic offences and accidents.

6. Conclusion

Based on the date obtained from conducted experiments, we intend to collect further data from future individual measurements in order to better identify the current condition of driver behaviour in road freight transport. Information retrieved from large amounts of data may be used to create effective alternative measurements of traffic exposure and help policymakers make better decisions. Although this study employs a very accurate simulator, it does not perfectly duplicate the experience of real traffic, therefore we want to compare two types of independent research from real traffic and the simulator to enhance the results. This future research should use information gathered from a variety of metrics and a total of at least 25 drivers. Driver behaviour is influenced by a variety of circumstances.

We are now unable to fully include personal characteristics of drivers (e.g., age, gender, education...) and psychological problems due to the limited sample size of drivers in individual measures (e.g., agitation, depression...). Despite these disadvantages, this study will offer a systematic strategy to evaluating possible dangers among various truck drivers. One point of contention is which approach and algorithm should be utilized to assess the outcomes. The use of a driving simulator, the research's methodology, the extent of the test results, and the scope of the core fact regarding sleepiness are all limitations of this study. These future results could be combined with safety interventions to address the causes of driving drowsiness, such as incorrect working patterns and sleep disorders, in order to create comprehensive systems that will improve driving safety [20].

References

[1] Ghole, U., Chavan, P., Gandhi, S., Gawde, R., & Fakir, K. (2020). DROWSINESS DETECTION AND MONITORING SYSTEM. In ITM Web of Conferences (Vol. 32, p. 03045). EDP Sciences. DOI:10.1051/itmconf/20203203045.

[2] Transport Accident Commission, (2021). Available Online at: https://www.tac.vic.gov.au/road-safety/statistics/summaries/fatigue-statistics.

[3] Apoorva, A., & Vali, D. (2020). Review on Drowsiness Detection. EAI Endorsed Transactions on Smart Cities, 20(12). DOI: 10.4108/eai.13-7-2018.165517.

[4] Regulation (EC) No 561/2006 of the European Parliament and of the Council of 15 March 2006 on the harmonization of certain social legislation relating to road transport and amending Council Regulations (EEC) No 3821/85 and (EC) No 2135/98 and repealing Council Regulation (EEC) No 3820/85. https://eur-lex.europa.eu/legal-content/SK/TXT/?uri=CELEX%3A02006R0561-20200820&qid=1648449087711.

[5] Furman, G. D., Baharav, A., Cahan, C., & Akselrod, S. (2008, September). Early detection of falling asleep at the wheel: A heart rate variability approach. In 2008 Computers in Cardiology (pp. 1109-1112). IEEE. DOI: 10.1109/CIC.2008.4749240.

[6] Dinges, D. F., Mallis, M. M., Maislin, G., & Powell, J. W. (1998). Evaluation of techniques for ocular measurement as an index of fatigue and as the basis for alertness management (No. DOT-HS-808-762). United States. National Highway Traffic Safety Administration. https://rosap.ntl.bts.gov/view/dot/2518.

[7] Lal, S. K., Craig, A., Boord, P., Kirkup, L., & Nguyen, H. (2003). Development of an algorithm for an EEG-based driver fatigue countermeasure. Journal of safety Research, 34(3), 321-328. DOI: 10.1016/S0022-4375(03)00027-6.

[8] Ji, Q., Zhu, Z., & Lan, P. (2004). Real-time nonintrusive monitoring and prediction of driver fatigue. IEEE transactions on vehicular technology, 53(4), 1052-1068. DOI: 10.1109/TVT.2004.830974

[9] Wierwille, W. W., Wreggit, S. S., Kirn, C. L., Ellsworth, L. A., & Fairbanks, R. J. (1994). Research on vehicle-based driver status/performance monitoring; development, validation, and refinement of algorithms for detection of driver drowsiness. final report (No. HS-808 247). https://rosap.ntl.bts.gov/view/dot/2578.

[10] Hanowski, R. J., Bowman, D., Alden, A., Wierwille, W. W., & Carroll, R. (2008, November). PERCLOS+: Development of a robust field measure of driver drowsiness. In 15th World Congress on Intelligent Transport Systems and ITS America's 2008 Annual Meeting, New York, NY. ISBN: 978-161567756-6.

[11] Eskandarian, A., & Mortazavi, A. (2007, June). Evaluation of a smart algorithm for commercial vehicle driver drowsiness detection. In 2007 IEEE Intelligent Vehicles Symposium (pp. 553-559). IEEE. DOI: 10.1109/IVS.2007.4290173.

[12] Krajewski, J., Golz, M., & Sommer, D. (2009). Detecting sleepy drivers by pattern recognition based analysis of steering wheel behaviour. Der Mensch im Mittelpunkt technischer Systeme, VDI-Verlag: Düsseldorf, Germany, 2009; Volume 29, 288-291.

[13] Forsman, P. M., Vila, B. J., Short, R. A., Mott, C. G., & Van Dongen, H. P. (2013). Efficient driver drowsiness detection at moderate levels of drowsiness. Accident Analysis & Prevention, 50, 341-350. DOI: 10.1016/j.aap.2012.05.005.

[14] Zilberg, E., Xu, Z. M., Burton, D., Karrar, M., & Lal, S. (2007, August). Methodology and initial analysis results for development of non-invasive and hybrid driver drowsiness detection systems. In The 2nd International Conference on Wireless Broadband and Ultra-Wideband Communications (AusWireless 2007) (pp. 16-16). IEEE. DOI: 10.1109/AUSWIRELESS.2007.44.

[15] Balkin, T. J., Horrey, W. J., Graeber, R. C., Czeisler, C. A., & Dinges, D. F. (2011). The challenges and opportunities of technological approaches to fatigue management. Accident Analysis & Prevention, 43(2), 565-572. DOI: 10.1016/j.aap.2009.12.006.

[16] McDonald, A. D., Lee, J. D., Schwarz, C., & Brown, T. L. (2018). A contextual and temporal algorithm for driver drowsiness detection. Accident Analysis & Prevention, 113, 25-37. DOI: 10.1016/j.aap.2018.01.005.

[17] Hoddes, E. et al., (1973). Quantification of Sleepiness: A New Approach. Psychophysiology, 10(4). DOI: 10.1111/j.1469-8986.1973.tb00801.x.

[18] Åkerstedt, T. & Gillberg, M., (1990). Subjective and objective sleepiness in the active individual. International Journal of Neuroscience, 52(1-2). DOI: 10.3109/00207459008994241.

[19] Putilov, A.A. & Donskaya, O.G., (2013). Construction and validation of the EEG analogues of the Karolinska sleepiness scale based on the Karolinska drowsiness test. Clinical Neurophysiology, 124(7), pp.1346 – 1352. Available at: http://dx.doi.org/10.1016/j.clinph.2013.01.018.

[20] Zhou, T., & Zhang, J. (2019). Analysis of commercial truck drivers' potentially dangerous driving behaviors based on 11-month digital tachograph data and multilevel modeling approach. Accident Analysis & Prevention, 132, 105256. DOI: 10.1016/j.aap.2019.105256

[21] Suhandi J., Habibullah A., (2018, September). Driver drowsiness detection based on face feature and PERCLOS. In Journal of Physics: Conference Series (Vol. 1090, No. 1, p. 012037). IOP Publishing. DOI: 10.1088/1742-6596/1090/1/012037.

[22] Bamidele, A., Kamardin, K., Syazarin, N., Mohd, S., Shafi, I., Azizan, A., & Mad, H. (2019). Non-intrusive driver drowsiness detection based on face and eye tracking. Int J. Adv. Comput. Sci. Appl, 10, 549-569. DOI: 10.14569/IJACSA.2019.0100775.

[23] World Health Organization, (2015). Global status report on road safety 2015. , pp.1–16. Available at: file:///C:/Users/AUTOSK~1/AppData/Local/Temp/9789241565066_eng.pdf

[24] Williamson, A. et al., (2014). Are drivers aware of sleepiness and increasing crash risk while driving? Accident Analysis and Prevention, 70, pp.225–234. Available at: http://dx.doi.org/10.1016/jaap.2014.04.007.

[25] Kubanova, J., Senko, S., Duranova, A., Bielikova, A., & Faturikova, E. (2020, October). Safe and secure truck parking places in EMEA region. In 2020 XII International Science-Technical Conference AUTOMOTIVE SAFETY (pp. 1-6). IEEE. DOI: 10.1109/AUTOMOTIVESAFETY47494.2020.9293518.

[26] Kubáňová, J., Kubasáková, I., & Dočkalik, M. (2021). Analysis of the Vehicle Fleet in the EU with Regard to Emissions Standards. Transportation Research Procedia, 53, 180-187. DOI: 10.1016/j.trpro.2021.02.024.

IOP Conf. Series: Materials Science and Engineering 1247 (2022) 012035 doi:10.1088/1757-899X/1247/1/012035

[27] Kalašová, Alica, et al. Analysis and Proposed Solutions for Parking in Rajecké Teplice. In: International Conference on Transport Systems Telematics. Springer, Cham, 2020. p. 259-271. DOI: 10.1007/978-3-030-59270-7_19.
[28] Gnap, J., & Dočkalik, M. (2021). Impact of the operation of LNG trucks on the environment. Open Engineering, 11(1), 937-947. DOI: 10.1515/eng-2021-0096.
[29] Konečný, V., Brídziková, M., & Semanová, Š. (2020, October). Assessing the Impact of Transport Service Quality on the Demand of Passengers for Bus Transport. Application of the Assessment Methodology in the Žilina Self-governing Region in the SR. In 2020 XII International Science-Technical Conference AUTOMOTIVE SAFETY (pp. 1-9). IEEE. DOI: 10.1109/AUTOMOTIVESAFETY47494.2020.9293508.
[30] Loman, M., Šarkan, B., & Skrúcaný, T. (2021). Comparison of fuel consumption of a passenger car depending on the driving style of the driver. Transportation research procedia, 55, 458-465. DOI: 10.1016/j.trpro.2021.07.009.
[31] Lawoyin, S.A., Fei, D.-Y. & Bai, O., 2014. A Novel Application of Inertial Measurement Units (IMUs) as Vehicular Technologies for Drowsy Driving Detection via Steering Wheel Movement. Open Journal of Safety Science and Technology, 04(04), pp.166–177. Available at: http://www.scirp.org/journal/PaperDownload.aspx; DOI: 10.4236/ojsst.2014.44018.
[32] Jaśkiewicz, M., Frej, D., & Poliak, M. (2022). Simulation of a Dummy Crash Test in ADAMS. Communications-Scientific letters of the University of Zilina, 24(1), B20-B28. DOI: 10.26552/COM.C.2022.1.B20-B28.
[33] Federal Motor Carrier Safety Administration (FMCSA), September 28, 2020; https://www.fmcsa.dot.gov/regulations/hours-service/summary-hours-service-regulations.
[34] Kapusta, J., & Kalašová, A. (2016). A comparison of truck driver safety between the EU and the USA. Zeszyty Naukowe. Transport/Politechnika Śląska. DOI: 10.20858/sjsutst.2016.93.6.
[35] Jurecki, R., & Jaskiewicz, M. (2012) Analysis of road accidents in Poland over the last ten years, Scientific Journals of the Maritime University of Szczecin-Zeszyty Naukowe Akademii Morskiej w Szczecinie 32 (104) Available at: https://repository.am.szczecin.pl/handle/123456789/71
[36] Szumska, E., Frej, D., & Grabski, P. (2020) Analysis of the Causes of Vehicle Accidents in Poland in 2009-2019 LOGI – Scientific Journal on Transport and Logistics, 11(2), DOI: https://doi.org/10.2478/logi-2020-0017.

Acknowledgments
This paper has been developed under support of project: VEGA No. 1/0245/20 Poliak: Identification of the impact of a change in transport related legislation on the competitiveness of carriers and carriage safety

Knowledge centre for transport accessibility and mobility of people with special needs using individual transport modes

J Jackowski [1], M Gidlewski[1,2*] and P Posuniak[2]

[1] Military University of Technology (WAT), Poland
[2] ŁUKASIEWICZ Research Network – Automotive Industry Institute (ŁUKASIEWICZ-PIMOT), Poland

* miroslaw.gidlewski@wat.edu.pl

Abstract. At the Military University of Technology in Faculty of Mechanical Engineering has been estabilished a Knowledge centre for transport accessibility and mobility of people with special needs (PwSN). The main aim of the Centre's activities is to extend issues of universal design and to develop, promote and disseminate innovative product solutions and universal service standards leading to increased mobility and improved road safety for PwSN using personal transport. An additional aim is to support teaching staff in preparation for education and in the implementation of training activities in the field of universal design, counteracting the exclusion of PwSN. The activities of the Centre involve the Faculty of Mechanical Engineering, the Institute of Psychology at Cardinal Stefan Wyszyński University and the Motor Transport Institute. The article presents in detail the intentions and expected results of the activities carried out by the Centre. It is assumed that the Centre's activities will make it possible to develop and disseminate original solutions of systems and devices facilitating and supporting the use of individual means of transport by PwSN.

Key words: disability, automotive safety, mobility

1. Introduction

More than a century of dynamic development of the motor vehicle industry has made vehicles of various forms and designs and construction have become a common good and an irreplaceable means of facilitating the movement of people and the transport of goods. Nowadays, the lack of free access to means of transport or its restriction is practically equivalent to exclusion from normal social and economic life and social isolation. The group most vulnerable to discrimination and marginalisation in terms of free and safe movement are people with special needs (PwSN).

According to [1], PwSN are people who, because of their external or internal characteristics or because of the circumstances in which they find themselves, have to take additional measures or apply additional means to overcome a barrier in order to participate in various spheres of life on an equal basis with other people. PwSN are nowadays a very large social group, which will steadily increase in the coming years. This group includes seniors (60/65+ people), people with disabilities (PwD) with permanent or temporary dysfunctions, sick people, people taking medication that prevents them from driving, people with unusual body structures, pregnant women, veterans and road accident victims with post-traumatic stress disorder. All of the above are or will become elderly people (60/65+). In the EU countries (and Poland) we observe a steady increase in the number of the elderly [2]. This is a result of increasing life expectancy and the baby boomers entering retirement age. According to the Central Statistical Office (CSO), at the end of 2020 people aged 60+ constituted about 25% of the population in Poland and it is projected that in 2030 1/3 of the population will be people over 60 years of age, and in 2050 the share of elderly people in the Polish population may reach 40%. It is also predicted that areas with the highest

percentage of people 60+ will be non-urban, away from the agglomerations and relatively sparsely populated, which also have a lower percentage of working age population [3].
For shorter trips, PwSN mainly use individual transport, i.e. bicycles, mopeds, motorbikes and cars. For longer trips they more often use public transport. The main criteria for mode choice are accessibility and safety [4]. Elderly people identify their own car as the best means of transport. The personal car, as opposed to public transport, is familiar to these people, as they have used this means of transport for many years. For many of them it is a symbol of freedom and independence as well as a higher material status and therefore they are reluctant to give it up [5], [6].
Unfortunately, a natural phenomenon accompanying the process of aging of PwSN is deterioration (often significant) of physical and psychomotor efficiency, which causes difficulties with the reception and rapid assessment of information provided by the environment in road traffic and as a result, slows down the reaction to the emerging threats [7], [8]. This makes results in a deterioration of the safety of PwSN actively participating in road traffic and other road users [9]. Nevertheless, with the continuous increase in the offer of different means of transport, PwSN will make more and longer trips in the future, if this offer includes solutions that take into account the specific needs of this group of people [10]. Car drivers will travel more frequently than current elderly drivers - this is especially true for women. Furthermore, it must be assumed that the transport and mobility requirements of PwSN in the future will be much higher than those of the current generation [11]. Thus, ensuring mobility for this group of society is a fundamental condition for creating and maintaining their active and high standard of living.
In order to keep up with demographic change and the needs of PwSN, anticipatory measures are needed to increase the mobility of this group of people and at the same time not to deteriorate road safety. One of such actions is the improvement of and adaptation of individual means of transport to specific needs and requirements of PwSN, in effect of using universal design of vehicles and their additional equipment taking into account needs of all their users [12].
The research carried out so far has identified the simplest but important examples of small vehicle adaptations in terms of components such as seats, doors, handles, brackets, supports, steps, knobs, steering wheel or seat belts. Many of these modifications can be made to standard vehicles at little cost and time. These adaptations should increase the overall availability of vehicles for PwSN [13]. At this point, it should be stressed that currently in Poland almost 80% of the passenger cars in operation are vehicles older than 10 years and nothing indicates a change in the age structure of the vehicles in operation in the coming years. Therefore, innovative projects and products facilitating the use of passenger cars by PwSN should also, or perhaps first of all, concern vehicles already in use.
At the Military University of Technology (MUT) in Faculty of Mechanical Engineering (FME) has been estabilished a Knowledge centre for transport accessibility and mobility of PwSN (Centre). This was made possible by obtaining for implementation the project No. POWR.03.05.00-00-CW07/20 co-financed by the European Social Fund under the Operational Programme Knowledge Education Development 2014-2020. The established Centre is the first in Poland operating in the field of transport [14]. It is assumed that the Centre's activities will at least partially contribute to improving the free and safe movement of PwSN in the near and distant future. The article presents the aims, objectives and expected effects of the activities carried out by the Centre.

2. The aims and the structure of the Centre

The main aim of the Centre is to develop, in cooperation with the wider socio-economic environment, issues of universal design and to create, promote and disseminate innovative product solutions and universal service standards leading to increased mobility and improved road safety for PwSN using means of individual transport. The meaning of individual transport are considered to be: cars and vans, motorbikes, mopeds, bicycles and wheelchairs to facilitate the movement of PwSN.
The indirect aim, leading to the achievement of the main aim, is to support the teaching staff of the Military University of Technology (MUT) and Cardinal Stefan Wyszyński University (CSWU) in their preparation for education and in the implementation of training activities in the field of universal design, counteracting the exclusion of PwSN from normal use of individual means of transport.
Three entities are involved in the Centre's activities:

- Military University of Technology (FME) MUT – leader;

- Institute of Psychology (IP) CSWU – partner no. 1;
- Motor Transport Institute (MTI)– partner no. 2.

Establishing a partnership with CSWU and MTI was primarily due to the need to build interdisciplinarity so important in developing issues related to universal design particularly focused on the needs of PwSN. The Faculty of Mechanical Engineering has a large number of research and teaching staff who have professional experience in working with PwD and with organisations working on behalf of PwSN. The department conducts extensive research in the fields of mechanical design innovation, biomedical engineering and road safety. The knowledge and experience of the faculty gained while conducting research on modelling of complex engineering structures, innovative remote-controlled structures, ergonomic assessment of structures, orthoses, biomechatronics of human movement, road traffic hazards, safety of drivers and passengers of vehicles, and behaviour of people with disabilities in vehicles will be particularly helpful in the Centre's activities.

The Institute of Psychology conducts extensive research in all fundamental areas of psychology. The experience gained in conducting research on building mathematical models of psychological processes, psychometric analyses, cognitive processes (e.g. perception, memory, etc.) and the effects of medication and fatigue on cognitive functioning will be used in the Centre's activities.

Motor Transport Institute (MTI) occupies a high position as a leading scientific-research institution in Poland in the field of motor transport. The Polish Road Traffic Safety Observatory operates at MTI. People with musculoskeletal dysfunctions occupy an important place in the Institute's research activity. As a result of the work carried out by MTI, in 2015 the Centre for Automotive Services (CAS) for People with Disabilities at MTI was opened - the first institution in Poland to provide a comprehensive mobility support system for people with mobility impairments.

All activities conducted by the Centre will be implemented by interdisciplinary teams consisting of representatives of each of the above mentioned entities. in the work of the Centre participate 52 people (29 peple from FME, 12 people from IP, 7 people from MTI and 4 external experts). These people are organised in 7 teams.

3. Activities carried out by the Centre

It is assumed that the activities carried out by the Centre will, at least in part, contribute to solving the important problems facing our society in the future in relation to transport and mobility of PwSN. The Centre's activities will be multi-pronged.

The first will be technical support for the design and production of devices and components to adapt vehicles under development, in production or already in use to the needs and requirements of people with special needs. The development of innovative vehicle systems and equipment in the context of the needs of users with special needs should be aimed at:

- practicality, comfort and ease of use of the vehicle (in particular, ease of getting on, off and moving within the vehicle, ease of packing and unpacking luggage) ease of boarding, alighting and moving about inside the vehicle, ease of packing and unpacking luggage) and ease of use of all equipment used in the vehicle;
- safety in road traffic, in pedestrian, bicycle and other traffic lanes (in particular: introduction of devices warning against traffic hazards and systems assisting the driver in difficult road manoeuvres).

In order for this to be effectively realised, it is necessary to adequately prepare academic research and teaching staff who, themselves fully informed, will be able to pass on to students - future and current engineers and designers - a sensitivity to PwSN problems together with the principles of universal design.

The second measure will be showing the problems of PwSN widely known to the public and to raise awareness that manual dexterity, responsiveness to external stimuli and decision-making decrease with age. Improving this awareness among the target group and decision-makers at various levels, which is a long-term process, should lead to creating demand for products and services resulting from the first measure. Another social aspect will be the provision of reliable knowledge resulting from analyses and research, which will facilitate individual decision-making for the people concerned (regarding research,

training, equipment purchases), and for decision-makers in creating policies and long-term social programmes. These activities must be supported by PwSN education to raise awareness of road traffic risks.
The third activity will be the coordinated, systematic research and analysis of the introduced technical and technological improvements, quality of service standards and their verification in terms of their usefulness and effectiveness in reducing adverse road incidents involving PwSN. Information on both useful and unhelpful solutions should be published along with reliable justification, preferably confirmed by industry authorities. The process approach, in a longer perspective, has a chance to create high quality expertise, which should translate into public confidence. It is also important that verified knowledge in this area is regularly updated and concentrated in one place to facilitate access to it by all interested parties.
In particular, the Centre's activities will consist in:

1) acquiring, collecting, analysing, developing, exchanging and making available (to all stakeholders) knowledge covering the problems and barriers related to the mobility of PwSN using means of personal transport;
2) to develop, update and promote universal design standards in the field of transport and mobility for PwSN, with particular reference to ergonomics and methods of selection of vehicle control aids and adaptive devices;
3) provide psychological and therapeutic support to PwSN in the context of their mobility and transport;
4) analysing PwSN transport and mobility regulations and legislation and proposing legislative amendments;
5) conducting surveys of PwSN on current topics developed by the Centre's teams;
6) monitoring the usefulness and effectiveness of new systems and equipment in vehicles in the context of the needs of their users with special needs;
7) organising national competitions for proposals and concepts of devices, objects and services to improve mobility and road safety of PwSN;
8) developing project concepts for innovative products and service standards friendly to PwSN in the area of transport;
9) preparing own curricula and providing opinions on other curricula for teaching staff, students, volunteers and other people and institutions interested in universal transport design;
10) preparation of training courses for the teaching staff of FME and IP;
11) giving opinions on documents concerning special needs and requirements of PwSN;
12) initiating cooperation with social and economic organisations working for the benefit of PwSN;
13) international cooperation in order to exchange information and best practices in the field of universal design and mobility support for PwSN;
14) organising and participating in conferences, seminars, and scientific and training workshops devoted to the promotion and development of universal design in the field of transport and mobility of PwSN;
15) creation of the Centre's website and its ongoing updating with current and important information on the research conducted by the Centre and the results obtained;
16) running two information and consultation points (one each in FME and MTI) supporting and advising individuals and socio-economic organisations cooperating with PwSN, as well as mediating in direct contacts with the Centre's specialists.

4. Expected results of the Centre's activities

As a result of the work carried out by the Centre:

1. Six original scientific studies will be created:
   - an analysis of the difficulties and problems of PwSN in movement related to transport and mobility;
   - a catalogue of special needs relevant to the transport and mobility of PwSN;

- universal design standards in the field of transport and mobility with particular reference to ergonomics and methods of selecting adaptive equipment;
- an analysis of the legislation in the field of PwSN transport and mobility and recommendations for legislative changes;
- guidelines for psychological and therapeutic support for PwSN in the context of transport and mobility;
- an analysis of the effectiveness of new in-vehicle systems and devices introduced in the context of the needs of users with special needs.

2. Cooperation with external experts and with the socio-economic environment related to transport and mobility of PwSN will be established and developed in order to carry out research and development work for the benefit of PwSN (including soldiers and war veterans with disabilities) and entrepreneurs producing vehicles and equipment for PwSN.
3. Model training programmes will be developed for academic staff of FME and IP to prepare for teaching students of universal design focused on the transport and mobility of PwSN using individual means of transport. The training programmes will be developed in cooperation with the socio-economic environment, including the use of knowledge and experience of leading experts from Polish and foreign universities.
4. Model curricula for six specialised training courses will be developed. These will be addressed to PwSN, volunteers and organisations servicing and cooperating with PwSN, as well as to economic entities involved in the design, production and adaptation of vehicles and their equipment to the needs of PwSN.
5. Curricula for selected courses of studies of the first and second degree conducted at FME and for uniform master's studies conducted at the faculty of Psychology will be modified. The curricula will be supplemented with new subjects promoting knowledge of accessibility and universal design in transport and mobility of PwSN. Information sheets on the new subjects will be developed separately for FME and IP.
6. A nationwide pilot competition will be held for innovative proposals and concepts for devices, objects and services to improve mobility and road safety for people with special needs using personal transport. The competition will be aimed at students and teachers of secondary vocational schools and student research clubs at universities and technical colleges. The goal of the competition is to identify innovative ideas and concepts of solutions supporting mobility and safe transport of PwSN, which will be a source of inspiration for the Centre's project team and other entities cooperating with PwSN. It is assumed that the competition will be held periodically.
7. At least 10 innovative product concepts and standards for services supporting transport and mobility of PwSN will be developed.
8. The competences of 50 academic staff members of FME and IP will be improved in the field of universal design principles in two six-day training cycles (one each for FME and IP).
9. Twelve specialist trainings will be carried out on the basis of previously prepared programmes (2 trainings per one subject). It is assumed that over 120 people will participate in specialist training courses.
10. Part of the building of the Institute of Vehicles and Transport (FME) and its immediate surroundings will be reconstructed in order to adapt them to the needs of the disabled. The part of the building adapted to the needs of the Centre's stakeholders will house the Centre's secretariat, information and consultation point, toilets and a didactic laboratory for teaching universal design. The laboratory will be equipped with specialised apparatus (driving simulator, elderly simulators, demonstrators of driving systems for people with different dysfunctions, semi-active and active wheelchairs, etc.). The laboratory will be equipped with new apparatus as the Centre develops.
11. Two information and consultation points will be established (run by PwD with extensive knowledge and experience) aimed at helping PwSN and organisations cooperating with them. The activities of the consultants will be supported by four substantive advisors of seven MTI

specialists, whose knowledge and experience has been consolidated while participating in the work of the Motor Service Centre. The consulting points will operate also after project completion.

All products developed and produced as part of the Centre's activities will be publicly available (e.g. on the Centre's website [14]) and can be reviewed, analysed, discussed and, as a result of information exchange, improved and used by external entities. It is assumed that the Centre's activities will become a permanent feature on the map of Poland.

5. Summary

It is assumed that the activities of the Centre described in the article will make it possible to develop, promote and disseminate innovative solutions for systems, products and universal service standards facilitating and supporting the use of vehicles by PwSN. As a result, PwSN will remain mobile and road safety will not deteriorate.

Currently, many devices and systems for driver assistance are being developed, but without a clear focus on the PwSN. Therefore, it is high time to design user-friendly vehicles and equipment. Recommended actions are:

- improving simple and advanced systems of driving support by including PwSN in the process of vehicle and road use and design;
- introducing a procedure for testing new road infrastructure solutions and new vehicle technologies with the participation of PwSN;
- educating PwSN to raise awareness of the road traffic risks arising from the limitations of their physical and mental fitness and information on the benefits and use of systems supporting vehicle users.

Transport is the second area, besides architecture, that is reported by PwSN area as requiring immediate improvement in terms of accessibility. Comments made during the public consultation on the Accessibility Plus Programme pointed to the risk of marginalisation, through all kinds of barriers to access of transport and mobility, which are the basis for the functioning of a modern society. In terms of accessibility of transport, a lot is changing for better, but most of the actions so far have concerned of transport and the environment of public transport.

The Centre's activities described in this article relate to means of transport. It demonstrates the originality and innovation of the Centre's activities.

Acknowledgments

The article was prepared as part of the project no. POWR.03.05.00-00-CW07/20-00 entitled: "Knowledge Centre on Accessibility to Transport and Mobility of People with Special Needs" co-financed by the European Social Fund under the Operational Program Knowledge Education Development 2014-2020.

Bibliography

[1] Act of 19 July 2019 on ensuring accessibility for people with special needs (Dz.Uz 2019, item 1696).
[2] Demographic situation of the elderly and consequences of population ageing in Poland in the light of projections for 2014-2050. Central Statistical Office, Warsaw 2014.
[3] Thematic study OT-662 Population ageing in the European Union - current status and projection. Chancellery of the Senate. Warsaw 2018.
[4] Raczyńska-Buława E. Mobilność osób starszych. Dlaczego nie transport publiczny? Technika transportu szynowego 1-2/2017.
[5] Solecka K. Potrzeby osób starszych w zakresie mobilności w mieście. *Autobusy* 6, 2018.
[6] Shaheen S. A., Niemeier D. A.: Integrating vehicle design and human factors: minimizing elderly driving constraints. *Transportation Research Part C* 9, pp. 155-174, 2001.
[7] Rottermund J., Knapik A., Szyszka M. Aktywność fizyczna a jakość życia osób starszych. *Społeczeństwo i Rodzina*, 42 (1), 2015.

[8] Zasadzka E., Pawlaczyk M., Wieczorowska-Tobis K. Test Short Physical Performance Battery jako narzędzie służące do oceny sprawności fizycznej osób starszych. *Gerontologia Polska*, 4, 2013.
[9] Łuczak A. Wiek a sprawność kierowców w zakresie cech warunkujących bezpieczne uczestnictwo w ruchu drogowym. *Bezpieczeństwo Pracy,* 3, 20215.
[10] Prochowski L., Gidlewski M., Dąbrowski F. Use of Regression Analysis for Comparative Evaluation of Accident Hazard in Poland, by Regions. *Journal of Kones*, 24(4), 2017.
[11] Ociepka A., Pytel S. Aktywność turystyczna seniorów w Polsce. *Ekonomiczne Problemy Turystyki* 2 (34), 2016.
[12] National Centre for Research and Development Modules of universal design classes within selected areas of education. Warsaw 2020.
[13] Ministry of Investment and Development Government Programme Accessibility Plus 2018-2025. Warsaw 2018.
[14] Website Accessibility Knowledge Centre: https://cwod.wim.wat.edu.pl/ [Access: 31.03.2022].

Lateral acceleration of passenger vehicle in roundabouts in term of cargo securing

O. Kuba[1*], J Jagelcak[1]

[1] Department of Road and Urban Transport, Faculty of Operation and Economics of Transport and Communication, University of Zilina, SK-010 26 Žilina, Slovakia

* kuba17@stud.uniza.sk

Abstract. In today's modern infrastructure, roundabouts are found in almost every city, which makes them impossible to avoid. Roundabouts with a radius of up to 15 m generate high accelerations even when traveling at average speeds. For this reason, it is essential to analyze the events that occur when passing through such roundabouts. In the research, we have performed a complex analysis of the accelerations on the x, y and z-axis for five passes through five different roundabouts, with a focus on the vehicle dynamics. This helped us to understand how the vehicle behaves when crossing a roundabout, as well as what accelerations and forces are affecting the cargo. Therefore, the paper is comparing data from two sensors that were mounted on various parts of the vehicle. The first sensor was mounted on the dashboard and the second on the roof of the vehicle. What is more, we found that the Y-axis accelerations, which are the most important for the analysis of vehicle dynamics in roundabouts, are not affected by the change in sensor location. In addition, we performed an acceleration calculation based on the data recorded from the sensor and radius, which can serve as a basis for further research.

Key words: acceleration, sensor, vehicle dynamics, cargo securing

1. Introduction

Accidents and damage of cargo occurs on European roads every day, mostly because of the incorrect storage or securing of loads on vehicles. The European Agency for Safety and Health at Work estimates that up to 25% of heavy vehicles accidents are caused by insufficient load securing [1].

Accelerometers based on a micro-mechanical system (MEMS accelerometers) are not new and have been serving different customer needs for many years [2]. With the introduction of MEMS technology, a number of sensor systems have emerged to record phenomena that would require professional technology only a few years ago [3]. Authors developed in [4] system for measuring pavement roughness. Low-cost MEMS accelerometer and global positioning system (GPS) was used for data measurements. They demonstrated that this solution is less expensive than previously known as [5], [6], [7]. MEMS sensors can also be used for saving lives and preventing accidents [8]. Research [9] describes system, that will in case of incident connect to nearest hospital and inform staff. Comparable system was developed [10] but rather than contacting officials, authors focused on early warning system for accidents that will provide location data and data from MEMS sensor. Low-cost navigation applications depend mainly on satellite navigation systems, especially the Global Positioning System (GPS). In [11], [12] authors discovered, that inaccurate coordinates can occur in urban areas or in tunnels. To solve this problem, author of [13] proposed improving the accuracy of the MEMS UMI stochastic model. Other solution was to use initialization of Kalman filtering for sensor parameters and including a body velocity constraint and improved zero velocity updates [13], [14]. Measurements with GPS device were performed in [15]

In this paper, we will be focusing on vehicle dynamics when passing roundabout, mainly lateral acceleration. Authors of [16], [17] also studied and analysed lateral acceleration of vehicles. In [18], [19], [20] analyse of vehicle breaking was described. This topic is connected to road safety and can also help us to better understand vehicle dynamics. Authors of [21] proposed the methodology of measurement for a vehicle's centre of gravity, which is important for precisely describing complex vehicle dynamics. Breaking performance and vehicles centre of gravity calculation was also performed in [22]. Research for cargo securing, which can prevent damage to cargo or even road accidents was carried out in [23], [24].

2. Definition of physical phenomena and methodology

Anyone who ever travelled knows, that speed of vehicle is constantly changing. That is why we are using average speed. The average speed is calculated as the ratio of the total distance s and the time t interval in which the movement took place: [25]

$$v = \frac{s}{t} \tag{1}$$

The unit of speed in the international system of units is determined according to the equation: [25]

$$v = \frac{\mathrm{s}}{\mathrm{t}} = \frac{m}{s} = m \cdot s^{-1} \tag{2}$$

When describing traffic phenomena, speed is usually expressed in kilometres per hour ($\mathrm{km} \cdot \mathrm{h}^{-1}$) [25]. The speed of movement may vary or may be constant. The movement in which the speed of movement changes is called acceleration. Accelerated movement is not only when the magnitude of the speed changes, but also when the direction of the speed changes. Acceleration is introduced as a measure of the change in speed per unit of time. The average (mean) acceleration $\bar{a}$ in the time interval Δt is defined as: [25]

$$\bar{a} = \frac{\Delta v}{\Delta t} \tag{3}$$

If the vehicle is moving with an acceleration of 1 m/s^2, this means that every 1 second the speed increases by 1 m/s. Unit of acceleration is m/s^2, but unit g is widely used in cargo securing instead, where 1g = 9,81 m/s^2 [26].

The instantaneous acceleration a and steady movement in the circle is caused by a change in the direction of the velocity and is directed to its centre at each point of the trajectory. We call it centripetal acceleration a_d. We calculate the magnitude of the centripetal acceleration from the relation: [26]

$$a_d = \frac{v^2}{r} \tag{4}$$

The vehicle can enter a curve at such a speed that two basic conditions are met, which are lateral skid resistance and rollover resistance. As the vehicle moves in a circular path, it is subjected to the centrifugal force F_o and the mass of the vehicle G. The centrifugal force tries to push the vehicle out of the circular path away from the centre of turn [27].

The sensor is a functional element forming the input block of the measuring chain, which is in direct contact with the measured environment [28]. For purpose of measuring vehicle dynamics, we used device that consist of three MEMES sensors. The BOSCH BHA250 is a small, low-power MEMS sensor with an integrated three-axis accelerometer. The BOSCH BMG250 is a three-axial gyroscope consisting of a state-of-the-art lowpower 3-axis gyroscope. The UBX-M8030 high-performance standard precision GNSS chip provides exceptional sensitivity and acquisition times for all GNSS systems.

In the research, one respectively two identical sensors were used to measure vehicle dynamics. We performed two groups of measurements. In first group, two roundabouts (measurements 1, 2) were recorded. To record the data, we used one sensor (A) that was placed in vehicle on dashboard. Both roundabouts were recorded within the same conditions. In second group three roundabouts (measurements 3, 4, 5) were recorded. This time we used two sensors, sensor A which was placed in vehicle on dashboard and sensor B which was placed on roof of the vehicle. All three roundabouts were recorded within the same conditions.

In all measurements raw data were recorded. Therefore, it was necessary to evaluate them. The maximum (minimum) average acceleration at 80 ms is applied in relation to the standard STN EN 12642: 2016 [29], which is used in strength tests of superstructures or load securing by dynamic driving tests. For evaluation it was essential to use average time. According to the standard EN 17321:2021 [30], when vehicle dynamic acceleration is tested, the minimum average time should be 300 ms. For the dynamic driving test, the minimum average time should be 1000 ms. Based on the above, all raw data from tests were evaluated at 80 ms, 300 ms and 1000 ms.

To describe physical processes, it is necessary to introduce a unified system that can clearly define the physical quantities. For this purpose, coordinate systems have been introduced. In describing three-dimensional events such as vehicle dynamic, it is best to use cartesian coordinate system, that contains three axes [31]. On axis X is shown horizontal acceleration, on axis Y lateral acceleration and on axis Z vertical acceleration. For simplification, we will be using shortcuts for average times and axis, e.g., y1000 which means that data were measured on axis Y and evaluated with time 1000 ms. When two sensors were used, data from sensor A has shortcut $A_c y$, while data from sensor B has shortcut $B_{sc} y$. In accordance to [25, 26], secured cargo should withstand following acceleration when transported in heavy vehicle: on axis X to the front 0.8 g, to the back 0.5 g; on axis Y 0.5 g and on axis Z 1 g.

For uniformity of data, we selected data at each roundabout from the same area that we called evaluated area (EA). Selected data are starting at value 0.0 g (y1000) and are ending at value 0.0 g (y1000). That is because at every entrance to roundabout there are positive values of acceleration, while at every roundabout there are negative values of acceleration (positive and negative value means if curve is to the right or to the left). We selected data beginning 0.0 g and ending 0.0 g to ensure, only negative values will be considered in calculations. We can see evaluated areas at figures 1, 2, 3, 4 and 5 between "start" and "end". This area is also background on graphs in figure 1, 2, 3, 4 and 5. On graphs that are showing axis Y, there are highlighted values 0.0 g from start and end of roundabout with red line. These are same places as point start and end in visualization of roundabouts. That means from every roundabout we will only be comparing data where acceleration happened. This guarantees uniformity of data from every roundabout. But it is important to mention, that every passing of roundabout is different, and it is impossible to achieve completely equal passing. In figures 1, 2, 3, 4 and 5, every graph has on axis X number of samples, while 200 samples are 1 second.

All measurements were performed with passenger vehicle. We wanted to achieve lateral acceleration 0.5 g or more. Because of required acceleration, it will be very dangerous to perform measurements with heavier vehicles. In such a high acceleration, slide or oven turnover can occur.

3. Evaluation of measured acceleration, comparison of sensor position and calculation of lateral acceleration

3.1 Evaluation of measured data

In figure 1 we can see roundabout 1 (R1) with a radius of 12.53 m. We observe two areas where the lateral acceleration was the highest. The first area is located after entering the roundabout at the first sharp turn. There was an acceleration of approximately -0.4 g to -0.45 g (y1000). Acceleration between 0.3 g and -0.4 g (y1000) was achieved throughout the majority of roundabout. The highest acceleration was recorded at the second area with increased values which was located at the exit of the roundabout, where a maximum value of -0.51 g (y1000) was also recorded. This was mainly due to the increase in speed V. We can clearly see the increase in value of the lateral acceleration as the speed increases in graphs in figure 1. Overall, the average speed was 24.02 km/h, with the lowest speed of 19.52 km/h recorded at the entrance of R1 and the highest speed of 33.03 km/h recorded at the exit of the R1. Since this was a long-term acceleration, we do not observe significant differences in the values of acceleration between the individual times. As for the accelerations on the X and Z axes, we did not observe any significant values on the X axis, with the highest value being -0.13 g, respectively 0.11 g (x80). On the

IOP Conf. Series: Materials Science and Engineering 1247 (2022) 012037 doi:10.1088/1757-899X/1247/1/012037

Z axis we can see increased values during the whole transit time. These accelerations were caused by the tilt of the vehicle. The accelerations on the Z axis are directly related to the accelerations on the Y axis and also to roll of the vehicle. The highest recorded acceleration value on the Z axis was -0.27 g (z80).

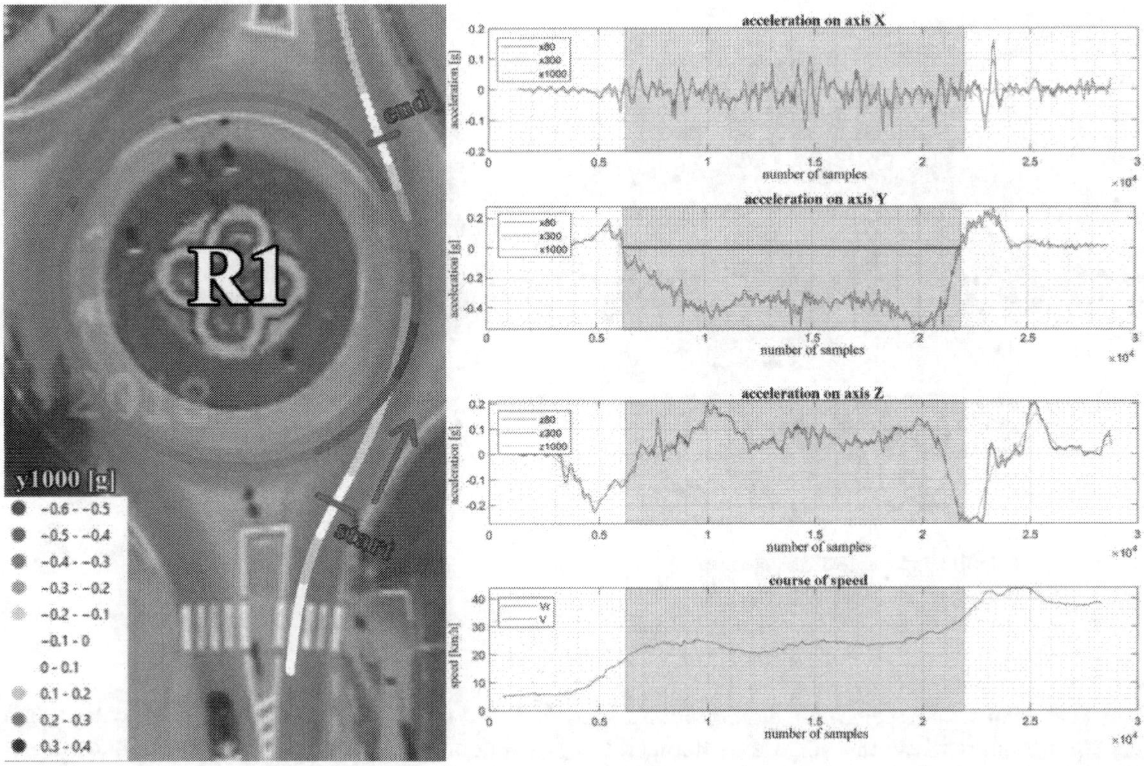

Figure 1. Analysis of accelerations at R1 with sensor A

Measurement at R2 were carried out at the same time as the measurement 1. Both roundabouts also have the same entry and exit points. However, the difference was in the radius, with the first having a radius of 12.53 m and the second 13.66 m. However, we see smaller differences in lateral acceleration, as there are 3 instead of 2 areas with increased values. The first part where accelerations of -0.40 g to -0.45 g (y1000) were achieved occurred earlier than in R1. This was followed by a second part with lower accelerations from -0.3 g to -0.4 g (y1000), followed by increased accelerations of -0.4 g to -0.45 g (y1000) in the third part. In fourth part acceleration decreased to values from -0.3 g to -0.4 g (y1000). The exit, and the maximum values took place in the same place as in the previous case. The driving style of the vehicle affects course of acceleration. Therefore, it is not possible to pass two roundabouts in the same way, even if they are similar. Overall, we see that the average (25.37 km/h), minimum (21.32 km/h) and also maximum (36.04 km/h) speeds were higher than for R1. This was caused not only by the driving style but also by the larger radius of the R2. This is also confirmed by the average lateral acceleration value of -0.38 g (y1000) and the maximum acceleration value of -0.54 g (y1000). Again, we do not observe a difference between the average times. We did not notice any significant values on the X axis. As in the previous case, the Z axis is dependent on the Y axis and on the roll, with the maximum measured acceleration value being -0.50 g (z80) at the exit of the roundabout, where roll the vehicle was very high. The average acceleration on axis Z was -0.11 g, which was the same at all average times.

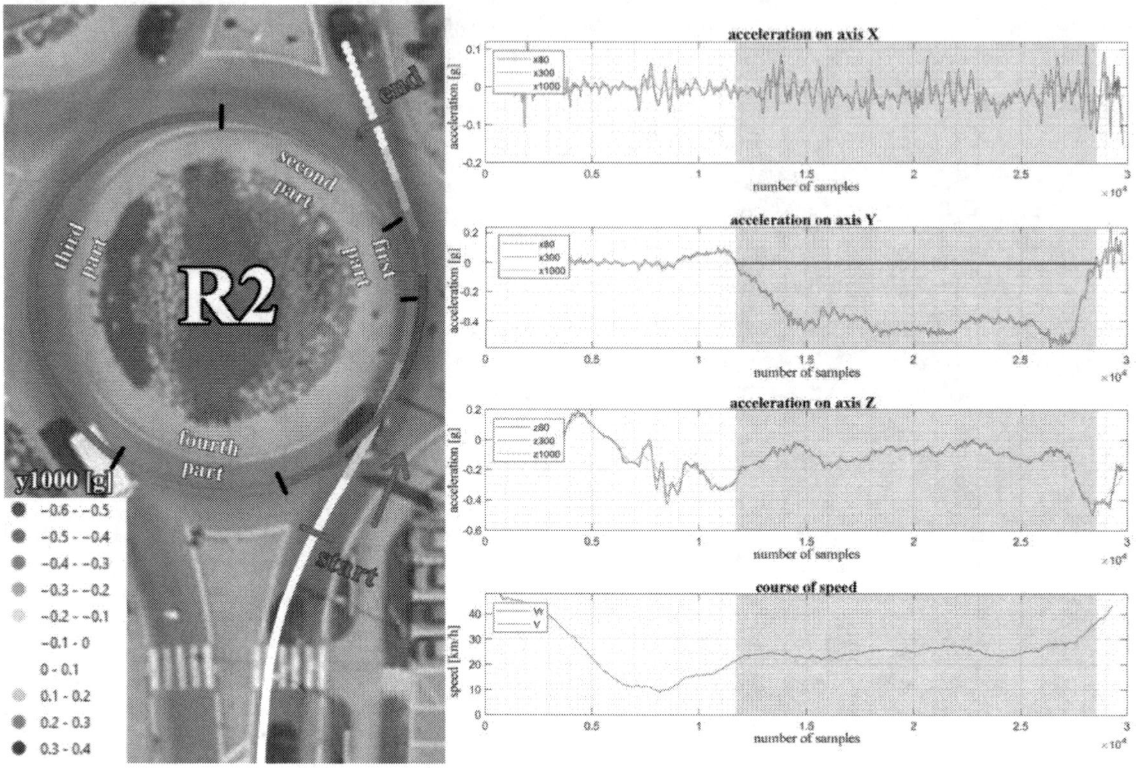

Figure 2. Analysis of accelerations at R2 with sensor A

Measurement of acceleration at R3 with a radius of 11.86 m took place at a different location than R1 and R2 but with similar conditions (temperature, road conditions) and with sensor that had similar position (sensor A). Data from sensor B were only used in comparison of sensors. Overall, the course of the accelerations on the Y axis was slightly different than in previous cases. In figure 3 we can observe a faster beginning of lateral acceleration. After beginning, values reached from -0.4 g to -0.5 g (A_cy1000) during the entire R3. The only exception was before the departure, where we can observe a reduction in accelerations to values just above -0.4 g (A_cy1000). During the departure from R3, we again observe an increase in the values of acceleration, at which we also recorded the maximum value of the lateral acceleration -0.54 g (A_cy1000). The increase in values in this case occurred earlier than in previous roundabouts. This could be due to smaller radius or the driver's driving style. The average value of the acceleration on the Y axis during the roundabout was -0.41 g (A_cy1000). This was a slightly higher value than in previous cases, which is confirmed by the graphs (figure 3), where a significant part of the passage was in the values from -0.4 g to -0.5 g (A_cy1000). We also observe the difference in values of the speed, where at the roundabout 3 we see the entrance at a higher speed (31.81 km/h), which was also the maximum speed. Subsequently, the speed decreased and during the passage reached values between 22.5 km/h and 26 km/h, while the average speed was 25.50 km/h. Before the exit, the speed increased and reached 30 km/h on the way out. On the X axis, we again observe negligible values, which were related to the smoothness of the ride. At Z axis accelerations, we observe a correlation with Y axis accelerations, with the highest value recorded being -0.20 g (A_cz80) and the average value during the pass being 0.05 g at all average times. The difference between the previous roundabouts occurred at the beginning and end of the lateral accelerations, where in R3 we observe a later beginning and an earlier end, so the area of action was smaller.

IOP Conf. Series: Materials Science and Engineering 1247 (2022) 012037 doi:10.1088/1757-899X/1247/1/012037

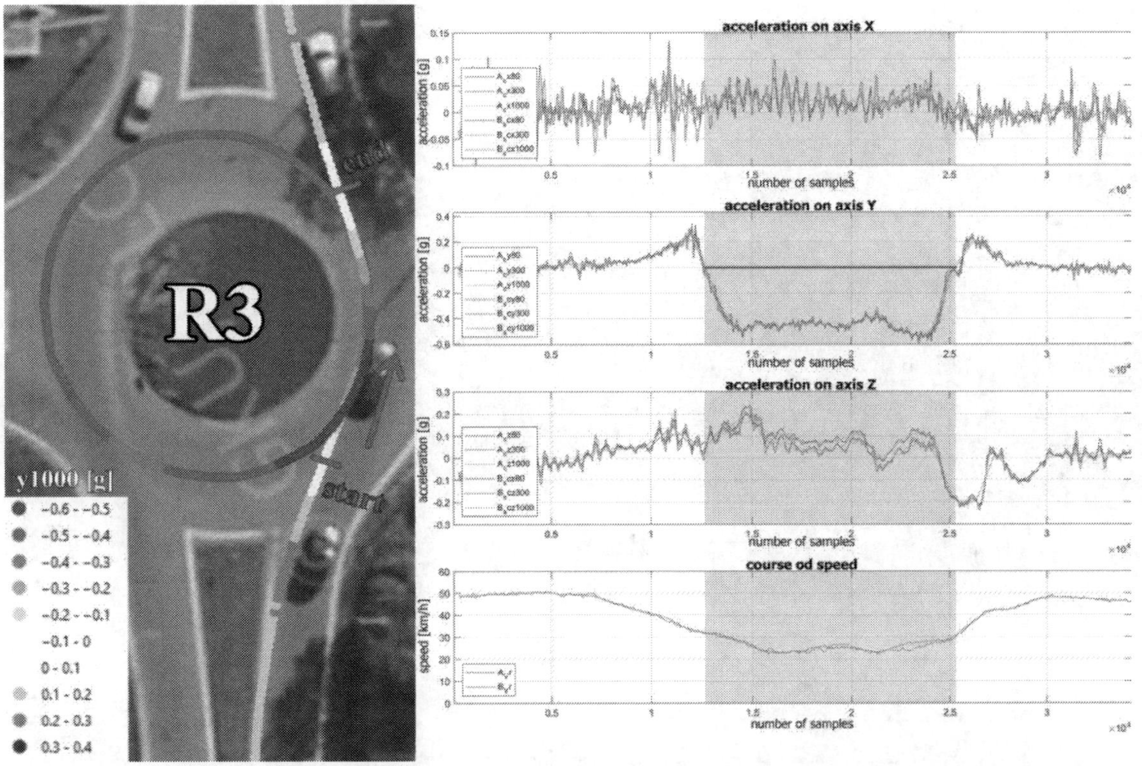

Figure 3. Analysis of accelerations at R3 with sensor A

Measurements at R4 with a radius of 13.82 m took place under the same conditions and at the same location as R3. However, we observe a slightly different course, with the highest value of 0.50 g (A_cy1000) being reached in the first sharp turn right after the entrance. Subsequently, during the entire transit period, we see values ranging from -0.4 g to -0.5 g (A_cy1000), with one exception. As with R3, the exception occurred before the departure, where there was a slight decrease in speed and also a slight decrease in values that reached slightly above -0.40 g (A_cy1000). Although the exit speed increased, there was no increase in lateral acceleration due to the larger radius, with an average acceleration of -0.40 g (A_cy1000). The development of speed was almost identical to the R3, with a top speed of 32.61 km/h recorded at entry while the average speed was 26.59 km/h. As in the previous cases, the values on the X axis were negligible. On the Z axis, we see the correlation with the Y axis as in all previous cases. The highest recorded acceleration on the Z axis was -0.25 g (A_cz80) while the average acceleration was 0.03 g at all average times. The beginning and end of the action of accelerations on the Y axis in R4 was at similar places as in R3.

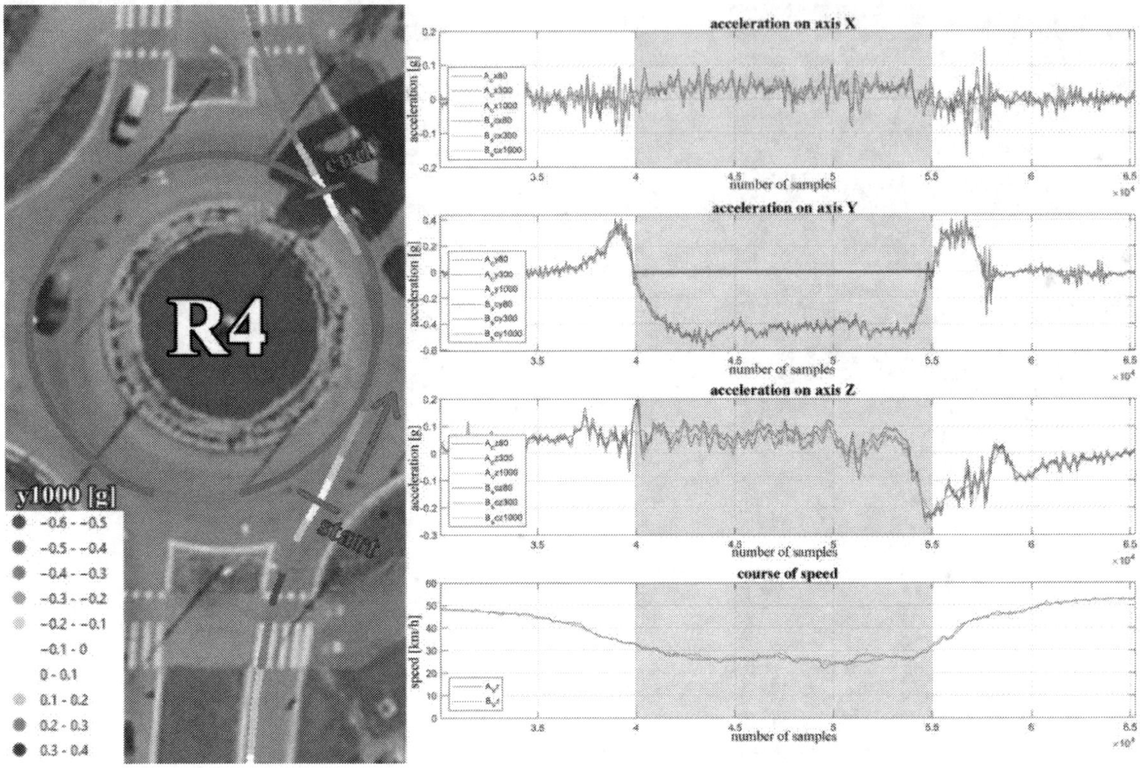

Figure 4. Analysis of accelerations at R4 with sensor A

Measurement in R5 with a radius of 11.19 m was carried out at the same location and under the same conditions as the R3 and R4, but the trajectory of the passing was different. In figure 5 we see that the whole 360 ° arc was not made. This is also indicated by the acceleration course, with the driver starting to turn earlier than in previous cases. As a result, the increased values of the lateral accelerations began earlier than in the previous case. Just after entering the arc, we see that the acceleration values reach from -0.4 g to -0.5 g (A_cy1000). Subsequently, we observe a difference in the driving trajectory, where the driver drove the vehicle more along the edge of R5. In this section, we also observe the highest values of accelerations, which were in the range from -0.5 to -0.56 g (A_cy1000), while the value of -0.56 g (A_cy1000) was the highest recorded value. After this action, there was departure from the roundabout, whereas the values of acceleration decreased. The average value of the lateral acceleration was -0.43 g (A_cy1000). The course of the speed was similar to the R3 and R4. However, we see the difference in the section where the highest values were achieved. The speed decreased slightly, but due to the small radius and sharper turning, the values of the lateral accelerations increased. The speed at the entrance was 29 km/h, but the highest recorded speed 31.35 km/h was at the exit while the average speed was 25.78 km/h. The values on the X axis were again negligible. On the Z axis, we can see once again correlation with the Y axis accelerations, with the highest value recorded being -0.24 g (B_{sc}z80) and the average value being 0.03 g at all average times. From the graph in figure 5 we can see that the maximum value did not occur when passing the roundabout, but when exiting it. In this moment, the values of the lateral acceleration changed direction what caused the vehicle to tilt. Similar action happened at every other roundabout. The beginning of the lateral acceleration at the crossing occurred at a similar place as in R3 and R4. However, the end of the operation was in a different place due to the fact that it was not a passage of the whole arch.

IOP Conf. Series: Materials Science and Engineering 1247 (2022) 012037 doi:10.1088/1757-899X/1247/1/012037

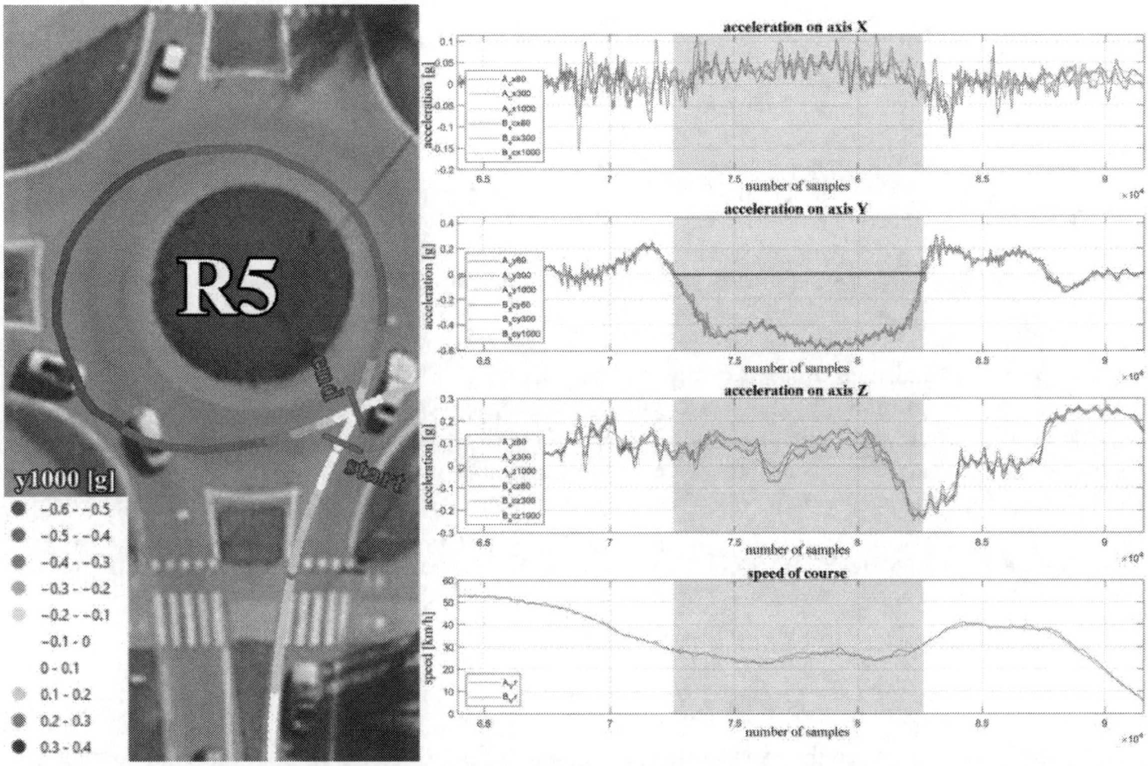

Figure 5. Analysis of accelerations at R5 with sensor A

3.2 Comparison of sensors with different positions

As mentioned earlier, two sensors denoted as sensor A and B were used in measurements R3, R4 and R5. Sensor A was located on the dashboard and Sensor B on the roof of vehicle. Both sensors were placed in the center of the vehicle. The height difference on the Z axis was 0.6 m and the vertical difference on the X axis was 0.7 m.

Table 1. Results comparisons for sensors A and B from EA

		Sensor	cx80 [g]	cx300 [g]	cx1000 [g]	cy80 [g]	cy300 [g]	cy1000 [g]	cz80 [g]	cz300 [g]	cz1000 [g]	V [km/h]
R1	Min	A	-0.130	-0.081	-0.042	0.027	0.025	-0.007	-0.265	-0.253	-0.236	19.52
	Max	A	0.101	0.054	0.014	-0.542	-0.531	-0.510	0.195	0.181	0.168	33.03
	Avg	A	-0.013	-0.013	-0.013	-0.346	-0.345	-0.343	0.054	0.053	0.053	24.02
R2	Min	A	-0.112	-0.069	-0.046	0.028	0.016	0.008	-0.500	-0.466	-0.422	21.32
	Max	A	0.113	0.048	0.017	-0.585	-0.545	-0.539	0.000	-0.007	-0.022	36.04
	Avg	A	-0.016	-0.017	-0.016	-0.379	-0.379	-0.378	-0.113	-0.113	-0.113	25.37
R3	Min	A	-0.037	-0.022	-0.018	-0.596	-0.549	-0.541	-0.194	-0.195	-0.191	22.28
		B	-0.031	-0.011	0.001	-0.571	-0.547	-0.537	-0.198	-0.198	-0.194	22.65
	Max	A	0.071	0.05	0.037	0.002	-0.02	0.007	0.208	0.197	0.17	31.81
		B	0.097	0.066	0.043	0.002	-0.006	0.036	0.238	0.234	0.203	32.13
	Avg	A	0.018	0.018	0.018	-0.414	-0.414	-0.412	0.047	0.046	0.046	25.5
		B	0.023	0.023	0.023	-0.416	-0.416	-0.414	0.08	0.08	0.079	26.18
R4	Min	A	-0.047	-0.026	-0.019	-0.587	-0.537	-0.504	-0.252	-0.236	-0.221	21.89
		B	-0.079	-0.041	0.007	-0.559	-0.525	-0.505	-0.238	-0.232	-0.221	24.21

	Max	A	0.09	0.057	0.038	0.046	-0.01	0.02	0.197	0.182	0.083	32.61
		B	0.101	0.067	0.047	0.068	0.018	0.063	0.188	0.17	0.1	32.8
	Avg	A	0.025	0.025	0.025	-0.407	-0.407	-0.404	0.029	0.029	0.029	26.59
		B	0.032	0.032	0.032	-0.409	-0.409	-0.406	0.056	0.056	0.056	27.25
	Min	A	-0.041	-0.02	-0.017	-0.597	-0.584	-0.562	-0.246	-0.227	-0.203	22.4
		B	-0.053	-0.031	-0.007	-0.598	-0.582	-0.567	-0.229	-0.224	-0.203	22.47
R5	Max	A	0.085	0.061	0.047	0.048	0.039	0.01	0.144	0.123	0.105	31.35
		B	0.114	0.081	0.057	0.098	0.108	0.059	0.167	0.161	0.152	31.96
	Avg	A	0.027	0.027	0.027	-0.435	-0.435	-0.431	0.028	0.028	0.029	25.78
		B	0.035	0.035	0.035	-0.437	-0.436	-0.433	0.063	0.063	0.063	25.96

As we can see in table 1, the differences in the measured values between the sensors were minimal. On axis Y, which values are the most important for evaluation of vehicle dynamics, we observe the difference from the maximum values only 0.009 g on average. On axis X, average difference between maximum values is 0.016 g while slightly larger average differences 0.027 g occurred in the values measured on the axis Z. In table 1 we can also see maximum, minimum and average values from R1 and R2. However, these measurements were performed with only one sensor, so we don't have data to compare.

Better than judging the maximum values will be to judge the average values. Overall, we will consider the average values for each axis and each time from all three roundabouts. This gives us a total of 27 average values for each sensor. Higher average values were recorded at sensor B for all cases. The largest recorded difference between the average values was 0.035 g at cz300/cz1000 in R5. This difference was mainly due to the fact that the acceleration on the Z axis was directly related to the vehicle's roll when crossing the R5, with sensor B being positioned higher than A. Looking at the average differences between the axes, on the X axis the difference from all times and roundabouts was 0.007 g, on the Y axis it was 0.002 g and on the Z axis it was 0.032 g. Thus, we can conclude that the sensor position in vertical axis when driving through the roundabout has the greatest effect on the accelerations on the Z axis. However, it should be added that the accelerations on the Z axis when crossing the roundabout are mainly caused by the vehicle roll and not by driving through the bumps. However, a better quantity to accurately determine vehicle tilt is unit roll, which can be recorded by more accurate sensors, but the sensor we used was not able to record this unit. Average speeds on sensor B were 0.503 km/h higher than A. This was mainly due to the accuracy of the GPS coordinates, since the speed is calculated from GPS coordinates, which do not reach the same accuracy as the accelerometer data and thus, cannot be directly compared.

Because accelerations on axis X and axis Z could get positive and negative values, we calculated average accelerations in absolute value (abs), that are shown in table 2. On axis Y we performed calculations only from 0.0 g to 0.0 g that means abs is not necessary. From results we can see, that average difference between sensor A and B on axis X is 0.006 g for 300 and 1000 ms and 0.007 g for 80 ms. On axis Z, average difference is 0.024 g for 80 and 300 ms and 0.025 g for 1000 ms. For values on Axis X we get similar results as in previous case. That also applied for axis Z, but with slightly smaller difference as in previous case. In all case, average acceleration on sensor B had higher values.

Table 2. Comparisons of average accelerations in absolute value

	Sensor	cx80_abs [g]	cx300_abs [g]	cx1000_abs [g]	cz80_abs [g]	cz300_abs [g]	cz1000_abs [g]
R3	A	0.021	0.020	0.019	0.072	0.072	0.068
	B	0.026	0.024	0.023	0.100	0.100	0.097
R4	A	0.029	0.027	0.026	0.059	0.057	0.054
	B	0.037	0.034	0.032	0.080	0.079	0.077
R5	A	0.030	0.029	0.028	0.083	0.082	0.076
	B	0.038	0.036	0.035	0.105	0.104	0.099

The acceleration on the Y axis, which is the most important for us when monitoring lateral dynamics when passing a roundabout, was almost completely unaffected by the location of the sensor.

3.3 Calculation of lateral acceleration
Due to the fact that roundabouts are circular ride, we decided to calculate the average lateral acceleration using equation (4). The average lateral acceleration for R1 is calculated in equation (5), while the other roundabouts are shown in table 3.

In order to calculate the average lateral acceleration from the centripetal acceleration a_d, the average velocity v and the radius of the circle r are needed. The instantaneous velocity was recorded by the sensor. From the instantaneous velocity we then calculated the average velocity. The average velocity was calculated in the region that is located between the "start" and "end" markers in figures 1, 2, 3, 4 and 5. We calculated the radius r using the map data from the route recorded by the GPS device. Thus, the calculation of the average lateral acceleration is as follows:

$$a_d = \frac{v^2}{r} = \frac{6.67^2}{12,53} = 0,36\ g \tag{5}$$

In table 3 we see the calculation of the average lateral acceleration along with the measured values. The direction of action, i.e. whether it is a positive or negative value, is not relevant in this calculation. Overall, we can conclude that the calculated values agree with the recorded values. The largest deviation is 0.04 g. Taking into account the fact that the vehicle was not driven exactly in a circle, the deviations of the GPS sensor and also the deviations in the speed data, the calculated values can be considered satisfactory.

Table 3. Calculation of lateral acceleration

Roundabout	Radius [m]	Average speed [km/h]	Calculated average lateral acceleration [g]	Measured average lateral acceleration [g]
R1	12.53	24.01	0.362	0.343
R2	13.66	25.38	0.371	0.378
R3	11.86	26.17	0.454	0.412
R4	13.82	27.25	0.423	0.404
R5	11.19	25.96	0.474	0.431

4. Conclusion

The analysis of the dynamics of a vehicle crossing a roundabout is an important topic. It can help us to prevent accidents in roundabouts such as sliding or even rollover of vehicle. Furthermore, the vehicle dynamics can help us to understand the movement of cargo better, which could help us with its securing. From our analysis, we have found that if the crossing is performed along the same trajectory, the highest accelerations always occur at the same places, namely at the beginning and at the end of the trajectory. We also found that even when the roundabout is passed at normal speed, accelerations greater than 0.5 g can occur on the Y axis at 1000 ms.

Comparing the two sensors, one on the dashboard and the other on the roof of the vehicle, we found that the difference in position mainly affects the accelerations on the Z axis, due to the vehicle's tilt. For the Y axis, which we found to be the most useful in analyzing the dynamics of roundabout crossing, we observed a minimal deviation, which was on average 0.002 g. Thus, we can claim that even if it will not be possible to maintain the same sensor placement height for different measurements, the accelerations on the Y axis will not be affected for passenger vehicle.

Although the calculation of the radius based on the data obtained from the sensor gave us values that are very similar to the measured values, we cannot declare our findings to be completely accurate. To test this theory, it will be necessary to make a larger number of measurements and develop a more sophisticated method of determining the radius to be used for the calculation of lateral accelerations.

However, our calculation can serve as a suggestion for further research. The results of the study are particularly important in terms of vehicle safety [31, 32].

Acknowledgments
This paper was realized with support of Operational Program Integrated Infrastructure 2014-2020 of the project: Innovative Solutions for Propulsion, Power and Safety Components of Transport Vehicles, code ITMS 313011V334, co-financed by the European Regional Development Fund and the project of institutional research of the Faculty of Operation and Economics of Transport and Communications, University of Žilina no. 2/KCMD/2021 Research on the impact of city logistics on the environment.

References
[1] Gnap J, Jagelčák J, Marienka P, Frančák M and Kostrzewski, M 2021 Application of mems sensors for evaluation of the dynamics for cargo securing on road vehicles *Sensors* **21**
[2] Noureldin A, Karamat T.B, Eberts and El-shafie, A 2009 performance enhancement of mems-based ins/gps integration for low-cost navigation applications *IEEE Transactions on Vehicular Technology* **58** 1077–1096
[3] Rivas J, Wunderlich R and Heinen, S.J 2012. A mems acceleration sensor for traffic condition detection *PRIME 2012; 8th Conf. on Ph.D. Research in Microelectronics and Electronics (Aachen)* 321–324
[4] Mirtabar Z, Golro A, Mahmoudzadeh A and Barazandeh F 2022 Development of a crowdsourcing-based system for computing the international roughness index *Int. J. Pavement Eng.* **23** 489–498
[5] Chen K, Lu M, Fan X, Wei M and Wu J 2011 Road condition monitoring using on-board three-axis accelerometer and gps sensor *Proc. of the 6th Int. ICST Conf. on Communications and Networking in China (Harbin)* 1032–1037
[6] Du Y, Liu C, Wu D and Li S 2016 Application of vehicle mounted accelerometers to measure pavement roughness *Int. J. Distrib. Sens. Networks* **2016**
[7] Anitha R, Deepak M and Snehesh, T Road quality and condition detection using gps and a raspberry pi *Int. J. Innov. Res. Creat. Technol.* **1** 527–529
[8] Mikusova M 2017 Crash avoidance systems and collision safety devices for vehicle occupants *MATEC Web of Conf. (Trstena)* 107
[9] Parmar P and Sapkal A.M 2017 Real time detection and reporting of vehicle collision *Proceedings of the 2017 Int. Conf. on Trends in Elect. and Informatics (Tirunelveli)* 1029–1034
[10] Mantoro T, Suryasa I.N, Moedjiono S and Nugroho M.R 2016 automatic early warning for vehicles accidents based on user location *Adv. Sci. Lett.* **22** 3065–3070
[11] Cossaboom M, Georgy J, Karamat T.B and Noureldin A 2012. Augmented kalman filter and map matching for 3d riss/gps integration for land vehicles *Int. J. Nav. Obser.*
[12] Meiling W, Guoqiang F, Huachao Y, Yafeng L, Yi Y and Xuan X 2017 A loosely coupled mems-sins/gnss integrated system for land vehicle navigation in urban areas *Proceedings of the 2017 IEEE Int. Conf. on Vehicular Elect. and Safety (Vienna)* 103–108
[13] Allan D 1966 Statistics of atomic frequency standards *Proceedings of the IEEE* **54** 221–230
[14] Li X, Xu Q, Li B and Song X 2016 A Highly reliable and cost-efficient multi-sensor system for land vehicle positioning *Sensors* **16**
[15] Mikusova M, Abdunazarov J, Zukowska J and Jagelcak J 2020 Designing of parking spaces on parking taking into account the parameters of design vehicles *Computation* **8**
[16] Visar B, Odhisea K and Doçi I 2017 The influence of the lateral acceleration on vehicle velocity moving on curved road *Int. J. Civil Eng. Tech.* **8** 414–420
[17] Xu J, Yang K, Shao Y and Lu G 2015 An Experimental study on lateral acceleration of cars in different environments in sichuan, southwest china *Discrete Dynamics in Nature and Society* **2015**
[18] Ondruš J and Hockicko P 2015 Braking deceleration measurement using the video analysis of motions by Sw tracker *Transport and telecommunication journal* **16** 127–137

[19] Ondruš J, Vrábel J and Kolla, E 2018 The influence of the vehicle weight on the selected vehicle braking characteristics *Transport Means-Proceedings of the International Conference (Trakai)* 384–390

[20] Ondruš J and Kolla E 2017 practical use of the braking attributes measurements results *MATEC Web of Conferences* (Ceske Budejovice) 134

[21] Skrúcaný T, Synák F, Semanová S, Ondruš J and Rievaj V 2018 Detection of road vehicle's centre of gravity *Proc. of the 11th Int. Scientific and Technical Conference on Automotive Safety (Častá)* 1–7

[22] Vlkovský M, Šmerek M and Michálek J 2017 Cargo securing during transport depending on the type of a road *IOP Conf. Ser.: Materials Science and Engineering 2017 (Prague)* 245

[23] Vlkovský M., Neubauer J, Malíšek J and Michálek J 2021 Improvement of road safety through appropriate cargo securing using outliers *Sustainability* **13** 1–16

[24] Vlkovský M, Ivanuša T, Neumann V, Foltin P and Vlachová H 2017 Optimizating cargo security during transport using data-loggers *J. Transp. Secur.* **10** 63–71

[25] Hockicko P 2015 *Fyzikálna videoanalýza reálnych dejov (Physical video analysis of real events)* (Žilina: EDIS) p 195

[26] Gonda J 1966 *Dynamika pre inžinierov (Dynamics for engineers)* (Bratislava: Vydavateľstvo SAV) p 454

[27] Rievaj V 2013 *Automobil a jeho dynamika (Automobile and its dynamics)* (Žilina: EDIS) p 224

[28] Ripka P 2011 *Senzory a převodníky (Sensors and transmitters)* (Praha: České vysoké učení technické) p 136

[29] EN 12642:2016 Securing of cargo on road vehicles - Body structure of commercial vehicles - Minimum requirements

[30] EN 17321:2021 Intermodal loading units and commercial vehicles - Transport stability of packages - Minimum requirements and tests

[31] Jurecki RS, Jaskiewicz M, Zuska A 2013 The variety of the behaviour of drivers at risk accident situations **35 (107)** 38-46

[32] Jurecki RS, Stanczyk TL 2018 Analyzing driver response times for pedestrian intrusions in crash-imminent situations, *Proc. of the 11th Int. Scientific and Technical Conference on Automotive Safety (Častá)* 1–7

IOP Conf. Series: Materials Science and Engineering 1247 (2022) 012038 doi:10.1088/1757-899X/1247/1/012038

Influence of harsh braking parameters on safety of load transport

A Zuska*, P Grabski, M Jaśkiewicz, R Jurecki, D Kurczyński, P Łagowski, T Stańczyk, E Szumska

Department of Automotive Engineering and Transport, Kielce University of Technology, Avenue Tysiąclecia Państwa Polskiego 7, 25-314 Kielce, Poland.

* Correspondence: a.zuska@tu.kielce.pl

Abstract. Loads acting on transported cargo are the result of the inertia forces associated with the movement of the motor vehicle. Longitudinal inertia forces acting on the goods transported by a vehicle may result from accelerating or braking the car. The maximum values of acceleration acting on cargo are important in terms of the selection of methods and tools used to secure the transported goods.

The paper presents the results of experimental tests of harsh braking of a tractor unit with a semi-trailer. The experiment was divided into three measurement series. Two of them were carried out on asphalt pavement, while the last one was carried out on concrete pavement. One series consisted of ten braking tests. The values of maximum decelerations recorded during the tests were subjected to statistical analysis and then compared with the values of the average full deceleration of braking determined for individual braking cases.
Keywords: road transport safety, cargo transport safety, cargo securing, harsh braking

1. Introduction

Due to the wide availability of infrastructure and high versatility, road transport is currently the most popular means of transport and considered the most important branch of European transport. It enables the execution of transports tasks virtually from any place and at any time, day or night. The widespread use of road transport is also the result of the high availability of cargo space, enabling the transport of virtually any cargo.

The benefits resulting from the development of road transport are also accompanied by hazards to the environment and road transport safety. This is manifested, inter alia, by the increased emission of pollutants and the increased number of various types of road accidents. It is estimated that almost a quarter of the accidents involving heavy goods vehicles are the result of improper cargo securing [1, 2]. The growing demand for transport services leads to increased safety risks for people involved in the transport process and other road users, as well as for vehicles and the loads they transport.

The main element affecting the safety of cargo transport is proper securing by selecting the appropriate methods and devices to secure the goods transported in the vehicle's cargo space. The cargo may be secured by blocking, lashings, increasing the friction force against the floor of the body or surface securing [3, 4, 5]. The selection of the appropriate method and the selection of tools for securing loads during transport depends on the forces acting on the cargo while the vehicle is moving. The loads acting on the cargo during transport result from inertia forces related to the movement of the motor vehicle, vibrations and inclination of the vehicle. The inertia forces acting on the load while the vehicle is moving are longitudinal force occurring during acceleration and braking, lateral forces occurring due to bendiness, vertical forces generated when driving on uneven roads [6, 7, 8].

The longitudinal inertia force is generated when the speed of the vehicle is changing, i.e. when the vehicle is braking or accelerating. The force can be expressed as the product of the mass of the cargo and **acceleration** of the mass centre. The maximum values of acceleration acting on the load are important in terms of selecting methods and tools used to secure the transported loads.

A significant percentage of road incidents caused by improper securing of loads was the reason to analyse the accelerations recorded during a series of harsh braking and acceleration tests of an articulated vehicle. The paper describes the object of the test, the test conditions, as well as the results of tests, showing the properties of the longitudinal dynamics of an unloaded truck.

2. Methodology of field tests

2.1. Research vehicle

In the years 2019-2021, the Department of Motor Vehicles and Transport of the Kielce University of Technology conducted research in the field of longitudinal dynamics (acceleration and braking) and transverse dynamics of vehicles of the following category: M1, M2, M3, as well as N1 and N3.

The present study presents and analyses the results of road tests of an articulated vehicle, consisting of a tractor unit and a semi-trailer (Fig. 1), classified as category N3. The N3 category includes the vehicles adapted to transport cargo of the maximum mass exceeding 12 tonnes. The total length of the vehicle was 12.3 m. The tractor unit had two single axles with a wheelbase of 4 meters, while the semi-trailer had three axles. The test weight, i.e. the tare weight of the tractor unit and semi-trailer, as well as the test apparatus and two people, was 13,450 kg in total. The tractor unit was equipped with 315/70/22.5 tyres, while the semi-trailer with 385/65/22.5 tyres.

Figure 1. View of the tested vehicle

2.2. Measuring apparatus

All the results obtained and presented in this study were recorded using the constructed measurement track that consisted of:

- the S-350 ™ optoelectronic head of the Correvit system for non-contact measurement of velocity vector components (speed measurement range 0.5 ÷ 250 km/h, measurement range of the lateral drift angle ± 40°, measurement resolution of the lateral drift angle <± 0.1°) [9],
- TANS ™ gyro sensor for measuring linear accelerations (measuring range ± 3 g) and angular velocities (yaw speed range ± 150 deg/s) [10],
- data acquisition stations µEEP-12 [11],
- laptop and software.

The arrangement of the constructed measurement track allowed to record:

- longitudinal and lateral speed, as well as the vehicle's chassis drift angle,
- longitudinal, lateral and vertical accelerations and angular velocities of the body (driver's cab).

2.3. The course of the test

The longitudinal dynamics tests were carried out on two types of pavement (dry asphalt and dry concrete). The asphalt pavement was in good technical condition, while the concrete pavement was covered with some small pieces of concrete debris. In this study, three series of measurements were analysed during which the following was performed:

- harsh braking of the vehicle on dry concrete pavement at a speed of 30 km/h,
- harsh braking of a vehicle on a dry asphalt pavement at a speed of 30 km/h and 50 km/h.

Each measurement series consisted of ten braking tests, during which the following time curves were recorded: the longitudinal and lateral velocity, as well as the vehicle's chassis sideslip angle; longitudinal, lateral and vertical accelerations and angular velocities of the driver's cab. The time curves of the above-mentioned values recorded during braking on a dry asphalt pavement at the initial speed of 50 km/h, are shown in Figure 2. For each of the braking tests, the value of the maximum braking deceleration a_{max} and the mean full developed deceleration value (MFDD) were determined.

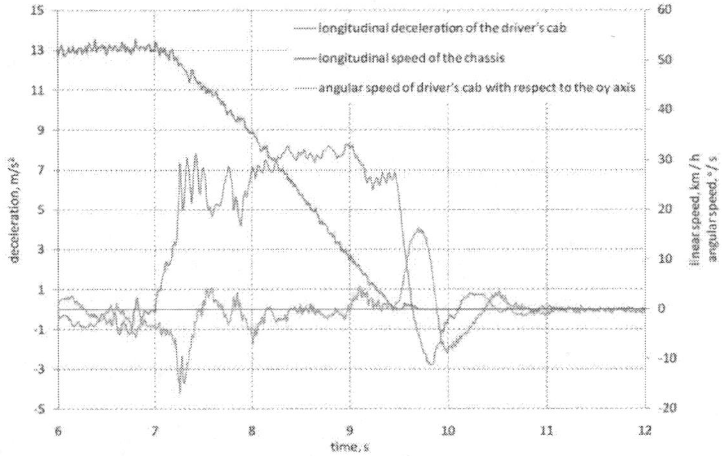

Figure 2. Sample time curves of selected values recorded during the braking test

3. Results and analysis of the tests

In the course of the research, the longitudinal acceleration of the vehicle was recorded using the acceleration sensor installed in the driver's cab. The time curves of these accelerations are shown in Figure 3. For each of the braking tests, the value of the maximum braking deceleration a_{max} was determined. Based on these values, for each series consisting of ten braking tests, the following values were calculated: the highest (a_{max}) and the lowest (a_{min}) maximum deceleration, a mean value of maximum decelerations $\bar{a}_{max}$ and its standard deviation $SD\bar{a}_{max}$. The calculation results are presented in Table 1.

Table 1. Maximum decelerations statistics

Type of road pavement	Initial speed, km/h	Statistics of maximum decelerations, m/s²					
		a_{max}	a_{min}	$\bar{a}_{max}$	$SD\bar{a}_{max}$	Q0,1	Q0,9
Dry asphalt pavement	30	10.64	6.55	8.66	1.24	10.15	7.16
	50	9.61	8.11	8.83	0.45	9.28	8.28
Dry concrete pavement	30	9.59	7.57	8.31	0.65	9.22	7.72

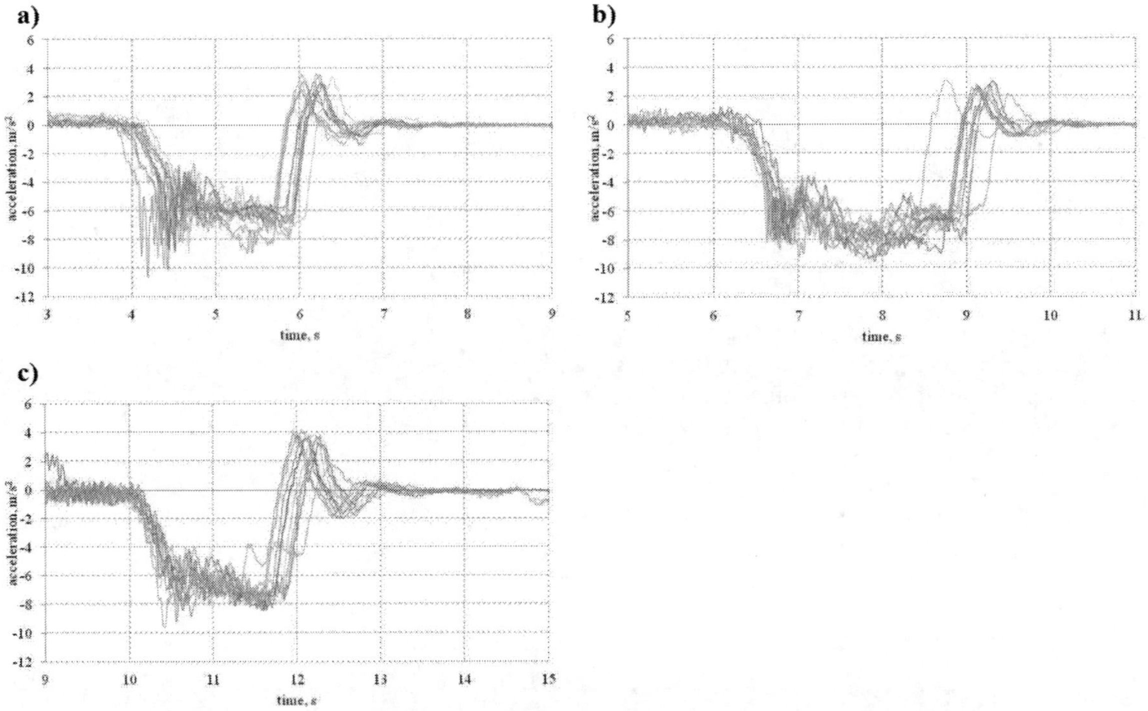

Figure 3. Longitudinal accelerations curves recorded during braking tests, carried out on the following pavement: a) dry asphalt at a speed of 30 km/h, b) dry asphalt at a speed of 50 km/h, c) dry concrete at a speed of 30 km/h.

A preliminary analysis of the above diagrams and the values summarised in Table 1 show that, in the individual measurement series, there is a high qualitative and quantitative similarity between the deceleration curves. All the graphs clearly show the effect of the ABS system activation and the tractor cabin rocking effect. The rocking of the body (cab) is most evident in the final stage of the braking (just before and immediately after stopping the vehicle).

The highest mean value of the maximum decelerations $\bar{a}_{max}$ amounting to 8.83 m/s^2 was recorded during the series of tests carried out on an asphalt pavement at the initial vehicle speed of about 50 km/h. It was only 6.6% higher than $\bar{a}_{max}$ recorded for the series of tests carried out on dry concrete pavement at the initial vehicle speed of about 30 km/h.

For the tests carried out at the initial speed of 30 km/h, the average value of the maximum decelerations for the tests carried out on the concrete pavement was lower than the average value of the maximum decelerations for the tests carried out on the asphalt pavement.

The international requirements for braking systems are regulated by the provisions of the United Nations Economic Commission for Europe. They are included in several regulations on the approval of braking systems. One of them is UN/ECE Regulation No 13 "Uniform provisions concerning the approval of vehicles of categories M, N and O with regard to braking". As regards the approval of braking systems, the national regulations are harmonised with the international ones.

According to the regulation, the effectiveness of the braking systems is determined by measuring the stopping distance of the vehicle from a given initial speed or from the mean fully developed deceleration.

The stopping distance is the distance travelled by the vehicle from the moment the driver begins to depress the brake pedal until the vehicle comes to a stop, and can be determined using the formula:

$$S_z = S_R + S_H = V_0 \cdot \left(t_r + \frac{t_n}{2}\right) + \frac{V_0^2}{2 \cdot MFDD} \ [m] \quad (1)$$

IOP Conf. Series: Materials Science and Engineering 1247 (2022) 012038 doi:10.1088/1757-899X/1247/1/012038

while the braking distance can be determined based on the following dependence:

$$S_H = \frac{V_0^2}{2 \cdot MFDD} \ [m] \tag{2}$$

Initial speed (V_0) means vehicle speed at the moment the driver begins to depress the brake pedal Whereas the mean fully developed deceleration ($MFDD$) is the average deceleration on the road in the speed range of V_b to V_e and they were calculated based on dependence number 3 [12]:

$$MFDD = \frac{V_b^2 - V_e^2}{25{,}92 * (S_e - S_b)} \ [m/s^2] \tag{3}$$

where:

- V_o- initial vehicle speed in [km/h],
- V_b - vehicle speed corresponding to 0.8 V_o in [km/h],
- V_e - vehicle speed corresponding to 0.1 V_o in [km/h],
- S_b - distance travelled between V_o and V_b in [m],
- S_e - distance travelled between V_o and V_e in [m].

Using the dependence 2 and the recorded time curves of velocity (examples of which are presented in Fig. 4) for each of the tests, the following were determined: V_o, V_b, V_e, S_b, V_b, MFDD, and S_H and the results of the calculations are given in Table 2.

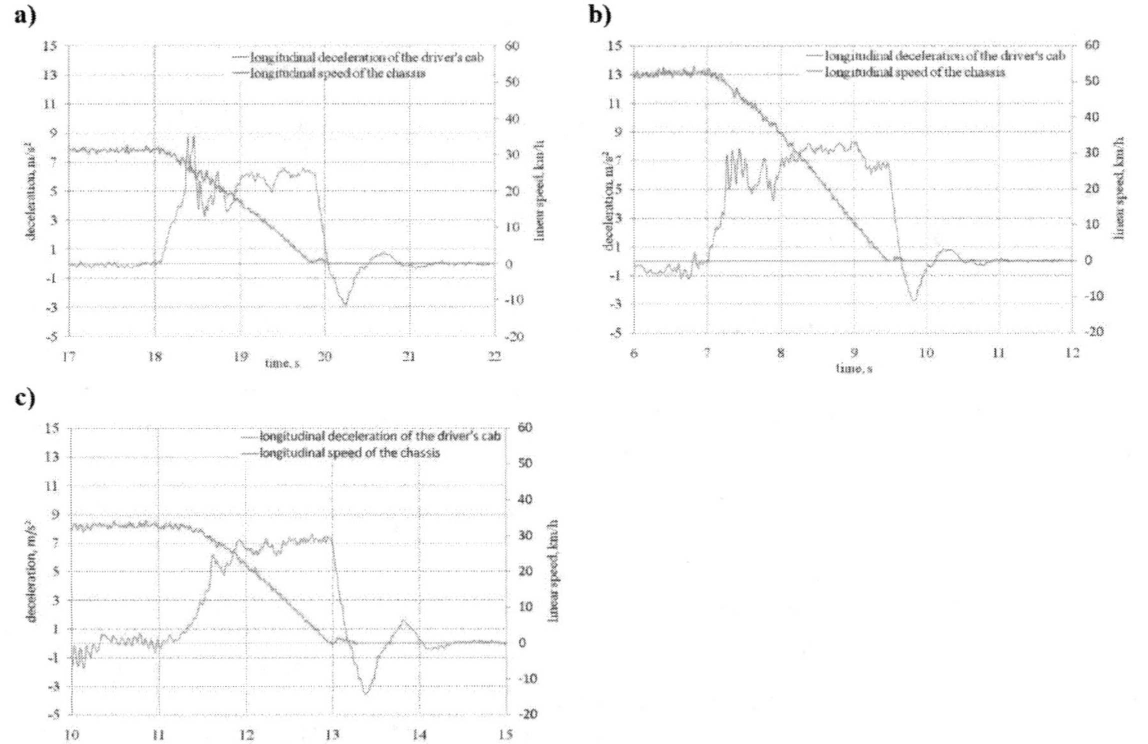

Figure 4. Examples of time curves of longitudinal deceleration and longitudinal speed recorded during braking on the following pavements: a) asphalt, at a speed of 30 km/h, b) asphalt, at a speed of 50 km/h, c) concrete, at a speed of 30 km/h.

Table 2. Parameters of the truck braking process course

Test no.	Initial vehicle speed, V_o, km/h		Maximum deceleration, a_{max}, m/s^2	Mean fully developed deceleration, MFDD, m/s^2	$\overline{MFDD}$, m/s^2	Braking distance, S_H, m
	Assumed	Actual				
Asphalt pavement						
1		31.91	6.55	5.08		7.74
2		31.27	8.83	5.17		7.30
3	30	30.88	10.61	4.73	5.16	7.78
4		31.49	7.16	5.52		6.94
5		31.89	8.72	5.29		7.42
1		52.33	8.29	6.49		16.29
2		52.61	9.00	6.88		15.35
3	50	51.22	8.63	6.76	6.78	14.98
4		51.66	9.60	6.79		15.18
5		50.62	8.66	6.96		14.20
Concrete pavement						
1		32.86	7.57	6.20		6.72
2		32.90	7.98	5.24		7.96
3	30	31.43	9.59	6.00	5.914	6.35
4		32.33	8.43	6.16		6.55
5		31.81	8.80	5.97		6.53

A detailed analysis of the time curves of accelerations and longitudinal velocities, as well as the values listed in Table 2 enables us to conclude that:

- The greatest value $MFDD$ amounting to 6.96 m/s^2 was recorded for the sudden braking test carried out on the asphalt pavement from the initial speed of the vehicle, amounting to 50.62 km/h. This value was over 47% higher than $MFDD$ determined during the test carried out on the same pavement but with an initial speed of 30.88 km/h.
- The initial speed of the vehicle had a significant impact on the values of $\overline{MFDD}$ For tests carried out on dry asphalt pavement, the initial speed increase from the range 30.88 - 31.98 km/h to the range of 50.62 - 52.61 km/h caused the increase of $\overline{MFDD}$from 5.16 m/s^2 to 6.78m/s^2.

When selecting the cargo securing elements, take into account the values of forces that may act on the cargo while driving. According to the standard, the maximum values of forces that should be taken into account in the selection of fastening elements are part of the cargo gravity force (Q_L) expressed by coefficient $C_{x,y,z}$ [13, 14].

$$C_x = \frac{a_x}{g} \quad (4)$$

Taking into account the dependence (4), the maximum deceleration values and the determined mean fully developed decelerations MFDD, $C_x$was determined for each of the tests. The results of the C_x calculations are given in Table 3.

Table 3. Values of the C_x coefficient for selected (example) braking tests.

Test no.	Initial vehicle speed, V_o, km/h		Maximum deceleration, a_{max}, m/s²	C_x factor for maximum deceleration	Mean fully developed deceleration, MFDD, m/s²	C_x factor for fully developed deceleration
	Assumed	Actual				
Asphalt pavement						
1	30	31.91	6.55	0.67	5.08	0.52
2	30	31.27	8.83	0.90	5.17	0.53
3	30	30.88	10.61	1.08	4.73	0.48
4	30	31.49	7.16	0.73	5.52	0.56
5	30	31.89	8.72	0.89	5.29	0.54
1	50	52.33	8.29	0.85	6.49	0.66
2	50	52.61	9.00	0.92	6.88	0.70
3	50	51.22	8.63	0.88	6.76	0.69
4	50	51.66	9.60	0.98	6.79	0.69
5	50	50.62	8.66	1.24	6.96	0.71
Concrete pavement						
1	30	32.86	7.57	0.77	6.20	0.63
2	30	32.9	7.98	0.81	5.24	0.53
3	30	31.43	9.59	0.98	6.00	0.61
4	30	32.33	8.43	0.86	6.16	0.63
5	30	31.81	8.8	0.90	5.97	0.61

4. Conclusions

The analysis of the tractor unit harsh braking road test results leads to the following conclusions:

- The average value of the maximum deceleration was 8.6 m/s² and it is more than 45% higher than the mean value of fully developed deceleration.
- The values of $MFDD$ for a series of braking tests carried out on the asphalt pavement at the initial speed of 50 km/h are within a fairly narrow range of 6.49 m/s² to 6.96 m/s². For the other two series, these ranges are much wider.
- Significant impact on the values of $MFDD$ had the initial speed of the vehicle and the type of road pavement.
- The highest value of $MFDD$ was recorded for the braking performed on the asphalt pavement at the initial speed ranging from 50.62 km/h to 52.61 km/h, while the lowest value of $MFDD$ was recorded during braking on the asphalt pavement at the initial speed ranging from 30.88 km/h to 31.91 km/h.
- The values of Cx determined based on $MFDD$ range from 0.48 to 0.71 and are definitely lower than the standard value of the Cx index, which is 0.8. This dependence is favourable because the actual values of the forces acting on the cargo during harsh braking will be much lower than the forces determined based on the standard Cx value. The fastening equipment selected on this basis will fully protect the cargo against moving on the floor of the trailer.

Literature:

[1] EUROPEAN COMMISSION "Traffic Safety Basic Facts 2018" https://ec.europa.eu/transport/road_safety/sites/roadsafety/files/pdf/statistics/dacota/bfs20xx_hgvs.pdf.

[2] EUROPEAN COMMISION "European Best Practice Guidelines on Cargo Securing for Road Transport" www.uirr.com/fr/component/ downloads/downloads/ 302.html.

[3] 2002 European best practice guidelines on cargo securing for road transport. European Commission Directorate – General for Energy and Transport.

[4] 2003 Driver's Handbook on Cargo Securement - A Guide to the North American Cargo Securement Standard.

[5] Vlkovsky M, Pochobradska K and Foltin P. 2016, The cargo securing based on european standards and its applicability in off-road transport conditions. *3rd International Conference on Traffic and Transport Engineering (ICTTE)*, pp 603-607.
[6] Vlkovsky M, Smerek M and Michalek J. 2017, Cargo securing during transport depending on the type of a road. *World Multidisciplinary Civil Engineering-Architecture-Urban Planning Symposium (WMCAUS)*, Book Series: IOP Conference Series-Materials Science and Engineering, vol. 245.
[7] Zong Ch, Zhang H and Huang Ch. 2017, Research on the Influence of Cargo Securing Force with Typical Road Alignments and Vehicle Working Conditions. *4th International Conference on Transportation Information and Safety (ICTIS)*, pp 27-32.
[8] Stokłosa J, Koszałka G and Gil L. 2012, Analysis of forces In elements used for fastening loads on Road vehicles. *Postępy Nauki i Techniki,* nr 12, Lublin, pp 94-100
[9] Correvit® S-350 Non-Contact Optical Sensors, Type CS350A... Kistler Patent No. DE 43 13 497 C2 https://www.kistler.com/files/document/000-807e.pdf.
[10] Acceleration modules SAA (1-axis), DAA (2-axis), TAA (3-axis), Type KCD15760, KCD15951, KCD15911. Kistler https://www.kistler.com/files/document/000-913e.pdf?callee=frontend.
[11] µEEP-12 Data Acquisition and Evaluation Typ CMEP1A... Kistler. https://www.techtarget.com.br/site/wp-content/uploads/2016/09/CMEP12A_000-819e.pdf
[12] Janczur R and Zawałeń J. 2021, Braking parameters of trucks and buses in the reconstruction and analysis of road accidents. *XVII Conference Problems of Reconstruction of Road Accidents*, (Zakopane), pp 105-20.
[13] EN 12195-1:2003 Load restraint assemblies on road vehicles — Safety — Part 1: Calculation of lashing forces. CEN 2003.
[14] EN 12195-1:2010 Load restraining on road vehicles — Safety — Part 1: Calculation of Securing forces. CEN 2010.

Acknowledgements

The research was carried out as part of the Innovative system research project supporting the motor vehicle insurance risk assessment dedicated to UBI (Usage Based Insurance) No. POIR.04.01.04 00 0004/19 00 financed by the National Centre for Research and Development.

A three-wheeled vehicle for the disabled people

K Stańko-Pająk[1,3], B Bursa[1], J Seńko[2], T Detka[1], S Korczak[4], R Nowak[4], K Popiołek[1], J Lisiecki[1], A Paczkowski[1]

[1] Research Group for New Automotive Technologies, Łukasiewicz Research Network-Automotive Industry Institute, Warszawa, ul. Jagiellońska 55, Polska
[2] Institute of Vehicles and Construction Engineering, Faculty of Automotive and Construction Machinery Engineering, Warsaw University of Technology, Warszawa, ul. Narbutta 84, Polska
[3] Doctoral School no 2, Warsaw University of Technology, Warszawa, ul. Narbutta 84, Polska
[4] Institute of Machine Design Fundamentals, Faculty of Automotive and Construction Machinery Engineering, Warsaw University of Technology, Warszawa, ul. Narbutta 84, Polska

* katarzyna.stanko-pajak@pimot.lukasiewicz.gov.pl

Abstract. This article presents the construction of a prototype three-wheeled vehicle for people with disabilities, approved in the L2e category. The vehicle is equipped with a special tilt mechanism that allows people with disabilities, who cannot balance their body, to return to an upright position after completing a driving maneuver on a curve of the road. The tilt angle of the vehicle depends on its speed and steering angle. When designing the vehicle, the availability and price of spare parts for scooters and motorcycles were taken into account to make the vehicle as cheap as possible.

Keywords: three-wheeled vehicle, disable people, tilt mechanism, electric drive, tilting frame, vehicle

1. Introduction

Currently, there are no systemic solutions and standardization of services in the area of mobility for people with disabilities. So far, support for people with disabilities in the area of mobility have been provided only by isolated activities of other people, driving schools or various non-governmental organizations, associations and foundations. The lack of a uniform, coordinated platform for activities in this area is the main problem and an obstacle in gaining a driving license by people with disabilities, and thus social and professional activation. The professional potential of people with disabilities, who often have higher education and great qualifications, is not used due to limited possibilities of commuting to work. One of the main reasons why people with physical disabilities are not engaging in social activities or have difficulties in finding employment and staying on the labor market is the limitation in individual mobility [1]. This problem is also magnified by insufficient urban infrastructure, especially in the field of public transport. Unadopted vehicles, bus stops, broken lifts or lack thereof significantly impede the ability of people with disabilities to move independently [2]. Currently, there are organized transports for people with disabilities, but even the best and most efficient transport systems will not fully meet the needs of independent travel at any time and over any distance.
In most cases, three-wheeled vehicles are characterized by a low curb weight, and because of this they are often homologated in the L category [3]. Solutions for three-wheeled vehicles with a different Delta

1-2 arrangement, i.e. one front wheel and two rear wheels, have been known all over the world for a long time. Wheels at the rear and Tadpole 2-1, i.e. one rear wheel and two front wheels. The most famous example of tricycles today is the Piaggio MP3, where the tilt is achieved by push arms. The vehicle also has a tilt lock at a speed of 115 km / h.

2. Results

The structure of the vehicle in the Delta 1-2 system is a support structure in a form of a frame that is tilted together with the driver when driving along a curve of the road [4]. For the design of the frame, round tubes were used, which are normally used in production of scooter frames [4,5] . The maximum dimensions of the vehicle are 1700 mm high, 900 mm wide, and the center distance is 1500 mm. A particularly important parameter is the width of the vehicle, because the assumption is that the vehicle will fit in a door frame, which will allow it to move around closed facilities such as shopping centers, airports, etc. A concept of the vehicle is presented in Figure 1. Mostly readily available parts were used for the construction of the vehicle, especially spare parts for scooters and motorcycles, e.g. the front suspension has been completely adapted from the scooter and the braking system was adapted from a quad vehicle. Other ready-made elements are: wheel rims, tires, brake discs, steering wheel, gas shifters, display, as well as wipers, washers, etc.

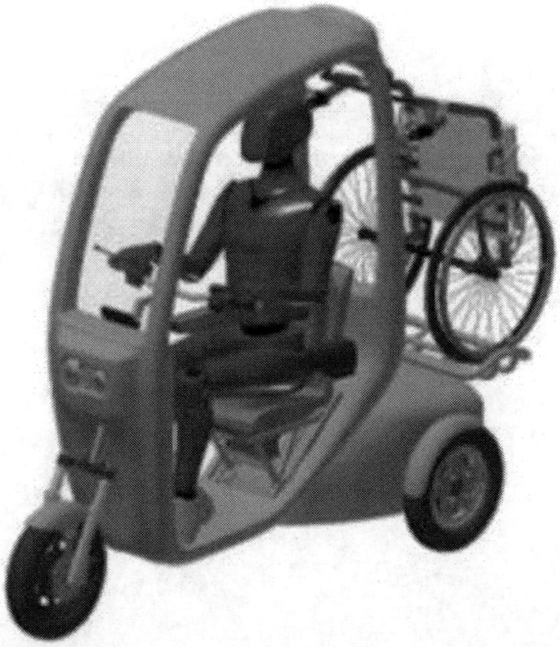

Figure 1. Three- wheeled vehicle – 3D CAD model.

The vehicle is driven by two electric motors located in its rear wheels. The use of an electric drive makes it a zero-emission vehicle, enabling it to move around landscape parks and city parks. The engine power is 3 kW. Thanks to bidirectional engines, reverse gear was obtained - required in wheelchair vehicles. Rear wheels are rigidly connected to each other, preventing them from moving relative to each other, thus creating a dependent rear suspension. Movement of rear suspension is possible by using a wishbone connection with the vehicle frame. An existing solution from a Honda Gyro tilting tricycle was used to design the swingarm. Appropriate modifications were made to ensure driving safety and for the vehicle to be allowed in land traffic. The rocker arm is an integral part of the tilt mechanism. The swivel arm allows the vehicle frame to tilt by a limited angle in relation to the rigid rear bogie [3].

IOP Conf. Series: Materials Science and Engineering 1247 (2022) 012039 doi:10.1088/1757-899X/1247/1/012039

2.1. Tilt Forcing Mechanism

The purpose of the vehicle is to facilitate a return to a vertical position after taking a bend or other curve of the road. For this purpose, a special tilting mechanism was designed, that consists of suitably selected elements, based on analytical calculations, such as: DC electric motor, belt transmission, worm gear, the aforementioned pivoting rocker arm and a three-arm lever [6]. The tilt mechanism also features a simple belt tensioning mechanism that is pulled out during use. It is easy to use and therefore even the driver can do it independently. The detailed structure of the mechanism is shown in Figure 2. The mechanism is protected by patent no. 239880. The principle of operation of the mechanisms is shown in Figure 3.

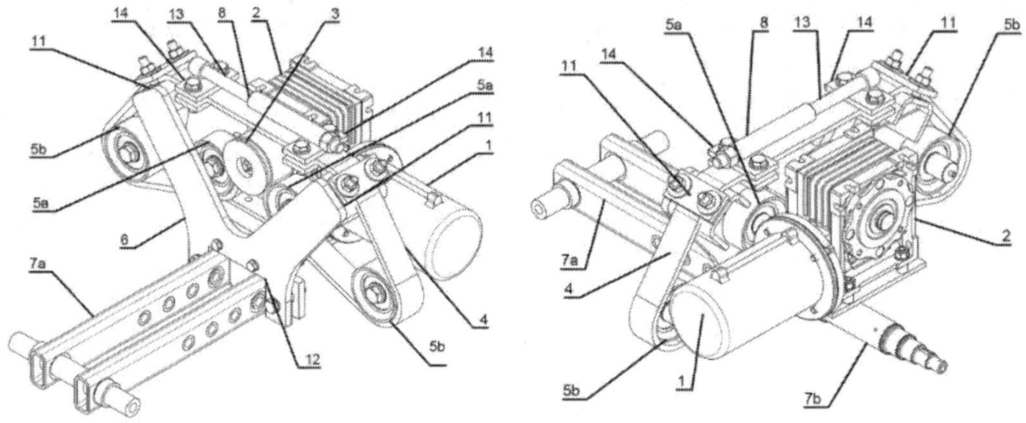

Figure 2. Tilt mechanism. 1- DC motor; 2- worm gear; 3- toothed gear wheel; 4- toothed belt; 5a- internal guide rollers; 5b- external guide rollers; 6- three-arm lever; 7a- swing side rocker arm (connected with frame of vehicle); 7b- swing rocker arm (connected with rigid trolley of vehicle); 8- tensioner; 11- upper arm three-arm lever; 12- down arm three-arm lever; 13- adjusting screw; 14- tensioner handle.

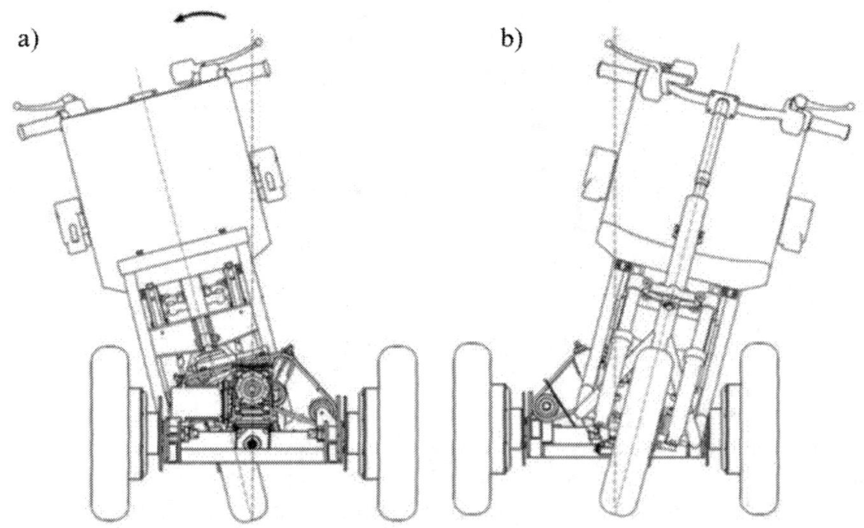

Figure 3. Example of tilt operation to the left side with the tilt mechanism visible a) - rear view, b) - front view.

The method for controlling the tilt of the vehicle frame works as follows: based on the signals from the steering wheel angle sensor and the vehicle speed sensor, the controller transmits the tilt angle signal,

IOP Conf. Series: Materials Science and Engineering 1247 (2022) 012039 doi:10.1088/1757-899X/1247/1/012039

which is corrected by the value of the feedback signal from the vehicle frame angle sensor, and then processed by the position controller, then corrected again for the current measured between the motor control module and the electric motor. This signal is sent to the current regulator. Based on this signal, the current regulator generates a control signal for the motor control system, which is converted into a PWM signal in the PWM signal generator (Pulse Width Modulation) and the engine control system powers the electric motor to drive the tilt function. A Hall effect sensor or a shunt connected to an analog-to-digital converter is used to measure the current between the motor control module and the electric motor [7]. The block diagram of the control system is shown in Figure 4. The tilt control method is protected by patent No. 239393.

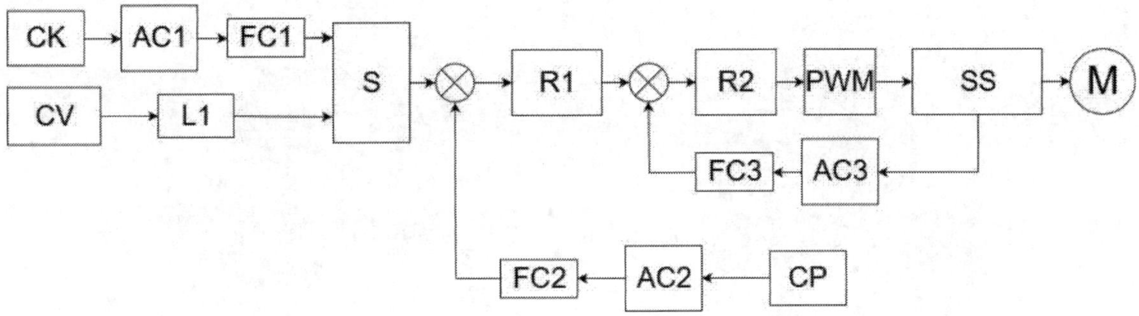

Figure 4. Schemat Block diagram of the vehicle tilt control system: CV- vehicle speed sensor; CK- steering angle sensor; CP- vehicle frame tilt angle sensor; M- electric motor; AC1, AC2, AC3- analog-to-digital converters; L1- counter circuit (timer); R1- position controller; R2- current regulator; FC1, FC2, FC3- digital filters; PWM- PWM signal generator; SS- engine controller; S- controller.

Value of the set angle of tilt of the vehicle's frame with the driver inside is selected from the tilt angle table, based on two data values - the steering angle and the vehicle's speed. The set angle is therefore a discrete-valued function. Vehicle's speed is obtained from its speed sensor located in the propulsion engine. Due to the presence of two identical motors, and the same two variable frequency signals, the system uses a counter circuit, which enables a frequency proportional to the frequency of the signal to be determined [6]. The arrangement of the sensors on the vehicle is shown in Figure 5.

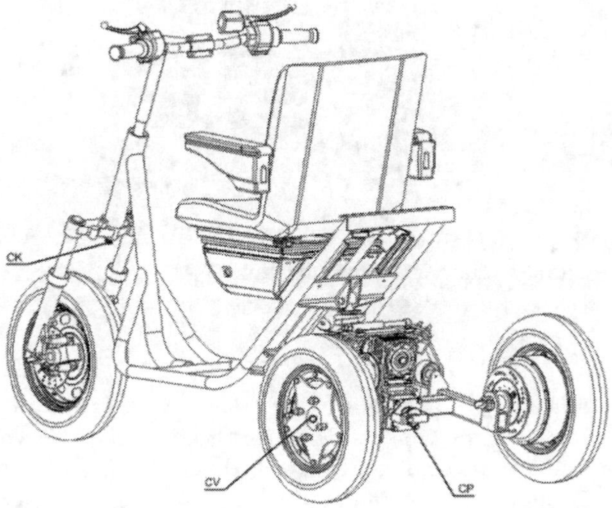

Figure 5. Location of sensors in the vehicle. CV- speed sensor; Frame angle sensor; CK- steering angle sensor.

2.2. Facilities for disabled people

The vehicle has been designed so that it has an open body, thus making it easier to transfer from a wheelchair to the driver's seat and to mount the wheelchair on the rear luggage compartment on your own (Figure 7). An ergonomic, adjustable seat with five-point seat belts was also used. The chair is adjustable in height, it lowers to a height of 550 mm from the ground, i.e. the height of the standard wheelchair seat. It is also possible to adjust the seat's distance from the steering wheel. In addition, the seat is rotatable by 360 ° with the possibility of locking every 45 °. Additional leg protection in the form of supports are also attached to the chair to prevent uncontrolled leg movement. The method of transfer is shown in Figure 6.

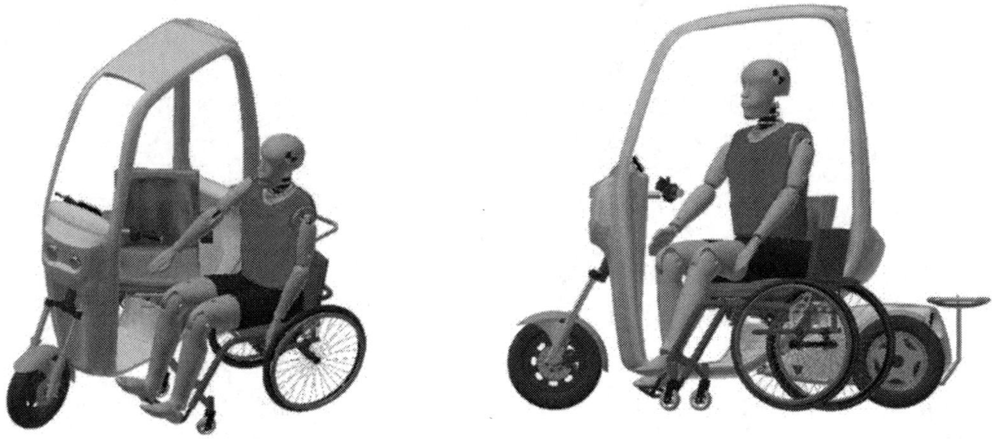

Figure 6. The method of transferring the driver to the driver's seat

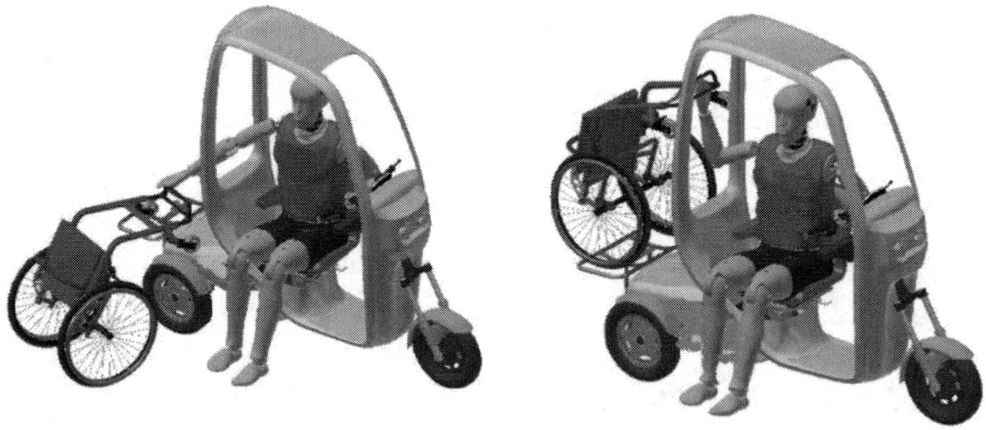

Figure 7. The way of loading a wheelchair on the trunk

3. Conclusion

The article presents the concept, model and application of a three-wheeled vehicle for people with disabilities. Table 1 presents the most important vehicle parameters which, apart from the requirements for adaptation for people with disabilities, also meet the vehicle approval guidelines in L2e category, i.e. three-wheeled mopeds. This category was chosen due to the fact that it is not necessary to have a category B driving license to be able to drive such a vehicle. This is due to the fact that some people

IOP Conf. Series: Materials Science and Engineering 1247 (2022) 012039 doi:10.1088/1757-899X/1247/1/012039

with disabilities, for reasons such as claustrophobia, cannot drive ordinary cars adapted for them. In addition, the vehicle can be used by people without disabilities and minors over 16 years of age.

Table 1. Technical parameters of the vehicle

Parameter	Value
Max. speed	45 km/h
Vehicle weight	190 kg
Engines power	3 kW
Max tilt angle	10 °
Range on one charging	100 km
Height vehicle	1700 mm
Axis distance	1500 mm
Width vehicle	900 mm

The modular structure of the vehicle allows the removal of luggage compartment for a wheelchair and installation of a trunk, creating a small cargo space, which enables it to be used as a small delivery vehicle or in carsharing. The use of many ready-made elements in the form of spare parts for scooters or motorcycles means, that the target price of the vehicle will be much lower than other solutions currently available. It also makes the servicing of the vehicle much easier. Work is currently underway on the construction of a prototype. It is planned to conduct tests related to the safety, driveability and stability of the vehicle

References

[1] Kamińska E S 2021 Możliwość wykorzystania pojazdów zautomatyzowanych w celu zapewniania wsparcia osobom niepełnosprawnym w transporcie indywidualnym *GospodarkaMateriałowa i Logistyka* **2** pp 14–20

[2] Kacperczyk K and Rymsza B 2021 Niepełnosprawni na przejściach dla pieszych- problemy z krawężnikami *Drogownictwo* **2** pp 35–40

[3] Nowak T, Seńko J, Jasiński P and Paczkowski A 2018 Selected issues relates to the research on Hybrid tree-wheeler for disabled *Archiwum motoryzacji* **80(2)** pp 95-113

[4] Caban J, Droździel P and Seńko J 2014 Wybrane materiały konstrukcyjne w budowie pojazdówSamochodowych *Logistyka* **3:3** pp 946-953

[5] Fligiel M 2014 Present-day trends in designing the supporting structures of special vehicles of Trike "REWACO" type *Autobusy: technika, Eksploatacja, systemy transportowe* **15(6)** pp 118-126

[6] Stańko-Pająk K, Lisiecki J and Paczkowski A 2018 *Mechanizm wymuszenia przechyłu* Pat.239880

[7] Bursa B, Popiołek K, Stańko-Pająk K, Paczkowski A, Lisiecki J and Detka T 2018 *Układ sterowania przechyłem ramy pojazdu oraz sposób sterowania przechyłem ramy pojazdu, zwłaszcza trójkołowego* Pat.239393

IOP Conf. Series: Materials Science and Engineering 1247 (2022) 012040 doi:10.1088/1757-899X/1247/1/012040

Evaluation of the safety of overweight and/or oversized traffic based on the analysis of lateral acceleration during transport

J. Vrábel[1*], M. Loman[1], J. Paľo[1], B. Šarkan [2], A. Hudcovský [2]

[1] Department of Road and Urban Transport, University of Žilina, 01026 Žilina, Slovakia
[2] Department of Road and Urban Transport, University of Žilina, 01026 Žilina, Slovakia

[*] jan.vrabel@fpedas.uniza.sk

Abstract: Ensuring the safety and smoothness of road traffic in transports, which by their parameters may cause a special situation in road traffic, is the primary task of the team of workers accompanying overweight and/or oversized vehicles. In order to successfully carry out such an activity, it is necessary not only the appropriate technical equipment but also the knowledge of accompanying staff on the issue of critical points during transport and knowledge of how to prevent the danger in the most appropriate way. For these reasons, there is a need to focus the interest of companies on the course of such transport and to increase the safety of all road users through trained staff to accompany vehicles performing overweight and/or oversized transport.
Keywords: safety, vehicle movement, road safety

1. Introduction

We encounter overweight and/or oversized traffic almost every day on the road. Usually, we do not even realize that he is an exceptional participant in road traffic. Only when the vehicle exceeds certain parameters (mostly dimensional) does it become interesting for other road users. Such an approach is not entirely correct because it is not always what is large and dangerous. There are situations where excessively heavy loads can cause great danger during transport and may not be dimensionally abnormal to capture our attention.

Therefore, to avoid road users having to estimate whether a given vehicle is dangerous, there are accompanying vehicles and the whole system of overweight and/or oversized traffic. This system should ensure, first and foremost, the safety but also the flow of road traffic during such services. Traffic safety conditions are also determined by drivers, vehicles, and the driving environment [1]. It is not enough to have an experienced escort team to eliminate the danger. Vehicle drivers also have a share in the level of safety during such transport. From the point of view of safety, it is also necessary to take into account the fastening and movement of the goods on the loading surface of the vehicle. While driving, various forces act on the vehicle and affect the movement of the goods. A study in which Gnap et al. deals with the measurement of these dynamic events using MEMS sensors at selected locations of the vehicle loaded with cargo and changes in dynamics after certain events that could occur during regular road haulage in order to analyse the possibilities of monitoring accelerations and related forces. acting on the load during transport [2].

Ensuring the safety and smoothness of road traffic in transports, which by their parameters may cause a special situation in road traffic, is the primary task of a team of workers accompanying overweight and/or oversized vehicles.

In order to successfully carry out such an activity, it is necessary not only the appropriate technical equipment, but also the knowledge of accompanying staff on the issue of critical points during transport and knowledge of how to prevent the danger in the most appropriate way. For these reasons, there is a need to focus the interest of companies on the course of such transport and to increase the safety of all road users through trained staff to accompany vehicles performing overweight and/or oversized transport.

This paper deals in more detail with the process of transporting overweight and/or oversized loads by measuring transverse and longitudinal accelerations, analysing the measured values, and identifying critical points on the observed routes.

2. State of knowledge

The field of overweight and/or oversized transport, which includes taking into account its safety, the emergence of many events during such transport, the simplification of the authorization process, and other issues in this area are addressed worldwide by many experts not only in the field of road transport. The JRelčák, J. Vrábel and M. Kiktová in the article Using of different road vehicle combinations for abnormal transports based on the assessment of load rating of bridges in Slovakia for the most critical vehicle combination from 2018 [3]. By identifying and assessing such a critical combination, the carrier would be able to carry out an unusual transport with all the vehicles in the fleet on the route under assessment without further evaluation of the route for another vehicle/vehicle combinations. This paper explains and compares several ways to obtain the most critical vehicle/vehicle combination and its purpose.

In their article [4] of 2015, Mr. P. Godavarthy, Mr. Russell, and Mr. Landman deal with the issue of the passage of roundabouts for vehicles with special traffic and, in particular, for oversized vehicles. In the study, they designed six standard roundabouts through which oversized vehicle crossings were subsequently simulated. This study evaluated the effectiveness of the respective vehicles. The disadvantageous proposals resulting from the studies are the impossibility of modifying the middle island by planting plants and greenery.

In 2017, J. Vrábel, J. Jagelčák, O. Stopka, M. Kiktová, and J. Caban dealt with the transport of various types of cargo with different dimensions exceeding the maximum permissible vehicle dimensions. This contribution concerns the main load whose dimensions exceed those of the semi-trailer. The paper describes the maneuverability circle of the vehicle, which states that each vehicle must be able to maneuver in the inner circle of 5.3 m as a minimum criterion and must be able to maneuver in the outer circle of 12.5 m as a maximum criterion. The paper also describes the effect of the position of the fifth wheel on the truck on the total length of the articulated vehicle, which is the maximum allowed length of the load in the rearward direction. This research study indicates the exact length of the maximum load in terms of its width and its location on the loading area of the semi-trailer [5].

The system described in the article [6] by D. Bazaras, N. Batarliene, R. Palšaitis, and A. Petraška is suitable for use in the planning and design of excessive or excessive cargo transport. This system allows objectively the most suitable sections of the route in the existing road network. It is also a means of comparing different layers in a given area, which allows you to objectively select the most suitable one according to certain criteria using mathematical calculations. The paper states that the system is necessary and can be used to compare different modes of transport for oversized and oversized loads for different modes of transport.

The issue of the generation of vibrations, their measurement, and analysis in road transport in India as a result of road inequalities are addressed in the article [7]. Its authors find that the most common road bumps, such as potholes and bumps, disrupt travel and cause great damage to vehicles. Recording road gaps and road roughness levels are key to monitoring road conditions that have an impact on traffic safety and driving comfort. This paper proposes a method that aims to monitor road surfaces, detect potholes and road bumps, and predict their severity by analysing the vertical vibrations of the signals produced by the vehicle as it moves.

Road irregularities, their position recording, and analysis are dealt with in the article [8], which elaborates on the issue of potholes, manhole covers and faulty cracks in the roadway. They say that current ways of monitoring the road surface in many countries depend on driver reports. However, these

methods are usually accompanied by long delays, making it difficult to obtain up-to-date information on road conditions. They tested the system in real conditions on the road. They found the detected conditions on the road on online maps.
Similar to the paper [9], the authors deal with the collection of data on road inequalities through the use of automatic detection of potholes and vehicle speed, which were recorded together with their coordinates and their severity using the built-in Android accelerometer system. A three-axis accelerometer was used along with GPS sensors. They performed such data collection on a 4 km straight road at different speeds and evaluated the data. Subsequently, they divided the potholes into several categories according to the severity of the deviation in the Z axis. The accuracy of the proposed record in comparison with the manual recording of inequalities was 93.75%. The advantage of this method of data acquisition is cost-effectiveness and ease of use.
Authors from China address the analysis and discussion of dimensional, axle loads and weights for road vehicles in their paper [10]. The authors of this article discuss, using comparative analysis of dimensions, axle loads and vehicle weight, which standard would be most suitable for the real needs of road transport. They compare five completely different countries. continents with completely different legislation. They compare China, Japan, Europe, America and Australia. Their comparison shall include the width of the lanes and the maximum permitted width of the combination, the maximum permitted lengths of the combinations according to the composition of the vehicles in the combination and the maximum permitted heights of the combinations. They included the axle load for the given number of axles and the total weight in the comparison of the maximum permissible weights.
The authors deal with a similar issue in their articles [11,12], where they comprehensively discuss the issue of overweight and oversized traffic in the given study, primarily focusing on crossing bridges during such transports. In this study, an analysis of the legislation in individual foreign countries was performed. This was followed by an in-depth analysis of the issues associated with crossing the bridges, also by measuring the emerging phenomena. The results of their study will enable them to make informed decisions issued when issuing permits for the performance of special transport. This should improve and unify current permitting procedures.
The area of overweight and oversized transport, specified in more detail for the transport of wind turbines and their parts, is addressed by the authors in their contribution [13]. The analysis found that the road infrastructure had to accommodate a wide range of complex vehicle configurations to transport wind turbine components, which are often considered overweight and/or oversized by transport authorities in terms of size and weight. The result of the authors' activities and their analysis is an overview of the identified difficulties that occurred during the road transport of wind turbine components. The authors' suggestions, after analysing the issue, include an operational planning strategy based on the maximum width required for the passage of the set transporting the turbine rotor blades.
The authors performed a detailed analysis of vehicle dimensions and weights [14,15]. During the processing of the transport, an abnormal transport of up to 60 tons was evaluated in relation to the combined transport.

3. Data and Methods

During the curve, it is possible to equivalently replace the action of the centrifugal force F_o in the center of gravity of the vehicle T (Figure 1), which is caused by lateral reactions in the contact surfaces of the tires, i.e. the lateral guiding forces F_{bpl}, F_{bpp}, F_{bzl} and F_{bzp}. With these forces, the tires counteract the effects of centrifugal force.

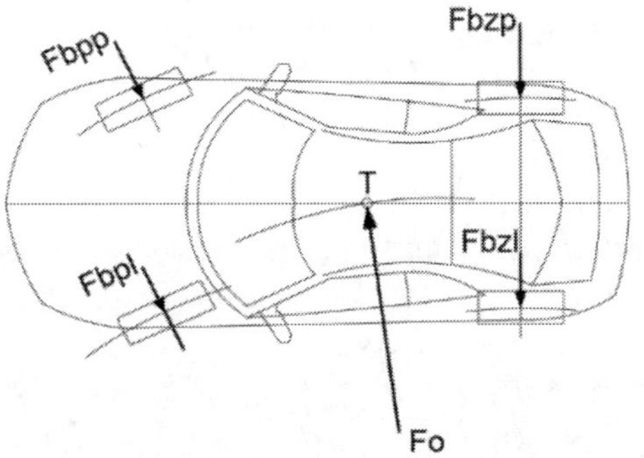

Figure 1. The effect of centrifugal force and lateral forces on the car when passing through a curve [16]

If we assume that the forces in Figure 1 in the direction of the x and y axes of the intended coordinate system are in equilibrium with each other, then the equation applies:

$$F_o = F_{bpl} + F_{bpp} + F_{bzl} + F_{bzp} \qquad (1)$$

3.1. Driving in a curve with a transverse slope of the road
Driving in a cross-country curve is divided into:
- driving in a curve that is inclined inwards,
- driving on a bend tilted to the outside.

3.2. Riding in a curve that is inclined inwards
When driving in a curve that is inclined inwards (to the inside), there is a lower risk of overturning or skidding out of a curve compared to driving on a curve without a transverse slope, with the same vehicle input parameters, since point A - Figure 2, which is the tipping point, shifts higher, and therefore the resultant forces R acting on the vehicle can reach a greater angle with respect to the weight of the vehicle, and this means that the centrifugal force F_o can have a higher value.

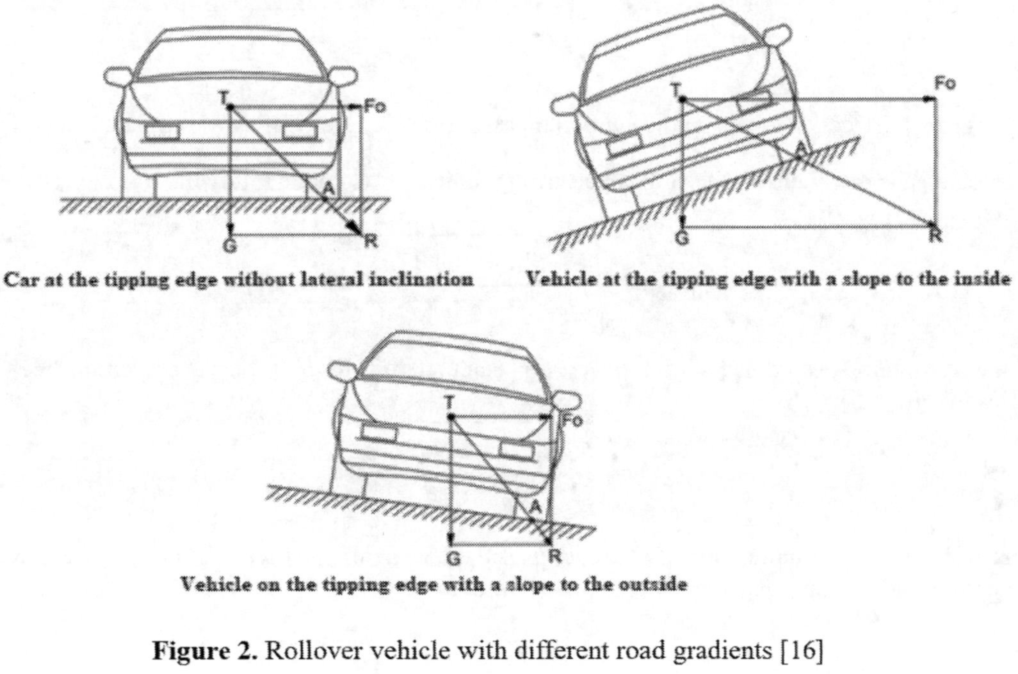

Figure 2. Rollover vehicle with different road gradients [16]

For this reason, it is sufficient to take into account the slope of the road to the outside (so-called out of the curve) in the calculations. In this case, the centrifugal force Fo is less than when passing the curve on without a lateral inclination, and it is sufficient for the vehicle to roll with less transverse acceleration, ie the speed of the combination was lower.
A problem when driving in a curve that is inclined inwards can occur when there is a surface with a reduced coefficient of shear friction in the curve. Then the car can skid not only at a very low speed, which we calculate:

$$v_{2min} = \sqrt{\frac{tg\beta - \mu_b}{1 + \mu_b * tg\beta} * g * r} \qquad (2)$$

Where β is the angle of inclination of the road, μb is the coefficient of shear friction, g is the gravitational acceleration in $m.s^{-2}$ and r is the radius of a curve in meters.

3.3. Riding in a curve tilted to the outside
Compared to a curve tilted inwards, in a curve tilted to the outside, the component of gravity G.sinβ acts in the same direction as the centrifugal force, so that the vehicle is more prone to overturning (Figure 3). The ability to "resist" lateral skid is also less because the centrifugal force component Fo.cosβ "lightens" the vehicle and reduces the frictional force on the contact surfaces of the tires.

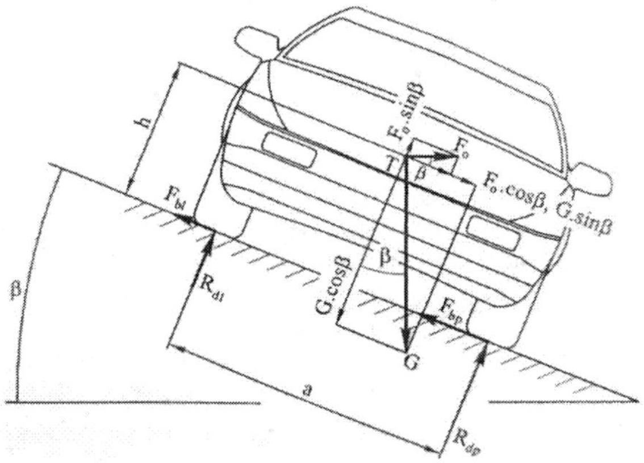

Figure 3. Effect of forces on the car when passing on a curve without a transverse slope [16]

We calculate the maximum speed at the overturning limit v_{1max} from the following equation (the equation applies to $tg\beta \leq a / 2h_T$):

$$v_{1max} = \sqrt{\frac{a - 2h_T * tg\beta}{2h_T + a * tg\beta} * g * r} \qquad (3)$$

and the maximum speed at the slide limit v_{2max} is calculated from the following equation (the equation applies to $tg\beta \leq \mu b$):

$$v_{2max} = \sqrt{\frac{\mu_b - tg\beta}{1 + \mu_b * tg\beta} * g * r} \qquad (4)$$

where β is the angle of inclination of the road, μ_b is the coefficient of slide friction, g is the gravitational acceleration in $m.s^{-2}$ and r is the radius of the curve in meters.

IOP Conf. Series: Materials Science and Engineering 1247 (2022) 012040 doi:10.1088/1757-899X/1247/1/012040

4. Measurements

Overweight and oversized transport - measurement on the section Skalica - Trnava

On 14.2.2021 and 15.2.2021, measurements were performed on one of the parts of overweight and oversized turbine transport on the Plzeň, Doosan Škoda Power (CZ) - Mochovce (SK) transport route. The cargo was transported by a company from the Czech Republic and in the territory of the Slovak Republic this transport was accompanied by a company operating in the conditions of the Slovak Republic. The vehicle set had the following parameters:

- Truck-Type: Mercedes Arocs 4 axles,
- semi-trailer -Type: Goldhofer,
- push - Type: Volvo,
- number of axles: 16 pcs tractor + semitrailer, + 4 pcs push,
- maximum width: 3.22 m,
- maximum height (increased): 5.15 m (5.40) m,
- total length: 39.80 m,
- total weight: 110.6 + 41.7 t.

Visual representation of the tractor (front view) and the loaded load on the trailer (rear view is like a technical drawing prepared in the AutoCAD software in Figure 4.

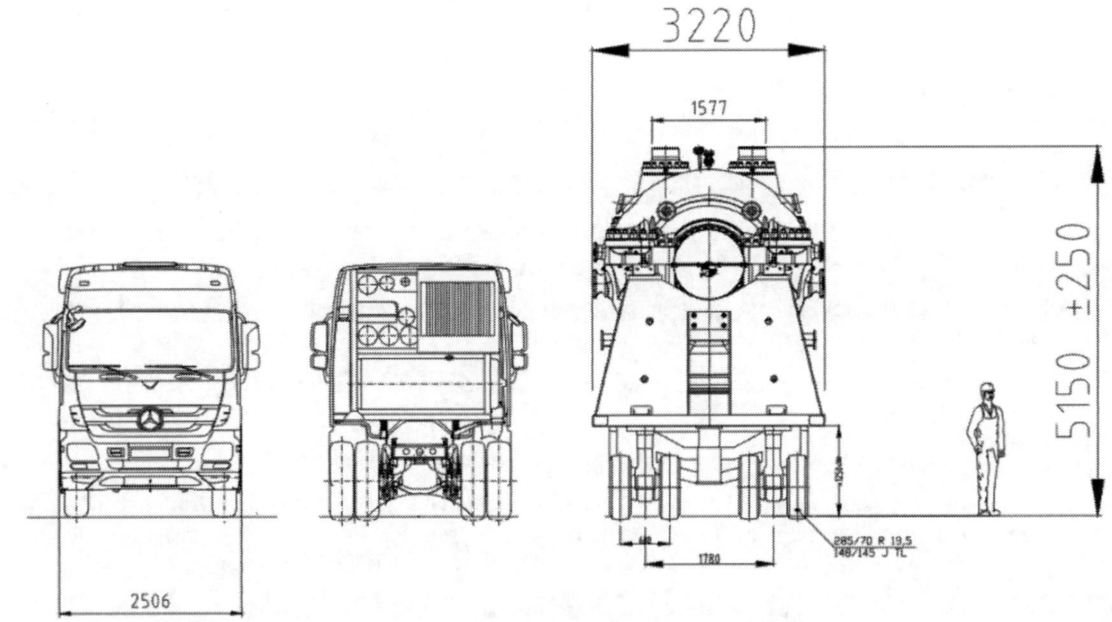

Figure 4. Technical drawing of the combination with a load - front and rear view [authors]

The longitudinal, transverse and vertical acceleration of the combination was recorded on the route using a recording device, which was then evaluated using an appropriate program. The device was placed just in front of the load on the chassis in the horizontal direction of the axis of the centre of gravity (centre of the width of the loading surface of the chassis). In FIG. 5 is indicated by a red arrow the location of the recording equipment on the technical drawing of the side view of the push-mounted combination.

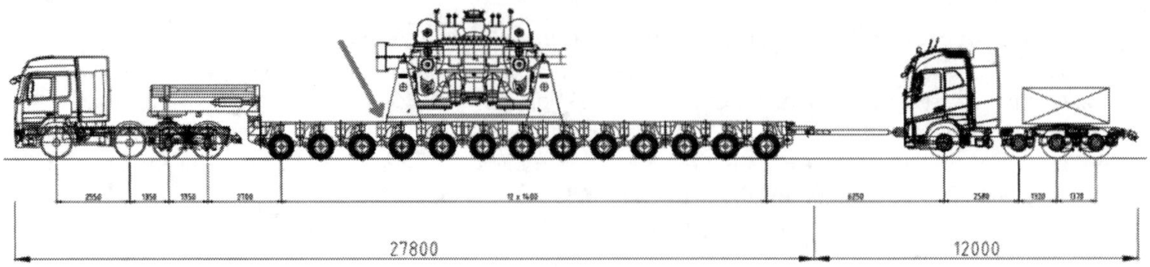

Figure 5. Technical drawing of the vehicle combination - side view with the location of the recording equipment [authors]

4.1. Measurement procedure

An important part of research is the measurement process. The following table (Table 1) shows the characteristics of the measurement steps performed.

Table 1. Elaboration of measurement procedure during research [authors]

measurement step	Measurement procedure
1	Measurement of the width, height and length of the combination, including the load, by means of a laser distance meter.
2	Location of the recording equipment together with the power supply on the combination chassis at a possible mounting point on the metal part of the chassis as close as possible to the center of gravity of the load.
3	Check the longitudinal and transverse parallelism of the installed equipment with the combination chassis using a spirit level and a length gauge.
4	Connecting the device to the power supply immediately before the start of the ride - start of data recording.
5	Visual inspection of device functionality and data logging via LEDs.
6	After a maximum of 3 hours 30 minutes, download the recorded data from the device (time determined during the test measurements, the limiting factor is the memory of the device). Possibility to download data even earlier according to the route itinerary.
7	Data evaluation and creation of graphical output.
8	Repetition according to the required number of shipments

4.2. Analysis of curve from route measurement data

Calculation of the centre of gravity height and determination of the lateral shear coefficient for the combination with the load

Calculation of the centre of gravity of the set with the load on the route Skalica - Trnava:

$$h_{T\,S-T} = \frac{m_P\,.\,v_{TP} + m_N\,.\,v_{TN}}{m_P + m_N} = \frac{31\,200kg\ \cdot\ 0.625m + 11\,0600kg\ \cdot\ 3.438m}{31\,200kg + 110\,600kg} = 2.819\ m \qquad (5)$$

Sample calculation for a left-hand curve in the village of Jablonica (route: Skalica - Trnava)

The measured values at the maximum lateral acceleration during transport in a curve are as follows:

- truck speed: 4.05 m.s^{-1}

- gravitational acceleration multiple for lateral acceleration (k): 0.112

Calculation of turning radius using orthophoto map in a given locality:

$$r_M = \frac{t^2 + 4\,.\,h^2}{8\,.\,h} = \frac{67^2 + 4 \cdot 11^2}{8\,.\,11} = 56.51\,m \qquad (6)$$

Calculation of the speed limit with respect to overturning without lateral inclination:

$$v_p = \sqrt{\frac{g\,.\,a\,.\,r_M}{2\,.\,h_T}} = \sqrt{\frac{9.81 \cdot 1{,}78 \cdot 56.51}{2 \cdot 2.819}} = 13.22\,m.s^{-1} \qquad (7)$$

Calculation of the speed limit with respect to overturning without lateral inclination:

$$a_p = \frac{{v_p}^2}{r_M} = \frac{13.22^2}{56.51} = 3{,}09\,ms^{-2} \rightarrow k_p = \frac{a_p}{g} = \frac{3.09}{9.81} = 0.31 \qquad (8)$$

Calculation of the speed limit value with respect to shear without lateral inclination:

$$v_2 = \sqrt{\mu_b\,.\,g\,.\,r} = \sqrt{0.8 \cdot 9{,}81 \cdot 56.51} = 21.05\,m.s^{-1} \qquad (9)$$

Sample calculation for a right-hand curve village Jablonica (route: Skalica - Trnava)

The measured values at the maximum lateral acceleration during transport in a given curve are as follows:

- truck speed: 3.25 $m.s^{-1}$
- gravitational acceleration multiple for lateral acceleration (k): 0.116

Calculation of turning radius using orthophoto map in a given locality:

$$r_M = \frac{t^2 + 4\,.\,h^2}{8\,.\,h} = \frac{57^2 + 4 \cdot 8^2}{8 \cdot 8} = 54.77m \qquad (10)$$

Calculation of the speed limit with respect to overturning without lateral inclination:

$$v_p = \sqrt{\frac{g\,.\,a\,.\,r_M}{2\,.\,h_T}} = \sqrt{\frac{9.81 \cdot 1{,}78 \cdot 54{,}77}{2 \cdot 2{,}819}} = 13.02\,m.s^{-1} \qquad (11)$$

Calculation of the speed limit with respect to overturning without lateral inclination:

$$a_p = \frac{{v_p}^2}{r_M} = \frac{13.2^2}{54.77} = 3.095\,ms^{-2} \rightarrow k_p = \frac{a_p}{g} = \frac{3.095}{9.81} = 0.31 \qquad (12)$$

Calculation of the speed limit value with respect to shear without lateral inclination:

$$v_2 = \sqrt{\mu_b\,.\,g\,.\,r} = \sqrt{0.8 \cdot 9.81 \cdot 56{,}51} = 20.73\,m.s^{-1} \qquad (13)$$

Table 2. Summary of calculations in selected curve with extreme values on the route from Skalica to Trnava [author]

		A curve	Jablonica - left curve	Jablonica - right curve
Measured valueas		Measured speed [m.s^{-1}]:	4.050	3.250
		Measured value of the lateral acceleration multiple [m.s^{-2}]:	0.133	0.116
		Radius according to ortho-photo map [m]:	56.511	54.766
Riding in a curve	No inclination	Limiting speed due to overturning [m.s^{-1}]:	13.230	13.024
		Limiting lateral acceleration for overturning [m.s^{-2}]:	3.097	3.097
		Limiting factor of lateral acceleration for overturning [-]:	0.316	0.316
		Limit speed due to skidding [m.s^{-1}]:	21.059	20.732
		Boundary lateral acceleration to skid [m.s^{-1}]:	7.848	7.848
		Limiting factor of lateral acceleration per skid [m.s^{-1}]:	0.800	0.800
	Transverse inclination inwards 5°	Limiting speed due to overturning [m.s^{-1}]:	15.162	14.926
		Limiting lateral acceleration for overturning [m.s^{-2}]:	4.068	4.068
		Limiting factor of lateral acceleration for overturning [-]:	0.415	0.415
		Limit speed due to skidding [m.s^{-1}]:	23.001	22.643
		Boundary lateral acceleration to skid [m.s^{-1}]:	9.361	9.361
		Limiting factor of lateral acceleration per skid [m.s^{-1}]:	0.954	0.954
		Minimálna rýchlosť šmyk dovnútra [m.s^{-1}]	nonexisting	nonexisting
		Size of lateral friction coefficient for inward shear at speed v_1 =1 m.s^{-1} (upper cell) a v_2 =2 m.s^{-1} [-]	0.086 0.071	0.086 0.071
	Transverse inclination out of the curve 5 °	Limiting speed due to overturning [m.s^{-1}]:	11.096	10.923
		Limiting lateral acceleration for overturning [m.s^{-2}]:	2.179	2.179
		Limiting factor of lateral acceleration for overturning [-]:	0.222	0.222
		Limit speed due to skidding [m.s^{-1}]:	19.214	18.914
		Boundary lateral acceleration to skid [m.s^{-1}]:	7	7
		Limiting factor of lateral acceleration per skid [m.s^{-1}]:	0.666	0.666
Driving on a straight road	Inclined to a ditch of 5 °	Limiting lateral acceleration for overturning [m.s^{-2}]:	3	3
		Limiting factor of lateral acceleration for overturning [-]:	0.259	0.259

Table 2 shows that the following relationship was verified:

$$a_p = \frac{v_p^2}{r_M} = \frac{\frac{g \cdot a \cdot r_M}{2 \cdot h_T}}{r_M} = \frac{g \cdot a \cdot r_M}{2 \cdot r_M \cdot h_T} = \frac{g \cdot a}{2 \cdot h_T} \quad (14)$$

IOP Conf. Series: Materials Science and Engineering 1247 (2022) 012040 doi:10.1088/1757-899X/1247/1/012040

namely that, at different curve radii, the lateral acceleration due to overturning of the combination has the same value at a given lateral inclination, as it depends only on the gravitational acceleration g (m.s-²), the wheelbase an (m), and the height of the centre of gravity when driving in a curve without transverse inclination. This case also applies to driving in a curve with a transverse slope, but the acceleration value changes the angle of the transverse slope along with the previous variables. This verified that the created calculation model is functional and calculates correctly.

5. Determination of critical points on the measured route Skalica - Trnava

In order to be able to determine the conclusions from the measurements and at the same time meet the idea of the applicability of the measurement outputs in practice, it is necessary to mark the critical points on the individual routes on a map base. The lowest calculated critical value of the transverse acceleration is 0.222 g = 2.179m.s^{-2} when considering the transverse slope of the bend outwards at a 5 ° angle. Due to the size of the recorded values, those that reach a value higher than or equal to half of this critical value, i.e. 0.111 g, were considered critical points. During the measured transport on the route Skalica - Trnava, the set value was reached by the vehicle at the following places:

- Jablonica left curve 0.113g,
- Jablonica right curve: 0.116g,

The marking of these places in Google maps with a note of the size of the value of the achieved lateral acceleration is shown in Fig. 6.

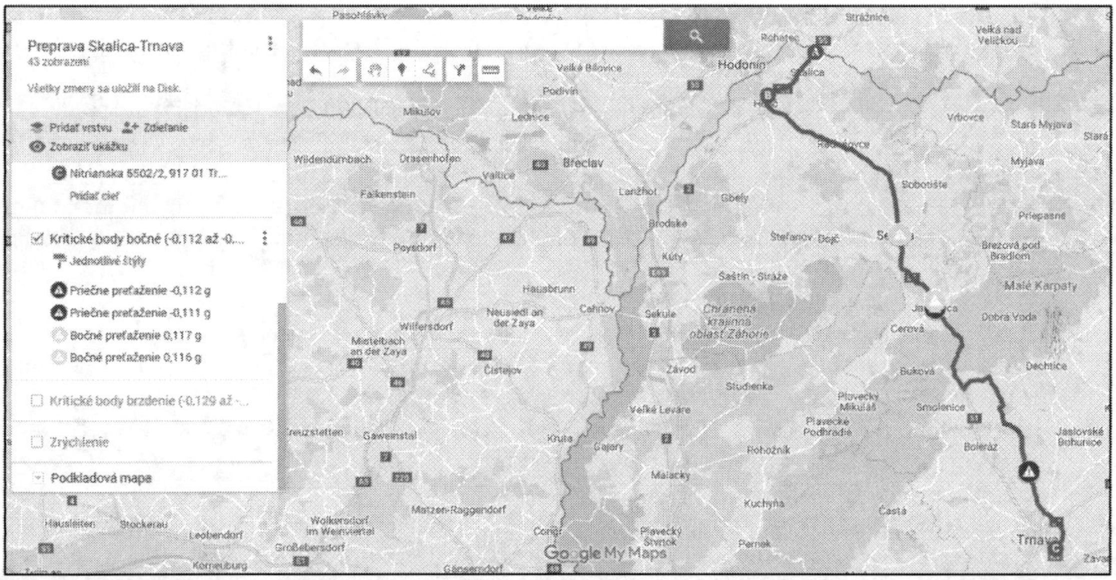

Figure 6. Designation of critical points due to transverse acceleration on the route Skalica - Trnava [authors]

6. Conclusion

After the overall route analysis, which included the calculation of the vehicle's centre of gravity, the calculations of the lateral acceleration limits with respect to overturning, and the analysis and determination of critical points and sections on the routes based on the measured values, several facts were found, which directly or indirectly affects the occurrence of these extreme values.

In the analysis of transverse acceleration, it was found that one of the causes of transverse forces on the vehicle during travel is the poor condition of the bridges, more precisely the unevenness of the road in the places of bridge expansion, but also the unevenness of the road on the bridge itself. When driving through such unevenness, the vehicle tends to sway when lateral forces begin to act on it and is more prone to tipping over. One of the other causes of extremes of lateral acceleration when passing a vehicle around a curve is unevenness on the road. When curving at low speeds, there is a low probability that this factor could affect the overturning of the combination. A critical situation at low speeds could occur due to extreme unevenness on the road and a very high centre of gravity of the vehicle combination. A critical situation is much more likely to occur when cornering with a transverse slope out of the curve when the vehicle will reach a relatively high speed, the centre of gravity will be relatively high and the transverse unevenness (poor technical condition of the road) may cause the vehicle to rock, thus reducing the speed limit (lateral acceleration) required to roll the combination to the side. When the addition of an extreme in a given curve in longitudinal acceleration is added to these parameters (most often braking occurs in a curve according to the measured values), such a curve becomes critical due to the forces acting on the combination and the load and on their tipping sideways.

7. References

[1] Poliak, M., Beňuš, J., & Nica, E. (2021). Analysis and research plan of commercial truck drivers' potentially dangerous driving behaviors caused by the changes of Regulation (EC) No. 561/2006. *Transportation Research Procedia*, *55*, 79-86.DOI: 10.1016/j.trpro.2021.07.172

[2] Gnap, J., Jagelčák, J., Marienka, P., Frančák, M., & Kostrzewski, M. (2021). Application of MEMS sensors for evaluation of the dynamics for cargo securing on road vehicles. *Sensors*, *21*(8), 2881.DOI: https://doi.org/10.3390/s21082881

[3] Jagelčák, J., Vrábel, J., KikováÁ, M. Using of different road vehicle combinations for abnormal transports based on the assessment of load rating of bridges in Slovakia for the most critical vehicle combination. In: Road and Rail Infrastructure. 2018, DOI: https://doi.org/10.5592/CO/cetra.2018.961

[4] Godavarthy, R.P., Russell, E., Landman, D. Using vehicle simulations to understand strategies for accommodating oversize, overweight vehicles at roundabouts. In: Transportation Research. 2016, vol 87, pp.41-50, DOI: https://doi.org/10.1016/j.tra.2016.03.002

[5] Vrábel, J., Jagelčák, J., Stopka, O., Kiktová, M., Caban, J. Determination of the maximum permissible cargo length exceeding the articulated vehicle length in order to detect its critical rotation radius. In: AUTOMOTIVE SAFETY . 2018, pp. 1-7. DOI: https://doi.org/10.1109/AUTOSAFE.2018.8373347

[6] Bazaras, D., Batarliene, N., Plašaitis, R., Petraška, A. Optimal road route selection criteria system for oversize goods transportation. In: Baltic Journal of Road and Bridge Engineering. 2013, vol. 8, pp. 19-24. DOI: https://doi.org/10.3846/bjrbe.2013.03,

[7] Harikrishnan, P.M., Gopi, V.P. Vehicle Vibration Signal Processing for Road Surface Monitoring. In: IEEE Sensors Journal. 2017, vol. 17, art. no. 7959046, pp. 5192-5197. DOI: https://doi.org/10.1109/JSEN.2017.2719865

[8] Jang, J., Smyth, A.W., Yang, Y., Cavalcanti, D. Road surface condition monitoring via multiple sensor-equipped vehicles. In: Proceedings - IEEE INFOCOM. 2015, art. no. 7179334, pp. 43-44. DOI: https://doi.org/10.1109/INFCOMW.2015.7179334 ,

[9] Rishwil, V., Khan, H. Automatic pothole and speed breaker detection using android system. In: 2016 39th International Convention on Information and Communication Technology, Electronics and Microelectronics, MIPRO 2016 – Proceedings. 2016, art. no. 7522334, pp. 1270-1273. DOI: https://doi.org/10.1109/MIPRO.2016.7522334,

[10] Zhang, X., Zhang, H., Zong, C., Zhang, H. Analysis and discussion on limits of dimensions, axle load and masses for road vehicles. In: ICTIS 2013: Improving multimodal transportation systems - information, safety, and integration - proceedings of the 2nd international conference on transportation information and safety. 2013. DOI: https://doi.org/10.1061/9780784413036.082

[11] Corbally, R., Cahill, F., O'conor, A. Administration of abnormal vehicles in Ireland. In: Proceedings of the Institution of Civil Engineers: Bridge Engineering. 2017, vol. 170 (4), pp. 235-247. DOI: https://doi.org/10.1680/jbren.15.00011,

[12] Gnap, J., Jagelčák, J., Marienka, P., Frančák, M., & Vojteková, M. (2022). Global Assessment of Bridge Passage in Relation to Oversized and Excessive Transport: Case Study Intended for Slovakia. *Applied Sciences*, *12*(4), 1931. DOI: https://doi.org/10.3390/app12041931

[13] Autelitano, F., Garilli, E., Ggiuliani, F. Road route planning for transporting wind turbines in europe. In: Transport Infrastructure and Systems - Proceedings of the AIIT International Congress on Transport Infrastructure and Systems, 2017, pp. 301-308. DOI: https://doi.org/10.1201/9781315281896-41

[14] Jagelčák, J., Kiktová, M., & Kubáňová, J. (2017). Legislative conditions for intermodal transport in Slovakia. In *MATEC Web of Conferences* (Vol. 134, p. 00020). EDP Sciences.DOI: https://doi.org/10.1051/matecconf/201713400020

[15] Jagelčák, J., Kiktová, M., & Stopková, M. (2018). The application of the verified gross mass of intermodal loading units in the conditions of the Slovak Republic. *NAŠE MORE: znanstveni časopis za more i pomorstvo*, *65*(4 Special issue), 218-223. DOI: https://doi.org/10.17818/NM/2018/4SI.10

[16] Kasanický, G., Kohút, P., Lukašik, M. Theory of motion and shock in the analysis and simulation of an accident. EDIS- vydavateľstvo ŽU v Žiline, 2001. ISBN 80-7100-597-5

Design of an electro-pneumatic driving simulator

Piotr Woś*, **Ryszard Dindorf** [1]

[1] Department of Mechatronics and Armament, Faculty of Mechatronics and Mechanical Engineering, Kielce University of Technology, 25-314 Kielce, Poland

* wos@tu.kielce.pl

Abstract. The paper discusses the construction of a car driving simulator prototype. The basic design assumptions for the mechanical structure of the simulator platform as well as the drives and control system are presented. A solid model of the device was presented and a solution to its kinematics was proposed to determine the stroke of the driving cylinders as a function of the angular displacement of its members. The issues related to virtual prototyping to determine the trajectory of movement and dynamic quantities characterizing the simulator were presented. The construction of the control system with the use of a dedicated computer control card is discussed. Such a solution enables testing the functions of the simulator control system before its connection with the real simulator. At the same time, the design process has been significantly improved and accelerated.

Keywords: pneumatic drive, driving simulator, mechatronic system design

1. Introduction

The task of the car driving simulator is to influence a person in such a way as to evoke the impression of movement. This effect is achieved by combining visual and audio effects with the movement of the simulation platform. The simulator is used as a research tool to study the driver's reactions in various simulated road conditions. A person driving the simulator is subjected to external stimuli, which in turn affect the human senses through receptors [3]. The simulator platform should generate motion in such a way that the driver is subjected to appropriate physical overloads. The most popular solutions of this type are flight simulators and driving simulators. The main challenge for simulator constructors is to induce comparable behavior of the simulator user and the real vehicle. To achieve this goal, simulator constructors place real cabins of various types of vehicles on special motion platforms with many degrees of freedom. Such a driving simulator is a tool that allows simulation participants to drive a vehicle on a virtual road, with the simulated conditions being carried out in safe conditions. Currently, driving simulators intended for research and training tasks usually use a platform with 6 degrees of freedom. The most advanced versions of simulators are used for training both military and civilian pilots [7], [9],. Their construction is usually based on the Stewart platform. These solutions are very expensive due to the use of complex propulsion mechanisms and control systems [4].

Hydraulic drives, pneumatic drives, linear actuators and electric motors are used to move the simulator platform [6]. Hydraulic actuators are mainly used in professional training simulators due to their ability to achieve high forces and accuracy. On the other hand, the main advantages of electric drives are their

IOP Conf. Series: Materials Science and Engineering 1247 (2022) 012041 doi:10.1088/1757-899X/1247/1/012041

ease of control, high efficiency and low noise level. It is the most popular solution in simulators created for entertainment. The most advanced actuators of this type are displacing hydraulic drives in professional training simulators [5].

Due to the simplicity of design and ease of installation, the least popular pneumatic actuators are used in low-budget simulators. Their disadvantages are low load carrying capacity and lower precision compared to other actuators. Thanks to the development of mechatronic systems, the construction of such devices has become increasingly popular. In recent years, there have been many companies offering simulators, which while simplifying the design (reduction of degrees of freedom, use of electric and pneumatic actuators/motors) in comparison to professional, typically training solutions, offer equally high level of reproduction of forces acting on humans [10].

2. Simulator design

A model of a prototype electropneumatic car driving simulator is shown in Figure 1. In this simulator, pneumatic actuators (4) are used to drive the moving members [1]. Movements of the actuators cause movement of the forward tilt frame (2) and lateral tilt (3), which in turn allows for the corresponding inertial forces acting on the driver. The two frames form an articulated kinematic linkage with two degrees of freedom. The driver's seat (6) is mounted directly on the tilt frame (2). The simulator is equipped with a steering wheel, control sticks (7) and a monitor (9).

Figure1. General view of the simulator model: 1 - base, 2- steering frame, 3 - acceleration and braking frame, 4 - pneumatic actuators, 5 - moving frame bearing, 6 - seat, 7- steering wheel, 8 - control pedals, 9 - projection screen [8].

Such a construction of the device is to provide real-time imitation of all overloads occurring in a moving car, i.e.: acceleration, braking, turning, ascending and descending.

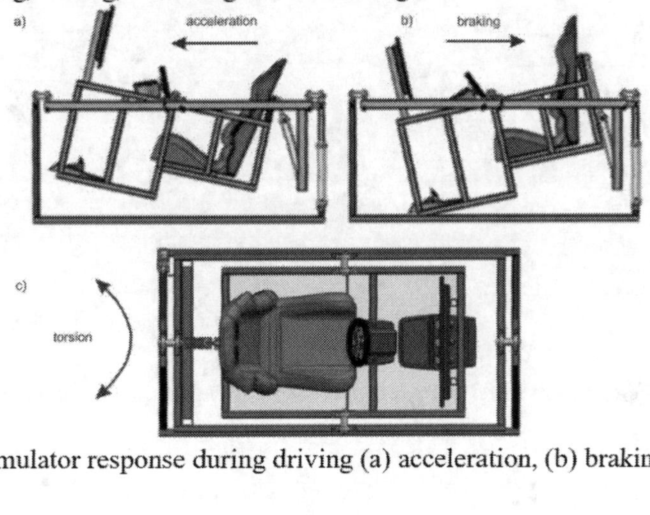

Figure 2. Simulator response during driving (a) acceleration, (b) braking, (c) turning.

2.1. Kinematic structure of the simulator

Figure 3 shows the kinematic structure of the driving simulator. The diagram shows the mutual relationships of the kinematic pairs in the kinematic chain [11].

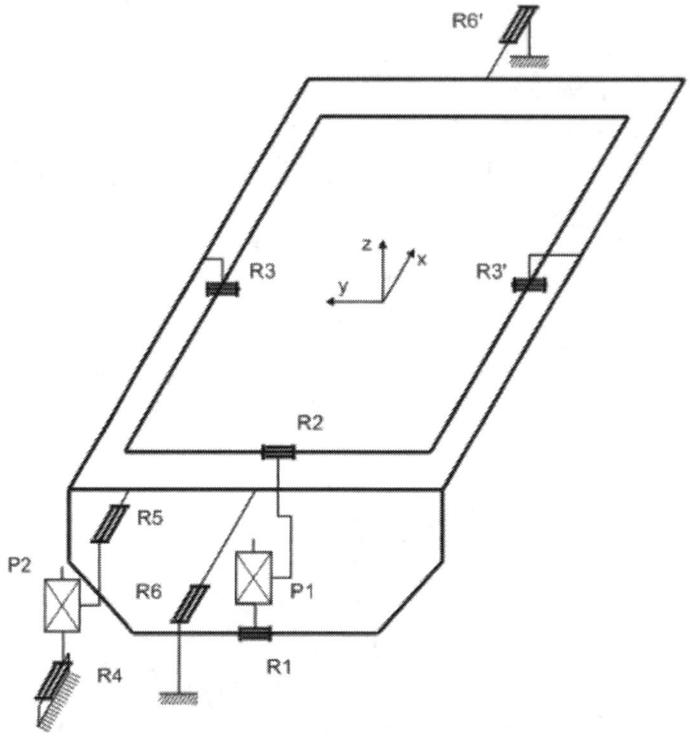

Figure 3. Kinematic diagram of the simulator.

The kinematic structure of the simulator contains 3 rotational pairs R_1, R_2, R_3providing movement during acceleration and braking, 3 rotational pairs R4, R5, R6 responsible for tilting during turning. Propulsion in the designed simulator is provided by two pneumatic actuators ($P_1$and P pairs$_2$).

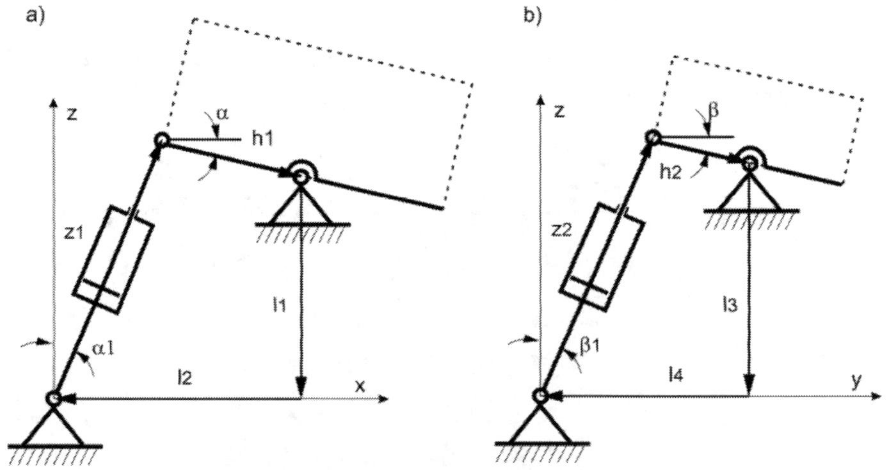

Figure 4: Flat kinematic diagram of the simulator axis s) *xz* plane, b) *yz* plane.

The purpose of the kinematic analysis is to determine the value of displacements of piston rods of pneumatic cylinders acting in two perpendicular planes responsible for acceleration-braking Z_1 and torsion Z_2. This analysis is necessary for the proper work of the control system of the simulator drives. The whole system can be simplified and considered as two independent plane mechanisms (fig.4). Figure 4 shows the kinematic diagram of the forward-backward swing mechanism (Fig. 4a) and the lateral tilt mechanism (Fig. 4b).
The mobility in the *xz* and *yz* planes of the simulator is:

$$W = 3 * n - 2p_1 - p_2 = 3 * 3 - 2 * 4 = 1 \quad (1)$$

Where: p_1 and p_2 are kinematic pairs of class one and two, respectively. The *p pairs* $_1$in the plane are pairs with rotational or sliding motion, *n the* number of moving members of the kinematic chain.
In order to determine the displacement of the pneumatic actuators Z_1, Z_2 providing forward tilt (Z_1) and side tilt (Z_2) using the analytical method, the equations are obtained:

$$h_1 * \cos \propto - l_2 = -z_1 * \sin (\propto_1) \quad (2)$$

$$-h_1 * \sin \propto - l_1 = -z_1 * \cos (\propto_1) \quad (3)$$

$$h_2 * \cos\beta - l_4 = -z_2 * \sin (\beta_1) \quad (4)$$

$$-h_2 * \sin\beta - l_3 = -z_2 * \cos (\beta_1) \quad (5)$$

$$z_1 = d_1 + s_1 \quad (6)$$

$$z_2 = d_2 + s_2 \quad (7)$$

where: d_1, d_2 constant value.
The corresponding stroke of the Z_1 and Z_2 actuator is:

$$s_1 = \sqrt{h^2 + l_2^2 + l_1^2 + 2 * h_1 * (l_1 * \sin \propto - l_2 * \cos \propto)} - d_1 \quad (8)$$

$$s_2 = \sqrt{h_2^2 + l_4^2 + l_3^2 + 2 * h_2 * (l_3 * \sin\beta - l_4 * \cos\beta)} - d_2 \quad (9)$$

3. Mechatronic solution of kinematics tasks

To verify the kinematic model, a series of simulation studies were conducted using *SolidWorks* (physical model) and *Matlab/Simulink* (implementation and simulation studies) software. This software makes it possible to design a virtual model and check its behavior under fixed initial conditions. With this solution, the design process is significantly streamlined and accelerated. With the help of *SimMechanics* toolkit it is possible to import a ready solid model from *SolidWorks* to *Matlab/Simulink*. Figure 5 shows the virtual simulator model and selected parameters of the driver's seat motion for the given duty cycle of the pneumatic actuators, determined during simulation in the *SimMechanics* program.

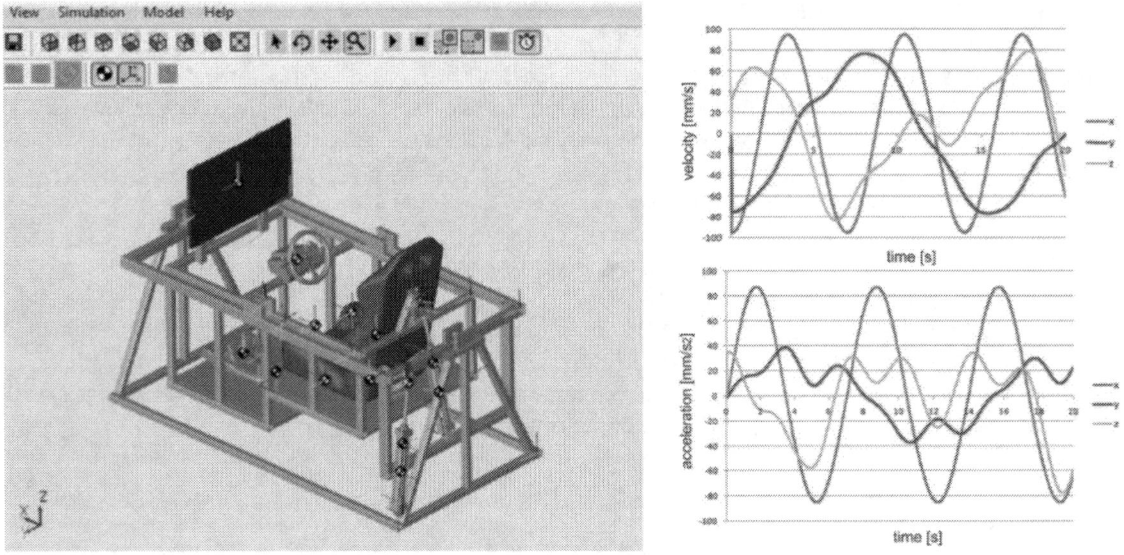

Figure 5. Virtual model of the simulator.

4. Control system

The manipulator control block system is shown in Figure 6. This system was developed assuming the use of electrically controlled pneumatic drives, where the control unit is a PC computer, working with a dedicated Velleman K8055 card. Additionally, *X-Sim* software was installed on the PC.

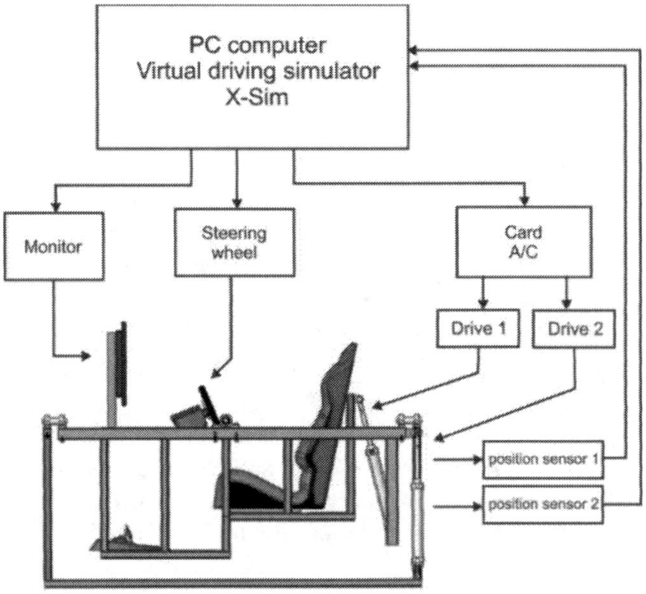

Figure 6. Block diagram of the simulator operation.

Figure 7 shows a schematic of the control system. The actuators (1, 2) that move the simulator are controlled by pneumatic distributor valves 5/3 [2]. The position of the simulator is read by two potentiometers (one for each drive axis). The whole is connected to the Velleman K8055 control card: valves digital outputs, potentiometers analog inputs. The card is connected to PC via USB interface. *X-Sim* software [8] was used to generate the simulator control signals.

IOP Conf. Series: Materials Science and Engineering 1247 (2022) 012041 doi:10.1088/1757-899X/1247/1/012041

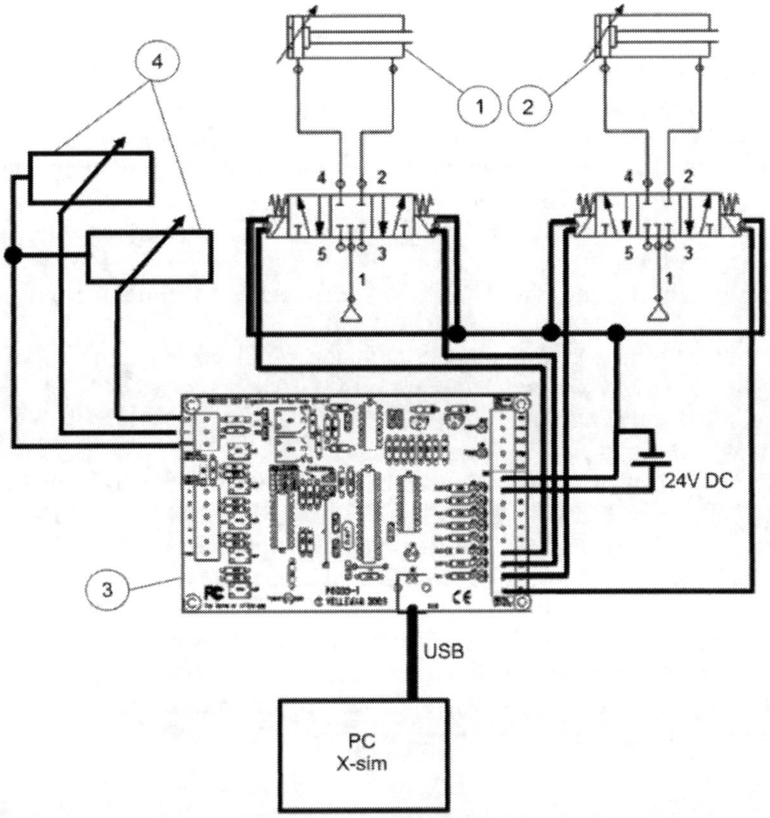

Figure 7. Schematic of the control system: 1, 2 - pneumatic actuators, 3 – Velleman K8055 control card, 4 - resistive position measurement of actuators.

X-Sim software works with most of the virtual game simulators available on the market. This is its main advantage because each game manufacturer uses a different method to export real-time telemetry data [2]. The software consists of two main modules *X-sim extractor* and *X-sim converter*. The first one has "plugins" and tools needed to read data from the game, which are then used to control the simulator [8]. The read values such as longitudinal, transverse, vertical overloads, among others, are sent to the *X-sim converter* module. The *X-sim converter module* is used to convert game variables such as velocity or acceleration into actual simulator motion by converting game input data into actuator control data [1].

5. Conclusion

The car driving simulator solution presented in the article is a dedicated solution covering the area of application of the mechatronic system as a uniform, functionally integrated system in which an original solution of mechanics and control was used. The car driving simulator project presented in the work is a simple and relatively inexpensive solution compared to commercial solutions. The original design of the supporting and controlled frames allows the use of a minimum number of drives - actuators. The main problem of the designed mechatronic simulator system with a multi-drive system is to obtain full synchronization of the pneumatic drives. The manipulator uses pneumatic actuators as an effective solution, but in the absence of a pneumatic installation, electric linear actuators can be successfully used to build the simulator. Another improvement for the simulator will be the use of electro-pneumatic or electro-hydraulic positioning servos [1], which will result in improved dynamics and smooth movement of the driver's seat.

References

[1] Dindorf R. and Woś P., Distributed real-time control system for fluid power drives, Hydraulika a Pneumatika No.**1**, pp.1-4, Slovakia, 2011.

[2] Dindorf R., Woś P., Rapid prototyping of an electrohydraulic drive control system; Drives and Controls - Technical and Information Monthly; No. **10**, pp.18-22, 2005.

[3] Fang Z. and Kemeny A., Motion cueing algorithms for a real-time automobile driving simulator, in Proc. Driving Simul. Conf., 2012, pp. 159–174.

[4] Haycock B, Campos J., Koenraad N., Potter M., and Advani S., Creating headlight glare in a driving simulator, Transp. Res. Part F: Traffic Psychol. Behav., vol. **61**, pp. 93–106, 2019.

[5] Kharrazi S., Augusto B., and Fröjd N., Vehicle dynamics testing in motion based driving simulators, Veh. Syst. Dyn., **58(1)**, pp. 92–107, 2020.

[6] Yuriy V, Yuriy V, Sergey A. Computer Simulation of Electro-pneumatic Drives for Vertical Motion Mobile Robots. Procedia Eng., vol. **100**, pp. 1003-1012., , 2015, DOI: 10.1016/j.proeng.2015.01.460. 5.)

[7] Lozia Z., Driving simulators, WKŁ. Warsaw. 2008.

[8] Pazdur M., Development of an electropneumatic car driving simulator, Master's thesis, Kielce University of Technology, 2013.

[9] Pitz J., Nguyen M.-T., Baumann G., and Reuss H.-C., Combined motion of a hexapod with a xy table system for lateral movements, in DSC 2014 Europe. Driving Simulation Association, pp. 06.1–06.2, 2014.

[10] Schramm D., Hiller M., and Bardini R., Vehicle Dynamics: Modeling and Simulation. Berlin, Germany: Springer, 2014.

[11] Zhang C. and Zhang L., Kinematics analysis and workspace investigation of a novel 2-DoF parallel manipulator applied in vehicle driving simulator, Robot. Comput.-Integr. Manuf., **29(4)**, pp. 113–120, 2013.

Carpooling and the reduction of urban traffic

Andrzej Szymanek[1], **Józef Stokłosa**[2*], **Tomasz Zawisza**[3]

[1] Kazimierz Pulaski University of Technology and Humanities in Radom

[2] University of Economics and Innovation in Lublin

[3]Industry Person

* jozef.stoklosa@wsei.lublin.pl

Abstract. The article presents the results of research that quantitatively confirm the well-known thesis on the impact of carpooling (measured by the declared carpooling indicator, DCI) on a significant reduction in the intensity of urban traffic flows. Of course, it is about joint journeys by passenger cars, which (especially in medium-sized and large cities) are an important source of transport congestion, and the connection with the intensity of traffic is obvious. The working hypothesis was verified by conducting observations and measurements in the central area of a medium-sized city in Poland. It is the city of Radom, 100 km south of Warsaw, with a population of 215,000. residents. To verify the hypothesis, the average daily traffic volume, ADTV were compared with the average reduced average daily traffic volume, RADTV. These were determined - by simulating carpooling, in which the main control variable (although not the only one) was the indicator, DCI, i.e. the number of additional passengers in a passenger car. It turned out that RADTV values decreased significantly (therefore congestion decreased) when DCI values increased. This is illustrated by the DCI - RADTV dependence graphs for different days of the week and different measurement locations. This relationship was confirmed by calculating the Spearman's rank correlation index between the DCI and RADTV variables.

Keywords: carpooling, transport congestion, declared carpooling indicator /DCI/, reduced average daily traffic volume /RADTV

1. Introduction

About 90 % commuting to work in France do it on their own. Increasing average occupancy rate of passenger cars by 12% from the existing value of 1.08 person would solve the majority of problems with traffic jams [1]. If these estimates are fine, then it shows what repair potential there is behind using of carpooling.

Transport congestion has been the problem known for ages. J. M. Thomson wrote that: „jams were the feature characteristic to London already long before inventing the car and in Rome – long before London was settled" [2]. But no earlier than in 20th century, in the period of intense development of automotive industry negative results of congestion became a big socio-economic problem. Congestion is a relatively long-lasting phenomenon when the level of traffic intensity of different means of transport is higher than the capacity of the infrastructure used. The effects of transport congestion significantly influence economic, environmental and social aspect of transport. It concerns mainly road transport, and most of all, transport in the cities. Among the most important disadvantages of transport congestion there are: time losses due to sitting in traffic jams, problems in finding parking places, degradation of the natural

environment (congestion is an important source of smog), impact on human health and life due to exposure to taking part in incidents and road collisions.
The concept that deserves particular attention is "carpooling". It was created in the USA in the 1940s. Carpooling is currently experiencing great development in Europe, also in Poland, mainly due to the Internet network and social media. Carpooling can be described as a displacement system consisting in increasing the use of a passenger car, for example in city trips. The principle is simple and it assumes increasing the number of passengers while traveling by car. This is mainly done by matching commuters for work or study destinations on the same routes. The benefits are for both the driver and passengers traveling with them.

The impact of carpooling on the level of transport congestion is important. This issue has been studied many times, but few studies have explicitly concerned empirical verification of the basic research thesis in this area. And the thesis concerns the question: does increasing the carpooling coefficient really reduce the level of transport congestion in urban traffic? And what are the quantitative relationships here? Therefore, it would be a question of estimating the hypothetical correlation between "congestion level" and "carpooling coefficient". The authors attempted such a verification using the developed simulation model and the results of field traffic research in the city of Radom in Poland.
Congestion has been defined in various ways, see e.g. works [3], [4]. Congestion increases transportation costs considerably; it also generates other consequences - it has been written about these issues, among others, in the paper [5].
Many sources of transport congestion are known. They were shown, among others, in [6]. There are many methods known to reduce the level of transport congestion; here are some of them:
- increasing the capacity of roads and streets which is realized by expanding and modernizing of transport infrastructure [7], [8],
- implementation of methods of efficient use of the capacity of the existing transport system [9];
- managing transport demand [10],
- introducing changes in the spatial development system of cities and urban space management [11].

The positive effects of sharing means of transport are particularly visible in reducing congestion in medium-sized and large cities, especially where traffic participants use a passenger car together. There are many known vehicle sharing methods, including: carpooling [12 - 14]; carsharing [15-17]; ridesharing [18], [19].
The essence of carpooling has been described – among others - in the works [12], [20], [21]. Definitions and a review of the literature on carpooling systems are included, for example, in [22]. Factors that encourage the use of carpooling are described, inter alia, in [23].
A more recent review of the literature in this area, as well as in the field of carpooling studies, is presented in [24]. It is especially about:
- costs of commuting to work [25], [26],
- travel time savings and environmental awareness [26 - 28],
- travel mode selection [29], [30],
- access to potential carpoolers (people who are willing to join and travel with another person) [12], [31 - [36].

Other important factors influencing carpooling are: demographic factors (gender, race, education, profession), economic (household income, vehicle availability) and others. In particular, the general, economic and legal issues of carpooling as well as its relationship with congestion - have a rich bibliography. More recent works in this field include [8], [37], [38], [39]. Motivating factors for traveling together, also innovations and trends in mobility, also have extensive literature - just as an example we mention two interesting studies from UC Berkeley: Transportation Sustainability Research Center [40], [41].
It was also written about how Government and employers can support carpooling – for instance at work [42].
Also various methods and research techniques used in carpooling research are worth mentioning. Let us focus here on the study of carpooling behavior on the basis of online surveys [43], and simulation

modeling of carpooling [44]. On the other hand, theoretical and experimental aspects of "carpooling dynamics" are discussed for example, in [45].
The following works cover the impact of carpooling on the congestion level: [2 - 4], [6], [8], [46].

2. Materials and Methods

2.1.Research object and research concept.
The subject of the research were the streams of road traffic in the central area of Radom city, which is the city of medium size and which can be included in the group of "road cities", namely these, where road congestion requires urgent reduction.
In 2018, the city of Radom had around 214,000. inhabitants, and 109 301 vehicles were registered here. The automotive index was 511 [vehicles / 1000 inhabitants] and it showed an upward trend. The city's street network was 291 km long, and the length of bicycle paths was 265; There were 496 taxis in the city. There were 265 road accidents that year with 13 fatalities and 322 injured.
The relatively large and still growing motorisation index heralds an increase in the transport congestion in Radom. In order to limit this phenomenon, it is necessary to introduce new forms of mobility - including planning some forms of carpooling.
Therefore, the main aim of the research (selected results of which are presented in this article) was the quantitative verification of the following working hypothesis: "increase in carpooling index means reduction of average daily traffic volume" (ADTV). And because the ADTV indicator indirectly shows the level of transport congestion - hence the reduction of ADTV, it is the reduction of transport congestion in urban traffic.
The verification of this research hypothesis has been carried out in the following stages:
1. ADTV measurements were made; in the area of the city where previously low levels of service traffic (LOS) were found, i.e. where congestion was particularly troublesome.
2. A model and a computer simulation program for carpooling was developed, where the main control variable of the simulation was the "declared carpooling indicator" (DCI) for passenger car journeys in Radom.
3. As a result of 100 simulations, the values of reduced average daily traffic volumes (RADTV), which indirectly characterize the level of transport congestion in the examined area of the city of Radom have been obtained.
4. ADTV results were compared with RADTV results calculated for different DCI values. These comparisons show that an increase in DCI implies a reduction in ADTV. Thus, the validity of the adopted working hypothesis has been demonstrated, which means that a quantitative correlation has been demonstrated between the increase in DCI and the reduction in ADTV, which indirectly describes the level of transport congestion in the analyzed measurement locations, ML.
5. The second (only preliminary) confirmation of the validity of the adopted working hypothesis was made - by calculating the Spearman's rank correlation index for the DCI and RADTV variables.
To sum up: the input data for the simulation program were the results of field traffic measurements in selected MLs of the city of Radom and the statistically compiled results of the questionnaire research on sharing places - only by people in passenger cars.

2.2.Levels' of service (LOS) analysis in the central area of Radom city
Field research / and survey research provided data on passenger car journeys in the city of Radom. There were three types of empirical data:
(1). Average daily traffic volume, ADTV - measured on the busiest sections of streets between selected intersections,
(2). Real (observed) fillings of passenger cars (FPC), in these measuring sections,
(3). Values of the declared (intentional) carpooling index (DCI).
ADTV and FPC values were measured in order to recreate the real urban traffic conditions in the simulation program. More precisely, the results of passenger car filling measurements were needed so that the simulation program could take into account the situations in which people from one vehicle

would fit into another vehicle going in the same direction. The measurement technique was simple: there were two people in the car, each of whom registered vehicles going in opposite directions of the analyzed inter-junction section.

Carpooling was simulated – with the change of DCI which is the number of people who want to take part in carpooling.

The map of measurements of ADTV and FPC parameters was developed using the known LOS values on the road and street network of the city of Radom. The most "jammed" sections between intersections (hubs), i.e. those where the highest ADTV values and the highest FPC values were previously recorded were selected for our own measurements. Traffic measurements in the city of Radom were carried out in the conditions of the highest daily traffic, i.e. from 7.00 - 10.00 a.m. and 2.00 - 5.00 p.m. The results of the LOS assessment concerned the so-called Radom Functional Area, [47], [48]. The LOS evaluations have been shown in Figure 1; individual LOSs are marked with colors: A - green; B - pink; C - yellow; D - orange; E - red; F - black.

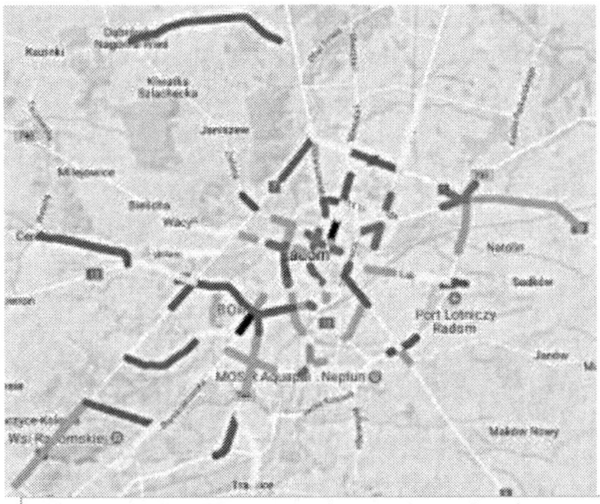

Figure 1. LOS map on the communication routes of the studied area of the city of Radom

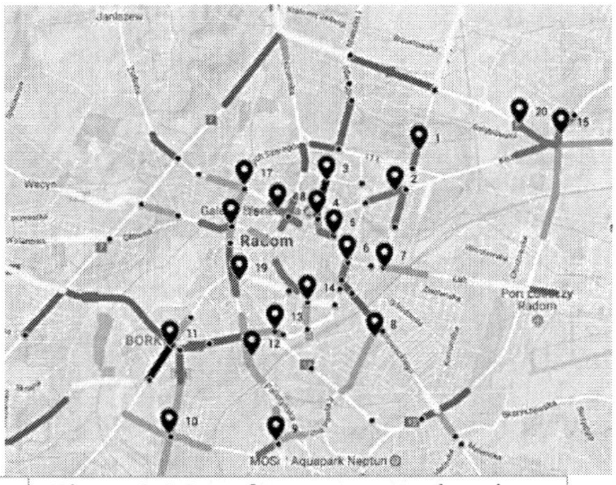

Figure 2. Map of measurements locations in the studied area of the city of Radom

Then, the LOS map was compared with the locations of traffic flow measurements in the city, which are carried out every year by the Municipal Road and Communication Authority in Radom. This comparison made it possible to select 20 measurement locations, in which own field research had to be carried out (in Fig. 2, black and white markers). It was in these locations that ADTV and FPC measurements were later carried out.

For particular test sites, the following was determined:

- traffic observer location points,
- measurement days: working day of the week, Saturday (before trading Sunday), Saturday (before Sunday without trading), trading Sunday, Sunday without trading,
- hourly range of measurements,
- measured values: ADTV, FPC.

2.3. ADTV measurements

Two-week measurements of traffic intensity were carried out on all previously established research sites (Fig. 2). This article presents some results of these traffic measurements. Figures 3 and 4 show the weekly measurement results for the test sites no. 18 (Wernera / Malczewskiego / Kelles-Krauza streets). The measurements were performed at 15-minute intervals.

IOP Conf. Series: Materials Science and Engineering 1247 (2022) 012042 doi:10.1088/1757-899X/1247/1/012042

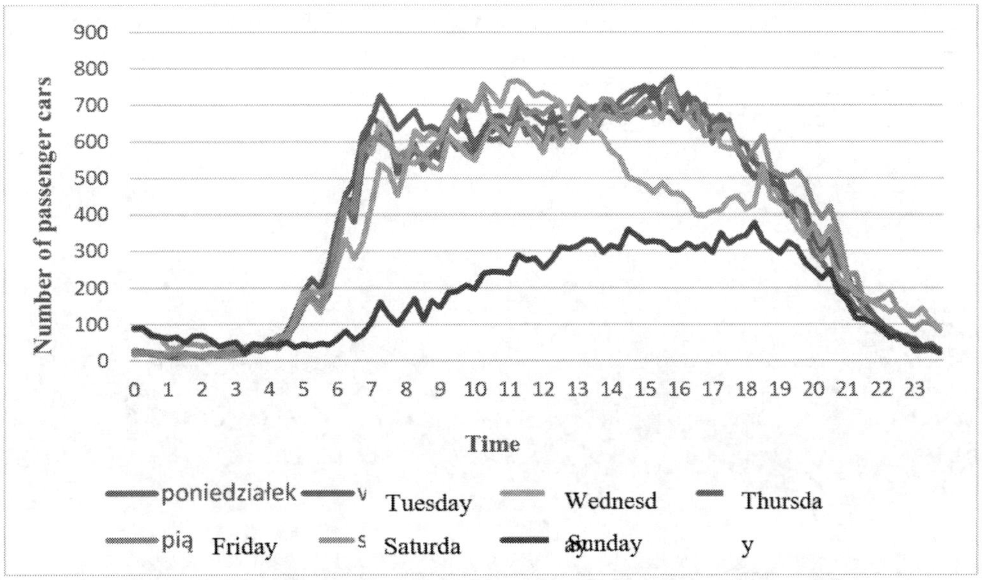

Figure 3. Weekly traffic in ML No. 18 with Sunday without trading

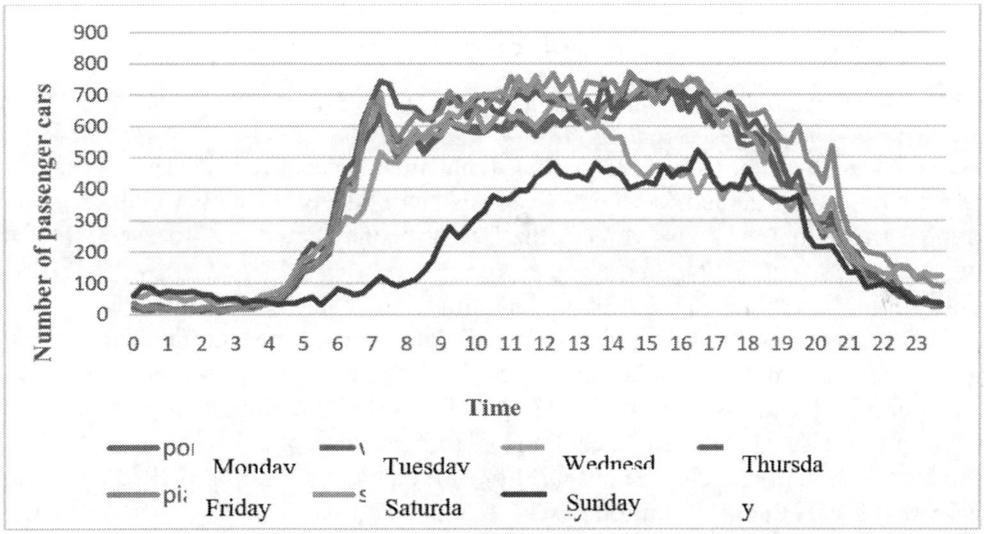

Figure 4. Weekly traffic in ML No. 18 with trading Sunday

The distributions of ADTV from Monday to Friday for both weeks are very similar and there is one peak of traffic, which indicates repeatability of traffic and therefore repeatability of vehicle filling. The traffic schedules for Saturdays and Sundays are different.

In line with the assumptions of this work - the results of the measured ADTV from all 20 MLs were used in the computer simulation to determine the RADTV value.

2.4.Measurements of filling passenger cars in city trips

Measurements were carried out in 20 street cross-sections of the central area of Radom. The filling of cars was recorded for both traffic flows in a given cross-section of the street, in 15-minute intervals. The following values were measured and recorded:

- filling passenger cars, FPC,;l7.:;l.;4
- number of 2-person cars,
- number of child seats – occupied and free.

Exemplary results of car filling measurements (for working days of the week) are shown in the diagram (Fig. 5). From Monday to Thursday, between 7.00 a.m. and 10.00 a.m., the FPC index fluctuated in the range 1.14 - 1.54, with the weighted arithmetic mean 1.29 and between 2 p.m. and 5 p.m., the values were 1.22 - 1.67 and the average 1.4, respectively.

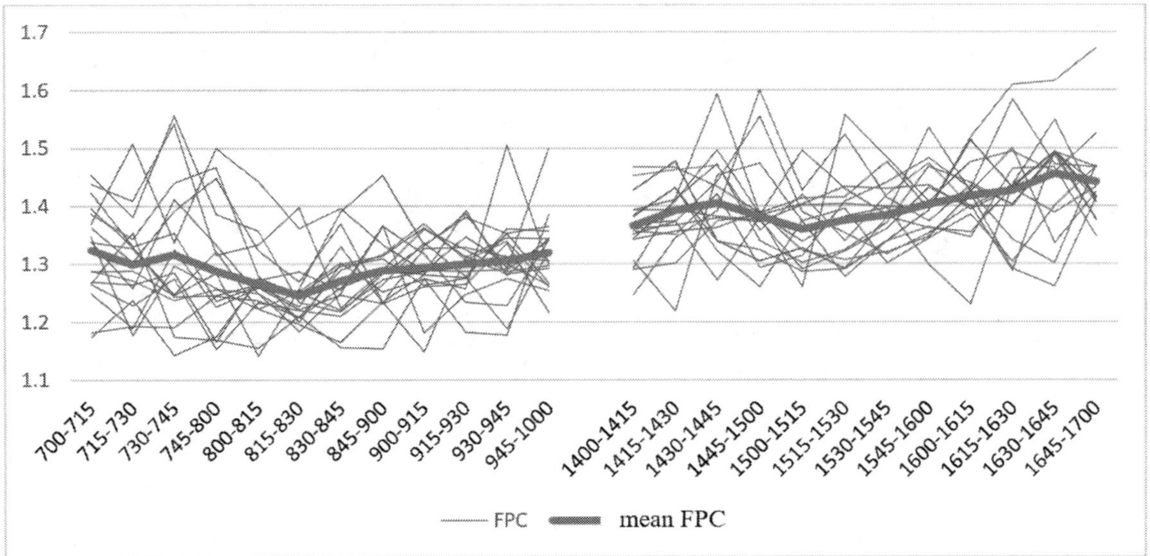

Figure 5 FPC index values (in all 20 measuring points). Days of measurement: Monday - Thursday

The following assumptions for creating carpooling simulation model have been developed:

- the 1st type driver, named "active" / "leading", can take passengers to his car,
- a type 2 driver named "passive" / "other" may become "active's" passenger, if there is one,
- in the simulation there is an option to select the driver as "active" and "passive",
- simulation runs simultaneously for empirical data from all measurement sites; this means that at the start of the simulation cycle, the number of cars is summed up from all measurement locations and at the end of the simulation cycle it will be 'weighted' according to the ADTV value at each measurement location,
- with each simulation cycle, the probability of the passenger being picked up increases,
- a two-seater car cannot be equipped with a child seat,
- full cars can be “active” even though they can no longer take any passengers,
- the car will be refilled to five passenger seats.

Following the above (also other) assumptions, a carpooling simulation model was developed. This simulation allowed to assess the quantitative influence of the DCI control variable - on the values of the RADTV index, which is the result of the simulation. The measured ADTV values and the observed FPC values were taken into account in the simulation.

The algorithm of the carpooling simulation program is as follows: Let NC denote the number of passenger cars observed in the 15-minute interval - located (in a given direction) on the in-between-hubs section. The carpooling simulation program performs the following operations:

1. selects from the pool of N_C vehicles - the f_S and f_D fractions of demand and supply places in passenger cars,
2. determines the f_S number of those cars where passengers want to join another vehicle,
3. determines the number f_D of these cars in which passengers agree to take people from another vehicle,
4. each of the N_C pool of cars has FPC values assigned,

5. each of the N_C pool of cars is assigned a fraction of p_E, p_F of cars with empty and full seats.

Therefore, in the simulation process, there are exchanges of "donors" and "recipients" of places in the analyzed N_C pool of passenger cars. These exchanges are registered at 15-minute intervals.
Circulation (transfer) of people becomes a fact (an event in the simulation process) if it takes place in no longer than WT_{max} (the maximum waiting time for a joint ride). One vehicle is then subtracted from the N_C number of vehicles, and the procedure continues until all the vehicles that have left the original pool due to sharing are subtracted from the original N_C pool. Fraction of vehicles that remains gives RADTV when counted.
A more detailed description of the assumptions and the carpooling simulation mechanism will be presented in the prepared, much more extensive work.
The RADTV index, i.e. the simulation output value, can be written symbolically as a function of seven factors:

$$RADTV = f\{N_c, f_s, f_p, FPC, p_E, p_F, WT_{max}\} \quad (1)$$

If we describe the carpooling simulation model developed here in the black-box convention, the main input to the simulation is DCI, while the output is represented by RADTV, (Fig.6). It should be added that the simulation is controlled by changing the supply and demand values of the DCI index.

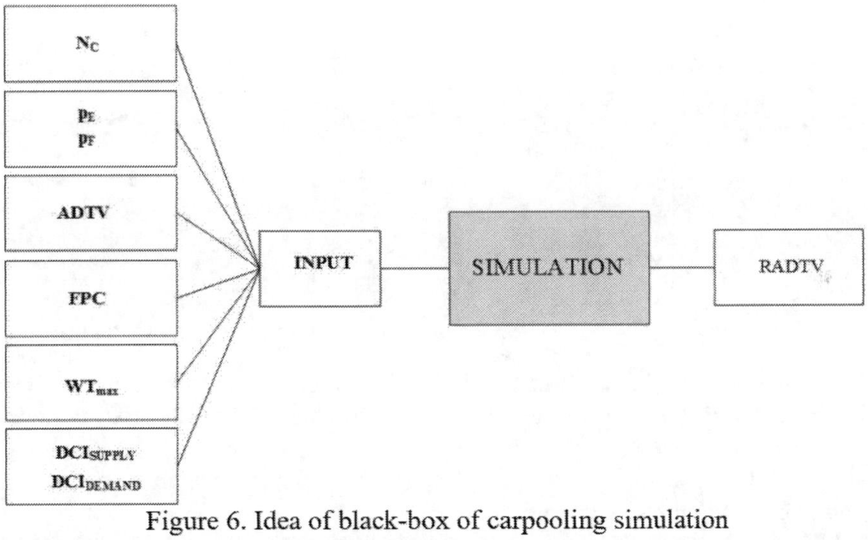

Figure 6. Idea of black-box of carpooling simulation

Carpooling simulation was written in C# language, project Windows Forms in the environment Visual Studio 2017 Community.

2.5. Data for simulation coming from questionnaire survey

In addition to the data obtained in traffic measurements in Radom, including the values of ADTV and FPC indexes, other data needed to perform a carpooling simulation was obtained in a questionnaire survey, on a random sample of 702 people. The aim of these studies was to obtain the empirical data needed to assess vehicle sharing preferences.
The respondents were residents of medium-sized cities in Poland (cities with 100,000-300,000 inhabitants). The questionnaire survey was not limited to the inhabitants of Radom, among other reasons, so that information could be gathered for a larger random sample. Moreover, the survey technique itself forced sending the questionnaires also outside the city of Radom.
The survey form contained only 8 questions. The first four questions concerned the respondent's demographic and social profile: gender, age, education, place of residence (here there were 5 categories

- ranging from rural areas to large cities). Question no. 5 concerned the professional status and the use of a private car as a work tool. Questions 6 and 7 concerned the SUPPLY of seats in the carpooling process. In question no. 6, the respondent answered whether he would like to participate in carpooling as a "donor" of a seat in the car, i.e. whether he would agree to take a person traveling in the same direction. In question no. 7, the respondent answered whether he would agree to be the "recipient", i.e. whether he would agree to get on another car going in the same direction, resigning from driving his own vehicle.

The declared total SUPPLY (answer - YES) of seats in the carpooling process was 60%, while the answer NO (32%). 8% of the respondents did not drive a car.

It turns out that 52% of respondents would not give up driving their own car. So 48% - it is the declared total DEMAND for "giving someone a lift", that is, the declared DEMAND in the hypothetical carpooling in Radom.

Question no. 8 also indirectly characterized the DEMAND for places in the carpooling process. The potential "recipient" of the place in the car answered the question: How many minutes maximum could you wait for another car that would go in the same direction and get on it, giving up driving your own car? The histogram of the WT_{max} value is shown in Figure 7.

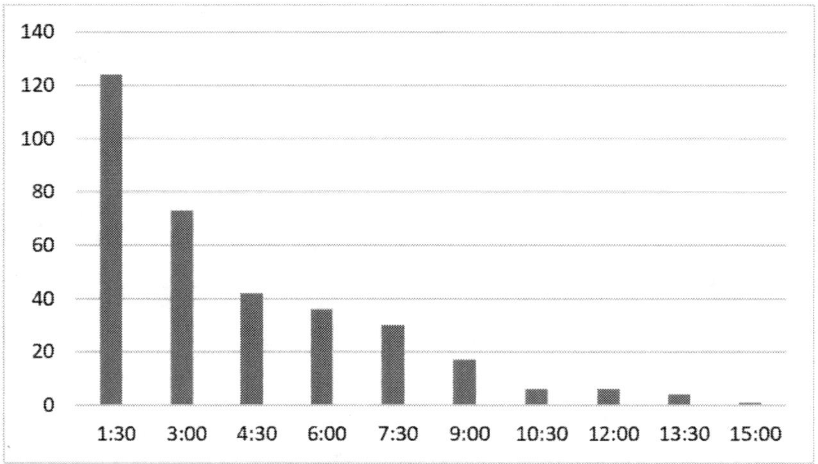

Figure.7. WT_{max} histogram of the declared maximum waiting time for a joint journey, i.e. to a place / vehicle in the carpooling system; on the vertical axis the number of people declaring their waiting time

The weighted arithmetic mean of the WT_{max} index is 4.08 minutes, which means that the statistical "demand" respondent was quite "impatient".

The shape of the above histogram is interesting. About 70% of the survey respondents (the first three bars of the histogram) declare slightly more than 30% of the WT_{max} value (1.5 - 4.5 minutes). Thus, the WT_{max} histogram resembles the Pareto diagram with the 70/30% rule, which is slightly weaker than the original Pareto observation (80/20%). Here, the proportion of 70/30% means that over 2/3 of the population of potential (declaring) carpoolers in a medium-sized Polish city is interested in the shortest "thirty percent" of the maximum time available to them to wait for a place in the carpooling system.

3. Results and discussion

The input data for the simulation was prepared and introduced on the basis of field tests and measurements. The percentage of "recipients" and "donors" was the same in the studies. The simulations were carried out consecutively for the following fractions: 10, 20, 30, 40, 50, 60, 70, 80, 90, 100% of people, where the fractions of "recipients" and donors "of carpooling seats were the same, which is a simplification at this stage of the simulation. Carpooling was simulated on: weekdays, Saturdays before trading Sunday, Saturday before Sunday without trading, trading Sundays and Sundays without trading.

IOP Conf. Series: Materials Science and Engineering 1247 (2022) 012042 doi:10.1088/1757-899X/1247/1/012042

Each of the simulated time periods was divided (it was to comply with field tests) into 15-minute intervals.

3.1. Illustration of DCI – RADTV dependence

The simulation confirmed the verified working hypothesis: an increase in carpooling (increase in DCI) significantly reduces the intensity of road traffic (RADTV).
Figure 8show exemplary simulation results - these are graphs showing the dependence of RADTV on the changing (every 10%) index controlling the DCI simulation.

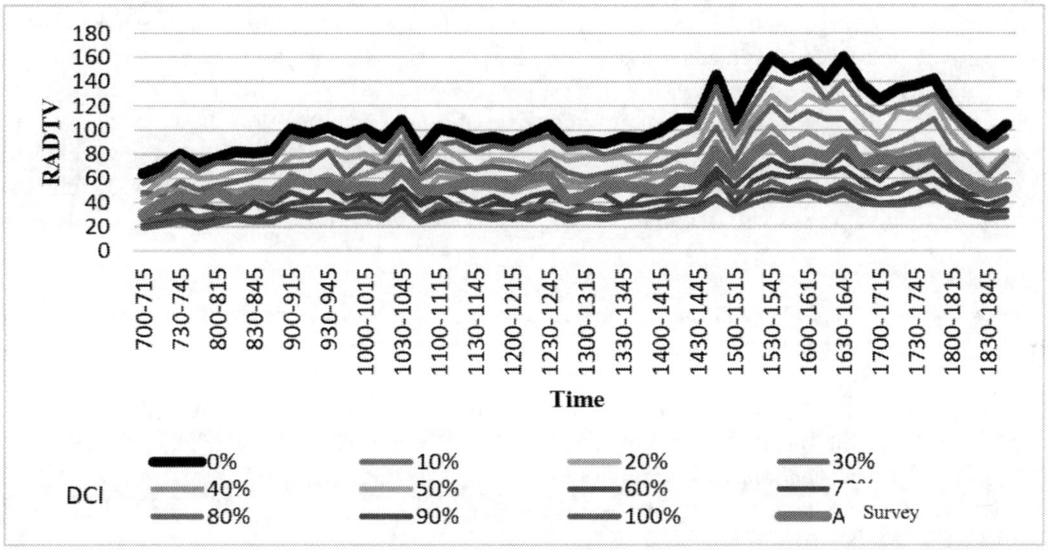

Figure 8. RADTV simulation (ML No. 4 - direction – Kusocińskiego Street; day of measurements – Thursday)

3.2. Spearman's rank correlation estimator in the study of dependences between DCI and RADTV indexes

The dependence between the DCI which controls the carpooling simulation and the RADTV which is the result of the simulation can also be confirmed differently and simply (see Figs. 11-26). It is enough to calculate the Spearman's rank correlation coefficient for these variables; it must be assumed that both variables can be treated as step random variables - although they are not. Both these variables are described by sequences of numbers determined for 15-minute intervals in the measurement (and simulations) interval, i.e. in the time interval [7.00 a.m. - 6.45 p.m.]. The RADTV variable is specified by the sequence of simulation result values. But, the values of the DCI variable are also non-random DCIk values of this variable (expressed in%), which control the carpooling simulation. Yet, each of these numerical values can be formally written as the value of a one-point probability distribution with the distribution function *pk = Pr(DCI = DCIk) =1, for k = 1, 2, ..., 11.*
Thus, the sequence of DCIk values controlling the carpooling simulation can be interpreted as a sequence of independent random variables with single point distribution. This sequence consists of *n = 11* elements, where n - the number of carpooling "levels of control" - into which the DCI variable is divided, (DCI = 0, 10, 20, ..., 100 [%]). Similarly, the sequence of RADTVk result values can be formalized - as a sequence of independent one-point random variables, where each of these values is assumed with a probability of 1.
When comparing the carpooling simulation results, it can be seen that increasing DCI causes a reduction in RADTV for all test time intervals and at all measurement locations. Then the maximum rank mismatch occurs. The Spearman's rank correlation coefficient *rs* (DCI, RADTV) of the DCI and

RADTV variables is calculated for $n = 11$ values obtained as a result of carpooling simulation (data to be calculated in the table below).

4. Conclusion

Creating new mobility in cities should ensure the possibility of using various methods of transfer realization in the transport space. It is also worthwhile to pay attention to various systems that allow for transport dedicated to people who prefer to use a passenger car for these purposes. Various activities can be undertaken on this occasion to encourage car sharing. This is particularly important, because, as indicated by numerous publicly available literature sources in the European Union, as many as 41% of people make their journeys with the use of a passenger car. If it is not possible to change habits regarding everyday movements, then travelers should at least be encouraged to use more car space.

The average filling of passenger cars varies depending on the place, day, time and direction of the traffic flow, though, the filling of all vehicles mainly consists of one or two people. The vast majority of cars on Radom, but also on Polish roads, can accommodate four or five people. Filling the vacancies with passengers from other cars will reduce the total number of vehicles in motion, which was confirmed by the simulation results. Roads will become less congested and travel times will be significantly shorter.

References

[1] https://www.rse-egis.fr/en/solution_egis/carpooling-solution-to-congestion/ (access:20.02.2022)

[2] Thomson, J. Michael (John Michael) 1974. *Modern transport economics*. Penguin Education, Harmondsworth.

[3] Alberto Bull (Ed.), Traffic Congestion. *The Problem and How To Deal With It*. Santiago, Chile, 2003 187 p; ISSN printed version: 0252-2195; ISSN online version: 1727-0413 ISBN: 92-1-121432-7; www.cepal.org/es/suscripciones.

[4] Managing urban traffic congestion.; ECMT, Paris 2007; ISBN 978-92-821-0128-5; DOI: 10.1787/9789282101506-en.

[5] Falcocchio J.C., Levinson H.S. 2015 The Costs and Other Consequences of Traffic Congestion. In: Road Traffic Congestion: A Concise Guide. Springer Tracts on Transportation and Traffic, vol **7**. Springer, Cham. https://doi.org/10.1007/978-3-319-15165-6_13.

[6] Kożlak, A. Economic, social and environmental effects of transport congestion. Gospodarcze, społeczne i ekologiczne skutki kongestii transportowej. *Research Papers of Wrocław University of Economics*, 402, 2015, Pages 153-164. doi:10.15611/pn.2015.402.15.

[7] Ungemah, David, et al. 2007. Examining Incentives and Preferential Treatment of Carpools on Managed Lane Facilities. *Journal of Public Transportation*, **10** (4): 151-169. DOI: http://doi.org/10.5038/2375-0901.10.4.8 Available at: https://scholarcommons.usf.edu/jpt/vol10/iss4/8. (access: 12.01.2022)

[8] Hideo Konishi, Se-il Mun, Carpooling and congestion pricing: HOV and HOT lanes. *Regional Science and Urban Economics*, **40**, Issue 4, 2010, Pages 173-186; ISSN 0166-0462, https://doi.org/10.1016/j.regsciurbeco.2010.03.009.

[9] Warberg, A., Larsen, J., & Jørgensen, R. M. 2008. Green Wave Traffic Optimization - A Survey. Informatics and Mathematical Modelling. D T U Compute. Technical Report, No. 2008-01.

[10] Gubins, Sergejs and Verhoef, Erik T., Teleworking and Congestion: A Dynamic Bottleneck Analysis (July 18, 2011). Tinbergen Institute Discussion Paper 11-096/3, Available at SSRN; https://ssrn.com/abstract=1888241 or http://dx.doi.org/10.2139/ssrn.1888241.

[11] Mark J. Nieuwenhuijsen, Haneen Khreis, Car free cities: Pathway to healthy urban living. *Environment International,* Volume **94**, 2016, Pages 251-262; ISSN 0160-4120; https://doi.org/10.1016/j.envint.2016.05.032.

[12] Teal R. F. Carpooling: Who, how and why. Transportation Research Part A: General. Volume 21, Issue 3, May 1987, pp. 203-214; ISSN 0191-2607; https://doi.org/10.1016/0191-2607(87)90014-8.

[13] Ahmad I., Dewan K.K., 2007. Carpooling: A Step To Reduce Congestion (A Case Study of Delhi), *Engineering Letters*, **14**:1, EL_14_1_12.

[14] Jurecki R., Jaskiewicz M., Zuska A., 2013. The variety of the behaviour of drivers at risk accident situations, *Scientific Journals of the Maritime University of Szczecin-Zeszyty Naukowe Akademii Morskiej w Szczecinie* **35 :107**, Pages 65-70.

[15] Cervero R., City CarShare: First-Year Travel Demand Impacts. *Transportation Research Record: Journal of the Transportation Research Board*, 2003, Volume: **1839** issue: 1, page(s): 159-166.

[16] Christine Celsor, Adam Millard-Ball, Where Does Carsharing Work?: Using Geographic Information Systems to Assess Market Potential. *Transportation Research Record: Journal of the Transportation Research Board,* 2007, Volume: **1992** issue: 1, page(s): 61-69; https://doi.org/10.3141/1992-08.

[17] Sebastian Kot, Chapter 9 - Carsharing concept implementation in relation to sustainability – Evidence from Poland. Editor(s): Manuela Tvaronavičienė, Beata Ślusarczyk, Energy Transformation Towards Sustainability, Elsevier, 2020, Pages 179-197, ISBN 9780128176887; https://doi.org/10.1016/B978-0-12-817688-7.00009-4; (http://www.sciencedirect.com/science/article/pii/B9780128176887000094).

[18] Masabumi Furuhata, Maged Dessouky, Fernando Ordóñez, Marc-Etienne Brunet, Xiaoqing Wang, Sven Koenig, Ridesharing: The state-of-the-art and future directions. *Transportation Research Part B: Methodological,* Volume **57**, 2013, Pages 28-46; ISSN 0191-2615; https://doi.org/10.1016/j.trb.2013.08.012, (http://www.sciencedirect.com/science/article/pii/S0191261513001483).

[19] Zhenning Dong, Mingming Leng, Managing on-demand ridesharing operations: Optimal pricing decisions for a ridesharing platform. *International Journal of Production Economics*, **2020**, 107958; ISSN 0925-5273; https://doi.org/10.1016/j.ijpe.2020.107958.

[20] Ferguson, E. The rise and fall of the American carpool: 1970–1990. *Transportation*, **24**, 349–376 (1997); https://doi.org/10.1023/A:1004928012320.

[21] Ron N. Buliung, Kalina Soltys, Catherine Habel, Ryan Lanyon, Driving Factors behind Successful Carpool Formation and Use. *Transportation Research Record: Journal of the Transportation Research Board,* 2009, Volume: **2118** issue: 1, page(s): 31-38; https://doi.org/10.3141/2118-05.

[22] Diana Jorge, Gonçalo Correia, Carsharing systems demand estimation and defined operations: a literature review. *EJTIR,* Issue **13(3),** 2013 pp. 201-220 ISSN: 1567-7141; www.ejtir.tbm.tudelft.nl.

[23] Ajzen I, The theory of planned behavior. *Organizational Behavior and Human Decision Processes*, Volume **50**, Issue 2, 1991, Pages 179-211; ISSN 0749-5978; https://doi.org/10.1016/0749-5978(91)90020-T. (http://www.sciencedirect.com/science/article/pii/074959789190020T).

[24] Jun Guan Neoh, Maxwell Chipulu, Alasdair Marshall, What encourages people to carpool? An evaluation of factors with meta-analysis. *Transportation* DOI 10.1007/s11116-015-9661-7. Springer Science+Business Media New York 2015.

[25] Levin, I.P. Measuring tradeoffs in carpool driving arrangement preferences. *Transportation* 11, 71–85 (1982). [https://doi.org/10.1007/BF00165595],

[26] Yochai Benkler, Sharing Nicely: On Shareable Goods and the Emergence of Sharing as a Modality of Economic Production. November 2004, *The Yale Law Journal* **114(2)**:273; DOI: 10.2307/4135731.

[27] Guanneoh J, Carpooling and the environment: directions for future research. *Proceedings of ISER* 5 th International Conference, Singapore, 5 th September 2015, ISBN: 978-93-85465-91-8.

[28] Hartl, B., Kamleitner, B., Holub, S. 2020 Take me on a ride: The role of environmentalist identity for carpooling. *Psychology & Marketing*, **37 (5).** pp. 663-676. ISSN 1520-6793; http://dx.doi.org/10.1002/mar.21340.

[29] Van Vugt, M., Van Lange, P. A. M., Meertens, R. M., & Joireman, J. A. 1996. How a structural solution to a real-world social dilemma failed: A field experiment on the first carpool lane in Europe. *Social Psychology Quarterly,* ***59***, 364–374.

[30] Ozanne L, Mollenkopf D., (1999) Understanding consumer intentions to carpool: a test of alternative models. Research paper from New Zealand: Lincoln University.

[31] Ferguson, E.: Demographics of carpooling. *Transportation Research Record* 1496, 142–150 1995],

[32] Kaufman, S: Why people (don't) carpool and change for the better: A social capital framework for investigating environmental behaviour. In: 2nd National Conference of Sustainable Campuses, RMIT, Melbourne, 2002.

[33] Buliung R. N., Soltys K., Habel C. Driving Factors behind Successful Carpool Formation and Use. Transportation Research Record: Journal of the Transportation Research Board, 2009, Volume: 2118 issue: 1, pages 31-38.

[34] Neoh, J.G., Chipulu, M. & Marshall, A. What encourages people to carpool? An evaluation of factors with meta-analysis. *Transportation* **44**, 423–447 (2017). https://doi.org/10.1007/s11116-015-9661-7.

[35] Kum Kum Dewan and Israr Ahmad , Carpooling: A Step To Reduce Congestion (A Case Study of Delhi). *Engineering Letters*, Vol. **14**, Issue 1, (Advance online publication: 12 February 2007).

[36] Heinrichs, H.: sharing economy: a potential new pathway to sustainability. *GAIA Ecological Perspectives for Science and Society,* Vol. **22**, No. 4, pp. 228–231 (2013); **DOI:** https://doi.org/10.14512/gaia.22.4.5.

[37] Codagnone, Cristiano and Abadie, Fabienne and Biagi, Federico, The Future of Work in the 'Sharing Economy'. Market Efficiency and Equitable Opportunities or Unfair Precarisation? (May 27, 2016). Institute for Prospective Technological Studies, Science for Policy report by the Joint Research Centre, 2016, Available at SSRN: https://ssrn.com/abstract=2784774

[38] Polkowski Z., Dysarz J., An analysis of the legal, economic and technical aspects of carpooling systems. *Scientific Bulletin – Economic Sciences*, Volume **15** Special Issue ETAEc 2016, pp. 76-88.

[39] Bresciani C., Colorni A., Costa F., Luè A. Carpooling: facts and new trends. Conference: 2018 International Conference of Electrical and Electronic Technologies for Automotive Conference Paper, July 2018 DOI: 10.23919/EETA.2018.8493206.

[40] Shaheen, S., Cohen, A., & Bayen, A. 2018. The Benefits of Carpooling. *UC Berkeley: Transportation Sustainability Research Center*. http://dx.doi.org/10.7922/G2DZ06GF

[41] Shaheen, S., & Cohen, A. 2020. Innovative Mobility: Carsharing Outlook; Carsharing Market Overview, Analysis, and Trends. *UC Berkeley: Transportation Sustainability Research Center*. http://dx.doi.org/10.7922/G2125QWJ Retrieved from https://escholarship.org/uc/item/61q03282.

[42] Parezanović T., Petrović M., Bojković N., Pejčić Tarle S., Carpooling as a measure for achieving sustainable urban mobility: European good practice examples. Scientific Proceedings XXIII International Scientific-Technical Conference "trans & MOTAUTO '15"; Year XXIII, Volume 3, pp. 82-85 (2015); ISSN 1310-3946.

[43] Suresh M. & Harish S. A study of carpooling behaviour using a stated preference web survey in selected cities of India. *Journal* . Volume **39**, 2016 - Issue 5, pp. 38-550.| Published online: 21 Apr 2016; https://doi.org/10.1080/03081060.2016.1174368.

[44] Galland S., Gaud N., Yasar Ansar-Ul-Haque Knapen L., Janssens D., Lamotte O. Simulation Model of Carpooling with the Janus Multiagent Platform. *Procedia Computer Science,* Volume **19**, 2013, Pages 860-866, ISSN 1877-0509; https://doi.org/10.1016/j.procs.2013.06.115.

[45] (http://www.sciencedirect.com/science/article/pii/S1877050913007230) (access: 15.01.2022)

[46] Mallus M., Colistra G., Atzori L., Murroni M. and Pilloni V. Dynamic Carpooling in Urban Areas: Design and Experimentation with a Multi-Objective Route Matching Algorithm. *Sustainability* **2017**, 9, 254; doi:10.3390/su9020254; www.mdpi.com/journal/sustainability.

[47] Schreier H., Grimm C., Kurz U., Schwieger B., Keßler S., Möser G. Analysis of the impacts of car-sharing in Bremen, Germany. Interreg North Sea Region Programme. „SHARE-North“ project . Final Report, 2018.

[48] Szymanek A. 2020. System Approach in Road Safety Studies. Communications - *Scientific Letters of the University of Zilina*, **22(4),** 201-210. https://doi.org/10.26552/com.C.2020.4.201-210.

Improvement of energy efficiency in the drive system of an electric car with PMSM

S German Galkin[1], D Tarnapowicz[1*], M Staude[1], T Zaleski[1] , M Jaśkiewicz[2]

[1] Faculty of Mechatronics and Electrical Engineering, Maritime University of Szczecin, 70-500 Szczecin, Poland
[2] Faculty of Mechatronics and Mechanical Engineering, Kielce University of Technology, Kielce, 25-314, Poland

*d.tarnapowicz@am.szczecin.pl

Abstract. Thanks to the advantages, i.e. high overload, high power density factor, wide range of rotational speed, low moment of inertia, as well as high torque-to-inertia ratio, PMSM AC motors are currently widely used in electric cars. The article presents a new method of PMSM control through the modification of the Clarke and Park transforms that enables to improve the efficiency of the mechatronic system due to the energy criterion. The article shows analytical tests confirming the possibility of maintaining the reactive power at zero level in the inverter-PMSM circuit. The analytical tests were supported by simulation tests with the use of the MATLAB-SIMULINK program.

Key words: active and reactive power control; energy optimization; control of the active converter, electric vehicle

1. Introduction

The main reasons for using an electric drive in cars is the fact that it does not generate toxic compounds – in contrast to internal combustion engines. Furthermore, it should be emphasized that electric motors are characterized by greater efficiency, easier maneuverability, lower failure rate and quieter operation compared to the commonly used combustion engines.

The first AC motors used in electric vehicles were squirrel-cage induction motors [1] - mainly due to their price and reliability. Taking into account electromechanical properties, in recent years, permanent magnet synchronous motors (PMSM) have been used. In relation to squirrel-cage motors, they have better properties [2 -4]:

- high value of the power density factor,
- wide range of rotational speeds,
- relatively high efficiency in a wide range of loads,
- high torque overload,
- low moment of inertia,
- high torque in relation to the moment of inertia.

Despite the much higher price of the PMSM in relation to the squirrel-cage motor, the above-mentioned advantages cause that it is more and more frequent use in drive systems of electric cars. The dynamics of changes in costs of energy production and the ecological aspect are the reason why engineers search for technical solutions that will improve the economic and environmental effects. One of the methods to increase the efficiency of the PMSM mechatronic drive system (in this respect) is energy optimization [5 - 8].

In order to benefit from all advantages of the PMSM, the motor should be powered from an electronic power converter with a microprocessor control. Commonly used methods regarding the control of the converter are field oriented control (FOC) and direct torque control (DTC) methods. The DTC method, compared to the FOC method, does not need continuous information about the rotor's position. However, it is characterized by the possibility of occurring the ripples [9], distortion of flux and reduction of flux (at low speeds).

The article presents the FOC method with control modification.

2. Analytical study of PMSM drive system

The control, according to the FOC method of the PMSM, in commonly used solutions, is connected with the orientation of the stator current vector with respect to the rotor flux. The decoupled control of the PMSM electromagnetic flux and the moment is realized thanks to the PMSM mathematical models (described in *d* and *q* coordinates). The *d* axis is related to the rotor flux, while the *q* axis is related to the stator current determining the electromagnetic moment (the control system scheme according to the FOC method is presented in Figure 1).

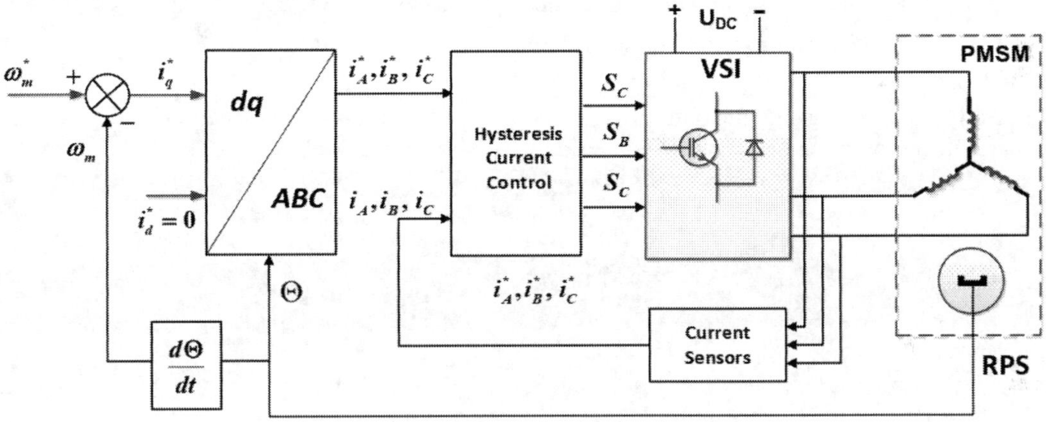

Figure 1. The diagram of the FOC system

For the ω_m speed equal to and smaller than the nominal speed (working area with a constant maximum torque), the current vector I1 is equal to the current component on the *q* axis. The component in the *d* axis is equal to zero. The mathematical description of the mechatronic system is conducted in a synchronously rotating *d,g* coordinate system with the use of the generalized vector method [10], as well as the theoretical analysis of mechatronic systems [11]. The equivalent scheme of the Inverter-PMSM circuit is shown in Figure 2.

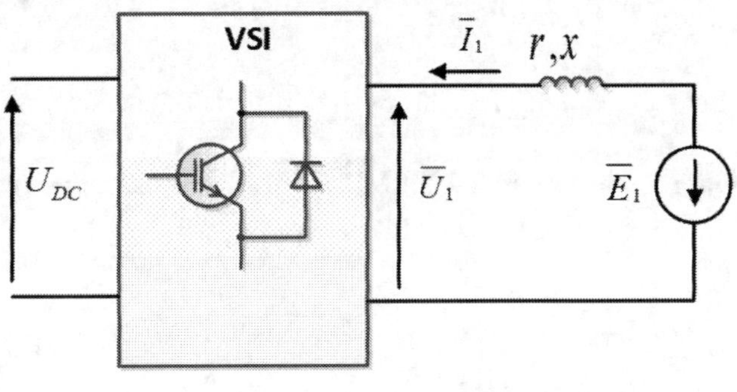

Figure 2. Equivalent scheme of the inverter-PMSM circuit

The mathematical description of the mechatronic system can be presented in the following way:

$$\begin{aligned} \bar{U}_1(t) &= r\,\bar{I}_1(t) + L\frac{d\bar{I}_1(t)_1}{dt} + j\bar{\psi}_0\omega, \\ T_e &= \frac{3}{2}p\cdot Mod[\bar{\psi}_0 \times \bar{I}_1(t)], \\ J\frac{d\omega_m}{dt} &= T_e - T_L. \end{aligned} \tag{1}$$

where:

$\bar{U}_1(t)$ - PMSM vector of the fundamental component of the stator voltage,
$\bar{I}_1(t)$ - PMSM vector of the fundamental component of the stator current,
$\bar{\psi}_0(t)$ - PMSM excitation flux vector,
p - number of pole pairs,
ω_m - angular speed of PMSM rotor's rotation,
ω - angular speed of sinusoidal state variables in the machine, $\omega = p\cdot\omega_m$
T_e T_L - PMSM electromagnetic torque and load torque,
J - moment of inertia in the machine's shaft,
r, L - resistance and inductance of stator windings.

In the steady state, taking into account the FOC control, assuming that Id = 0 and I1 = Iq – for a speed not greater than the nominal speed, the equation (1) can be determined as:

$$\begin{aligned} \bar{U}_1 &= r\bar{I}_q + jx\bar{I}_q + j\bar{E}_1, \\ T_e &= \frac{3}{2}p\cdot Mod(\bar{\Psi}_0 \times \bar{I}_q), \\ T_e &= T_L. \end{aligned} \tag{2}$$

where:

$\bar{E}_1 = \omega\bar{\psi}_0$ - PMSM electromotive force,

And in the d and q axes (respectively), taking into account the phasor diagram (Figure 3), the system of equations (2) can be presented in the following form:

$$\begin{aligned} U_q &= U_1\cos\Theta_0 = r\,I_q + \omega L\,I_d + E_1, \\ U_d &= -U_1\sin\Theta_0 = r\,I_d - \omega L\,I_q, \\ I_q &= \frac{2}{3}\frac{T_e}{p\Psi_0} = I_1 \\ I_d &= 0. \end{aligned} \tag{3}$$

where:

Θ_0 - PMSM load angle, at (Figure 1) for $I_d = 0$ this angle is determined in an automatic way

The space vector diagram in the "inverter-PMSM" circuit is shown in Figures 3 and 4.

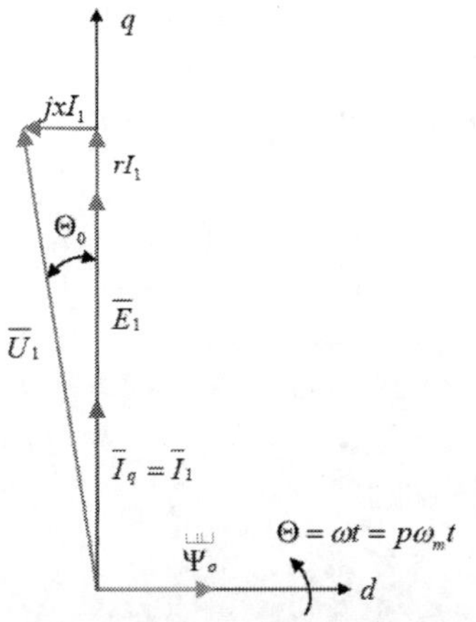

Figure 3. Space vector diagram of PMSM without optimization

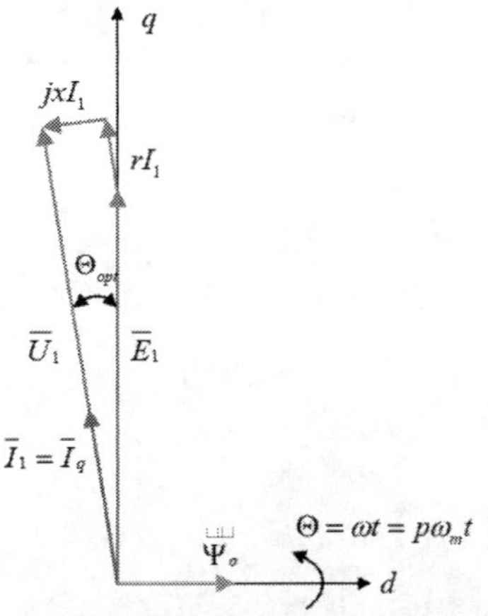

Figure 4. Space vector diagram of PMSM with optimization

The optimization proposed in this article, due to the energy criterion of PMSM drive control, consists in modifying the inverse Park transform by adding an Θ angle to the Θ_{opt} angle between $\overline{U}_1$ and $\overline{I}_1$ (Figure 1). The determination of the Θ_{opt} angle consists in the analysis of the vector diagram (Figure 3b) and indication of this angle from the trigonometric functions [13]:

$$\Theta_{opt} = \arcsin\left(\frac{LI_1}{\psi_0}\right) \tag{4}$$

where :

L, ψ_0 - PMSM parameters.

Adding of the Θ_{opt} angle to the angle (instantaneous angular position of the rotor's field) enables to obtain a zero phase shift between the I_1 current and the U_1 voltage – and thus the reactive power $Q = 0$. Active power and reactive power in the inverter-PMSM circuit can be presented as follows:

$$\begin{aligned} P_1 &= 1.5 \cdot U_q I_q, \\ Q_1 &= -1.5 \cdot U_d I_q. \end{aligned} \tag{5}$$

The PMSM parameters used in the tests are presented in Table 1.

Table 1. Parameters of the PMSM

Parameters of PMSM	symbol	unit	size
Stator phase resistance	r	Ω	0.05
Armature inductance	L	mH	635
Flux linkage	ψ_0	Vs	0.192
Nominal Torqe	T_{eN}	Nm	126
Nominal speed	ω_{mN}	RPM	3000
pole pairs	p	-	4

IOP Conf. Series: Materials Science and Engineering 1247 (2022) 012043 doi:10.1088/1757-899X/1247/1/012043

The presented analysis based on PMSM parameters (Table 1) is reflected in the energy characteristics [12] for the dependence of active and reactive power from external parameters – ω_m, T_e for the FOC control system without (Figures 5 and 6) and with (Figures 7 and 8) energy optimization.

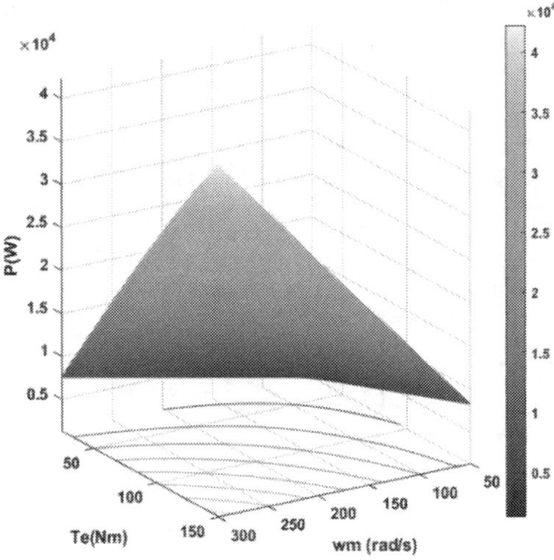

Figure 5. Energy characteristics (active power P) with FOC control without optimization

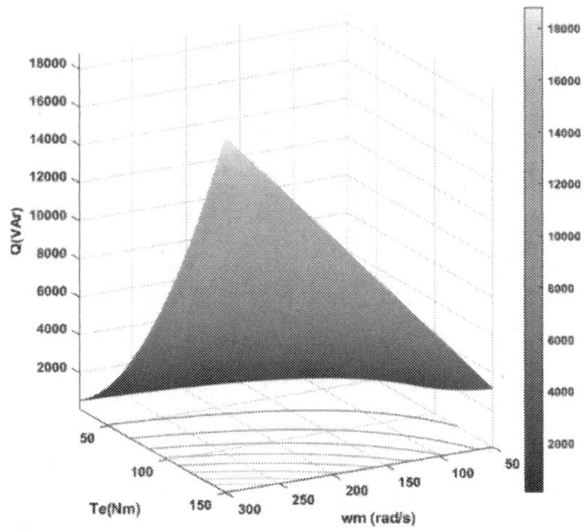

Figure 6. Energy characteristics (reactive power Q) without optimization

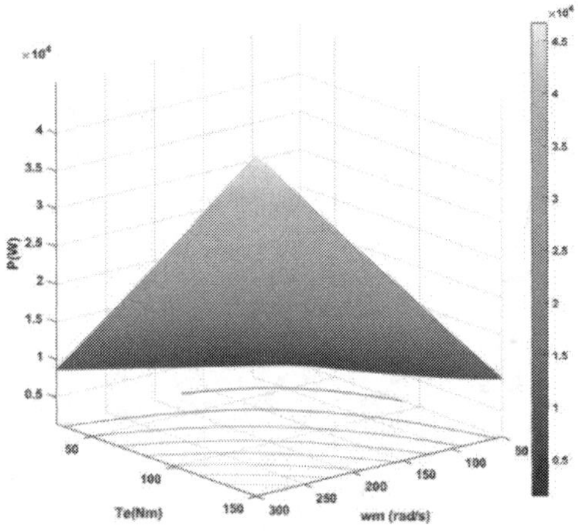

Figure 7. Energy characteristics (active power P) with FOC control with optimization

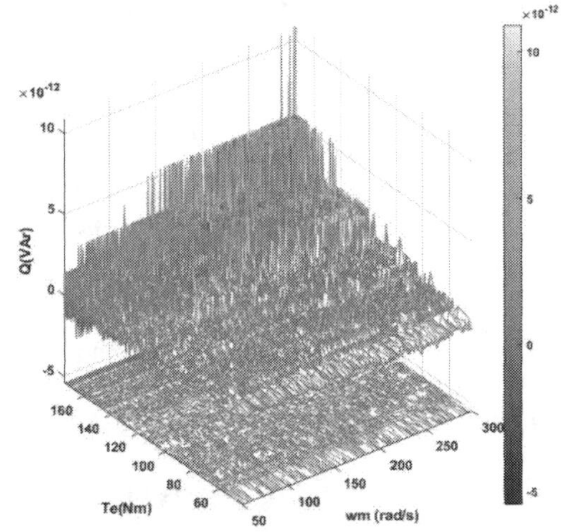

Figure 8. Energy characteristics (reactive power Q) with FOC control with optimization

3. Simulation tests of the electric car drive system with PMSM

The simulation tests of the PMSM drive system were conducted on the basis of the model implemented in the MATLAB-SIMULINK program (Figure 9). The PMSM parameters are presented in Table 1.

During the tests, torque, rotational speed, active power and reactive power were compared with the operation of the drive system (with and without energy optimization).

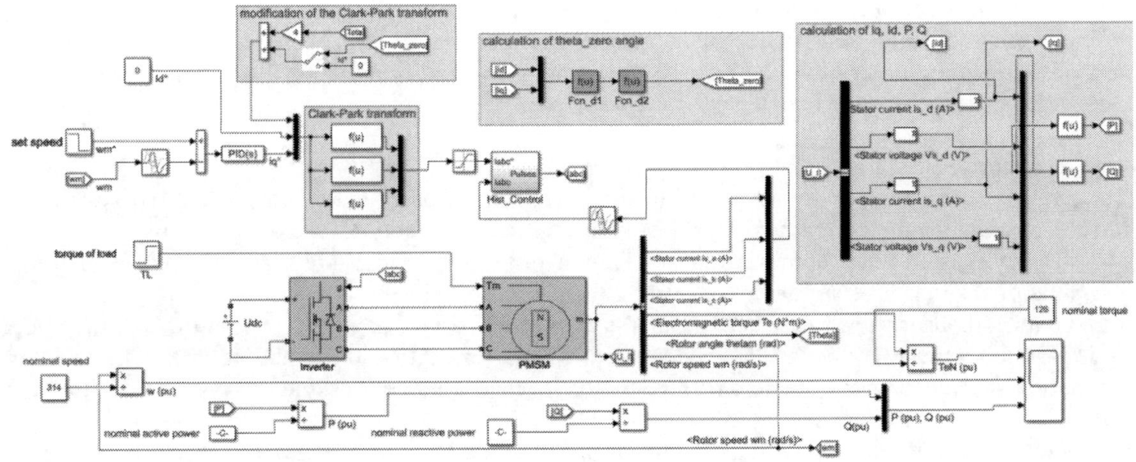

Figure 9. Simulation model of PMSM drive system with FOC control without and with optimization

Figures 10 and 11 show the results of simulation tests for both systems. The waveforms present the behavior of the T_e torque, ω_m speed, active power (P) and reactive power (Q) in the INV-PMSM system.

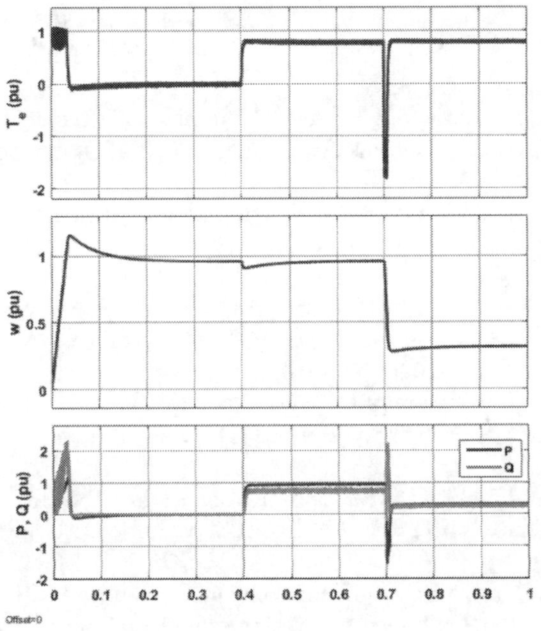

Figure 10. Energy processes (P, Q) in the inverter-PMSM drive system during speed and torque changes without optimization

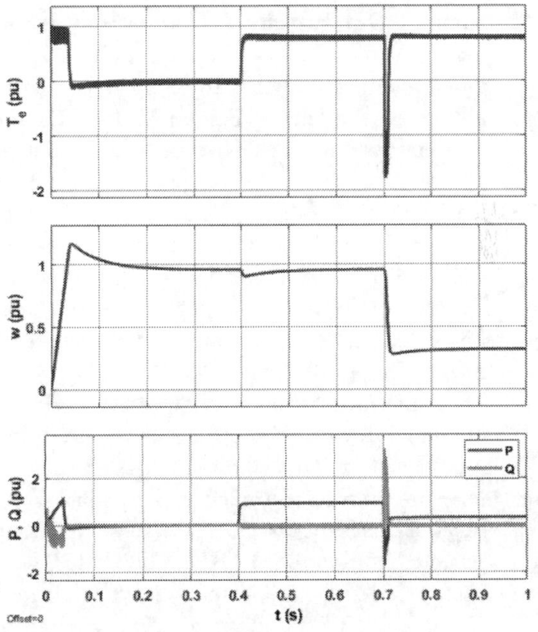

Figure 11. Energy processes (P, Q) in the inverter-PMSM drive system during speed and torque changes with optimization

In the steady state of operation (with a constant torque load), the waveform for the active power for both operating modes of the system (with and without optimization) is similar, while the waveform of the reactive power in the system with optimization is practically zero. During the change of speed from 1 (pu) to 0.3 (pu) and unchanged load, the system operates properly both in the mode with and without the optimization. The obtained results confirm the previously presented theoretical analysis.

At PMSM startup, the speed curve (in both cases) is aperiodic, and the power transient state is more favorable in the operating mode with optimization.

The results of simulation studies of the proposed control algorithm with reference to the reactive power optimization criterion in the INV-PMSM system by modifying the inverse Park transform show the effectiveness of this method.

4. Conclusions

The aim of the optimization with regard to the energy criterion of the electric car drive system with PMSM is to reduce the current and limit losses through the limitation of the reactive power.

Various methods of such optimization are known [13 - 15], which are characterized by extensive computational algorithms significantly loading the processor unit.

The control modification presented in this paper (which was patented [16]) allowed for energy optimization in the inverter-PMSM system with a small modification of the control. The results of simulation studies for the system confirmed the theoretical assumptions and the possibility of reactive power regulation and its reduction to zero.

Currently, advanced research work is being carried out on a laboratory stand, the first results of which are very promising.

The suggested solution can be used not only in electric car drives with PMSM, but also in other mechatronic drive systems with PMSM.

References

[1] Kenjo T and Takano Y, Comparison of torque-to-copper loss ratio in switched reluctance and induction motors for EV applications, 2012 15th International Conference on Electrical Machines and Systems (ICEMS), 2012, pp. 1-6. ISBN:978-1-4673-2327-7

[2] Hafner M, Finken T, Felden M and Hameyer K, Automated Virtual Prototyping of Permanent Magnet Synchronous Machines for HEVs, in IEEE Transactions on Magnetics, vol. 47, no. 5, pp. 1018-1021, May 2011, doi: 10.1109/TMAG.2010.2091675

[3] Renyuan Tang, Zhao Qing and Zhou Ting, Rare earth permanent magnet electrical machines stepping a new period of rapid development, Journal of Shenyang University of Technology, vol. 33, no. 1, pp. 1-8, 2011. ISSN 1000-1646

[4] German-Galkin S and Tarnapowicz D. Optimization of the electric car's drive system with PMSM, 2020 XII International Science-Technical Conference AUTOMOTIVE SAFETY, 2020, pp. 1-6, DOI: 10.1109/AUTOMOTIVESAFETY47494.2020.9293514.

[5] German-Galkin S and Tarnapowicz D. Energy optimization of the 'shore to ship' system — a universal power system for ships at berth in a port. SENSORS, Volume20, Issue14, Article Number 3815, DOI10.3390/s20143815

[6] German-Galkin S, Tarnapowicz D, Matuszak Z and Jaskiewicz M. Optimization to limit the effects of underloaded generator sets in stand-alone hybrid ship grids. ENERGIES, Volume13, Issue3, Article Number708, DOI10.3390/en13030708.

[7] Tarnapowicz D andGerman-Galkin S. Energy optimization of mechatronic systems with PMSG. Volume 4613 September 2018 Article number 000163rd International Conference on Energy and Environmental Protection, ENOS 2018Krakow13 September 2018. DOI 10.1051/e3sconf/20184600016

[8] German-Galkin S and Tarnapowicz D. Energy Optimization of Ship's Shaft Generator with Permanent Magnet Synchronous Generator. NASE MORE 2020, Volume67, Issue2, Page138-145, DOI10.17818/NM/2020/2.6

[9] Li Y, Gerling D, Ma J, Liu J and Yu Q. The Comparison of Control Strategies for the Interior PMSM Drive used in the Electric Vehicle. World Electric Vehicle Journal Vol. 4. DOI: 10.3390/wevj4030648

[10] Kovacs K P and Racz I. Transient processes in alternating current machines, Transiente Vorgänge in Wechselstrommaschinen. Verlag der Ungar., Budapest: Akad.der Wissenschaften, 1959.
[11] Bulgakov A A. New theory of controlled rectifiers. - M.: Science, 1970-326 p.: Il.
[12] German-Galkin S, Sakharov V and Tarnapowicz D. Energy characteristics of asynchronous electric drive. Management Systems in Production Engineering, Volume 27, Issue 1, Pages 51 - 541 March 2019. DOI 10.1515/mspe-2019-0009.
[13] Xiaotan Z, Chongjian L, Yaohua L and Chengsheng . Analysis of a large power PMSM using different control methods. 2005 International Conference on Electrical Machines and Systems, 2005, pp. 416-421 Vol. 1, doi: 10.1109/ICEMS.2005.202559.
[14] Jianbo c, Yuwen H, Wenxin, Yong L, Jianfei Y and Mingjin W. Direct active and reactive power control of PMSM, 2009 IEEE 6th International Power Electronics and Motion Control Conference, 2009, pp. 1808-1812, doi: 10.1109/IPEMC.2009.5157687.
[15] Wu F, Wan S and Huang S, Unity power factor control for PMSM position sensorless drive. 2008 International Conference on Electrical Machines and Systems, 2008, pp. 1618-1620.
[16] German-Galkin S, Tarnapowicz D and Staude M. Method and speed control system for a permanent magnet synchronous motor (PMSM). Polish Patent Office. Patent application: 439070, 00-950 Warsaw, Niepodległosci 188/192, Warsaw 29.09.2021.

IOP Conf. Series: Materials Science and Engineering 1247 (2022) 012044 doi:10.1088/1757-899X/1247/1/012044

Carports as an element of energy security for electric vehicles

A Małek[1*], Ł Wojciechowski[2]

[1] University of University of Economics and Innovation in Lublin, Department of Transportation and Informatics, Projektowa 4, 20-209 Lublin, Poland; Email: email: arkadiusz.malek@wsei.lublin.pl
[2] Lublin University of Technology, Department of Mechanical Engineering, Nadbystrzycka 36, 20-618 Lublin, Poland; Email: l.wojciechowski@pollub.pl

[*] Corresponding author e-mail address: arkadiusz.malek@wsei.lublin.pl

Abstract. Energy security with regard to electric vehicles consists in providing an appropriate infrastructure in the form of charging points for electric vehicles, which are able to generate and supply the traction batteries of vehicles with an appropriate amount of electricity. This role is played very well by photovoltaic carports, which are the structures designed to generate electricity and generate a shadow. The vehicle parked under the carport structure is protected against excessive heating. The shade for the parked vehicle is provided by a photovoltaic system mounted on the roof of the carport. Carports are, therefore, an ideal solution for charging electric vehicles. The article analyzes the problem that is related to the selection of peak power of the carport for the electric vehicle (or vehicles) that will be parked under it. An economic and energy analysis is carried out to ensure the fastest and cheapest charging of an electric vehicle. The authors present a study of two cases of carports of different sizes, which were used to charge electric vehicles of various brands with different sizes of traction batteries and different powers of on-board chargers. The results obtained from the conducted research and analysis can be generalized to a larger number of carports and vehicles available on the market.

Keywords – electric vehicle, traffic monitoring, carport

1. Introduction

Energy security in relation to electric vehicles consists in providing an appropriate number of charging points for electric vehicles that are able to generate and supply the vehicle's traction batteries with an appropriate amount of electricity. At present, i.e. in Europe and in the world, in the 21st century, almost all areas of our life have been mechatronized. At the beginning of the 20th century, they were electrified in order to obtain greater reliability and safety in relation to the previously present mechanical solutions. Electric drives turned out to be more reliable, more efficient (efficient) and more ecological than steam drives based on direct combustion of coal in a boiler [1]. Moreover, the electrification of towns and villages allowed for the transfer of emissions of carbon dioxide and other exhaust components generated during the combustion of coal, outside the clusters of people [2]. Alternating current makes it possible to transmit electricity over long distances with little losses. At the same time, mankind has started to use alternative [3] and renewable energy sources [4]. At the beginning, they were used in stationary mode, but with time, they were developed to such a degree

that they could find widespread mobile use [5]. For centuries, it has been a huge problem to store large amounts of non-fossil energy. However, the development of technology ensured the appearance on the market of traction batteries capable of storing the amount of energy necessary for a vehicle to travel several hundred kilometers [6]. Importantly, each year such battery packs are becoming cheaper and more reliable [7]. Due to the possibility of charging with high currents, the electric drive of vehicles is becoming more and more popular [8]. Some countries have already (2022) been declaring the complete withdrawal of diesel and even gasoline engines from production in a very short term (by 2028). Obviously, humanity faces many challenges, which are studies of risk, reliability and the impact of a given technology on human life [9].
The multitude of energy generating sources and their electrical receivers requires monitoring and supervision to ensure safe and reliable operation [10]. It would have been impossible on such a large scale if it had not been for the global Internet network introduced in the last three decades. In the 20s of the 21st century, most power generation sources and receivers are already Internet of Things devices and have Industry 4.0 functions [11]. Their monitoring, diagnostics and even servicing are carried out by means of sending and selected parameters to the data cloud and their on-line or off-line analysis [12]. Advanced internet platforms allow you to manage electricity over large areas. The types and interconnection of electricity generation and consumption infrastructure are shown in Figure 1 in the form of infographics [13][14].

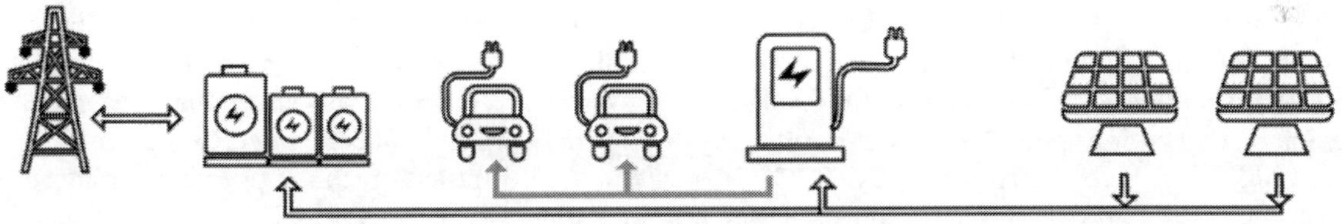

Figure 1. Power supply and power take-off infrastructure [own source]

Photovoltaic carports are a very effective way of obtaining electricity from RES and charging electric cars with it [15]. Photovoltaic carports are structures that generate electricity and shade. The vehicle parked under the carport structure is protected against excessive heating. The shade for the parked vehicle is provided by a photovoltaic system mounted on the roof of the carport. Carports are therefore an ideal solution for charging electric vehicles [16]. The article analyzes the problem related to the selection of peak power of the carport for the electric vehicle (or vehicles) that will be parked under it. Large parking spaces located in front of shopping centers or office complexes can be developed with carports. The vehicles will be protected from excessive sunlight and charged with electricity at the same time while their owners are shopping or working.

The safety of an electric vehicle can be considered in many ways:

- Safety of the driver and passengers traveling by electric vehicle. It involves, among others, safe reaching the destination, which is related to the range of the electric vehicle.
- Thermal safety when charging high power DC traction batteries. It can damage the batteries as a result of overheating and in extreme cases ignite them.
- The safety of an electric vehicle involved in a road collision or accident.
- Safety during diagnostics and repairs of selected components of the electric vehicle. Authorized persons should have appropriate qualifications and permits.
- Ecological safety resulting from zero emissions to the atmosphere at the place of use of the vehicle.

Safety in each of the above areas affects the road safety of this vehicle. If the vehicle has a battery charged with electricity, it will have good driving properties and will be able to move safely on the roads. If the battery is fully discharged, the electric vehicle will not be roadworthy and will present a traffic hazard, especially if it occurs on the motorway. Topping up a traction battery is not as simple as adding gasoline to an empty fuel tank. Thus, one of the main goals in the operation of an electric vehicle is to provide an adequate amount of electricity needed to cover the planned route. The infrastructure for charging electric vehicles, developing every year, reduces the risk of an electric vehicle being stopped in an undesirable place. Carports are part of such an energy security infrastructure.

2. Case 1: PV carport with a peak power of 1 kWp and a small electric vehicle

At the very beginning, the photovoltaic minicarport with a peak power of 1 kWp located at the University of Economics and Innovation in Lublin is analyzed. The supporting structure of the carport consists of two legs made of welded steel profiles. The steel elements have been protected against corrosion by galvanizing. Another part of the carport is the supporting structure of photovoltaic panels. The roof structure is made of anodized aluminium profiles. The structure was subjected to strength calculations due to its own weight, additional load of snow remaining in the winter and gusts of wind. Four photovoltaic panels in polycrystalline technology with a peak power of 250 Wp each were installed on the roof structure.

The design assumption was to make a compact structure of utility importance. The minicarport is to generate electricity needed to charge electric cars. If the vehicle does not receive energy, it is transferred to the power grid. An additional function of the minicarport is to be the exhibition capacity to promote the idea of electromobility, as well as the university itself, at various seminars, exhibitions, demonstartions and fairs. To fulfill this function, the minicarport must be easily and quickly assembled and disassembled. The appearance of the final version of the carport with the connected electric vehicle is shown in Figure 2. The photovoltaic system of the carport was connected to the power grid by means of an inverter. Additionally, 3 230 V sockets were installed on the supporting structure of the minicarport in order to power external receivers, including the electric vehicle. The author has the Renault Twizy electric vehicle, which is a small city vehicle with a traction battery with a nominal capacity of 6.1 kWh. The vehicle is designed to carry two people and has a range of 60 to 100 km on a single charge.

Figure 2. Charging a Renault Twizy vehicle from the WSEI photovoltaic minicarport [own source]

The time course of the current parameters during the process of full charging of the Renault Twizy vehicle from the WSEI PV minicarport is shown in Figure 3. The process of a fully discharged battery starts with a power level of 1200 W. After a few minutes, the power consumed by the on-board charger was about 1800 W (Pev). The power consumed by the vehicle slowly increases to around 2,000 W. After the battery has reached a level of approx. 95%, the charging current and, therefore, the power consumed have been reduced. The entire charging process took 210 minutes (3.5 hours). The charging process took place during the sunny noon between 11:45 a.m. and 3:15 p.m.. At that time, the

WSEI photovoltaic minicarport generated instantaneous power ranging from 600 to 800 W (Ppv). This power is lower than the power consumed by the vehicle during charging. For this reason, the photovoltaic minicarport has been connected to the power grid, which is an energy buffer for the vehicle being charged. This connection is called on-grid. In contrast to the off-grid connection, when the photovoltaic system is not connected to the power grid and the generated energy is stored in stationary or mobile energy storage [17].

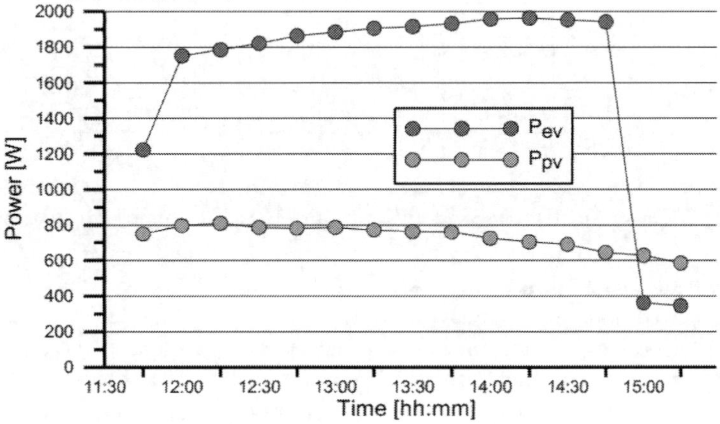

Figure 3. Mileage of the power generated by the carport and the power consumed during vehicle charging [own source]

The vehicle consumed almost 6 kWh of electricity to be fully charged. At that time, the photovoltaic minicarport generated 2.75 kWh of electricity, which is 46% of the demand.

The graphs in Figure 3 show that the WSEI minicarport is not able to cover the instantaneous power needed to charge the Renault Twizy vehicle. It has to get the missing power from the utility grid. It is worth noting that the charging process shown in Figure 4 took place in very good sunlight conditions which are rare. Usually, over the course of a year, the minicarport generates less instantaneous power and for a shorter period of time. The high uniqueness of the power generated by each photovoltaic system depends on weather conditions and is characterized by seasonality (seasons). Thus, the peak power of a carport cannot be a parameter on the basis of which its size can be selected for a specific electric vehicle.

In addition to the charging power, an important parameter of traction batteries in vehicles is their energy capacity. It is the amount of electricity stored in the battery and is expressed in kWh. The range of each electric vehicle depends on the amount of energy stored in the battery and the energy consumption. Renault Twizy has a nominal battery capacity of 6.1 kWh. This is the amount that corresponds to the daily electricity requirement of a single-family house. A fully charged battery provides the vehicle with a range of 50 to 100 km, depending on the speed and characteristics of the vehicle. Previous long-term studies conducted by the author show that the average range on a single full charge is 65 km [24].

Now let's analyze the daily amount of energy produced by the minicarport WSEI on individual days of the month of June in 2021. The amount of energy produced varies greatly from 1.5 to over 6 kWh. The maximum amount of energy produced in the sunniest month of the year is similar to the energy capacity of an electric vehicle battery.

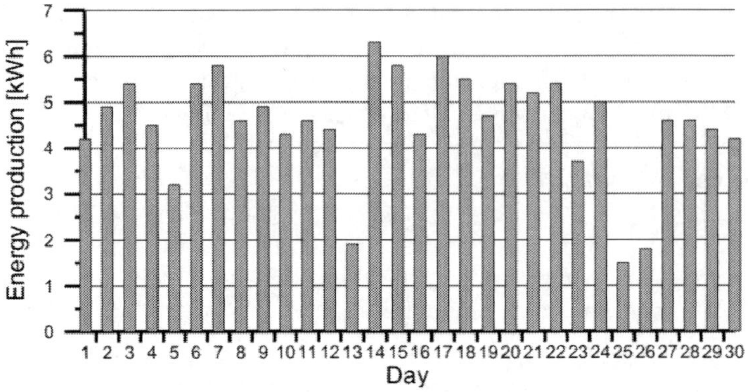

Figure 4. Daily electricity production by the WSEI minicarport in June 2021

However, the amount of energy produced in particular months of the year varies significantly [18]. In the winter months (December and January), in Polish climatic conditions, the amount of energy produced is 10 to 20 times lower than in the most sunny months (May, June and July). In the winter months, the amount of energy produced during the day is sufficient to drive a vehicle from several to several kilometers. However, the monthly amount of produced energy is an important parameter characterizing photovoltaic systems [19]. This is due to the fact that the customer's settlements with the electricity supplier and seller are usually monthly. Thus, the amount of monthly energy produced by the WSEI carport in the winter months corresponds to the need for one or more full charges of a Renault Twizy battery.

From the amount of energy produced in the summer months, you can fully charge the vehicle about 20 times. According to the author, the minimum size of the photovoltaic carport should be selected to ensure the desired range of the electric vehicle on summer days. With 20 fully charged batteries, Renault Twizy will travel about 1300 km. This is the real value of the monthly mileage of this small city vehicle. This monthly mileage has been achieved many times by the author who owns such a vehicle. The research results were also analyzed and published in a peer-reviewed article.

The amounts of electricity produced in the spring and winter months are smaller and ensure a monthly vehicle mileage of less than 1,000 km. Due to the semi-open structure of the Renault Twizy body, its use in the winter months in Poland is not possible. The size of the WSEI carport was selected in a proper way to the real electricity demand of the Renault Twizy vehicle. The basis for such selection is the monthly amount of energy produced by the minicarport needed to cover the actual demand of the vehicle.

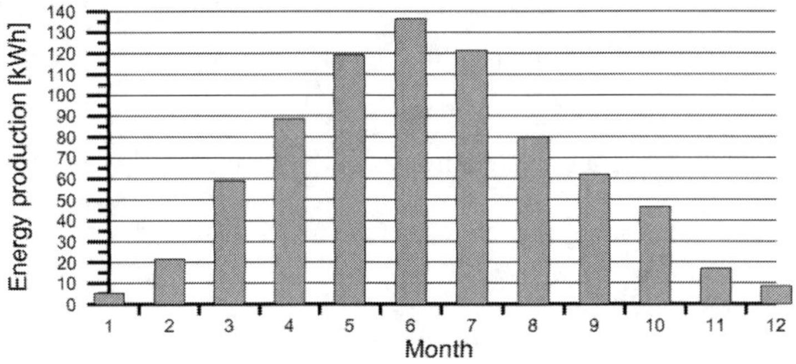

Figure 5. Monthly electricity production by the WSEI minicarport in 2021

3. Case 2: PV carport with 3 kWp peak power and electric city vehicle

Another carport designed and made by the author is the carport in the Lublin Science and Technology Park (LSTP). Carport LSTP consists of 12 monocrystalline photovoltaic panels made in the glass-glass technology, with an individual optimizer. Carport LSTP has a single-phase Schuko electricity receiving socket and is connected to Segment 4 LSTP. This means that when an electric vehicle is not charged from the carport, it transfers energy to the LSTP building. The on-grid approach used to connect the carport to the power grid ensures that the vehicle is charged in all sunlight conditions. If the power generated by the carport is not able to cover the power needed to charge the electric vehicle, the shortage will be taken from the power grid. On the other hand, if the power generated by the carport is greater than the power consumed by the vehicle, its excess will be returned to the power grid. The appearance of the LSTP carport as well as the charging socket and plug is shown in Figure 6.

Figure 6. Charging a Nissan Leaf vehicle from a photovoltaic carport in LSTP [own source]

The study used a Nissan Leaf vehicle manufactured in 2016, which had a traction battery with a capacity of 30 kWh. The vehicle is equipped with two battery charging sockets. The slow-charging socket is Type 1 and is connected to the 3 kW on-board charger. If the charging socket in the Type 1 vehicle is powered from a single-phase socket with a Schuko plug, the charging power is 2 kW. It is easy to calculate that the time to fully charge the battery will be approx. 15 hours. In the case of slow charging using the Type 2 socket, the charger can achieve a maximum charging power of 3 kW. Then the time to fully charge the battery will be approx. 10 hours. The fastest possible battery charging is via a Chademo socket and a 50 kW DC external charger. The charging time is then approx. 40 minutes.

Figure 7 shows depicts the course of the instantaneous power generated by the LSTP carport on several days in June 2020. The presented data shows that on a sunny summer day (June 21), the photovoltaic carport is able to cover the real power demand needed to slowly charge the Nissan Leaf vehicle with a capacity of 2 kW. The graph clearly shows that the instantaneous power generated by the LSTP carport is higher than 2 kW (2,000 W) for a period of approx. 6 hours (between 10.00 and 16.00). However, the size of the LSTP carport is not able to charge the Nissan Leaf in the off-grid mode, even in the process of slow charging with the lowest power.

IOP Conf. Series: Materials Science and Engineering 1247 (2022) 012044 doi:10.1088/1757-899X/1247/1/012044

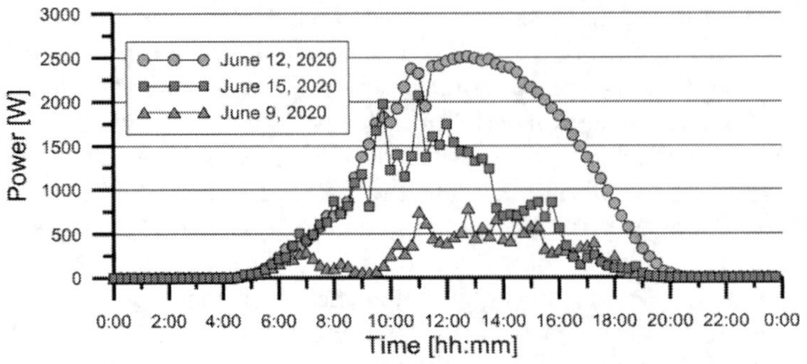

Figure 7. Course as a function of time of the power generated by the LSTP carport on 9, 12 and 15 June 2020

Figure 8 shows the daily production of electricity by carport LSTP in June 2020. The average daily amount of energy produced by carport LSTP in the sunniest month of the year is enough to charge half the battery of a Nissan Leaf vehicle. The actual mileage of this vehicle with a full battery is approx. 200 km. Thus, the amount of energy produced by the LSTP carport in the summer months ensures a daily vehicle mileage of about 100 km. This distance is traveled daily by people living in rural areas and commuting to a large city for work [20].

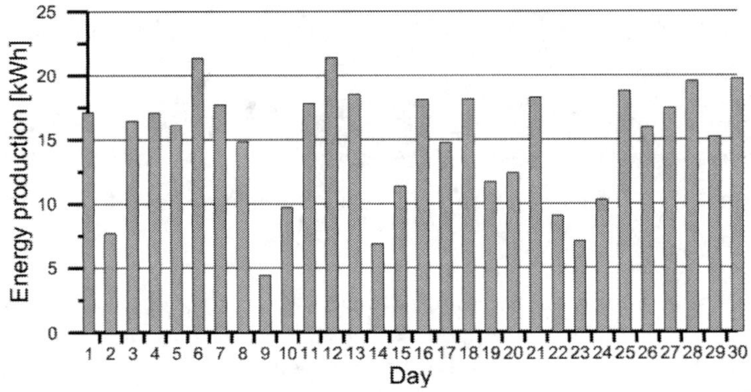

Figure 8. Daily electricity production by carport LSTP in June 2020

Figure 9 shows the monthly production of electricity by carport LSTP in 2020. The monthly amount of energy produced in the months from April to August is over 400 kWh [21]. This amount of energy gives the electric vehicle a monthly range of over 2,600 km. The amount of energy produced by carport LSTP in 2017-2021 was over 3 MWh per year. This amount of produced energy ensures that the vehicle's battery is fully charged 100 times. This in turn corresponds to the annual mileage of a Nissan Leaf vehicle of 20,000 km [22].

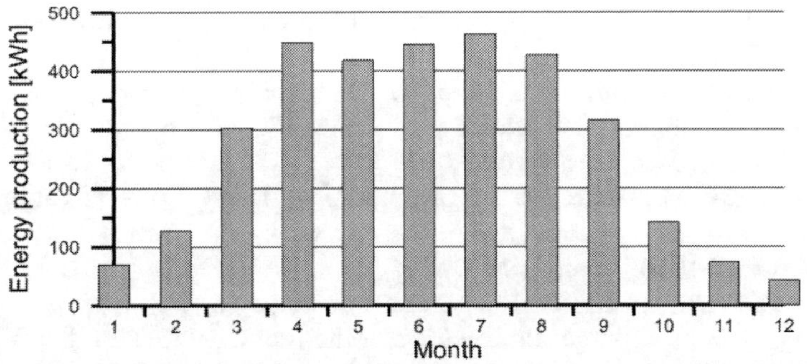

Figure 9. Monthly electricity production by carport LSTP in 2020

4. Conclusions

The analysis presented above show that it is possible to rationally select the size of a photovoltaic carport for an electric vehicle. According to the author, the minimum size of the photovoltaic carport should be selected to ensure the desired range of the electric vehicle on summer days. The nominal size of the carport should be selected to cover the annual electricity demand of the vehicle. The author proved that it is very difficult or impossible to charge an electric vehicle battery from a carport in off-grid mode. In each of the analyzed cases it was necessary to connect the carport to the power grid and take the missing power for charging from it.

The author intends to continue research on solar carports in the future. An interesting direction is adding a stationary energy storage and a hybrid inverter to the carport. The selection of the size of the energy storage is an interesting challenge due to the high purchase costs of the energy storage.

References

[1] Muha R., Perosa A.: Energy consumption and carbon footprint of an electric vehicle and a vehicle with an internal combustion engine. Transport Problems. 2018, **13(2)**, 49–58, DOI: 10.20858/tp.2018.13.2.5.

[2] Merkisz-Guranowska A., Daszkiewicz P.: Possibility of reducing CO_2 emissions for example electric vehicles. Journal of KONES Powertrain and Transport. 2014, **21(3)**, 211–217, DOI: 10.5604/12314005.1133215.

[3] Gis W, Gis M. Overview of the method and state of hydrogenization of road transport in the world and the resulting development prospects in Poland. *Open Engineering*. 2021, **11(1)**, 570-583. https://doi.org/10.1515/eng-2021-0039

[4] Colmenar-Santos A., Muñoz-Gómez A-M, Rosales-Asensio E., López-Rey Á., Electric vehicle charging strategy to support renewable energy sources in Europe 2050 low-carbon scenario, Energy, **183**, 15 September 2019, Pages 61-74, doi:10.3390/en12203834

[5] Mruzek M., Gajdáč I., Kučera L., Gajdošík T.: The Possibilities of Increasing the Electric Vehicle Range. Procedia Engineering. 2017; **192**: 621-625. https://doi.org/10.1016/j.proeng.2017.06.107

[6] Du J, Liu Y, Mo X, Li Y, Li J, Wu X, Ouyang M. Impact of high-power charging on the durability and safety of lithium batteries used in long-range battery electric vehicles. Applied Energy. 2019, **255**, 113793. DOI: https://doi.org/10.1016/j.apenergy.2019.113793.

[7] Ansari S, Ayob A, Hossain Lipu MS, Hussain A, Saad MHM. Multi-Channel Profile Based Artificial Neural Network Approach for Remaining Useful Life Prediction of Electric Vehicle Lithium-Ion Batteries. Energies. 2021; **14(22)**, 7521. https://doi.org/10.3390/en14227521

[8] Tomaszewska A, Chu Z, Feng X, O'Kane S. Lithium-ion battery fast charging: A review. eTransportation. 2019, **1**, 1-28. DOI: https://doi.org/10.1016/j.etran.2019.100011

[9] Flasza J.: Electromobility in Poland : challenges and possibilities taking into account intelligent

installation RES. AUTOBUSY – Technika, Eksploatacja, Systemy Transportowe. 2017; **6**: 1196-1198.

[10] Madeti SR, Singh S. Monitoring system for photovoltaic plants: A review. Renewable and Sustainable Energy Reviews. 2017, **67**, 1180-1207. DOI: https://doi.org/10.1016/j.rser.2016.09.088

[11] Fragiacomo P, Piraino F, Genovese M. Insights for Industry 4.0 Applications into a Hydrogen Advanced Mobility. Procedia Manufacturing. 2020, **42**, 239-245. DOI: https://doi.org/10.1016/j.promfg.2020.02.077.

[12] Gan Y, Chen Z, Wu L, Cheng S, Lin P. Fault diagnosis of PV array using adaptive network based fuzzy inference system. Proceedings of the IOP Conf. Series: Earth and Environmental Science. 2020;467:012083. DOI:10.1088/1755-1315/467/1/012083

[13] Novoa L., Brouwer J., Dynamics of an integrated solar photovoltaic and battery storage nanogrid for electric vehicle charging, Journal of Power Sources, Volume **399(30)**, September 2018, Pages 166-178

[14] Seddig K., Jochem P., Fichtner W., Two-stage stochastic optimization for cost-minimal charging of electric vehicles at public charging stations with photovoltaics, Applied Energy, **242(15)**, May 2019, Pages 769-781, DOI: 10.1016/j.apenergy.2019.03.036

[15] Kulik AC, Tonolo ÉA, Scortegagna AK, da Silva JE, Urbanetz Junior J. Analysis of Scenarios for the Insertion of Electric Vehicles in Conjunction with a Solar Carport in the City of Curitiba, Paraná—Brazil. Energies. 2021; **14(16)**, 5027. https://doi.org/10.3390/en14165027

[16] Iringová A., Kovačic M, Design and optimization of photovoltaic systems in a parking garage - a case study, Transportation Research Procedia. 2021;55:1171-1179. https://doi.org/10.1016/j.trpro.2021.07.097.

[17] Mehrjerdi H., Off-grid solar powered charging station for electric and hydrogen vehicles including fuel cell and hydrogen storage, International Journal of Hydrogen Energy, **44(23)**, 3 May 2019, Pages 11574-11583, doi:10.1016/j.ijhydene.2019.03.158

[18] Małek A, Caban J, Wojciechowski Ł. Charging electric cars as a way to increase the use of energy produced from RES. Open Engineering. 2020, **10(1)**, 98-104. DOI:10.1515/eng-2020-0009

[19] Kostopoulos Em., Spyropoulos G., Christopoulos K., Kaldellis J. K., Solar energy contribution to an electric vehicle needs on the basis of long-term measurements, Procedia Structural Integrity, **10**, 2018, Pages 203-210, doi:10.1016/j.prostr.2018.09.029

[20] Wahid MR, Budiman BA, Joelianto E, Aziz M. A Review on Drive Train Technologies for Passenger Electric Vehicles. Energies. 2021; **14(20)** 6742. https://doi.org/10.3390/en14206742

[21] Małek A, Dudziak A, Stopka, O, Caban J, Marciniak A, Rybicka I. Charging electric vehicles from photovoltaic systems – statistical analyses of the small photovoltaic farm operation. Energies 2022, 15, 2137. https://doi.org/10.3390/en15062137

[22] Nait-Sidi-Moh A., Ruzmetov A., Bakhouya M., Naitmalek Y., Gaber J., A Prediction Model of Electric Vehicle Charging Requests, Procedia Computer Science, **141**, 2018, Pages 127-134, doi:10.1016/j.procs.2018.10.158

Determining the trajectory of the vehicle motion on the basis of the recorded physical quantities describing its dynamics during typical maneuvers in road traffic

T Pusty[1]

[1] Łukasiewicz Research Network – Automotive Industry Institute, Jagiellońska 55, 03-301 Warszawa, Poland, tomasz.pusty@pimot.lukasiewicz.gov.pl

Abstract: The paper compares the results of the vehicle motion trajectory calculations determined on the basis of the recorded waveforms of the acceleration vector and angular velocity components during the double lane change maneuver, taking into account the errors of acceleration sensors reset (zeroing). The differences in the results of the calculations were determined and possible causes of these differences were indicated. The test results indicate that special attention should be paid to zeroing the acceleration sensors, in particular the a_y component.

Keywords – vehicle movement, traffic safety problems, vehicle dynamics

1. Introduction

Obtaining data on physical quantities describing the movement of vehicles is of interest to many research entities: passive safety systems, active safety systems e.g. during the modeling of autonomous lane change algorithms [1, 2], compliance of technological solutions with applicable normative documents, e.g. LDWS (Lane Departure Warning System) or experts analyzing the records of UDS (Unfall Daten Speicher) or ADR (Accident Data Recorder) black boxes [3] or during modelling vehicle dynamics [4, 5, 6]. Determination of the trajectory of the vehicle motion takes place in the global reference system with the use of kinematic relationships in a situation where linear accelerations, angular velocities and linear velocities in the local system were measured directly. This is important both during the experimental verification of simulation programs and the reconstruction of road accidents. In the works [7] the possible errors in the processing of data recorded by EDR were subjected to simulation analysis. The research was based on the results of vehicle motion simulations and acceleration records of two mathematical EDR models. One records - shaft acceleration and angular velocities in relation to three axes, the other - longitudinal and transverse acceleration and yaw rate. In the latter case, significant errors are possible, in particular with regard to the reconstructed trajectory of the vehicle's movement. The paper [8] presents a comparative analysis of velocity and linear acceleration courses measured with analog devices with measurements carried out in GPS systems. The problem of reducing these values from the device mounting points to the center of masses is also discussed. An example is shown with the results of measurements carried out during the steady-state driving test. A four-wheeled flat model and the related theory of motion kinematics were used. Acquisition of data on physical quantities describing the movement may be subject to uncertainty due to transducer noise, zeroing and calibration errors, and limited data sampling frequency [9, 10]. Therefore, research focused on determining the size of the differences in the determined trajectory end positions at a different level of accuracy of acceleration sensors zeroing (zeroing at the level of tens and hundreds). The results of the work may be

IOP Conf. Series: Materials Science and Engineering 1247 (2022) 012045 doi:10.1088/1757-899X/1247/1/012045

a guide for people using sensor data to determine the uncertainty range of calculations of the motion trajectory based on the recorded signals from the acceleration and angular velocity sensors.

2. Results and Discussion

During the analysis of the results of the experimental research, the vehicle kinematics model was used, in which right-handed coordinate systems were used: global and local. Coordinate systems used: – local coordinate system, rigidly associated with the car body. Coordinate systems were used:

– local coordinate system, rigidly associated with the car body. The Oxyz coordinate system has an O origin at the center of the car's mass, the Ox axis is parallel to the longitudinal axis of the vehicle. In the local coordinate system Oxyz, the components of the acceleration vector (ax, ay, az) and the components of the angular velocity vector of the car body (P, Q, R) were recorded;

– $O_GX_GY_GZ_G$ global coordinate system related to road. The $O_GX_GY_GZ_G$ plane of this system is on a horizontal roadway, and the O_GZ_G axis is directed vertically upwards. The speed vector of the car is parallel to the O_GX_G axis before starting the measurements.

Vehicle movement is identified, among others. in relation to the road infrastructure, the location of which is described in the global coordinate system. Therefore, it is necessary to transform the measurement results of the local components of the acceleration vector (a_x, a_y, a_z) into the global system (a_X, a_Y, a_Z). The angular velocities of the car body in the global coordinate system are described by the expression:

$$\begin{bmatrix}\dot{\Phi}\\ \dot{\Theta}\\ \dot{\Psi}\end{bmatrix} = \begin{bmatrix}1 & \sin\Phi \,\mathrm{tg}\Theta & \cos\Phi \,\mathrm{tg}\Theta\\ 0 & \cos\Phi & -\sin\Phi\\ 0 & \frac{\sin\Phi}{\cos\Theta} & \frac{\cos\Phi}{\cos\Theta}\end{bmatrix}\begin{bmatrix}P\\ Q\\ R\end{bmatrix} \tag{1}$$

$$\begin{bmatrix}a_{xP}\\ a_{yP}\\ a_{zP}\end{bmatrix} = \begin{bmatrix}\cos\Theta & \sin\Theta\sin\Phi & \sin\Theta\cos\Phi\\ 0 & \cos\Phi & -\sin\Phi\\ -\sin\Theta & \cos\Theta\sin\Phi & \cos\Theta\cos\Phi\end{bmatrix}\begin{bmatrix}a_x\\ a_y\\ a_z\end{bmatrix} \tag{2}$$

$$\begin{bmatrix}a_{Gx}\\ a_{Gy}\\ a_{Gz}\end{bmatrix} = \begin{bmatrix}\cos\Psi & -\sin\Psi & 0\\ \sin\Psi & \cos\Psi & 0\\ 0 & 0 & 1\end{bmatrix}\begin{bmatrix}a_{xP}\\ a_{yP}\\ a_{zP}\end{bmatrix} \tag{3}$$

where:
Ψ, Θ, Φ – quasi-Euler angles, defining the orientation of the local system $\{C_S\}$ relative to the global system $\{0\}$;
P, Q, R – components of the angular velocity vector of the body in local coordinate systems.

To achieve the planned aim of this work, a double lane change maneuver was performed, during which the acceleration and angular velocity waveforms were recorded in the local reference system. For registration of the above-mentioned waveforms, the OXTS RT 3002 measurement set was used, which recorded signals at a frequency of 100 Hz. The waveforms of accelerations and angular velocities have been centered by removing the so-called offset and filtered with a 2nd order Butterworth lowpass filter with no phase shift with a cutoff frequency of 25 Hz. The OXTS RT 3002 measuring set was placed inside the car (test object) near the location of its center of mass (Figure 1a and Figure 1b). The tests were conducted during double lane change with a constant forward speed v_X=58 km/h. View from the vehicle during test and view from inside of the vehicle has been shown in the Figure 1c and 1d.

IOP Conf. Series: Materials Science and Engineering 1247 (2022) 012045 doi:10.1088/1757-899X/1247/1/012045

a) b)

c)

d)

Figure 1. Picture from the test object, a) and b) view of the OXTS RT 3002 apparatus installed in the car, c) view from the vehicle during test, d) view from inside of the vehicle

IOP Conf. Series: Materials Science and Engineering 1247 (2022) 012045 doi:10.1088/1757-899X/1247/1/012045

The determined acceleration of the center of mass of the vehicle in the global system a_{Gx}, a_{Gy}, a_{Gz} was used to determine the coordinates of the center of mass of the car by double numerical integration. The error of zeroing the acceleration sensors at the start of the maneuver was assumed to be:

– variant 1: a_{x0}=-0.4÷0.4 m/s^2 with a step of 0.1 m/s^2, a_{y0}=0 m/s^2;
– variant 2: a_{x0}=0 m/s^2, a_{y0}=-0.04÷0.04 m/s^2 with a step of 0.01 m/s^2;
– variant 3: a_{x0}=-0.04÷0.04 m/s^2 with a step of 0.01 m/s^2, a_{y0}=0 m/s^2.

The results of the calculations are presented in Figures 2a-2d in the form of the trajectories of the center of mass during lane change.

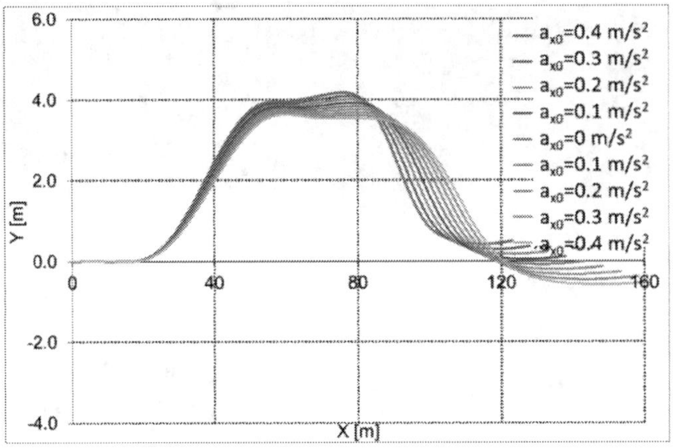

a)

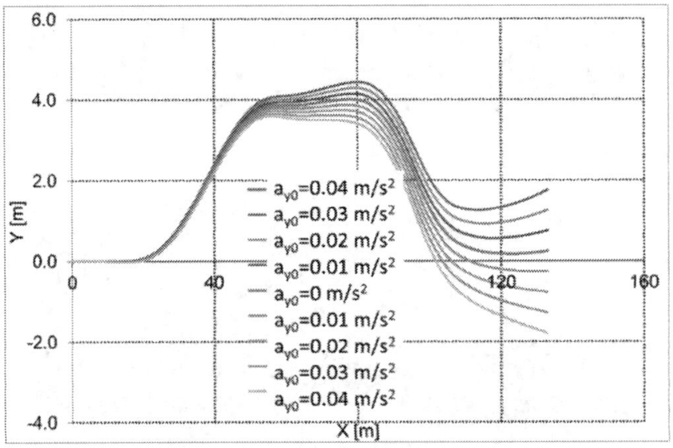

b)

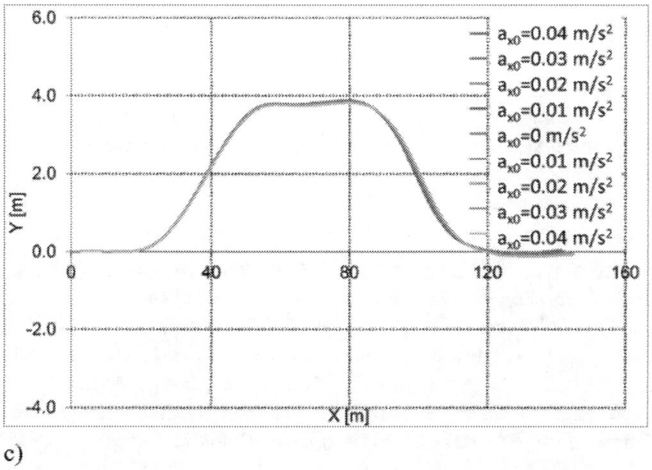

c)

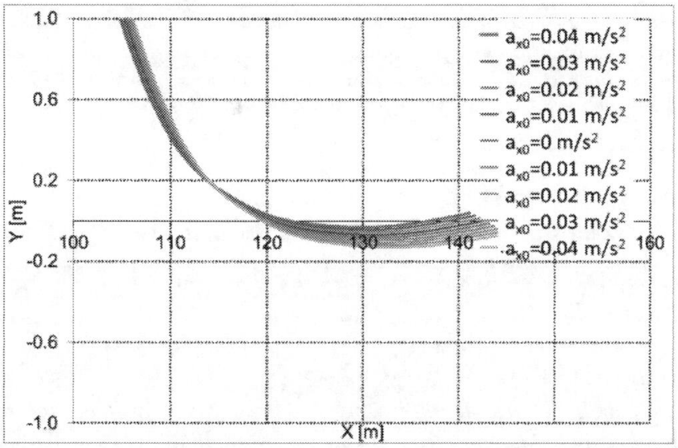

d)

Figure 2. Trajectories of movement of the car; a) variant 1, b) variant 2, c) variant 3, d) a fragment of variant 3 enlarged

The scope of the obtained calculation results for the variant

– 1 is: x_G from 122.93 m to 163.47 m and y_G = from 0.49 m to -0.51 m;
– 2 is: x_G from 133.07 m to 133.02 m, y_G = 1.76 m to -1.79 m;
– 3 is: x_G from 141.17 m to 145.23 m, y_G = 0.04 m to -0.06 m.

3. Concluding Remarks

In case of errors in resetting the longitudinal component of the acceleration sensor a_{x0}=-0.4÷0.4 m/s^2, the range of end positions was within the range Dx_G=40.54 m and Dy_G=1 m after covering the designated road section. In the case of errors in resetting the longitudinal component of the acceleration sensor a_{y0} =-0.04÷0.04 m/s^2, the range of end positions was in the range Dx_G=0.5 m and Dy_G=3.55 m after driving the designated distance. In the case of errors in resetting the transverse component of the acceleration sensor a_{x0}=-0.04÷0.04 m/s^2, the results of the calculations were Dx_G=4.06 m Dy_G=0.1 m after driving the designated road section. The test results indicate that special attention should be paid to zeroing the acceleration sensors, in particular the a_y component.

Acknowledge

The research was carried out as part of the Innovative system research project supporting the motor vehicle insurance risk assessment dedicated to UBI (Usage Based Insurance) No. POIR.04.01.04 00 0004/19 00 financed by the National Centre for Research and Development

References

[1] Prochowski L, Ziubiński M, Szwajkowski P, Gidlewski M, Pusty T, Stańczyk T L, *Impact of Control System Model Parameters on the Obstacle Avoidance by an Autonomous Car-Trailer Unit: Research Results*. Energies, **14**, 2958, 2021. https://doi.org/10.3390/en14102958

[2] Prochowski L, Pusty T, Gidlewski M, Jemioł J, *Experimental studies of the car-trailer system when passing by a suddenly appearing obstacle in the aspect of active safety of autonomous vehicles*, International Automotive Conference (KONMOT2018), IOP Conf. Series: Materials Science and Engineering **421** 032024 2018, doi:10.1088/1757-899X/421/3/032024

[3] Guzek M, Lozia Z, *Analiza dokładności rekonstrukcji wypadku drogowego wykorzystujących zapisy samochodowej „czarnej skrzynki"*, Materiały II konferencji naukowo-technicznej "Rozwój techniki samochodowej a ubezpieczenia komunikacyjne". Radom 2004.

[4] Lozia Z, *Vehicle dynamics and motion simulation versus experiment*. 1998 SAE International Congress and Exposition. Detroit, Michigan USA. SAE TP 980220, 1998.

[5] Jackowski J, Jakubowski J, *Recognition of Moving Tracked and Wheeled Vehicles Based on Sound Analysis and Machine Learning Algorithms*. International Journal of Automotive and Mechanical Engineering, **18(1)**, 2021.

[6] Żardecki D, Dębowski A, *Non-smooth models and simulation studies of the suspension system dynamics basing on piecewise linear luz(...) and tar(...) projections. Applied Mathematical Modelling*, **94**, 2021.

[7] Guzek M, Lozia Z, *Possible Errors occurring during Accident Reconstruction based on Car "Black Box" Records*, 2002 World SAE Congress and Exposition. Detroit, Michigan USA. March SAE TP 2002-01-0549 2002.

[8] Pieniążek W, Wolak S, *The problematic aspects of linear velocity and acceleration measurements in curvilinear motion of automobiles*, Proceedings of the Institute of Vehicles **5(114)**, 2017.

[9] Dziewiecki M, Gidlewski M, *Limitations of use of an inertial positioning system in a truck during a maneuver of avoiding a suddenly appearing obstacle*, ESV Conference Paper Number 15-0341. 24th ESV International Technical Conference on the Enhanced Safety of Vehicles. Goteborg, Szwecja 2015.

[10] Wach W, *Wiarygodność strukturalna rekonstrukcji wypadków drogowych*, Wydawnictwo Instytutu Ekspertyz Sądowych, Kraków 2014.

IOP Conf. Series: Materials Science and Engineering 1247 (2022) 012046 doi:10.1088/1757-899X/1247/1/012046

The use of probabilistic networks in the analysis of risks to the components of the bus power system with hydrogen fuel cells

D Kasperek, G Bartnik, A Marciniak, A Małek*, D Pieniak, L Gil

University of University of Economics and Innovation in Lublin, Department of Transportation and Informatics, Projektowa 4, 20-209 Lublin, Poland;

*arkadiusz.malek@wsei.lublin.pl

Abstract. Probabilistic networks can be a useful tool for reliability modelling and risk analysis in the operation of technical facilities. In the article, the authors present the method of using probabilistic networks in the analysis of threats to the elements of the electric bus power supply system with a hydrogen fuel cell. Both the use of probabilistic networks as a tool supporting decision-making processes and the use of the latest achievements in the field of non-emission propulsion indicate a new area of research. The article presents the construction of a hydrogen bus with particular emphasis on the hydrogen storage and transmission installation. Next one specific case is considered where there is a leak in the hydrogen system with the simultaneous appearance of an ignition initiator in the form of a short circuit in the electrical system. Authors considered Fault Tree-Event Tree method in context of Bayesian networks technology. The conducted research has shown that there is a real risk of hydrogen explosion in a hydrogen fuel cell-powered bus and that it can be accurately calculated.

Key words: alternative drive system, probabilistic networks, automotive safety

1. Introduction

Probabilistic networks can be a useful tool for reliability modeling and risk analysis in the operation of technical facilities [1]. The nodes of the network represent the states of the components of a complex technical object and the arcs of dependence between them. Thus, the entire network represents the state of the object as a whole. Each node of the network is associated with a conditional probability distribution of a given component (part, subassembly) in a given state, determined by the state of the elements functionally related to it.

The knowledge and information representation model in the form of a Bayesian network is a graph that consists of nodes and inter-node relations [2]. The nodes in this case represent events. An event is something that is associated with a change in the state of the process or object under study and takes place at some point in time, where the system changes from one state to another and these variables are set with accuracy to the probability distribution. If such a model is a representation of knowledge and information, questions may be asked of this representation of knowledge and information. The question is a sentence function, i.e. it is a sentence that contains variables and such variables in Polish are interrogative pronouns (e.g. how many, what, where, how, why, etc.).

The use of probabilistic measures results from the fact that changes in state may occur according to more or less unpredictable influence of the environment and the processes of wear and aging of

IOP Conf. Series: Materials Science and Engineering 1247 (2022) 012046 doi:10.1088/1757-899X/1247/1/012046

construction materials. Additionally, network nodes are introduced to represent operational activities, e.g. replacement of a damaged element [3].

In the article, the authors present the method of using probabilistic networks in the analysis of threats to the elements of the electric bus power supply system with a hydrogen fuel cell. Both the use of probabilistic networks as a tool supporting decision-making processes and the use of the latest achievements in the field of non-emission propulsion indicate a new area of research.

The constant striving to improve air quality in cities forces the need to look for new solutions aimed at reducing CO_2 emissions in the urban transport sector [4] [5]. An important activity in this area is the use of alternative fuels and drives [6]. Today they include electricity and hydrogen [7]. Polish and European city bus manufacturers are introducing new generations of ecological city buses [8] [9].

With the first commercial vehicles powered by hydrogen fuel cells entering the market, the first questions related to the safety of their use arose. The differences in operation in relation to traditional internal combustion drives or increasingly popular hybrid drives are also puzzling [10]. On the part of both the users and mechanics, there are questions about potential problems related to the servicing of such vehicles. In 2013, the Korean company Hyundai launched its ix35 Fuel Cell model first, with an ambitious plan to sell 1,000 units in the first two years and 10,000 per year from 2015. In 2014, another strong player appeared on the market - Toyota with its Mirai model. Immediately after the market launch, Toyota decided to popularize the new technology by making available free of charge over 5,000 patents developed by the company in the planning, development and production of the hydrogen Mirai model. This allows other manufacturers to take advantage of patents relating to fuel cell technology, high pressure hydrogen storage tanks, and advanced control systems. According to Toyota, this convenience will increase the interest of car companies and customers in the hydrogen drive. It is considered to be the most ecological type of vehicle propulsion. The first phase of the dissemination of hydrogen vehicles lasted until 2020. Its implementation included technical education at the level of vocational schools, technical universities and Authorized Service Stations. At the beginning, these were the basics of operation, and then the details related to the use, diagnostics and servicing of hydrogen vehicles. The task of scientists in the near future will be to see further development of individual components of hydrogen vehicles in order to reduce weight, cost and increase the range of such vehicles. An important issue significantly affecting the operation, diagnostics and servicing of vehicles powered by hydrogen cells is the type of fuel cells used [11][12]. Another is the method of storing hydrogen on board the vehicle and the related safety of using hydrogen vehicles.

2. Hydrogen power supply system for electric city bus - components and functions

In the 21st century, many components are available for building a bus's electric drive system. The most important of them are the electric motor and traction batteries, and in the case of hydrogen propulsion, a stack of fuel cells and hydrogen tanks [12][14]. The appropriate selection of each of them determines the suitability of the bus for a specific application (city, intercity) and its price. In addition to the correct selection of components, their integration is also important, which enables failure-free operation with high efficiency [15][16][17].

One of the problems with hydrogen vehicles is its storage. There are many ways to store hydrogen. It can be collected in a compressed or cryogenic version. The latter method enables large amounts of hydrogen to be stored in a small volume. However, it is very expensive due to the need to maintain very low temperatures at the level of production, distribution and transportation of hydrogen. There are also hybrid tanks with cryogenic and additionally cryo-compressed hydrogen [18]. The hydrogen can also be stored in solid form in metal hydrides. It is very easy to use and efficient method, however, it has one disadvantage - it is relatively heavy. One of the newest methods of collecting hydrogen is the use of carbon nanotubes [19]. However, metal or composite high-pressure tanks are most often used to store hydrogen on board vehicles. At present, composite tanks have almost completely replaced the originally used, very heavy, steel tanks. Lightweight and capable of collecting hydrogen under high pressure, the composite tanks are manufactured in a cylindrical version with various diameters and lengths. They have a capacity of several to several hundred liters. Several European and world producers produce tanks that are approved at the international level.

Alternative drive technologies used in city buses include hybrid electric (HEV), plug-in hybrid (PHEV), pure electric (EV) and fuel cell buses (FCEB).

These alternative technologies are the basis of the electrification of urban transport. The electrification strategy enables the improvement of air quality in cities (no local emissions), the diversification of primary energy sources (electricity / hydrogen can be produced from a wider variety of sources, not necessarily fossil sources) and the use of technologies that can improve energy efficiency [20][21].

Fuel Cell Electric Bus (FCEB) buses use an electric motor to drive. In a hydrogen vehicle, electricity to power the engine is generated on an ongoing basis in a hydrogen fuel cell powered by hydrogen stored in tanks usually located on the roof of the vehicle. A chemical reaction takes place in a hydrogen fuel cell whereby the hydrogen combined with oxygen in the air produces electricity. The reaction product is also steam and heat [22].

Electric buses using hydrogen fuel cells (abbreviated as "hydrogen buses") are the most technologically advanced solution among zero-emission vehicles (Fig. 1). During the operation of hydrogen buses, zero emission of pollutants is achieved at the place of use of the vehicle. As a result of the chemical reaction of hydrogen and oxygen in the fuel cell, electricity is generated and a by-product is water vapor.

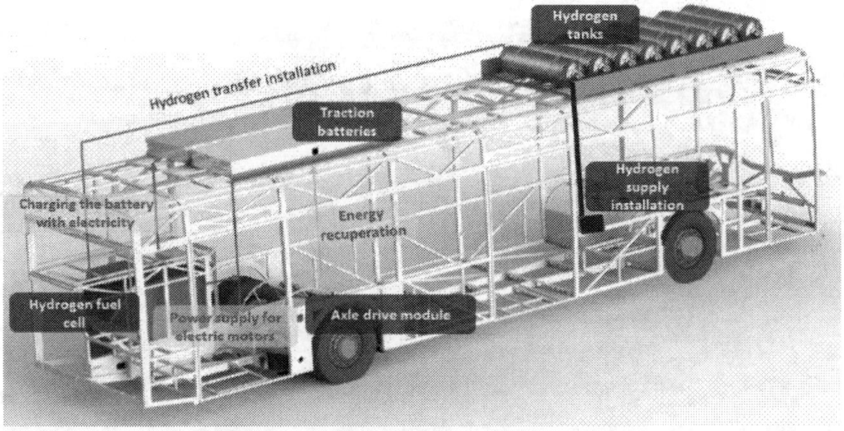

Figure 1. The hydrogen fuel cell bus

Buses powered by pure hydrogen energy differ from classic electric buses in that the source of electricity is a hydrogen fuel cell, while the batteries act as an energy buffer - they are powered by both the fuel cell and the energy recuperation system obtained during vehicle braking [23].

A hydrogen bus uses an average of 9 kg of hydrogen per 100 km. Consumption depends on the parameters of the built-in components and the number of passengers transported, but also the driver's driving technique, season, i.e. weather conditions and outside temperature. Greater consumption will occur in summer due to the need to supply air conditioning equipment in the passenger compartment, and in winter due to the need to use some of the electricity to heat the vehicle.

Hydrogen tanks placed on the roof of the bus have a capacity of approx. 35 kg of hydrogen, which, depending on the conditions, is enough for a distance of approx. 350-450 km, without the need to refill the route (as is the case with battery electric vehicles). Hydrogen is collected in tanks under a pressure of 35 MPa [8]. PEMFC (Proton Exchange Membrane Fuel Cell) hydrogen cells are successfully used in electromobility applications.

While the bus is stationary, the traction batteries can be charged with electricity from the power distribution network. In many European countries, a large proportion of electricity comes from renewable energy sources [24].

3. Methodology

The article presents the safety risk during the operation of electric bus power system components with a hydrogen cell, which has been considered the most dangerous. According to the theory of probability, dangerous events are rare and usually result from the fact that certain elements of the structure of a technical object accumulate considerable energy and the sudden release of this energy is the cause of an event defined as dangerous or dangerous.

In the bus power supply system with a hydrogen cell (Fig. 2), there is an element (there are of course more of them) in which there is a significant amount of energy - it is a hydrogen tank. The energy of the hydrogen tank, or actually the chemical energy stored in this hydrogen, if it is released in the form of an explosion, it will be this dangerous event.

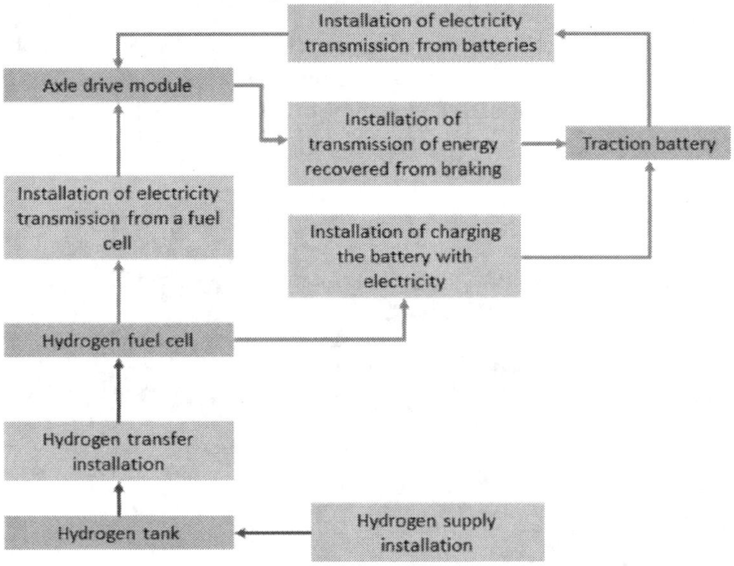

Figure 2. Diagram of the hydrogen supply system for the bus drive

The problem of the reliability of the hydrogen supply to fuel cells is not considered here, but one specific case is considered where there is a leak in the hydrogen system with the simultaneous appearance of an ignition initiator in the form of a short circuit in the electrical system. The occurrence of both events will lead to a hydrogen explosion (explosive combustion of hydrogen) (Fig. 3).

IOP Conf. Series: Materials Science and Engineering 1247 (2022) 012046 doi:10.1088/1757-899X/1247/1/012046

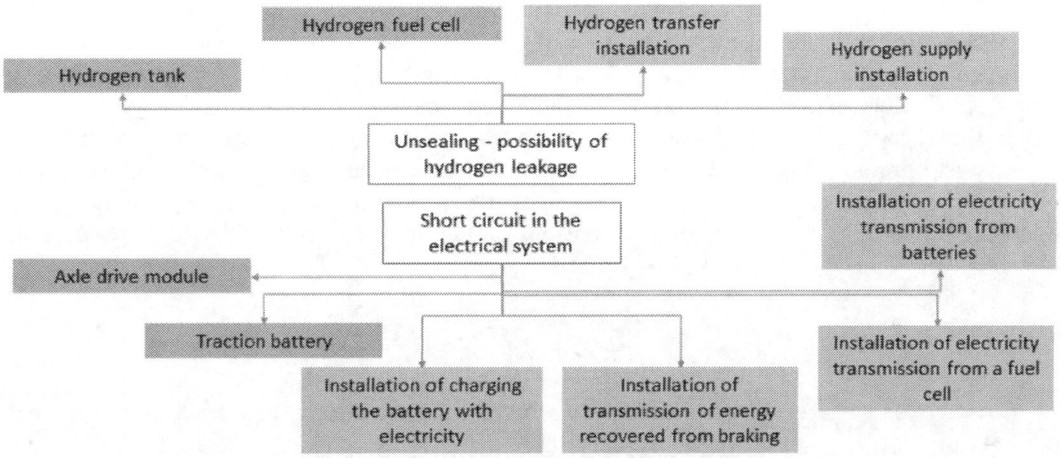

Figure 3. Potential locations of the initiators of the explosion

4. Bayesian networks modelling the operational risk of rare catastrophic events

Bayesian networks (BN) are graphs whose relations represent cause-and-effect interactions and nodes are propositional functions whose arguments are known with accuracy to the conditional probability distribution. BN allow you to infer value of unobserved variables and calculate the effect of observing the values of a subset of model variables on the probability distribution of other variables. Bayesian update, also called belief update, or probabilistic inference, is based on determining cumulative probability distributions.

Application of Bayesian networks in systems reliability and safety domain is well known. The aim of presented article is an introductory conceptualization and analysis of modeling methods of rare catastrophic events. We considered well known Fault Tree-Event Tree method in context of Bayesian networks technology. The FTA-ETA_BN is shown below as a modified "bow tie" graph (Fig. 4).

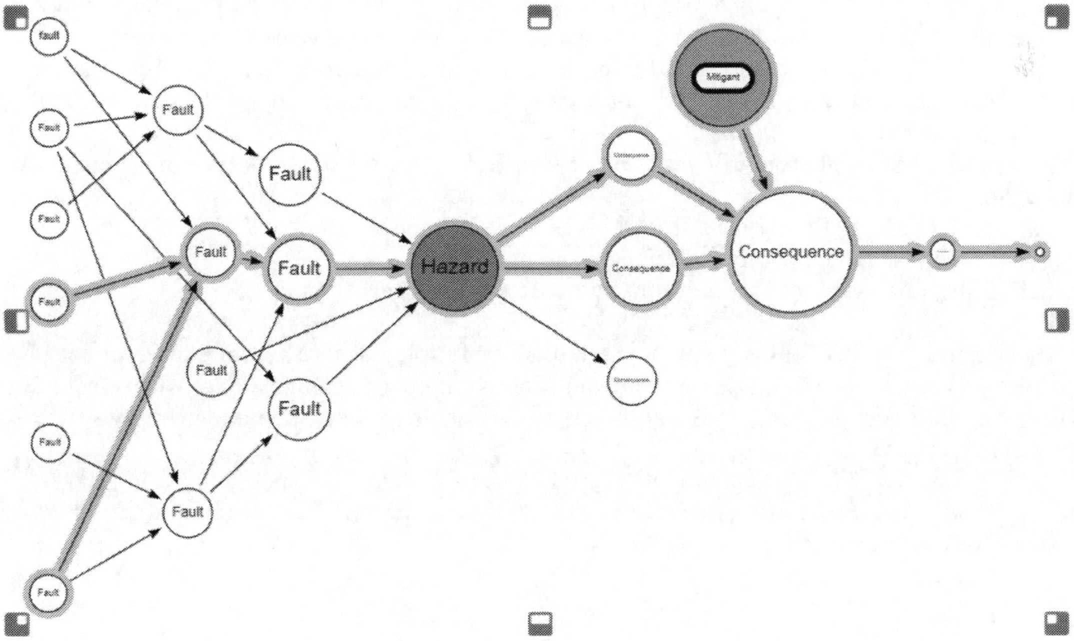

Figure 4. Fault–Consequences events graph – (modified "bow tie" graph)

In presented article we analysed two methods of risk modelling as Bayesian networks. The first one was invented by N. Fenton & M. Neil in "Risk Assessment and Decision Analysis with Bayesian Networks" [25]. Risk conceptualization as an risk event that can have negative or positive impact, involving the risk event, one at least consequence event which characterizes the impact, and additionally one or more trigger events, one or more control events, and one or more mitigating or inhibiting events (Fig. 5). "Control event" represents the prevention of the occurrence of danger and "Mitigant" represents the occurrence of interventions mitigating the effects of the "Risk event".

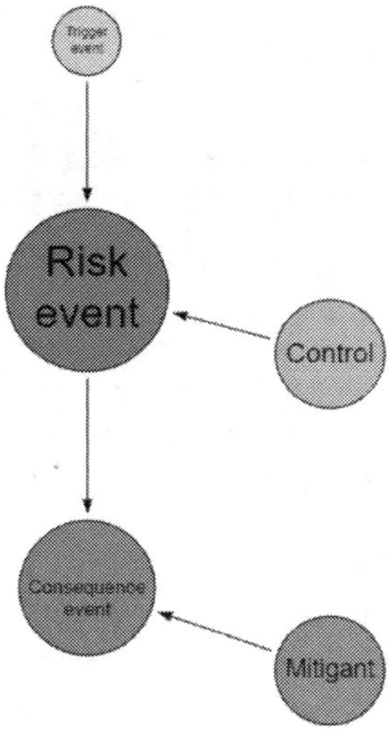

Figure 5. Bayesian network conceptualization of risk (Fenton, Neil 2013)

A risk characterized in Fig.5 has five categories of events. Each of them may have deeper specification as a sub network of detailed events of a given type. Events have a number of possible outcomes (in the simplest case it can be just two outcomes true and false). Every event and hence every object associated with risk has uncertainty that is characterized by the probabilities of the event's outcomes.

It is possible to build arbitrarily large risk-based BNs using the above approach. This involves the following steps:

1. Consider the set of risk events from a given subject domain
2. For each risk event identify any triggers and/or controls
3. For each risk event identify any consequences and mitigants

By chaining together different risks we can model multiple risks, risks from different perspectives, and common causes, consequences and mitigants, all within the same model. So, we can get the BN to generate quantified risk predictions and other useful functions like simulation, backward reasoning and what-if-analysis.

Another method of risk representation in Bayesian Networks is possible with Bayes Fusion product Genie [26]. Especially interested is its qualitative variant Queenie. It explicitly distinguishes four categories of causal relationships, Fig.6.

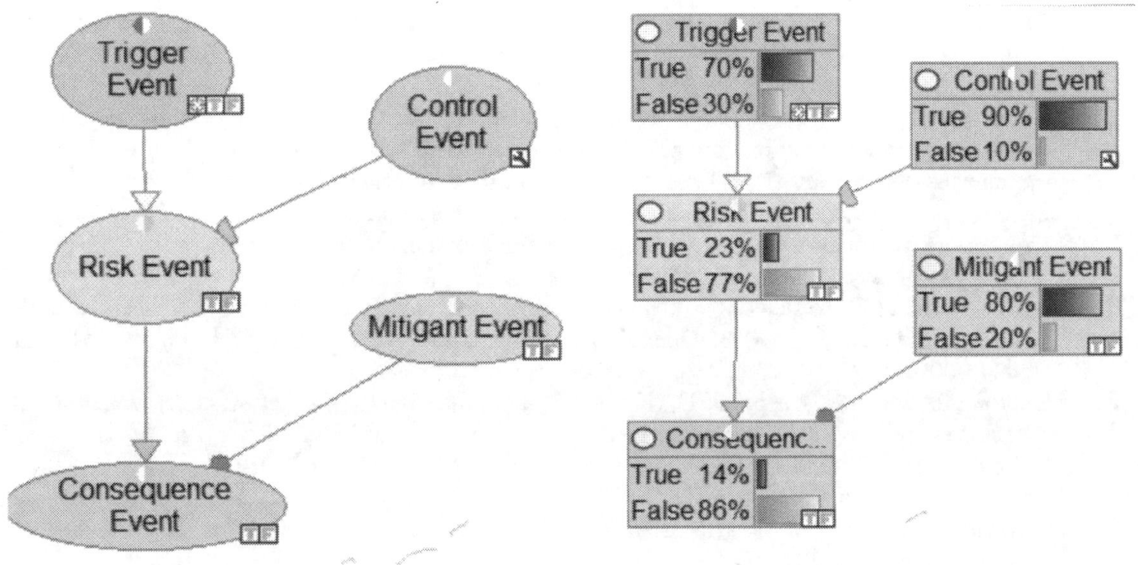

Figure 6. Representation of risk event in QGenie BN

The QGeNIe qualitative modeling environment has many applications, for example:

- Rapid prototyping – even if the goal is to build a fully quantified Bayesian network, models developed at QGeNIe can be exported to GeNIe for further refinement and learning of parameters.
- Modeling the reliability and safety of processes in which significant threats are very rare events and we have almost no hard data about them, only expert qualitative knowledge.

5. Conclusions

In the article, the authors present the method of using probabilistic networks in the analysis of threats to the elements of the electric bus power supply system with a hydrogen fuel cell. Both the use of probabilistic networks as a tool supporting decision-making processes and the use of the latest achievements in the field of non-emission propulsion indicate a new area of research. The article presents the construction of a hydrogen bus with particular emphasis on the hydrogen storage and transmission installation. Next one specific case is considered where there is a leak in the hydrogen system with the simultaneous appearance of an ignition initiator in the form of a short circuit in the electrical system. Authors considered Fault Tree-Event Tree method in context of Bayesian networks technology. The conducted research has shown that there is a real risk of hydrogen explosion in a hydrogen fuel cell-powered bus and that it can be accurately calculated.

References

[1] Bartnik G., Kusz A., Marciniak A. Modelling of operation process for engineering facilities using dynamic Bayesian networks. Inżynieria Rolnicza, Volume 12, Issue 87, Pages 9-16

[2] Yang Y, Sørensen JD. Probabilistic Availability Analysis for Marine Energy Transfer Subsystem Using Bayesian Network. *Energies*. 2020; 13(19):5108. https://doi.org/10.3390/en13195108

[3] Soltanali H, Khojastehpour M, Farinha JT, Pais JEdAe. An Integrated Fuzzy Fault Tree Model with Bayesian Network-Based Maintenance Optimization of Complex Equipment in Automotive Manufacturing. *Energies*. 2021; 14(22):7758. https://doi.org/10.3390/en14227758

[4] Berger R., Fuel Cell Electric Buses – Potential for Sustainable Public Transport in Europe. A

Study for the Fuel Cells and Hydrogen Joint Undertaking: http://www.fch.europa.eu/sites/default/files/150909_FINAL_Bus_Study_Report_OUT_0.PDF

[5] Skrúcaný T, Kendra M, Stopka O, Milojević S, Figlus T, Csiszár C. Impact of the Electric Mobility Implementation on the Greenhouse Gases Production in Central European Countries. Sustainability. 2019; 11(18):4948. https://doi.org/10.3390/su11184948

[6] Chang C. C., Liao Y. T., Chang Y. W., Life cycle assessment of alternative energy types – including hydrogen – for public city buses in Taiwan. International Journal of Hydrogen Energy, Volume 44, Issue 33, 5 July 2019, Pages 18472-18482

[7] Małek A., Caban J., Wojciechowski Ł., Charging electric cars as a way to increase the use of energy produced from RES. Open Engineering. 2020; 10(1): 98-104. doi:10.1515/eng-2020-0009

[8] Małek A., Taccani R., Kasperek D., Hunicz J., Optimization of Energy Management in a City Bus Powered by the Hydrogen Fuel Cells. Communications - Scientific letters of the University of Zilina. DOI: https://doi.org/10.26552/com.C.2021.4.E56-E67

[9] Correa G., Muñoz P., Falaguerra T., Rodriguez C. R., Performance comparison of conventional, hybrid, hydrogen and electric urban buses using well to wheel analysis. Energy, Volume 141, 15 December 2017, Pages 537-549

[10] de Miranda P. E. V., Carreira E. S., Icardi , Nunes G. S., Brazilian hybrid electric-hydrogen fuel cell bus: Improved on-board energy manage U. A. ment system. International Journal of Hydrogen Energy, Volume 42, Issue 19, 11 May 2017, Pages 13949-13959

[11] Zuliani N., Radu R., Taccani R., Design and experimental characterization of a 350 W high temperature PEM fuel cell stack. 9th YSESM, Trieste, Italy, July 7-9, 2010; ISBN 978-88-95940-30-4

[12] Wendeker M, Małek A, Czarnigowski J, Taccani R, Boulet P., Breaban F. Adaptive airflow control of the PEM fuel cell system. SAE Technical Papers. JSAE/SAE International Fuels & Lubricants Meeting 2007. https://doi.org/10.4271/2007-01-2012

[13] Jwa K., Lim O., Comparative life cycle assessment of lithium-ion battery electric bus and Diesel bus from well to wheel. Energy Procedia, Volume 145, July 2018, Pages 223-227

[14] Wang L., Wang Z., Ju Q., Characteristic Analysis of Lithium Titanate Battery, Energy Procedia, Volume 105, May 2017, Pages 4444-4449

[15] Xu L, Hua J, Bao L, Li J et al. Development of a Vehicle Control System for a Fuel Cell Hybrid Bus. 2008 SAE International Powertrains, Fuels and Lubricants Congress. https://doi.org/10.4271/2008-01-1574

[16] Lee D. Y., Elgowainy A., Vijayagopal R., Well-to-wheel environmental implications of fuel economy targets for hydrogen fuel cell electric buses in the United States. Energy Policy, Volume 128, May 2019, Pages 565-583

[17] Vepsäläinen J., Otto K., Lajunen A., Tammi K., Computationally efficient model for energy demand prediction of electric city bus in varying operating conditions. Energy, Volume 169, 15 February 2019, Pages 433-443

[18] Ahluwalia R. K., Peng J. K., Roh H. S., Hua T. Q., James B. D., Supercritical cryo-compressed hydrogen storage for fuel cell electric buses. International Journal of Hydrogen Energy, Volume 43, Issue 22, 31 May 2018, Pages 10215-10231

[19] Fazle Kibriaa A.K.M., Moa Y.H., Parka K.S., Nahma K.S., Yun M.H., Electrochemical hydrogen storage behaviors of CVD, AD and LA grown carbon nanotubes in KOH medium; International Journal of Hydrogen Energy 26/2001, Pages 823-829.

[20] Jereb B, Stopka O, Skrúcaný T. Methodology for Estimating the Effect of Traffic Flow Management on Fuel Consumption and CO2 Production: A Case Study of Celje, Slovenia. Energies. 2021; 14(6):1673. https://doi.org/10.3390/en14061673

[21] Laib F., Braun A., Rid W., Modelling noise reductions using electric buses in urban traffic. A case study from Stuttgart, Germany. Transportation Research Procedia, Volume 37, 2019,

Pages 377-384
[22] Zuliani N., TaccaniR., Microcogeneration system based on HTPEM fuel cell fueled with natural gas: Performance analysis, Applied Energy, Volume 97, September 2012, Pages 802-808
[23] Erd A, Stokłosa J. Main Design Guidelines for Battery Management Systems for Traction Purposes. Proceedings of the XI International Scientific and Technical Conference Automotive Safety 2018. Slovakia, Apr. 2018. DOI: 10.1109/AUTOSAFE.2018.8373345.
[24] Małek A, Dudziak A, Stopka, O, Caban J, Marciniak A, Rybicka I. Charging electric vehicles from photovoltaic systems – statistical analyses of the small photovoltaic farm operation. Energies 2022, 15, 2137. https://doi.org/10.3390/en15062137
[25] Fenton N., Neil M. Risk Assessment and Decision Analysis with Bayesian Networks - 2nd Edition. Published August 28, 2018 by Chapman and Hall/CRC
[26] https://www.bayesfusion.com

AUTHOR INDEX

Institute of Physics
Dirac House, Temple Back
Bristol BS1 6BE UK

ISSN: 1757-8981
ISBN 978-1-7138-5811-9

9 781713 858119

CURRAN ASSOCIATES INC.
proceedings.com